Biochemistry

Concise Medical Textbooks

Biochemistry

S. P. DATTA
B.Sc., M.B., B.S.
Professor of Medical Biochemistry, University College, London

J. H. OTTAWAY
B.Sc., PhD., A.R.I.C.
Senior Lecturer in Biochemistry, University of Edinburgh

THIRD EDITION

BAILLIÈRE TINDALL · LONDON

Baillière Tindall
7 & 8 Henrietta Street, London WC2E 8QE

Cassell & Collier Macmillan Publishers Ltd, London
35 Red Lion Square, London WC1R 4SG
Sydney, Auckland, Toronto, Johannesburg

The Macmillan Publishing Company Inc.
New York

First published 1965
Second edition 1969
Reprinted 1972, 1974
Third edition 1976

ISBN 0 7020 0512 6

Printed by J. W. Arrowsmith Ltd., Winterstoke Road, Bristol BS3 2NT

Preface to the Third Edition

Two changes which addicts may notice in this edition of the book are the absence of the calorie and of the Kenya blotched genet. The first has been replaced by the joule as a result of the very strong recommendation by the British Nutrition Society, although many of us will continue to be both figuratively and literally calorie conscious.

Genetta tigris erlangeri has disappeared because the complete chapter on inborn errors of metabolism has been omitted from this new edition. It is not that I feel that the subject is now unimportant; on the contrary, examples of congenital defects have accumulated at such a rate that it would have been impossible even to catalogue them adequately, let alone to have discussed such important topics as human polymorphism, or the still remote but exciting possibility of inserting suitable lengths of DNA into the chromosomes. Much of the material from this chapter has, however, been incorporated into the text elsewhere.

The space released has been used to extend the treatment of protein and nucleic acid synthesis, as I judge that these topics will become ever more important in medicine in the future. Elsewhere I have tried to bring the material up to date without greatly altering the basic structure. The book is indeed as complete as I can make it, and attempts that I have made to suggest excisions have brought cries of reproach from my colleagues. It is hoped, therefore, that the book in this new edition will once again have a warm welcome from students, doctors and teachers.

My thanks are due to my wife and Dr Ian Mason, for preparing the index. It is very thorough and should be of great assistance to all those who use this book.

Edinburgh, June 1975

J. H. OTTAWAY

From the Preface to The Second Edition

Fundamentally, this text is based on the courses of lectures given to medical students at University College, London and at Edinburgh University. Our experience of teaching students who are starting their training has, however, led us to expand our discussion of topics which are frequently puzzling to them, or which are in some way controversial.

The emphasis which is placed in the book on the biochemistry of blood, and particularly on acid–base status, probably makes it unsuitable for use as a text by students who are not in the medical sciences. Nevertheless, this comprehensive and, we hope, lucid account of this complex subject, whose clinical importance cannot be overemphasized, should help advance the spread of scientific medicine. We hope also that the book is still small enough to be of value to practising doctors whose days of full-time study are over.

J. H. OTTAWAY

S. P. DATTA

Contents

Definitions

Before starting the book proper it will be useful to define some units which frequently occur in biochemistry.

Mole. One mole of a substance is the mass of that substance in grammes that is numerically equal to its molecular weight.

Molar solution. A molar solution (abbreviation: M) of a substance is one which contains one mole of that substance in one litre of solution.

The terms mole and molar solution are extended to ions, thus a one molar solution of disodium hydrogen phosphate (Na_2HPO_4) is one molar with respect to phosphate ions and two molar with respect to sodium ions.

Equivalent. One equivalent (abbreviation: equiv.) of an ion is the weight of one mole of that ion divided by the number of its electric charge. Thus one equivalent is equal to one mole for singly charged ions (Na^+, Cl^-), to $\frac{1}{2} = 0.5$ moles for doubly charged ions (Ca^{2+}, HPO_4^{2-}), to $\frac{1}{3} = 0.33$ moles for triply charged ions (Fe^{3+}, $citrate^{3-}$), etc. Or, conversely, one mole of a singly charged ion contains one equivalent, one mole of a doubly charged ion contains two equivalents, one mole of a triply charged ion contains three equivalents, etc.

Multiples and submultiples of units are shown by prefixes as follows:

Multiplier	Prefix	Symbol	Example
$1\,000\,000 = 10^6$	mega	M	megaunit (Mu)
$1\,000 = 10^3$	kilo	k	kilogramme (kg)
$0.1 = 10^{-1}$	deci	d	decimetre (dm)
$0.01 = 10^{-2}$	centi	c	centimetre (cm)
$0.001 = 10^{-3}$	milli	m	millilitre (ml)
$0.000\,001 = 10^{-6}$	micro	μ	microlitre (μl)
10^{-9}	nano	n	nanometre (nm)
10^{-12}	pico	p	picogramme (pg)

Only one multiplying prefix is used at one time to a given unit. Thus one thousandth of a millimole is not called one millimillimole (1 mmmol) but is known as one micromole (1 μmol) and instead of 1 millimicrometre (0·001 μm) one writes 1 nanometre (nm).

Examples. The following examples will be found useful in the laboratory.

A 1 molar (M) solution contains 1 mole per litre (1 mol/l), 1 millimole per millilitre (1 mmol/ml), and 1 micromole per microlitre (1 μmol/μl).

A 1 millimolar (1 mM) solution contains 1 millimole per litre (1 mmol/l), 1 micromole per millilitre (1 μmol/ml), and one nanomole per microlitre (1 nmol/μl).

Each millilitre of a solution which is 1 millimolar with respect to *both* KH_2PO_4 and Na_2HPO_4 contains 1 μmole K^+, 2 μmoles Na^+ and 2 μmoles total phosphate ($H_2PO_4^- + HPO_4^{2-}$). *Each litre* of such a solution contains 1 meq. K^+, 2 meq. Na^+, and 3 meq. total phosphate, made up of 1 meq. $H_2PO_4^-$ and 2 meq. HPO_4^{2-} since each mole of HPO_4^{2-} is equal to 2 equivalents.

Dalton, the atomic mass unit equal to $\frac{1}{12}$ of the mass of an atom of carbon-12.

Joule. The joule is a unit of energy and is numerically equivalent to 0·239 calorie. One calorie is equal to 4·184 joules (J) and one kilocalorie is equal to 4·184 kilojoules (kJ).

Hydrogen Ion Concentration

Acids and Bases

An acid is a molecular species tending to lose a hydrogen ion while a base is a species tending to add on a hydrogen ion. The dissociation of a hydrogen ion from an acid may be represented by the equilibrium

$$A \rightleftharpoons B + H^+ \tag{1}$$

Since the dissociation is reversible the species B formed when A loses a hydrogen ion is in fact a base; when equilibrium (1) is displaced to the left B adds a hydrogen ion. Such a pair of species is known as a conjugate acid-base pair. Since an acid loses a hydrogen ion to form its conjugate base it follows that the acid must always have a charge which is one unit more positive than its conjugate base. These points are illustrated in the following equilibria:

$$HCl \rightleftharpoons H^+ + Cl^- \tag{2}$$

$$CH_3 \cdot COOH \rightleftharpoons H^+ + CH_3 \cdot COO^- \tag{3}$$

$$H_2PO_4 \rightleftharpoons H^+ + HPO_4^{2-} \tag{4}$$

$$NH_4^+ \rightleftharpoons H^+ + NH_3 \tag{5}$$

In all biological systems the solvent is water which can itself act as an acid or as a base as is shown by the following equilibria:

$$CH_3 \cdot COOH(A_1) + H_2O(B_2) \rightleftharpoons H_3O^+(A_2) \\ + CH_3 \cdot COO^-(B_1) \tag{6}$$

and

$$R \cdot NH_2(B_1) + H_2O(A_2) \rightleftharpoons OH^-(B_2) + R \cdot NH_3^+(A_1) \tag{7}$$

The Ion Product of Water

Because of the acidic and basic potentialities of water it follows that inter-actions between water molecules themselves will give rise to H_3O^+ and OH^- ions, thus

$$H_2O(A_1) + H_2O(B_2) \rightleftharpoons H_3O^+(A_2) + OH^-(B_1) \qquad (8)$$

and the ion product of water, K_w, is given by

$$K_w = [H_3O^+][OH^-] \qquad (9)$$

The constant, K_w, has a value of about 10^{-14} moles²/litre² at ordinary temperatures, and it follows from (9) that there is a reciprocal relation between $[H_3O^+]$ and $[OH^-]$. When $[H_3O^+] = [OH^-]$, the concentration of $H_3O^+ = \sqrt{K_w} = 10^{-7}$ moles/litre. This is the concentration of H_3O^+ ions at *neutrality*.

When acids dissociate in water, as in (6), they give rise to the hydronium ion H_3O^+ and not to the hydrogen ion H^+ though for simplicity we shall always refer to the hydrogen ion and write H^+.

Strong Acids

When strong mineral acids are dissolved in water the dissociation of the hydrogen ion may be considered to be complete. Thus HCl, $HClO_4$, HNO_3 and the first hydrogen of H_2SO_4 are completely ionized in dilute solution. In other words equilibrium (1) is completely over to the right.

Weak Acids

When weak acids such as $CH_3 \cdot COOH$, H_3PO_4, $H_2PO_4^-$, HPO_4^{2-}, HSO_4^- and $CH_3 \cdot NH_3^+$ are dissolved in water they are incompletely dissociated, that is to say both the acids and their conjugate bases are present in the solution in similar concentrations. All these dissociations may be represented by the general equilibrium

$$HA \rightleftharpoons A^- + H^+ \qquad (10)$$

where the charge on the conjugate base, A^-, is one unit less positive than on the conjugate acid, HA.

Acid Dissociation Constants

The Law of Mass Action may be applied to these equilibria giving (from 10)

$$K_{HA} = \frac{[A^-][H^+]}{[HA]} \qquad (11)$$

where K_{HA} is the equilibrium or acid dissociation constant of the acid HA. The constant K_{HA} has the dimensions of concentration and is a measure of the 'strength' of the acid, the larger the value of K_{HA} the 'stronger' the acid. The following acids are arranged in order of their 'strengths' at 25°C. H_3PO_4, $K = 8.91 \times 10^{-3}$; $CH_3 \cdot COOH$, $K = 2.24 \times 10^{-5}$; $H_2PO_4^-$, $K = 1.58 \times 10^{-7}$; $CH_3 \cdot NH_3^+$, $K = 2.40 \times 10^{-11}$, and $K_w = 10^{-14}$.

pH and pK

The numerical values of $[H^+]$ and K with which we have to deal are very small, such as the values of K listed above; $[H^+]$ at neutrality $= 10^{-7}$ moles (100 nanomoles)/litre. To simplify calculations the pH and pK scales are used; these are defined as the negative logarithms to the base 10 of the hydrogen ion concentration and the acid dissociation constant respectively.

$$pH = -\log [H^+] = \log \frac{1}{[H^+]} \qquad (12)$$

$$pK = -\log K = \log \frac{1}{K} \qquad (13)$$

The hydrogen ion concentration of the blood, which is kept fairly constant, can also conveniently be expressed as nanomoles of hydrogen ion per litre. Normally blood has a pH = 7.4 or $[H^+] = 40$ nmol/l (see Chapter 14).

It follows from these definitions that the functions pH and pK have the following important properties:

(a) The higher the hydrogen ion concentration $[H^+]$, in moles/litre, the lower the pH and vice versa, e.g. if $[H^+] = 3 \times 10^{-7}$, pH = 6.523 and if $[H^+] = 2 \times 10^{-4}$, pH = 3.699. Similarly the lower the pK the greater K and the 'stronger' the acid; thus at 25°C for H_3PO_4, pK = 2.05; $CH_3 \cdot COOH$, pK = 4.65; $H_2PO_4^-$, pK = 6.8; $CH_3 \cdot NH_3^+$, pK = 10.62, and $pK_w = 14$.

(b) A tenfold change in $[H^+]$ or K corresponds to a change of one unit in pH or pK; e.g.

$$[H^+] = 10^{-6} \text{ mol/l or } 1 \ \mu\text{mol/l, pH} = 6$$

$$[H^+] = 10^{-7} \text{ mol/l or } 100 \text{ nmol/l, pH} = 7$$

$$[H^+] = 10^{-8} \text{ mol/l or } 10 \text{ nmol/l, pH} = 8$$

The Relation between pH and pK

We can now rewrite equation (11) in terms of pH and pK to give the very important equations:

$$pK = pH - \log \frac{[A^-]}{[HA]} \qquad (14)$$

and

$$pH = pK + \log \frac{[A^-]}{[HA]} = pK + \log \frac{[\text{conjugate base}]}{[\text{conjugate acid}]} \quad (15)$$

From (15), the Henderson–Hasselbalch equation, it follows that $pH = pK$ when $[A^-] = [HA]$, i.e. when the acid is half neutralized.

The pH of a Solution of a Weak Acid

A weak acid dissociates in solution as shown in equations (10) and (11), and further it is necessary for the solution to remain electrically neutral, i.e. there must be the same *total* number of positive charges on the ions as there are negative charges. Then ignoring the small $[OH^-]$ in an acid solution and assuming HA is uncharged, the electroneutrality condition is:

$$[A^-] = [H^+] \quad (16)$$

Let the total concentration of acid be A_T, then from (10)

$$A_T = [A^-] + [HA] \quad (17)$$

and from (11)

$$[H^+][A^-] = K[HA] \quad (18)$$

Combining (16), (17) and (18) we have:

$$[H^+]^2 = K(A_T - [H^+]) \quad (19)$$

In a dilute solution of a weak acid $[H^+]$ may be assumed to be small compared with A_T, so (19) becomes

$$[H^+] \approx \sqrt{(KA_T)} \quad (20)$$

or

$$pH = \tfrac{1}{2}pK - \tfrac{1}{2}\log A_T \quad (21)$$

For example if we have solutions of acetic acid, $pK = 4.65$, at concentrations of 0.1, 0.01, and 0.001 molar their pH's will be given by:

$$pH = \tfrac{1}{2}(4.65) - \tfrac{1}{2}\log (0.1, 0.01, \text{ and } 0.001)$$

$$= 2.325 + 0.5, 1.0, \text{ and } 1.5$$

$$= 2.825, 3.325, \text{ and } 3.825 \text{ respectively.}$$

Buffers

These are solutions of weak acids, HA, and their salts, MA; such systems resist changes in the pH when acid or alkali is added to the solution.

The acid HA is by definition a weak acid so we may assume that it is only very slightly dissociated and the concentration [HA] is equal to the *total concentration of acid added.* Further we may assume that the salt MA, if it is an alkali metal salt, is completely dissociated into M^+ and A^- and hence the concentration of the conjugate base, $[A^-]$, is equal to the *total concentration of salt added.* Equation (15) then becomes:

$$pH = pK + \log \frac{[salt]}{[acid]} \tag{22}$$

Both monobasic and polybasic acids form buffers. Typical examples of the first type are the acetic acid–acetate buffers and of the second type, the phosphate buffers. When the pK's of the various groups of a polybasic acid are near each other (e.g. citric acid) the analysis of the buffer system is more complex and will not be considered.

Acetate buffers. The buffering action of mixtures of acetic acid (pK 4·65) and Na acetate is illustrated in Fig. 1.1 which shows the changes in pH when 100 ml 0·2 molar $CH_3 \cdot COOH$ are titrated with 2M–NaOH.

On each addition of 1 ml of 2M–NaOH, the pH rises sharply from A to B and then less rapidly past C, the point of half neutralization, to D. Then the pH rises very sharply as the equivalence point is reached at E.

Fig. 1.1 shows that buffers most strongly resist changes in pH near the point of half neutralization (C), that is when pH = pK. Further it is seen that the range over which buffers are effective is about 1 pH unit on either side of the pK, i.e. from [salt]/[acid] = 1/10 to 10/1 (equation (22)).

Putting this in another way, we can say that the rate of change of pH with titre is minimal when pH = pK, but it should be noted that the rate of change of hydrogen ion concentration, $[H^+]$, with titre is least at the

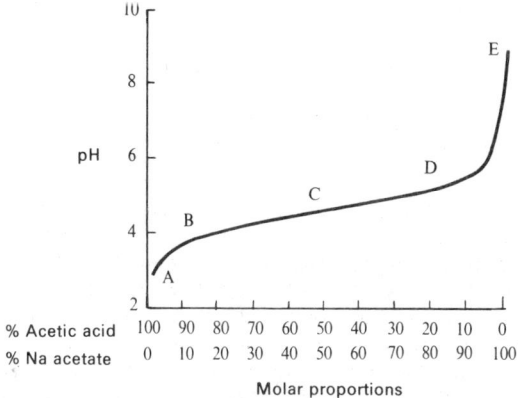

Fig. 1.1. pH titration curve of $CH_3 \cdot COOH$.

lowest values of $[H^+]$ (highest pHs). On the other hand, the rate of change of hydroxide ion concentration, $[OH^-]$, with titre is large at low $[H^+]$ (high pH) and low at high $[H^+]$ (low pH).

Phosphate buffers. In the titration curve of phosphoric acid shown in Fig. 1.2 three distinct regions of buffering can be distinguished; these correspond to the three dissociations of phosphoric acid. As phosphoric

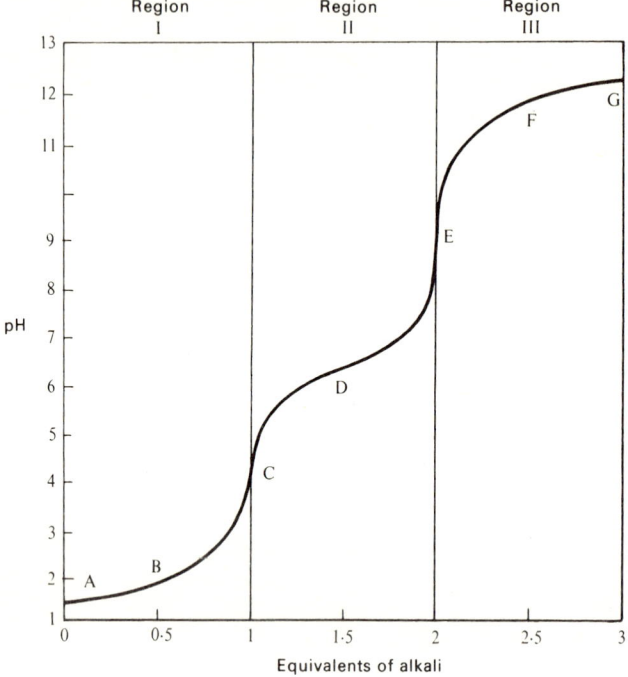

Fig. 1.2. pH titration curve of H_3PO_4.

acid, H_3PO_4, can lose 3 hydrogen ions per mole, 3 equivalents of alkali are required to neutralize it completely. There are, therefore, 3 sodium (or other metal) phosphates, NaH_2PO_4, Na_2HPO_4, and Na_3PO_4. The main features of the phosphate system are indicated below and in Fig. 1.2.

The pH of any solution of phosphoric acid and/or its alkali metal salts can be determined from equations (23), (24) or (25) below. The addition of alkali to any phosphate solution will cause the pH to rise along the curve in Fig. 1.2, while the addition of a strong acid will cause the pH to fall along the curve.

Region I
Equilibrium $\qquad H_3PO_4 \rightleftharpoons H_2PO_4^- + H^+$

Curve $\qquad\qquad$ A $\quad$ to $\quad$ C

Mid-point of equilibrium, B

$$pK_1 = 2{\cdot}0$$

$$pH = pK_1 + \log\frac{[H_2PO_4^-]}{[H_3PO_4]} \tag{23}$$

Region II
Equilibrium $\qquad H_2PO_4^- \rightleftharpoons HPO_4^{2-} + H^+$

Curve $\qquad\qquad$ C $\quad$ to $\quad$ E

Mid-point of equilibrium and point of maximum buffering, D

$$pK_2 = 6{\cdot}8$$

$$pH = pK_2 + \log\frac{[HPO_4^{2-}]}{[H_2PO_4^-]} \tag{24}$$

Region III
Equilibrium $\qquad HPO_4^2 \rightleftharpoons PO_4^{3-} + H^+$

Curve $\qquad\qquad$ E $\quad$ to $\quad$ G

Mid-point of equilibrium, F

$$pK_3 = 11{\cdot}7$$

$$pH - pK_3 + \log\frac{[PO_4^{3-}]}{[HPO_4^{2-}]} \tag{25}$$

The Measurement of pH

The fundamental instrument for the measurement of pH is the hydrogen electrode, though for routine use the glass electrode is more convenient.

The glass electrode. If a thin bulb of a special glass is placed in a solution it acquires a potential which depends on the pH in the same way as does that of a hydrogen electrode. In order to measure the potential of the glass membrane it is necessary to have a reference electrode (generally Ag · AgCl · HCl) *inside* the glass bulb as well as a reference electrode connected to the test solution by a salt bridge. Then the potential difference between the two reference electrodes is given by the equation:

$$E = E' + \frac{2{\cdot}303\mathbf{R}T}{\mathbf{F}} \times pH \tag{26}$$

where **R** is the gas constant, **F** the Faraday, T the Kelvin temperature, and E' is a constant for the system.

In practice it is always first necessary to measure the potential of the glass electrode system in a standard buffer of known pH and then in the test solution. If E_S is the potential of the electrode system in a standard buffer pH_S, then the pH of the test solution, pH_X, is given by:

$$pH_X = pH_S + \frac{(E_X - E_S)F}{2 \cdot 303 R T} \tag{27}$$

The potential of the hydrogen-saturated calomel electrode system can be measured with an ordinary potentiometer. The glass electrode system, on the other hand, has so high a resistance that the potential has to be measured with a high input impedance voltmeter usually arranged as a pH meter, that is to say the potentiometer is divided to read directly in pH units. In spite of the scale calibrated in pH units it must be emphasized that a calibration measurement in a buffer of known pH must always be made before measuring the pH of the test solution.

Indicators. Indicators are weak organic acids which change colour on ionization. Thus the acid form of methyl red is red while the conjugate base is yellow. Similarly with phenolphthalein, the acid is colourless while the base is pink. The dissociation of the indicator, HI, may be represented thus

$$HI \rightleftharpoons H^+ + I^- \tag{28}$$

with a dissociation constant K_I. Then equation (15) becomes:

$$pH = pK_I + \log \frac{[I^-]}{[HI]} \tag{29}$$

Since the species HI and I^- are of different colours (red and yellow for methyl red, colourless and pink for phenolphthalein) the colour of the solution depends on the ratio $[I^-]/[HI]$, i.e. it depends on the second term on the right of equation (29) and therefore on the pH of the solution.

Because of the difficulty in discriminating between small changes in one colour in a large excess of another colour, the useful range of an indicator is only about 1 pH unit, i.e. over the range $pK_I \pm 0 \cdot 5$ pH (see Table 1.1). The measurement of pH then resolves itself into the choice of an indicator with a suitable pK_I, near the pH to be measured, and the determination of the concentration ratio of the two colours. This is most easily done by comparing the colour of the unknown solution containing a little indicator, with the colour of standard buffers of known pH containing the same concentration of indicator. Indicators are not so reliable as glass or hydrogen electrodes for the measurement of pH, as pK_I is often affected by the presence of salts, proteins, etc. Indeed the 'protein error' of a paper strip soaked in

Table 1.1
Table of Indicators

Indicator	Useful range of pH and colour change
Thymol blue (acid range)	1·2 red–2·8 yellow
Tropaeolin—thymol blue	1·0 red–3·5 yellow
Methyl orange	3·0 red–4·4 yellow
Bromophenol blue	2·8 yellow–4·6 blue
Methyl red	4·2 red–6·3 yellow
Chlorophenol red	5·0 yellow–6·6 red
Bromothymol blue	6·0 yellow–7·6 blue
Phenol red	6·8 yellow–8·4 red
Phenolphthalein	8·3 colourless–10·0 violet-red
Thymol blue (alkali range)	8·0 yellow–9·6 blue
Thymol violet	9·0 yellow–13·0 violet

a suitable indicator can be used as a test for the presence of protein, e.g., in urine.

Physiological Buffers

The buffers important in vivo are those which are effective around pH 7·4, the pH of blood. The pH of urine, however, can vary between 4 and 9. The chief systems are listed below.

Bicarbonate. The pK_1 of carbonic acid is 6·1. The ratio of base/acid at pH 7·4 is, therefore, 20/1, which means that the bicarbonate system is a good buffer when blood is being acidified, but very poor if it is being made alkaline. The concentration of HCO_3^- ions in plasma is about 0·03M. Bicarbonate is also useful in buffering urine.

Phosphate. The pK of the equilibrium $H_2PO_4^- \rightleftharpoons HPO_4^{2-}$ is 6·8, i.e. the ratio $[HPO_4^{2-}]/[H_2PO_4^-]$ in plasma is 4/1. This makes phosphate a more efficient buffer than bicarbonate at physiological pHs, but its concentration in plasma is only 0·002M. In cells, the various phosphate esters, which have very roughly the same pK as inorganic phosphate, come to about 0·08M, and are therefore important buffers. Inorganic phosphate is the chief buffer in urine.

Amino acids. Most of these compounds are dibasic, i.e. in going from pH 1–pH 10 they lose two protons. The pKs of the COOH and NH_3^+ groups are, however, far removed from 7·4 and they are not important, except in buffering the HCl released in the gastric juice. The free amino acid concentration is also small.

Proteins. Many of the amino acids in peptide chains have acidic or basic groups not forming part of a peptide bond (e.g. glutamic acid, lysine). These groups can buffer solutions but, as with the amino acids, the pKs are far removed from 7·4, with one exception, *histidine* (pK 6·0). As haemoglobin is so concentrated in blood (14 g/100 ml ≡ 0·008M) and it contains a good deal of histidine, haemoglobin accounts for 60 per cent of the buffer capacity of whole blood. On the same principle, the plasma proteins account for another 20 per cent.

Chemistry of the Carbohydrates

Stereoisomerism

Stereoisomerism depends on the fact that the 4 valencies of carbon are given a direction *in space*. Many kinds of experiment have made it quite clear that the valencies point away from each other as much as possible, and this means that they are directed towards the angles of a regular tetrahedron, if the atom itself is supposed to be at the centre of the tetrahedron (Fig. 2.1).

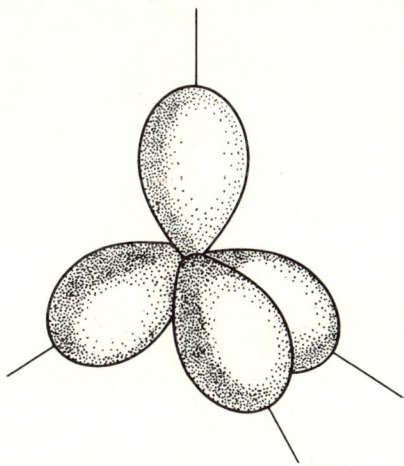

FIG. 2.1. The spatial arrangement of the valencies of carbon.

Optical Isomerism

The common factor in this type of isomerism is that all the compounds which show it are *optically active*; that is they, or their solutions, rotate plane-polarized light. A light ray can be thought of as made up of a number of waves of energy, pulsating at all angles to the direction of the ray. If a ray passes through a crystal of calcite it can be shown that in the emergent ray the waves of energy are pulsating in one direction only. Such light is called *polarized* (Fig. 2.2). The angle of the polarized ray may be determined with another calcite crystal. If the crystal lattice of this second crystal makes the same angle with the polarized ray or beam as did the first, the ray will pass through it. Otherwise no light will emerge from the second crystal. If a translucent solution of an optically active substance is placed between the two calcite crystals, the angle of the second crystal will have to be changed by an amount, α, characteristic for each compound at a given concentration. This is the principle of the *polarimeter*.

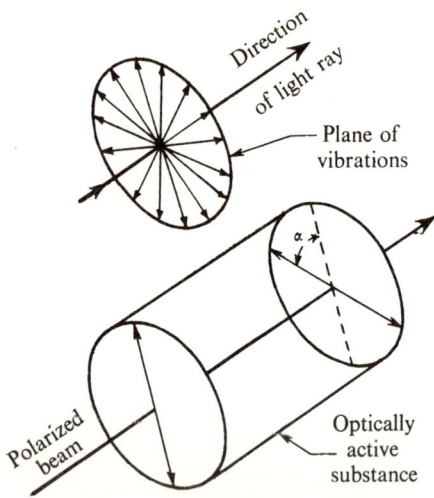

Direction of light ray

Plane of vibrations

Polarized beam

Optically active substance

FIG. 2.2. Diagrams of unpolarized (above) and polarized (below) light beams. The lower diagram shows the rotation of the plane of polarization by an optically active substance.

All optically active molecules are without a plane, or centre, of symmetry. An asymmetric molecule is not identical with its mirror image.

The simplest way in which an organic molecule can be asymmetric is for it to possess one carbon atom with *four different* radicals attached to it. The mirror image of such a carbon atom cannot be superimposed on the original (Fig. 2.3).

Fig. 2.4 represents D-glucose and its mirror image, L-glucose, and it will be seen that no matter how L-glucose is turned about in space, it cannot be superimposed on D-glucose.

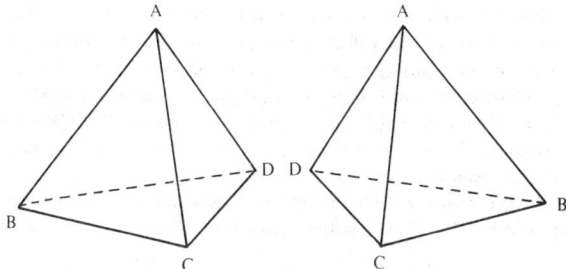

FIG. 2.3. Diagram of asymmetric molecules.

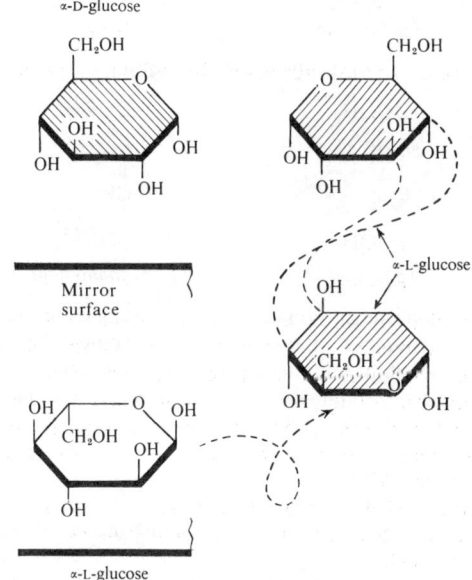

FIG. 2.4. Non-identity of D- and L-glucose. D-glucose is the mirror image of L-glucose.

Since the two 'mirror image' formulae of the optically active compounds cannot be superimposed, there are actually two molecular forms of each such compound, and furthermore the forms differ in their effect on polarized light. If one molecule rotates the plane to the right, its mirror image will rotate it equally to the left. Such pairs of mirror image compounds are called *enantiomorphs*. They have identical melting points and physical properties and, in all but a few particulars, identical chemical properties. Separation of the enantiomorphs is thus often a very difficult process. When a compound containing an asymmetric carbon atom is made by chemical synthesis, it is usually a mixture of equal amounts of the two enantiomorphs. This has no optical activity, it is a *racemic* mixture (*externally compensated*).

It will be appreciated that the optically active atom cannot take part in a double bond. Thus lactic acid is optically active, but not pyruvic:

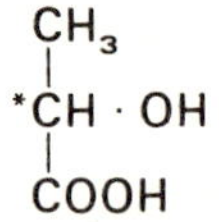

$$CH_3 - {}^*CH \cdot OH - COOH$$
Lactic acid

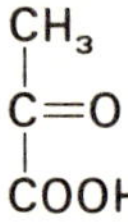

$$CH_3 - C{=}O - COOH$$
Pyruvic acid

Similarly malic acid is optically active, but not fumaric acid:

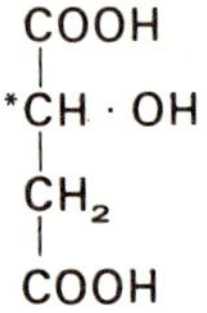

$$COOH - {}^*CH \cdot OH - CH_2 - COOH$$
Malic acid

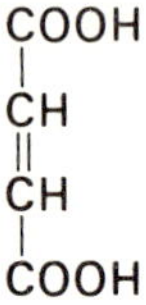

$$COOH - CH {=\!=} CH - COOH$$
Fumaric acid

When more than one carbon atom in a molecule is asymmetric, determination of the number of optically active isomers becomes difficult because of the possibility of *internal compensation*. Tartaric acid presents one of the simplest examples. L- and D-tartaric acids are enantiomorphs, and are optically active. A third isomer, however, has a plane of symmetry, and its mirror image is identical with itself. It is not optically active and is known as *meso*-tartaric acid.

Mesotartaric acid has a different melting point from D- or L-tartaric acids. A fourth variety of tartaric acid is the mixture of the D- and L-forms, *racemic* tartaric acid.

In principle, with n asymmetric carbon atoms there are 2^n possible isomers, i.e. 2^{n-1} pairs of enantiomorphs. However, the possibility of *meso*-forms often reduces this number considerably.

Nomenclature

The D- and L-forms of tartaric acid, for example, were so named because they are in fact dextro- and laevo-rotatory, i.e. they rotate the plane of polarized light to the right and the left respectively. An optically active lactic acid made from L-malic acid must of course be L-lactic acid; it is, however, dextrorotatory. This frequently occurs, and to avoid confusion the signs $(+)$ and $(-)$ are used to indicate rotations to the right and left. Thus L-$(+)$lactic acid means the dextrorotatory acid spatially (but not biologically) related to L-glyceraldehyde. Some ambiguity still exists when this convention is used, but a more precise convention (the Prelog, or R/S convention) has not yet been widely adopted by biochemists. For the *stereospecific notation* for glycerol, see p. 37.

There are some optically active molecules which do not have an asymmetric carbon atom, the whole molecule being asymmetric. Among such molecules are *inositol* and *benzene hexachloride* (the insecticide Gammexane).

Geometrical isomerism is touched on in Chapter 3.

Composition and Structure of the Monosaccharides

Simple carbohydrates, strictly speaking, have the empirical formula $(CH_2O)_n$, where n is usually 5 or 6 in the compounds discussed in this book. A number of related compounds such as deoxyribose, or glucuronic acid, which do not quite have this empirical formula, are always classed with the carbohydrates.

The monosaccharides are poly-alcohols which are also aldehydes or ketones, and their composition is best shown by the most simple of the structural formulae:

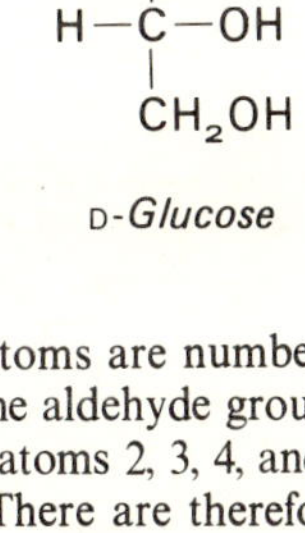

                D-*Glucose* 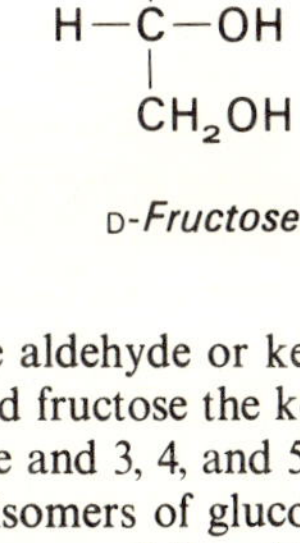   D-*Fructose*

The carbon atoms are numbered from the aldehyde or ketone end: thus glucose has the aldehyde group on C–1 and fructose the ketone group on C–2. Carbon atoms 2, 3, 4, and 5 of glucose and 3, 4, and 5 of fructose are asymmetric. There are therefore $2^4 = 16$ isomers of glucose, and $2^3 = 8$ isomers of fructose. In fact 8 distinct isomers of the aldohexoses exist, the other 8 being enantiomorphs (mirror images) of the first set, thus D- and L-glucose make a pair of enantiomorphs. The biologically important enantiomorphs are the D-forms: of the aldohexoses, glucose, galactose, and mannose; of the ketohexoses, fructose; and of the aldopentoses (which have 3 asymmetric carbon atoms, and hence 8 isomers), ribose, xylose, and arabinose.

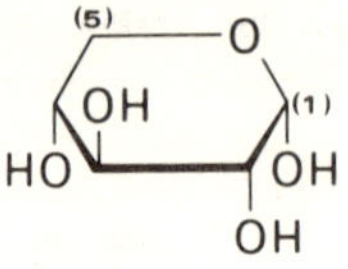

Xylose

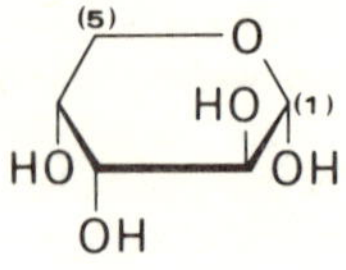

Arabinose

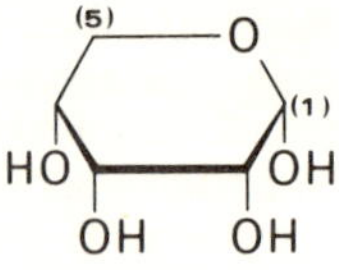

Ribose

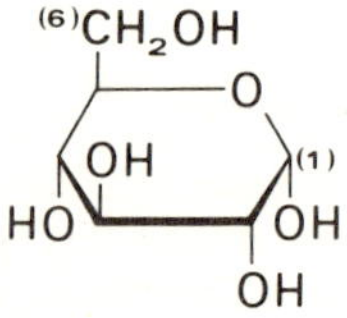

Glucose

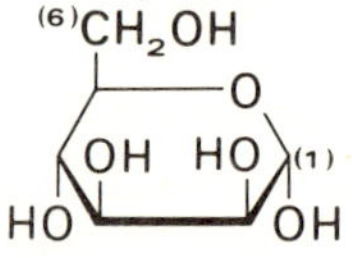

Mannose

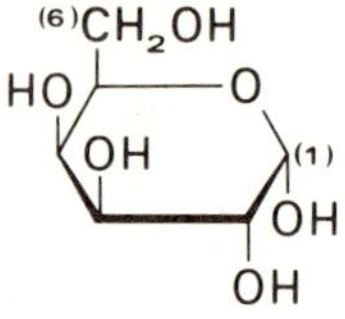

Galactose

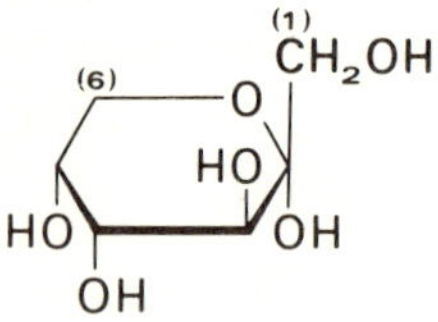

Fructose

α-Pyranose forms of some naturally occurring D-sugars. Carbon atoms 1 and 5 (pentoses) or 1 and 6 (hexoses) are numbered

The compounds are shown above in a *ring* form, because it is known that in solution, and in most of their compounds, the carbohydrates exist as rings and not as straight chains. The ring usually has 6 members, as

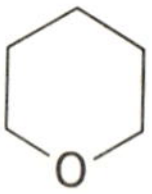

Pyran

Furan

above, and is then known as *pyranose* after the simplest similar ring *pyran*, but some compounds of fructose and the pentoses exist as 5-membered rings, shown below, and are known as *furanoses* after *furan*. The rings are conventionally shown flat, but are actually puckered, usually in a 'chair' form.

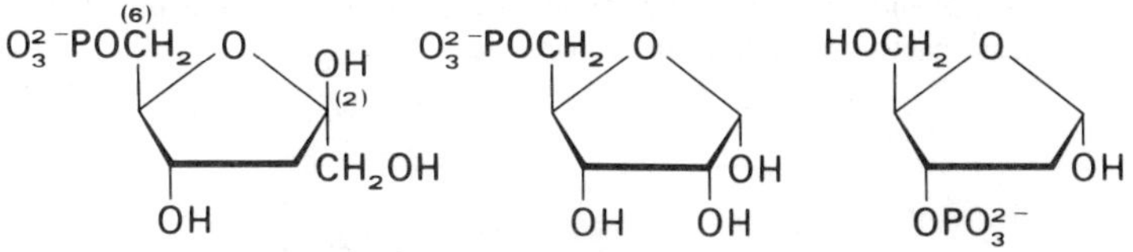

| β-D-*Fructose* 6-phosphate | α-D-*Ribose* 5-phosphate | α-D-2-*Deoxyribose* 3-*phosphate* |

Some Furanoside Esters

The ring structure introduces a further complication into the stereo-isomerism of carbohydrates. C–1 (or, in the ketoses, C–2) is asymmetric in ring structures because it is attached to 4 different groups as shown.

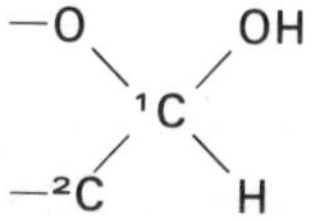

The number of isomers is thus doubled, making 32 in the aldohexose series. For example, D-glucose exists in 2 forms, known as α- and β-D-glucose, which are *not* mirror images. The α-form of D-glucose is shown on page 18, in the β-form the hydroxyl on C–1 is *above* the plane of the ring. The two forms have different optical activities, but are otherwise identical. The two forms are interconvertible in solution, a process known as *mutarotation*. However, when either anomer has been incorporated into a compound, it cannot be converted into the other. α- and β-glycosides are quite distinct compounds (see p. 21).

Properties of the Carbonyl Group

In solution there is very little of the free aldehyde (or ketone) form of the sugars, so that they do not give some of the more delicate reactions of aldehydes, such as Schiff's test or the formation of a bisulphite compound. Slightly stronger reagents will show the typical reactions.

1. *Oxidation.* By very mild oxidizing agents, e.g. ammoniacal $AgNO_3$, to an '-onic acid'. Thus glucose is oxidized to gluconic acid $CH_2OH \cdot (CHOH)_4 \cdot COOH$. Ketoses can only be oxidized by breaking the carbon chain, and are slightly less easily oxidized than aldoses (e.g. not by hypoiodite). Both aldoses and ketoses are more susceptible to *alkaline* oxidizing agents, and both may be disintegrated by vigorous oxidation. Periodic acid readily breaks the chain between two hydroxyls which are '*cis*' to each other.

2. *Condensation*. Carbonyl groups condense readily with compounds such as phenylhydrazine ($C_6H_5 \cdot NH \cdot NH_2$).

3. *Acetals and glycosides*. Almost all aldehydes and ketones in solution are at least partly in the hydrated form:

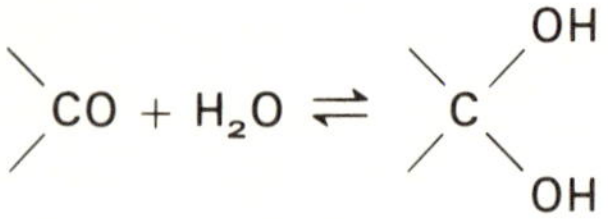

$$\backslash CO + H_2O \rightleftharpoons \begin{array}{c} \diagup OH \\ C \\ \diagdown OH \end{array}$$

This form reacts easily with alcohols to form *acetals*, e.g.

$$CH_3 \cdot CH(OH)_2 + 2\,C_2H_5OH \rightarrow CH_3 \cdot CH(OC_2H_5)_2 + 2\,H_2O$$

These compounds are perfectly distinct, though they are easily hydrolysed by acids back to the original components. If only one molecule of alcohol condenses, the compound is a hemi-acetal, e.g.

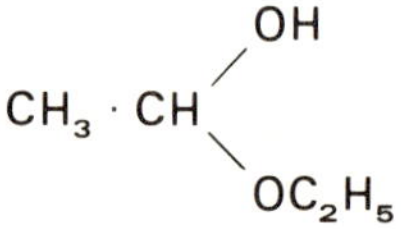

$$CH_3 \cdot CH \begin{array}{c} \diagup OH \\ \diagdown OC_2H_5 \end{array}$$

Ring forms of the monosaccharides are, by definition, hemi-acetals. They easily form the full acetal by reaction with alcohols, or with other sugars. These compounds are called, generically, *glycosides*; in particular, glucosides, galactosides, fructosides, etc. Some of the glycosides formed with complex alcohols, such as digitonin, with the sterol digoxin, are medically important. Glycosides are usually water-soluble even if the *aglycone* is insoluble in water. Sugars also form glycosides with nitrogenous compounds. The most important of these are the *nucleosides*, the ribosides of pyrimidines and purines (see Chapter 5):

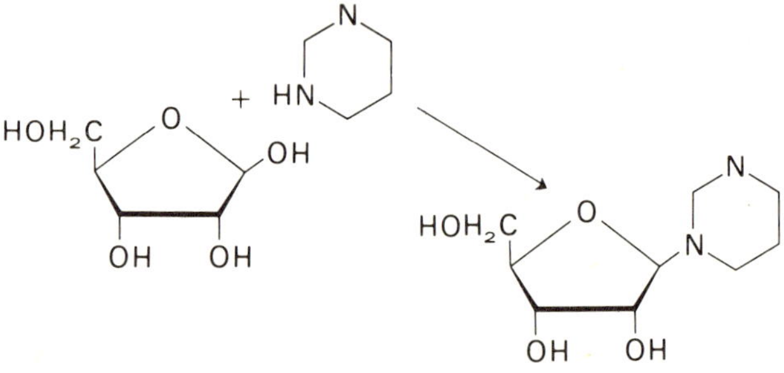

Schematic representation of nucleoside formation between ribose and a pyrimidine (the aglycone)

The glycosides do not react like esters. They are fairly stable to alkali but easily hydrolysed by acids. The $=CO$ group is protected from oxidation, so that glycosides are non-reducing (unless the aglycone contains a reducing group, as in some disaccharides).

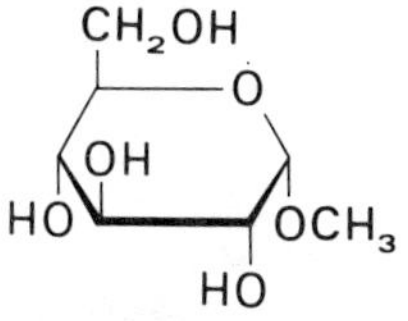

α-Methyl glucoside 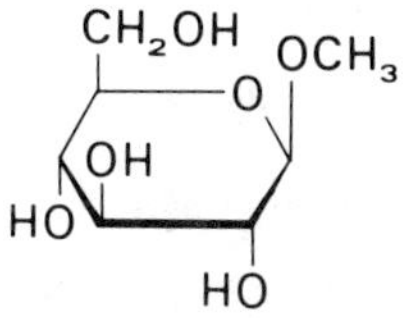   *β-Methyl glucoside*

α- and β-glycosides are separate entities, and are hydrolysed by different enzymes. The above discussion applies equally to aldoses and ketoses.

Properties of the Hydroxyl Group

Esters. All the hydroxyl groups of a monosaccharide can be esterified. Esters of the anomeric hydroxyl group are very readily hydrolysed. The most important esters biologically are the phosphates; sulphates occur in chondroid tissue.

Ethers. All $-OH$ groups other than that on the anomeric carbon can be transformed into ethers by reaction with alcohols. These ethers are not easily decomposed.

Acetals. As described above, the hydrated $=CO$ group of sugars forms acetals with alcohols. Any of the $-OH$ groups of a sugar can be used to form an acetal with another sugar. The product is a *disaccharide*.

Dehydration Reactions

Carbohydrates, on warming with concentrated H_2SO_4, are decomposed to carbon, the water being taken up by the acid. On boiling with mineral acids (usually concentrated or somewhat diluted HCl), monosaccharides are converted into *furfural*, or close relatives of it. Furfural is obviously derived from the furanose ring structures of the sugars, and ketoses and pentoses are converted into it more easily than aldoses. This is the basis of several tests for the former sugars, but these tests can be very misleading if it is not realized that *all* monosaccharides can be converted into furfural if they are boiled long enough with strong acid.

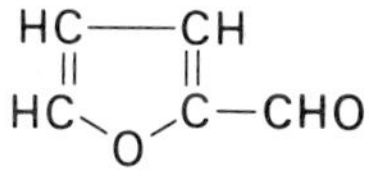

Furfural

Compounds Related to Monosaccharides

Amino sugars. Glucosamine (2-amino-D-glucose) and galactosamine (2-amino-D-galactose) are widely distributed in polysaccharides. Galactosamine is of particular interest as a component of chondroitin sulphate in cartilage. The amino sugars react very much like normal hexoses.

Deoxysugars. 2-deoxy-D-ribose is the carbohydrate of deoxyribose nucleic acid. It is more unstable than a normal sugar, and reacts with Schiff's reagent. This is the basis of the Feulgen stain for nucleic acid. *Fucose* (6-deoxy-L-galactose) is a characteristic sugar of blood-group substances (see p. 27).

Sugar acids. There are three possible sugar acids corresponding to the aldoses:

1. The dibasic *saccharic acids*, $COOH \cdot (CHOH)_4 \cdot COOH$. Mucic acid formed by the oxidation of galactose with dilute HNO_3 is relatively insoluble and can be used as a test for this sugar.

2. The *aldonic acids*, $CH_2OH \cdot (CHOH)_4 \cdot COOH$. Gluconic acid is formed by oxidizing glucose with bromine water. On further oxidation it loses CO_2 to form a pentose.

3. The *uronic acids*. These cannot easily be made chemically from sugars, but they are widely distributed in the body. Glucuronic acid, $COOH \cdot (CHOH)_4 \cdot CHO$ is a constituent of mucopolysaccharides and of chondroitin. Glucuronides, i.e.

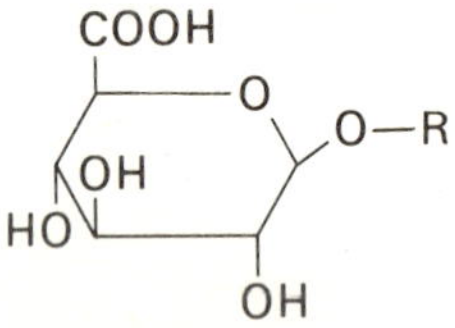

β-Glucuronide

of many compounds, especially steroids, are found in the urine. These glycosides are water-soluble and this is therefore an important way of detoxifying fat-soluble substances (see Chapter 14). With hot mineral acids, uronic acids lose CO_2 and form pentoses; they therefore give positive reactions in the pentose tests.

Sialic acids. This group of substances is formed by an aldol condensation between pyruvic acid and *N*-acylhexosamines. The unacylated compound is called *neuraminic acid* and 'sialic acid' is used as a group name for the acylated neutraminic acids. The acylating group on the amino nitrogen is frequently acetyl. The amino sugar moiety is mannosamine.

Sialic acids occur in mucoproteins of the submaxillary gland and in mucolipids of the brain; they are always accompanied by galactosamine.

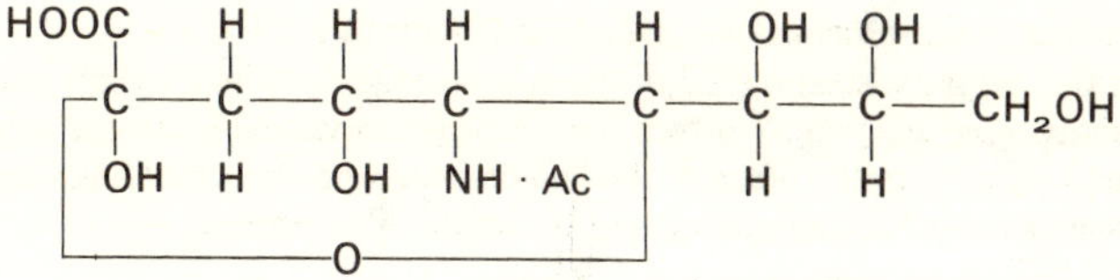

$Ac = CH_3 \cdot CO-$, or $HO \cdot CH_2 \cdot CO-$.

β-D(—)-N-*Acylneuraminic acid*
Mannosamine provided the right-hand six carbon atoms

Polysaccharides

1. *Oligosaccharides*

This means those carbohydrates composed of two, three, or four monosaccharide residues. Those of chief interest are the disaccharides maltose, lactose, and sucrose.

Maltose is α-D-glucosyl-1,4-D-glucose. One aldehyde group is left free and maltose is therefore a reducing sugar, although not so easily oxidized as the monosaccharides. It does not occur free in nature, but is one of the products of the action of amylase on starch.

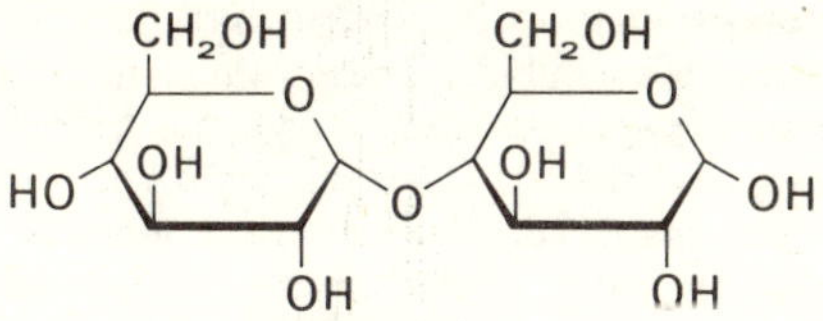

α-*Maltose*

Lactose is β-D-galactosyl-1,4-D-glucose; it is a reducing sugar appreciably less soluble and less sweet than most other sugars. It occurs to the extent of about 5 per cent in milk. Unlike sucrose and maltose it is not fermented by most yeasts.

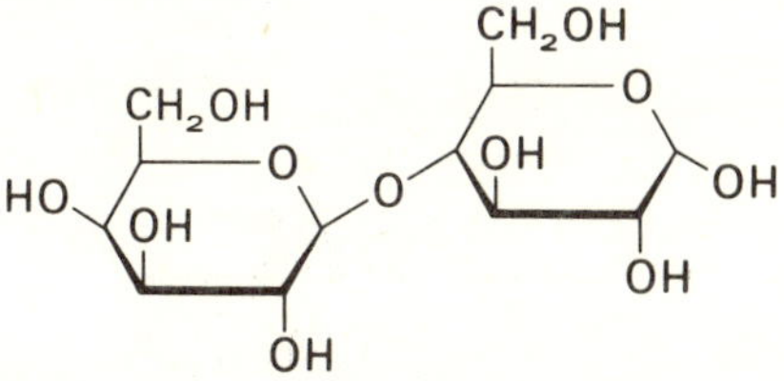

α-*Lactose*

Sucrose is α-D-glucosyl-β-D-fructoside. Note that it is a *double glycoside*; both C–1 of glucose and C–2 of fructose are involved in the link, and it is therefore *non-reducing*. Further the fructose molecule is in the *furanose* form. It occurs in the juice of the sugar-cane and the sugar-beet up to a concentration of 20 per cent and is the sugar of commerce.

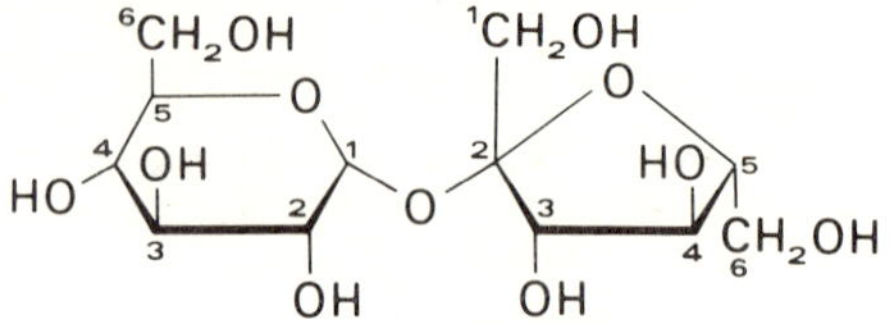

Sucrose

2. *True Polysaccharides*

These are macromolecules and form colloidal solutions. They are composed of repeating units of monosaccharides combined by glycosidic links; the patterns can be very complex, and although many of these compounds are very important in biology, only the simplest will be dealt with.

Glucosans (only glucose units)

Starch. This is a mixture of two main components, one soluble in boiling water and making up to 10–20 per cent of the total, called *amylose*; the other 80–90 per cent is insoluble in boiling water and is called *amylopectin*. Both are made up of D-glucose units. Amylose is unbranched, containing 200–2 000 glucose units linked α-1,4 in a straight line. Amylopectin, on the other hand, is highly branched. It has one end-group to 24–30 glucose units, which means that the outer chains are about 13–18 residues long. The molecule is very large, containing 250–5 000 units. The main linkage is α-1,4, but at the branch points a third molecule of glucose is joined in the 6 position, thus:

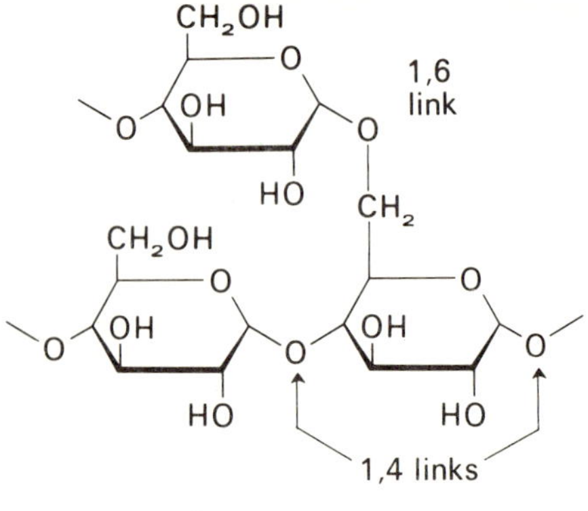

The mixture of amylose and amylopectin known as *starch* forms a gel when concentrated solutions cool. It gives a blue colour with iodine.

Dextrins. This is the generic term for the mixture of branched and unbranched soluble polysaccharides produced by the partial hydrolysis of starch by acids or amylases. The larger branched dextrins give a red colour with iodine (erythrodextrins), the smaller ones give no colour (achromodextrins).

Dextran. This is a polysaccharide, consisting of relatively unbranched 1,6 linked glucose molecules, produced by the bacterium, *Leuconostoc mesenteroides*, acting on sucrose. It is soluble, but colloidal, and can be used as a plasma substitute in the treatment of shock.

Glycogen. This polysaccharide is found only in animals. It has a structure very like that of amylopectin, except that it is even more highly branched. The average chain-length of the exterior chains is only 8 glucose units (13–18 in amylopectin), and in the main chains there is a branch point every 3 units on the average (every 5–6 units in amylopectin). The molecular weight is very high, about 5 000 000 ($\equiv$ 25 000 units). It gives a red colour with iodine (amylopectin gives a red-violet). Its structure is shown in Fig. 2.5.

Cellulose. This is a straight-chain polysaccharide consisting of β-1,4 linked glucose units. The β-configuration at C–1 ,which is apparently a trifling difference, is very important, as no vertebrate has a digestive

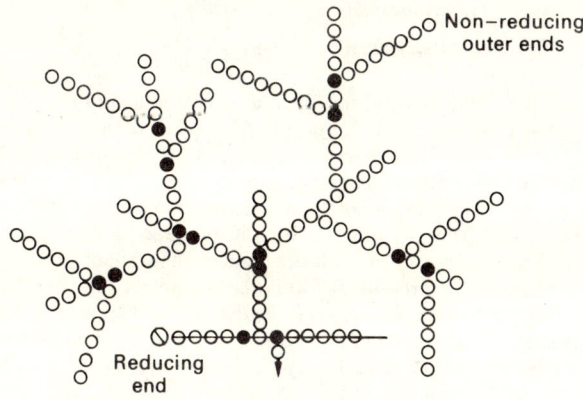

FIG. 2.5. Diagram of the structure of glycogen. The open rings represent glucose residues linked only at carbon atoms 1 and 4 while the black circles represent branch points where glucose residues are linked at carbon atoms 1, 4, and 6. The molecule is symmetrical about the original chain; the lower half of the molecule has been suppressed for the sake of clarity.

enzyme able to attack it. Ruminants digest cellulose with the aid of symbiotic micro-organisms. Cellulose is very insoluble. It may be solubilized by chemical modification to make, for example, the bulky but still indigestible carboxymethylcellulose.

Other Hexosans

Inulin is a polymer of fructose; it is inert when injected into the body.

Acid Polysaccharides (Glycosaminoglycans)

There is a wide range of polysaccharides containing acid groups, either as sulphate esters or as uronic acids. They appear to play a structural role, as distinct from the storage of carbohydrate for energy as with starch and glycogen. They form gels which can be very rigid, particularly if metal ions are present.

The most important glycosaminoglycans in the body are based on a repeating unit of the type shown in Fig. 2.6.

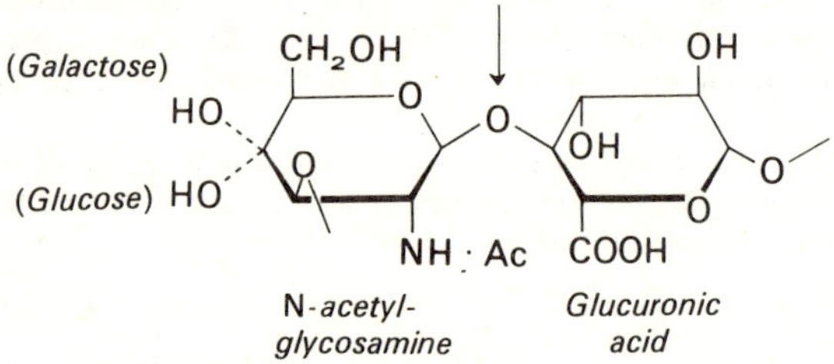

FIG. 2.6. The unit structure of many glycosaminoglycans.

Note that the bond between the glycosamine residue and the next unit is 1,3 and not the more usual 1,4 or 1,6. Also because the configuration at C–1 is always β, alternate pairs of residues will be 'upside down'. These polymers have very large water shells around them, which exclude other macromolecules. This plays a large part in their lubricating function.

Hyaluronic acid. This is a polymer of N-acetyl-glucosamine-glucuronic acid units as shown in Fig. 2.6 with the reducing end attached ultimately to a seryl residue of a protein. It forms the cement of interstitial tissue generally, and of hyalin. The enzyme *hyaluronidase* ('spreading factor') which is present in sperm and in some bacteria and in some snake venoms, depolymerizes hyaluronic acid by hydrolysing it at the bonds one of which is marked by an arrow in Fig. 2.6.

Mucoitin sulphate. This is a chain of repeating units like those in Fig. 2.6 but with a sulphate ester on one of the glucosamine hydroxyls. It is found in the lubricating mucoproteins, such as the mucin of saliva.

Heparin (the anticoagulant) is very similar to mucoitin sulphate, but more highly sulphated, and not of very high molecular weight.

Chondroitin sulphate. This differs from the previous polymers in that the glycosamine is N-acetyl-galactosamine. In chondroitin-4-sulphate (formerly chondroitin sulphate A) there is a sulphate ester on the 4-hydroxyl of this residue, and in chondroitin-6-sulphate (formerly chondroitin sulphate B) the sulphate is on the 6-OH. These polymers are again linked to a protein which is probably interpolated among collagen fibrils, as shown in Fig. 2.7. In rheumatoid arthritis chondroitin sulphate hydrolysed from this protein finds its way into synovial fluid.

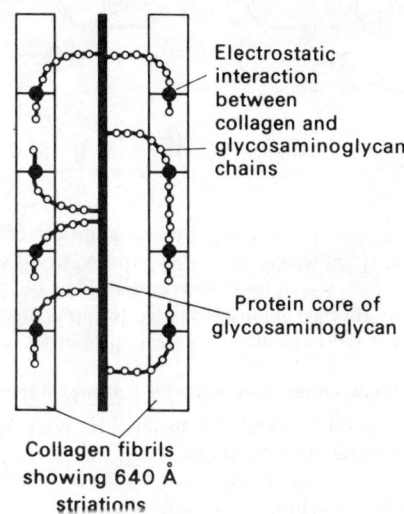

Electrostatic interaction between collagen and glycosaminoglycan chains

Protein core of glycosaminoglycan

Collagen fibrils showing 640 Å striations

FIG. 2.7. Schematized interaction of proteinpolysaccharide and collagen.

Seaweed polysaccharides. Agar and agarose are sulphated galactose polymers; alginic acid is a linear mannuronic acid polymer. All are widely used as food additives, or sometimes for adding 'roughage' to the diet.

Blood-group substances. These are glycoproteins, the peptide moiety of which contains many seryl and threonyl residues. Attached to the hydroxyl groups of these residues are as many as 300 oligopolysaccharides, all with a similar repeating pattern. The glycoprotein conferring 'H' specificity (most marked in group O individuals) terminates in the

α-fucosyl

↓1,2

β-galactosyl-(1 → 3)-N-acetyl-glucosaminyl

sequence shown in Fig. 2.8.

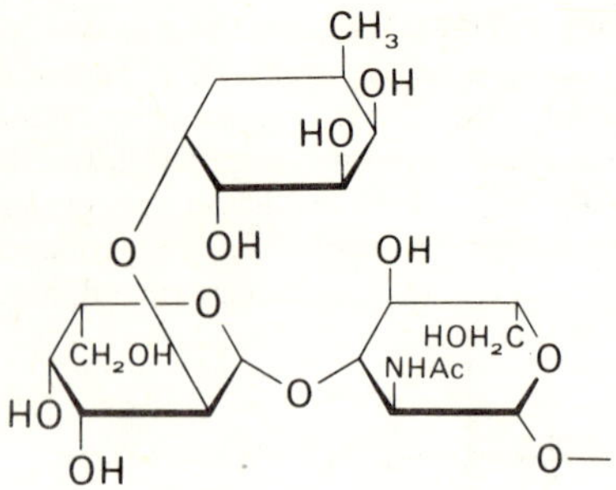

Fig. 2.8. Terminal sequences in the 'H'-specific
blood-group glycoprotein.

B group specificity is conferred by the addition of an α-galactosyl
residue at the point shown by the arrow, and A group specificity by the
addition of an N-acetyl-galactosamine residue. Thus the genetically determined presence or absence of closely related terminal glycosyl transferases
gives rise to marked alterations in antigenic properties.

Some Qualitative Tests for Carbohydrates

Molisch's test (α-naphthol and concentrated H_2SO_4). All carbohydrates,
and certain other reducing substances.

Iodine. Colours with starch, glycogen, and erythrodextrins. No colour
with achromodextrins, cellulose, or inulin.

Fehling's test (alkaline copper tartrate). All reducing sugars, uric acid
and other strong reducing agents.

Benedict's test (alkaline copper citrate). All reducing sugars and other
strong reducing agents. Not uric acid.

Fearon's test (methylamine and NaOH). Reducing disaccharides.

Barfoed's test (copper acetate in acetic acid). Monosaccharides.

Bromine water or alkaline KI. Aldoses react.

Selivanoff's test (resorcinol in HCl). Ketohexoses (aldohexoses react
slowly).

Foulger's test (urea in H_2SO_4 with $SnCl_2$). Ketohexoses.

Bial's test (orcinol in HCl + $FeCl_3$). Pentoses, glycuronic acids.

Tollen's test (naphthoresorcinol in HCl). Glucuronides.

Yeast fermentation. Glucose, mannose, fructose, sucrose, maltose.

Glucose oxidase. A specific test for free glucose.

Galactose oxidase. A specific test for free galactose.

3 Chemistry of Lipids and Steroids

Geometrical Isomerism

Geometrical isomers arise because of the arrangement in space of the bonds around a carbon atom, just as do optical isomers, but the molecules do not necessarily have a centre of symmetry, and are therefore not optically active.

Cis-Trans Isomerism of Double Bonds

This arises whenever there is a double bond between two carbon atoms. If a carbon atom is represented by a tetrahedron, the bond $C-C$ is represented by two tetrahedra attached by their apices. A double bond $C=C$ must be represented by two tetrahedra sharing an edge. If the two remaining bonds of each carbon atom are occupied by two different substituents, these latter may be arranged in two different ways. This is true even if the two pairs of substituents are identical as $XHC=CHX$, the simplest case. The two substituents X may be adjacent (*cis*), or on opposite sides of the double bond (*trans*). This is illustrated by Fig. 3.1, and by the case of maleic and

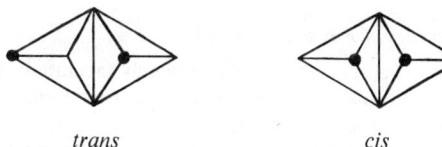

trans   *cis*

Fig. 3.1. Diagram of the two possible arrangements of two substituents attached to two carbon atoms linked by a $C=C$ double bond.

fumaric acids, below. The two forms are not interconvertible without breaking the double bond, but each form is identical with its mirror image, so there is no rotation of the plane of polarization of light.

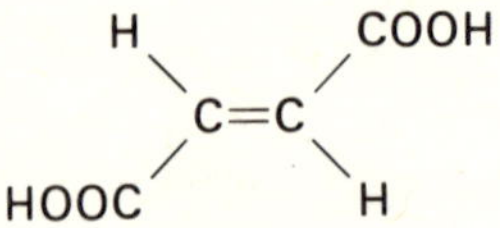

$$H\diagdown C{=}C\diagup COOH$$
$$HOOC\diagup \qquad \diagdown H$$

Fumaric acid
(trans)

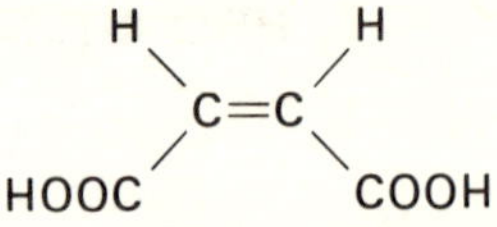

$$H\diagdown C{=}C\diagup H$$
$$HOOC\diagup \qquad \diagdown COOH$$

Maleic acid
(cis)

The physical properties of *cis*- and *trans*-isomers are almost always markedly different (contrast this with optical isomers). This is particularly true of the melting points (e.g. that of fumaric acid is 287°, and of maleic acid 130°), and solubility. The chemical properties may differ too; e.g. maleic acid easily forms an anhydride

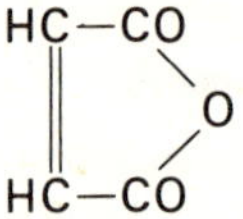

$$HC{-}CO\diagdown$$
$$\qquad\qquad O$$
$$HC{-}CO\diagup$$

fumaric acid does not. Maleic acid is a stronger acid than fumaric, and is less stable. Exposure to ultra-violet light, heat, or certain chemicals helps to convert it to fumaric acid. A lower melting point and greater chemical reactivity are characteristic of *cis*-isomers, although the chemical differences may be slight if the double bond is some distance from the reactive group.

The number of isomers of a compound containing more than one double bond may be large ($= 2^n$ for non-symmetrical molecules), as in some of the highly unsaturated lipids. Since enzymes are specific for geometrical, as well as optical, isomers, there is usually only one naturally occurring isomer of an unsaturated compound. A reservation must be made about compounds containing alternate (conjugated) double bonds, $-C{=}C-C{=}C-C{=}C-$, whose isomers are often more easily interconverted than those of simpler compounds, cf. retinene (p. 423).

Cis-Trans Isomerism of Fused Rings

Substituents on the carbon atoms of completely saturated rings stand out above or below the plane of the ring (cf. the ring forms of the sugars, Chapter 2). When two rings are fused together at *two* adjacent carbons, two isomers are formed, depending on the position of the substituents on *the atoms common in both rings* (Fig. 3.2). These isomers are also known as *cis*- and *trans*-; they are particularly important in the steroids. The diagram also shows that the rings do not lie completely flat, but fall into what are called 'bed' and 'chair' shapes. All the rings in Fig. 3.2 are drawn in the 'chair' shape, as the 'bed' shape is usually rather rare. The pyranose rings of the sugars (Chapter 2) also have this puckered conformation. Because

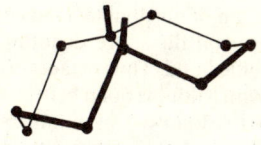   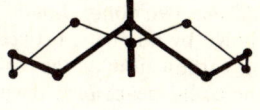

cis   *trans*

FIG. 3.2. Diagram of *cis* and *trans* isomerism of fused, non-aromatic rings. The position in space of the remaining bond on each carbon atom common to the two rings is indicated by a heavy line.

of the puckering, the substituents do not truly stand out above and below the plane of the ring. One substituent is usually at right angles to the local plane (*axial*), while the other is more or less in the plane (*equatorial*), and their reactivity may be different.

Fatty Acids and their Esters

Nomenclature

The fatty acids were originally given Greek names indicating their source. The systematic names are based on the Greek numbers, and indicate, besides the chain length (the carboxyl carbon atom is number 1), the degree of unsaturation of the compound by a suffix. Thus the suffix *-anoic* means a fully saturated compound, *-enoic*, containing one double bond,

Table 3.1
Some Common Fatty Acids

Common name	Formula	Systematic name	Melting point (°C)
Acetic	$CH_3 \cdot COOH$		16·6
Propionic	$C_2H_5 \cdot COOH$		−22
Butyric	$C_3H_7 \cdot COOH$		−19
Caproic	$C_5H_{11} \cdot COOH$	Hexanoic	−2
Caprylic	$C_7H_{15} \cdot COOH$	Octanoic	16
Capric	$C_9H_{19} \cdot COOH$	Decanoic	31
Lauric	$C_{11}H_{23} \cdot COOH$	Dodecanoic	44
Myristic	$C_{13}H_{27} \cdot COOH$	Tetradecanoic	54
Palmitic	$C_{15}H_{31} \cdot COOH$	Hexadecanoic	63
Stearic	$C_{17}H_{35} \cdot COOH$	Octadecanoic	69
Oleic	$C_{17}H_{33} \cdot COOH$	Octadec-9-enoic	4
Ricinoleic	$C_{17}H_{32}(OH) \cdot COOH$	12-Hydroxyoctadec-9-enoic	(*cis*) 6
Linoleic	$C_{17}H_{31} \cdot COOH$	Octadeca-9,12-dienoic	−12
Linolenic	$C_{17}H_{29} \cdot COOH$	Octadeca-9,12,15-trienoic	−16—17
Arachidic	$C_{19}H_{39} \cdot COOH$	Eicosanoic	77
Arachidonic	$C_{19}H_{31} \cdot COOH$	Eicosa-5,8,11,14-tetraenoic	−49

-dienoic, two double bonds, and so on. The position of the double bonds is shown by numbers, between the Greek number and the suffix, denoting the carbon atom on the carboxyl side of the double bond. Thus octadec-9-enoic acid means an acid with 18 carbon atoms with a double bond between carbon atoms 9 and 10. Sometimes the symbol Δ followed by numbers is used to indicate the position of the double bond, then the above compound is written Δ^9-octadecenoic acid. Table 3.1 lists some of the more common acids; for reasons outlined in Chapter 10, fatty acids containing an *odd* number of carbon atoms are rather rare. Acids with branched chains are also very unusual.

Saturated Fatty Acids

These are rather unreactive compounds, progressively less soluble in water as the chain lengthens. Only acetic, propionic, and butyric acids are miscible with water in all proportions. They are soluble in all fat solvents. They are rather weak acids (pK acetic acid 4·65, pK stearic acid 5·75). The alkali metal salts (soaps) of the acids are quite soluble; the heavy metal salts are very insoluble. The melting points rise steadily with increasing chain length (acetic acid is anomalous), and all the acids with 12 or more carbon atoms are solids at room temperature. The C_4 to C_{10} acids are volatile in steam, and have rather revolting rancid smells. All, with the exception of acetic acid, are less dense than water.

Saturated fatty acids can be oxidized with difficulty to acids with two atoms less in the carbon chain, together with $2CO_2$. They can be reduced to the corresponding alcohols; this is commercially important, but is not practicable in the student laboratory.

Unsaturated Fatty Acids

Short-chain unsaturated fatty acids are not often found, so that all the natural compounds are insoluble in water, and are not steam-volatile. They are *all* liquids (the effect of the double bond on the melting point is striking, see Table 3.1), and the common acids have the *cis*-configuration about the double bonds.

They are much more reactive compounds than their saturated congeners:

1. They can be *oxidized* very readily, for example by dilute $KMnO_4$ or Br_2 water. The acid usually breaks at the double bond(s) and yields two or more short-chain acids. The reaction with O_2 is more complex. The first product is a *peroxide*:

$$-CH{=}CH- \; + \; O_2 \; \longrightarrow \; \underset{\underset{OOH}{|}}{-C}{=}CH-$$

Such compounds are both formed and break down more readily in the light. They may give rise to short-chain aldehydes, or the peroxides may polymerize. This behaviour with oxygen is important both in *rancidity* and in the drying of paints.

2. *Addition* takes place at the double bond. The addition of hydrogen is, of course, reduction. The direct addition of H_2 may be catalysed by powdered nickel or other metal, and is commercially important in converting vegetable oils into edible fats.

Halogens, or water (see p. 188), may add across the double bond to form a saturated, but substituted, fatty acid, e.g.

$$-CH=CH- + HX \rightarrow -CH_2-CHX-$$

3. Unsaturated fatty acids tend to decompose at high temperatures.

Fatty Acid Esters (Neutral Fats)

Liquid esters of the higher fatty acids are called *oils*, solid esters, *fats*, or *waxes*. Fats are tri-esters of glycerol:

$$CH_2OH \cdot CHOH \cdot CH_2OH$$

Waxes, which have higher melting points than fats, are esters of high molecular weight alcohols, such as cholesterol or cetyl alcohol, $C_{15}H_{31} \cdot CH_2OH$.

Glycerol is closely related to the carbohydrates. It is a polyhydric alcohol, completely miscible with water, and sweet tasting. It can be readily oxidized to glyceric aldehyde,

$$CH_2OH \cdot CHOH \cdot CHO$$

and subsequently to glyceric acid. On heating, either alone or with a dehydrating agent such as $KHSO_4$, it forms the unsaturated aldehyde *acrolein*:

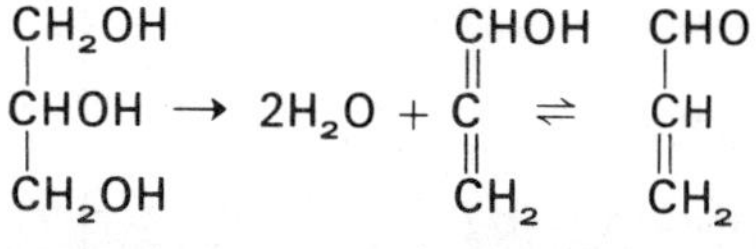

$$\begin{array}{ccc}
CH_2OH & CHOH & CHO \\
| & || & | \\
CHOH \rightarrow 2H_2O + & C & \rightleftharpoons & CH \\
| & || & || \\
CH_2OH & CH_2 & CH_2
\end{array}$$

which has a highly characteristic irritating odour.

Fats, oils, and waxes are neutral, water-insoluble compounds which can be hydrolysed by boiling with alkali to an alcohol and the salt of one or more fatty acids. Waxes are less easily hydrolysed than glycerides (note that *paraffin wax* is not an ester, but a mixture of long-chain hydrocarbons, and cannot be hydrolysed). All react to the hydroxamic acid test and are less dense than water.

Triglycerides

Most naturally occurring triglycerides, whether fats or oils, are mixed, that is to say, they contain at least two different fatty acid residues esterified to the same glycerol molecule. One of these residues is usually unsaturated (see Table 3.2).

Table 3.2
Glycerides of Depot Fat (Moles %)

	Pig	Sheep	Ox
Saturated:			
palmitic-stearic	5	5	17
Mono-oleic:			
di-palmitic	5	13	15
palmitic-stearic	27	28	32
di-stearic	0	1	2
Di-oleic:			
palmitic	53	46	23
stearic	7	7	11
Tri-oleic	3	0	0

From Table 3.2 it is seen that oleic acid is by far the most frequently occurring fatty acid in animals. This important point is further emphasized by the data in Table 3.3. Oleic acid is also very common in plant lipids.

Table 3.3
Component Acids of Fats (Moles %)

	Pig	Ox	Human
Saturated:			
$C_{12} + C_{14}$	1·3	2·6	2·6
C_{16} (palmitic)	29·0	33·4	24·7
C_{18} (stearic)	13·8	21·4	7·7
Unsaturated:			
$C_{14} + C_{16}$	2·7	2·5	7·7
C_{18} (oleic)	43·9	35·2	45·8
C_{18} (linoleic)	7·2	3·5	10·0
All others	2·1	1·4	1·5

The melting points of fats and oils are largely controlled by the proportion of unsaturated acids which they contain. Vegetable oils are more highly unsaturated even than animal fats. Butter fat is unusual in that its component acids are largely saturated. Its low melting point is a consequence of the fact that the proportion of short-chain (C_4–C_{10}) acids present is much higher than usual.

The Chemical Characterization of Fats and Oils

The routine use of gas chromatography, and of thin-layer chromatography, particularly on supports impregnated with $AgNO_3$, together with the possibility of rapidly hydrolysing fats with lipase, has made the characterization of lipids much more precise than older techniques could have hoped to achieve. Whenever necessary, the spectrum of chain lengths, and of the number and position of double bonds in the individual acids can be quoted for any sample.

Other Natural Esters

Medically, at least, the most important esters of the higher alcohols are those of cholesterol. Blood plasma contains a varying amount of these esters as well as free cholesterol. About 60 per cent of the acid in these plasma esters is linoleic acid; the significance of this is unknown. Wool fat, or lanolin, contains a large percentage of cholesterol and lanosterol esters; it has the property of absorbing a large amount of water without forming two phases. Hydrated lanolin is therefore used as an ointment base for carrying water-soluble drugs.

Rancidity

This is partly caused by the enzymic hydrolysis of glycerides to produce free fatty acids; any short-chain acids will be volatile, and have an objectionable odour. It is also due to the formation of peroxides by unsaturated fats exposed to air and light. The peroxides may break down to aldehydes or acids of rancid odour or taste. Purified glycerides become rancid much more easily than natural fats, because the latter contain substances, such as the tocopherols, which act as antioxidants (see vitamin E, Chapter 20).

Phosphatides

Phosphatides contain an esterified phosphate group and are not neutral fats. Every cell contains phospholipids; they appear to play an essential, probably structural role. In starvation the body's reserve of neutral fat may almost completely disappear, yet the amounts of phospholipid and cholesterol will remain unchanged. Various lines of evidence indicate that many membranes probably have a unit structure consisting of a double layer of lipid, including much polar lipid, and protein (Fig. 3.3). The most striking evidence comes from electron micrography and from the observation that if the lipid is removed by solvents, or if the phospholipids are attacked by phospholipases (p. 39), the functional and sometimes the physical integrity of the membrane disappears. It is certain that the membranes of animal cells, and of their constituent organelles, are not held together by covalent forces, unlike the components of bacterial cell walls. The stability of these membranes depends very largely on the hydrophobic attraction that brings together the fatty acyl groups of the constituent lipids, and segregates them from the aqueous environment (cf. protein conformation, p. 69).

The rigid pattern of Fig. 3.3 may be a fair approximation to reality for simple membranes such as the myelin sheath, but it is likely that it oversimplifies the picture for other membranes, especially those which contain

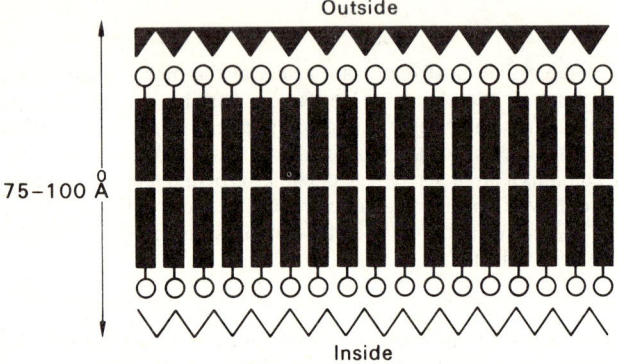

Outside

75–100 Å

Inside

Fig. 3.3. Schematized version of the arrangement of molecules in a unit membrane structure.

Circles represent the polar heads of lipid molecules; the dense bars, the nonpolar carbon chains; the zig-zag lines, monomolecular films of non-lipid (often protein or glycoprotein). The outer monolayer is filled to indicate that it is chemically different from the inner.

permeases or enzymes. For one thing, many proteins contain hydrophobic regions which may be folded into the lipid bi-layer. It is also probable that electron microscopy exaggerates the rigidity of the structures; some recent work suggests that membranes can have a mobile structure, with 'islands' of protein forming and dispersing in the lipid sheet. Indeed, evidence for pinocytotic ingestion into some cells continues to accumulate, while there is very strong evidence that proteins 'for export' which are synthesized in non-apocrine cells are first organized in the membrane-like Golgi bodies. These membranes then become contiguous with the cell membrane, so that the proteins are on the outside of the cell. It would be very difficult to interpret these phenomena without invoking a dynamic membrane structure. Phospholipids certainly play an important role in maintaining such structures, and also those of the lipoprotein complexes found in plasma.

In general, phospholipids are not soluble in water, although they may form emulsions or *micelles* (sub-microscopic aggregates). They are not soluble in acetone, which enables them to be separated from neutral fats and fatty acids.

The phosphate present is usually esterified twice, thus:

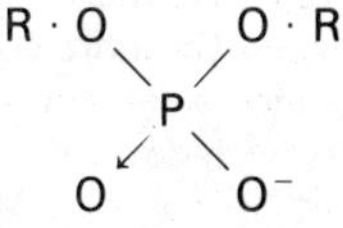

$$R \cdot O \diagdown \quad \diagup O \cdot R'$$
$$P$$
$$O \diagup \quad \diagdown O^-$$

where R and R' are groups derived from alcohols. The major alcohol, forming the 'backbone' of the phospholipids, is glycerol in the larger number of phospholipid types. The complex alcohol sphingosine forms the backbone of another important group.

Stereospecific Notation for Glycerol Derivatives

Glycerol itself is not optically active, but any derivative of even one of its —OH groups is bound to be so. In the older literature the L- and D- convention (see p. 16) was used, together with the convention β- for a function on the central —OH, and α- for one on either of the outer hydroxyls. This is, however, ambiguous, and since 1967 a *stereospecific (sn) notation* has been used. This specifies that if the central —OH is spread out to the *left* (in the Fischer notation), the top —OH is numbered 1-, and the bottom —OH is 3- (see Fig. 3.4).

$$\begin{array}{cc} CH_2OH & CH_2O \textcircled{P} \\ | & | \\ HO-C-H & IIO-C-H \\ | & | \\ CH_2O \textcircled{P} & CH_2OH \end{array}$$

3-sn-Phosphoglycerate   *1-sn-Phosphoglycerate*

Both these could be called L-α-glycerophosphate

FIG. 3.4. Stereospecific notation for glycerol esters.

The most common alcohols making up the other part of the phospho-diester group are serine, ethanolamine and choline, which are related as follows:

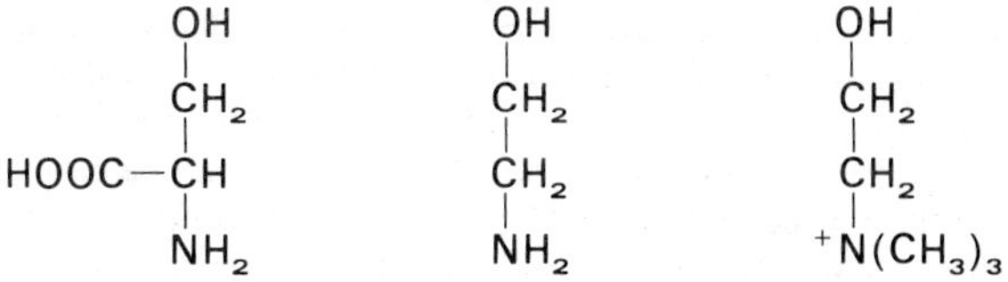

$$\begin{array}{ccc} OH & OH & OH \\ | & | & | \\ CH_2 & CH_2 & CH_2 \\ | & | & | \\ HOOC-CH & CH_2 & CH_2 \\ | & | & | \\ NH_2 & NH_2 & {}^+N(CH_3)_3 \end{array}$$

Serine   *Ethanolamine*   *Choline*

Choline is a strong water-soluble base, as strong an alkali as NaOH and forming neutral salts. On heating it gives rise to ethylene glycol and trimethylamine, which has a strong fishy odour.

The phosphatides were originally isolated from lipid-rich tissues, such as brain, by solvent fractionation, and given the trivial names shown in Table 3.4. When it was realized that the cephalins could contain either serine or ethanolamine, and when yet more complex lipids were isolated, it became clear that these names were not sufficiently accurate, and a more logical nomenclature has been devised.

Table 3.4
Phosphatides

First alcohol	Second alcohol	Trivial name	Refer to page
A. Diglyceride	None	Phosphatidic acids	38
Diglyceride	Choline	Lecithins	39
Diglyceride	Ethanolamine Serine Inositol	Cephalins	39
B. Monoglyceride ether	Choline Ethanolamine	Plasmalogens	40
C. Sphingosine	Choline	Sphingomyelins	40

A (Table 3.4). The common grouping of the lecithins and cephalins is that of *phosphatidic acids*:

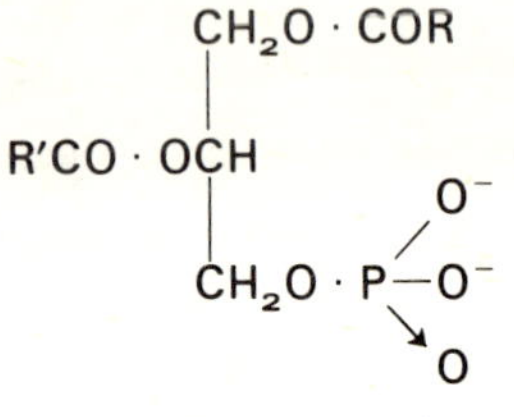

$$CH_2O \cdot COR$$
$$R'CO \cdot OCH$$
$$CH_2O \cdot P\text{—}O^-$$

3-sn-Phosphatidic acids

R and R′ are two long-chain fatty acid residues. Usually at least one residue is unsaturated, generally that esterifying the 2-hydroxyl. Only 3-*sn*-phosphatidic acids and their derivatives occur naturally. The acids themselves are not found in any quantity, but they may be easily formed from lecithins and cephalins by suitable enzymes. As shown in Chapter 10, phosphatidic acids are intermediates in triglyceride and phospholipid synthesis.

Lecithins are choline esters of the phosphatidic acids which are widely distributed in plants and animals; particularly good sources are egg-yolk and soya-bean. They are insoluble in water, although they hydrate to

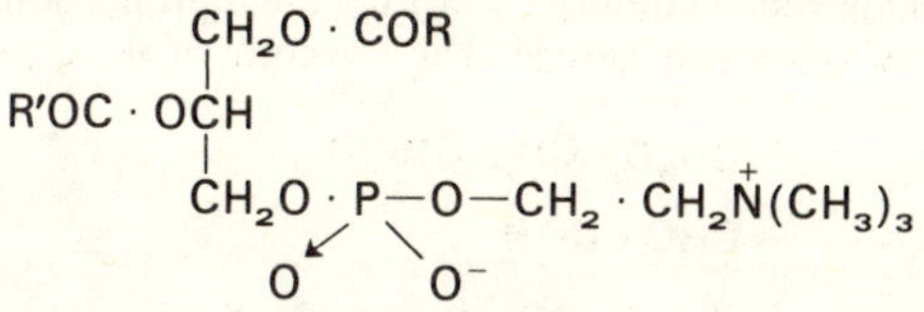

$$CH_2O \cdot COR$$
$$R'OC \cdot OCH$$
$$CH_2O \cdot P-O-CH_2 \cdot CH_2\overset{+}{N}(CH_3)_3$$
$$O \qquad O^-$$

produce the so-called 'lecithin forms', and may give a colloidal dispersion in water. At neutral pHs both the phosphate and choline groups are ionized, so that the molecule is a dipolar ion (see Chapter 4). Because of this polar grouping, and the lipid solubility conferred by the fatty acid residues, lecithins are very powerful emulsifying agents.

They are rather unstable substances. In particular the fatty acid residues are more easily hydrolysed and much more easily oxidized than the same acids in a neutral triglyceride. For this reason it is very difficult to purify and store lecithins, and indeed any phosphatide, without chemical changes taking place. Unoxidized lecithins are soluble in all fat solvents except acetone.

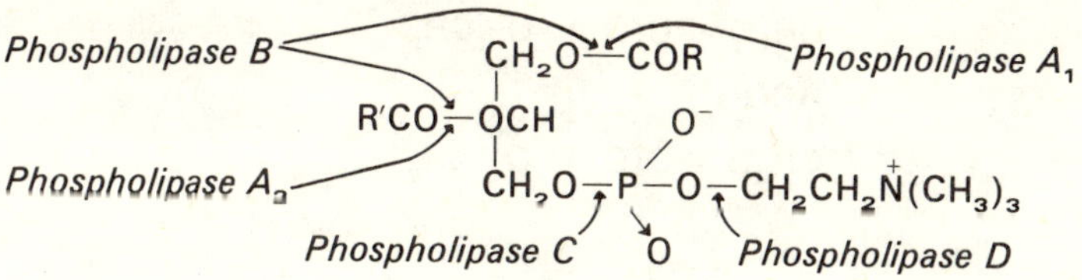

Phospholipase B CH_2O-COR *Phospholipase A*$_1$
                    $R'CO-OCH$ O^-
Phospholipase A$_2$ $CH_2O-P-O-CH_2CH_2\overset{+}{N}(CH_3)_3$
          Phospholipase C O *Phospholipase D*

There are many enzymes included in the class of *phosphatidases* (lecithinases). The general specificity which they show is indicated in the diagram above.

Phosphatidases A are found in snake venoms and bee stings. The products of their action are the so-called lysolecithins (monoglyceride phosphoryl cholines), which are very powerful haemolysing agents. Phosphatidases A are also found in pancreatic juice and elsewhere in the body. The existence of phosphotidase B is now doubtful. Phosphatidases C and D are almost entirely confined to bacteria and plants.

Cephalins are structurally very similar to lecithins, except that choline is replaced by ethanolamine, serine, inositol, and possibly other compounds. They are found most extensively in brain tissue.

One of the cephalins is almost certainly involved in the clotting reaction. It appears to be a pro-thromboplastin; that is, the cephalin is acted on enzymically to produce one of the components of the thromboplastin system.

B (Table 3.4). *Plasmalogens.* Very little is known about the function of these compounds which comprise a considerable proportion of the total phospholipid of brain and muscle. The structure of these compounds is

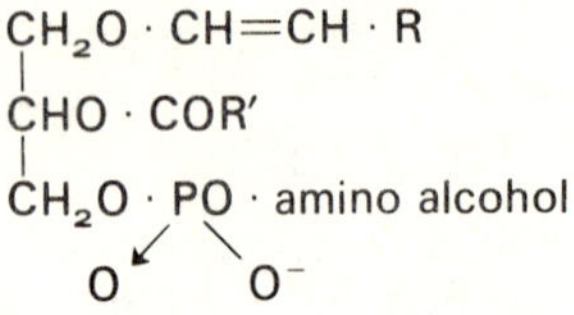

$$CH_2O \cdot CH{=}CH \cdot R$$
$$CHO \cdot COR'$$
$$CH_2O \cdot PO \cdot \text{amino alcohol}$$
$$O \overset{\nearrow}{} \overset{\nwarrow}{} O^-$$

The characteristic linkage at the 1-*sn*-carbon atom is that of a vinyl ether. This is a very unusual bond which has not been found elsewhere in lipids. The amino alcohol is either choline or ethanolamine.

Cardiolipins. These compounds were first isolated, as the name suggests, from heart, where they are found in the mitochondria. They have the general formula:

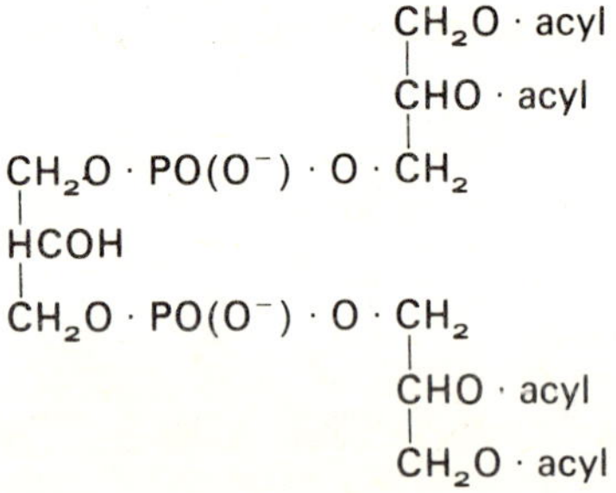

$$CH_2O \cdot \text{acyl}$$
$$CHO \cdot \text{acyl}$$
$$CH_2O \cdot PO(O^-) \cdot O \cdot CH_2$$
$$HCOH$$
$$CH_2O \cdot PO(O^-) \cdot O \cdot CH_2$$
$$CHO \cdot \text{acyl}$$
$$CH_2O \cdot \text{acyl}$$

C (Table 3.5). *Sphingomyelins.* These phospholipids do not contain glycerol, but instead a long-chain amino alcohol, sphingosine. This is

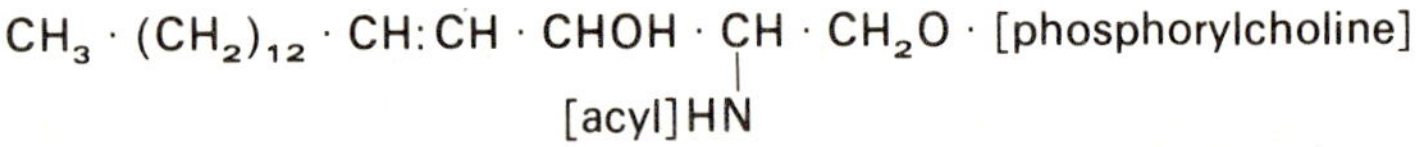

$$CH_3 \cdot (CH_2)_{12} \cdot CH{:}CH \cdot CHOH \cdot \underset{[acyl]HN}{CH} \cdot CH_2O \cdot [\text{phosphorylcholine}]$$

A sphingomyelin. Sphingosine lies outside the square brackets

combined by way of its terminal hydroxyl group with phosphorylcholine, and by way of its amino group with a fatty acid (through an *amide* linkage).

Sphingomyelins are found in brain and nerve, particularly in the myelin sheath (cf. Table 3.6). In spite of their complexity they are much more stable to air and light than are lecithins. They are rather insoluble compounds, usually extracted from brain with hot alcohol.

Glycolipids (Cerebrosides)

Large amounts are present, in brain, of compounds obviously allied to the sphingomyelins, but containing no phosphate and therefore not phospholipids. They are called glycolipids because they contain, in place of phosphorylcholine, a monosaccharide (usually galactose).

$$CH_3 \cdot (CH_2)_{12} \cdot CH{:}CH \cdot \underset{\underset{HO}{|}}{CH} \cdot \underset{\underset{NH \cdot CO \cdot R}{|}}{CH} \cdot CH_2O\text{-hexose}$$

Plasma Lipoproteins

A number of proteins, mostly in the globulin fraction of plasma, are conjugated with lipid or have lipid material strongly adsorbed. Cholesterol and phospholipid can form part of these lipoproteins. They are discussed in more detail in Chapter 10.

Detergents and Emulsifiers

Detergents work partly by lowering the surface tension between water and fat. It is thought that many detergents also form a monolayer at the surface of the fat or oil droplets, so that the detergent molecule is partly in lipid and partly in water (see Fig. 3.5). The *hydrophilic* part of the detergent molecules then prevents two or more fat droplets from coming into close contact and coalescing, i.e. it keeps the fat or grease as an emulsion in water.

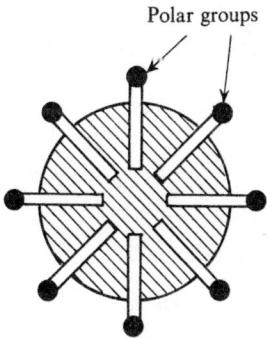

Polar groups

Fig. 3.5. Adsorption of a detergent on the surface of a fat droplet.

The soaps (the salts of fatty acids) are detergents of this type, in which the ionized $-COO^-$ radical is the hydrophilic group, and the fatty acid chain the *lipophilic* group. Many synthetic detergents have been made which are an improvement on simple soaps. Among them are *anionic detergents*, such as the sulphate esters of fatty alcohols; *cationic detergents*, such as quaternary ammonium compounds containing a long-chain paraffin radical; and *non-ionic detergents*, such as fatty acid esters of polymerized sugar alcohols. The quaternary ammonium compounds are often quite efficient bactericides, as well as detergents.

Emulsifying agents work in much the same way, but they need a more rigid structure, as an emulsion must remain stable indefinitely. Emulsions important in food products are often water-in-fat, so that it is an advantage for an emulsifying agent to be fat-soluble with hydrophilic groups, rather than the reverse, as in a detergent. Among the important natural emulsifying agents are proteins, lecithins, bile salts, and monoglycerides. Others may be based on the non-ionic detergents described above, or on acidic polysaccharides.

Prostaglandins

Prostaglandins are complex unsaturated fatty acids which contain a 5-membered ring. Fig. 3.6 shows two examples of the 'PGE series', together with eicosa-8,11,14-trienoic acid, which is the precursor of PGE1. In the 'PGF series' of prostaglandins, the keto group at C–9 has been reduced to an alcohol. *Cis-trans* isomerism in both the double bonds, and the

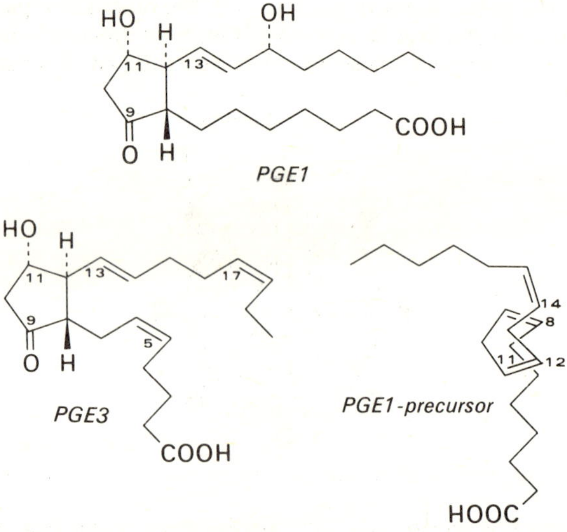

FIG. 3.6. Prostaglandins.

The PGE1-precursor, eicosa-8,11,14-trienoic acid, has been given a distorted three-dimensional representation to show how close the propinquity between C–8 and C–12 can be.

arrangement of hydroxyl groups, permits many variations in the overall shape of the molecules. As the name suggests, prostaglandins were first isolated from the prostate gland and are present in seminal plasma, but they have since been isolated from many other tissues. They have powerful physiological and biochemical effects, which are described, in relation to their possible hormonal role, in Chapter 18.

Steroids

A very large number of steroids have been identified in animals, plants, and micro-organisms. Some of them have very important physiological or pharmacological actions. They are all based on the 'cyclopentanoperhydrophenanthrene' nucleus (this means a 5-membered ring + reduced phenanthrene) shown below. The biological activity of these compounds often depends on the presence of a few groups attached to the nucleus, or rather it depends on the spatial relationship between these groups—

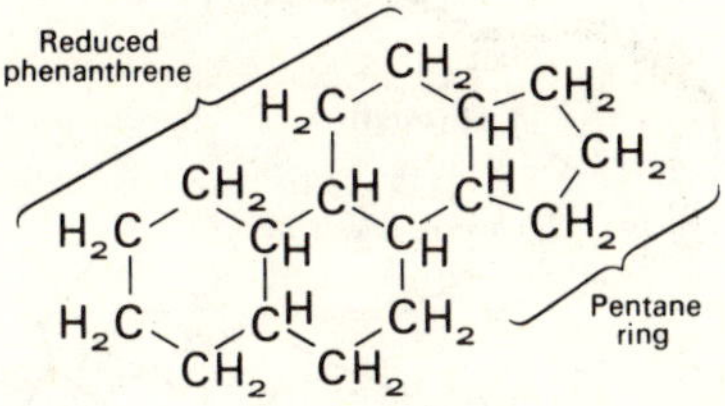

Cyclopentanoperhydrophenanthrene

whether they lie more or less in the plane of the nucleus, or project at right-angles to it. The nucleus is puckered, which can bring projecting groups quite close together. See, for example, *aldosterone*, p. 48, and Fig. 3.6. Naturally occurring steroids can be modified by the addition or replacement of groups, to make them more soluble in water, or less readily attacked by catabolic enzymes. There are consequently so many steroids in existence that only the requirements for biological activity in the main groups of natural steroids can be dealt with here.

The basic molecule of a steroid is shown on p. 44 with the ring lettering and numbering of the carbon atoms. There are no fewer than 8 asymmetric carbon atoms, which means that there are 256 isomers, including the *cis-trans* isomers of the fused rings which were described at the beginning of this chapter. Fortunately, most of the natural steroids have the shape either of dihydrocholesterol (rings A/B *trans*, B/C *trans*, C/D *trans*) or of its isomer coprostanol, also called coprosterol (rings A/B *cis*, B/C *trans*, C/D *trans*). These two arrangements are shown in Fig. 3.7. Whenever a group (e.g. an —OH radical) is attached to one of the carbon atoms it can project either 'behind' or 'in front' of the ring, if this is considered

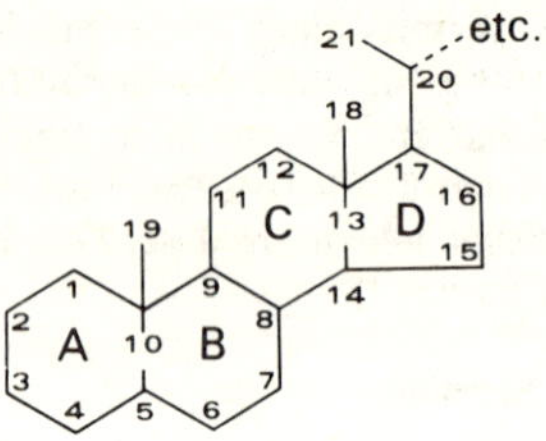

The steroid nucleus

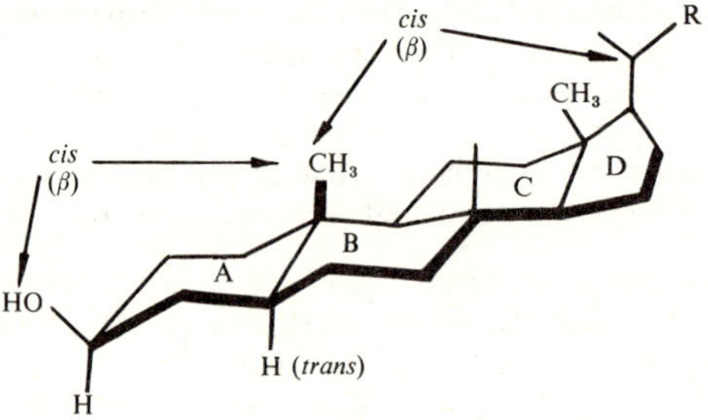

Dihydrocholesterol

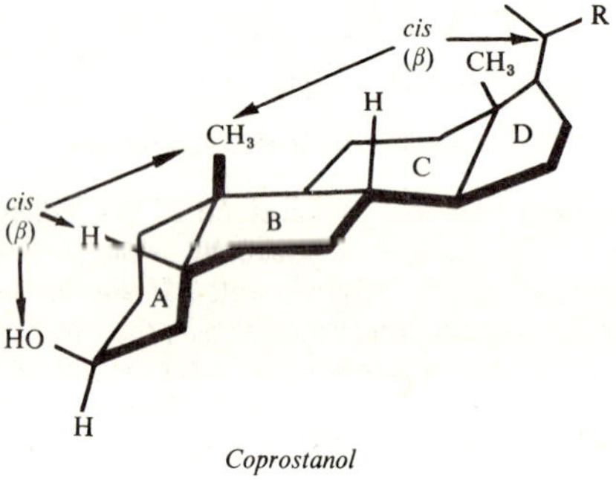

Coprostanol

FIG. 3.7. Configurational representations of dihydrocholesterol and coprostanol. The bond linking the 'reference' $-CH_3$ group to C–10 of the nucleus is indicated by a heavy line. It will be seen that the $-OH$ at C–3, the $-CH_3$ at C–13, and the side-chain at C–17 are all *cis* to the reference methyl group in both sterols, i.e. they all have the β configuration.

The H atom attached to C–5 is, however, *trans* to the reference $-CH_3$ in dihydrocholesterol (rings A/B *trans*), and *cis* in coprostanol (rings A/B *cis*).

The diagram also shows to some extent that, because of the crumpling of the rings into a 'chair' form, the $-OH$ at C–3 has the 'axial' configuration, i.e. is at right-angles to the plane running through the rings and is not 'equatorial' (more or less in this plane).

to be flat. Thus two isomers exist of all such compounds. This is indicated in Fig. 3.7, which shows a view of the two typical ring arrangements from the 'side'. For the purpose of identifying these stereo-isomers the methyl group on C–10 is always taken as a reference point. All groups projecting on the same side of the molecule as this methyl group have a '*cis*' or *β*-configuration: those projecting on the other side are '*trans*' or *α*.

There are 8 main groups of steroids, not all of which can be dealt with here. They have certain structural features in common:

(*a*) an —OH or =O group at C–3
(*b*) methyl groups at C–10 and C–13 (except for oestrogens)
(*c*) a side chain *or* an —OH or =O group at C–17.

Sterols have a hydroxyl group, and sometimes in addition a double bond, and a side chain at C–17. *Cholesterol* is the important sterol found in animals. It has one double bond (between C–5 and C–6) but is a remarkably stable compound, although it can be both oxidized and reduced. The hydroxyl is aliphatic, i.e. it forms esters, etc.; it is not phenolic. Cholesterol and other sterols are not soluble in water, but are soluble in most fat solvents except cold alcohol. *Sitosterol* is a common plant sterol; *ergosterol* is found in yeast and fungi. *Coprosterol* arises in the gut from the cholesterol excreted in bile. It does not have a double bond, and, moreover, the conformation of rings A/B has been changed from *trans* to *cis*.

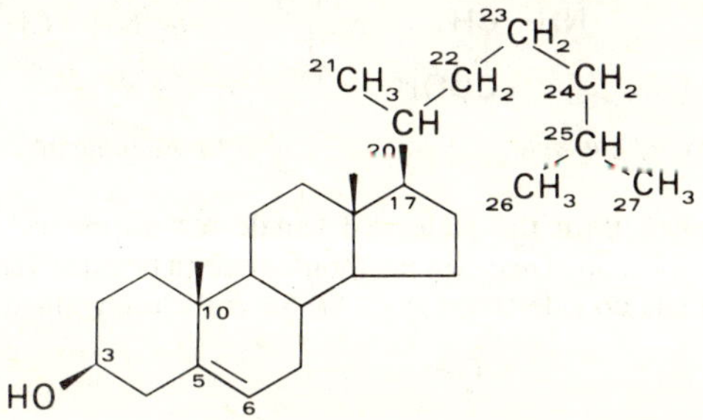

Cholesterol

In this and subsequent formulae, groups which project above the plane of the paper are joined by a heavy line to the nucleus (*β* configuration), those which project below the paper are joined to the nucleus by a dotted line (*α* configuration). The nucleus is considered to lie flat in the plane of the paper, and the reference methyl (at C–10) always projects upwards.

Bile acids. The bile of higher animals contains three important bile acids, lithocholic acid (3-hydroxy), deoxycholic (3,12-dihydroxy), and cholic

(3,7,12-trihydroxy). The last-named has the formula shown:

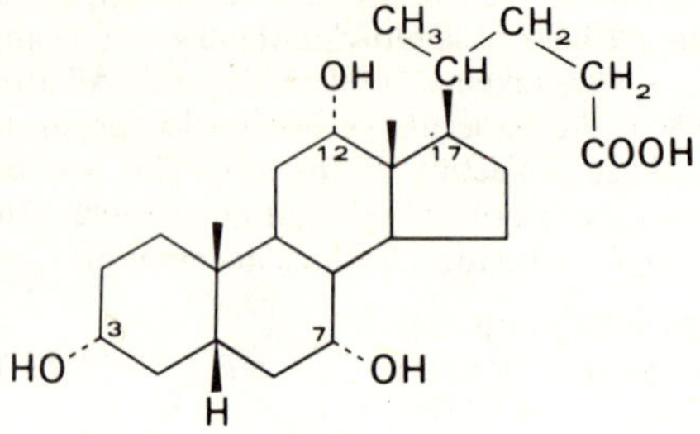

Cholic acid

Rings A/B are *cis* (coprosterol structure).

The side-chain on C–17 is shorter than that of cholesterol. Since it ends in a —COOH group, these steroids are soluble in alkalis. They are not excreted as such in bile, but as *conjugates* with glycine and taurine; these are very efficient emulsifying agents (see Chapter 8).

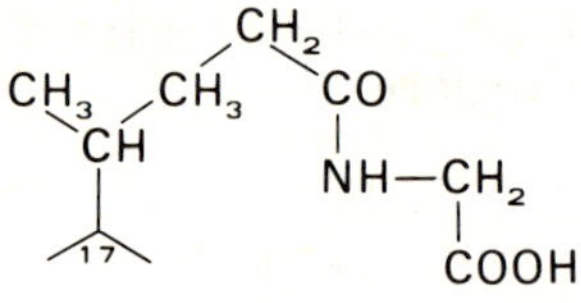

Glycocholic acid

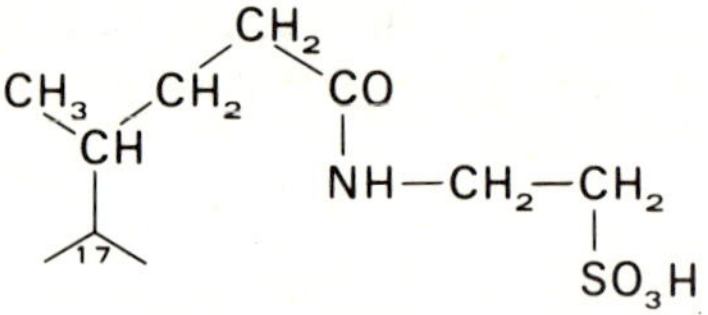

Taurocholic acid

Sex hormones. Both the male and female sex hormones are steroids. *Testosterone* is the chief *male* sex hormone, and can be extracted from testes. Note that it has no side-chain at C–17. Several closely allied compounds

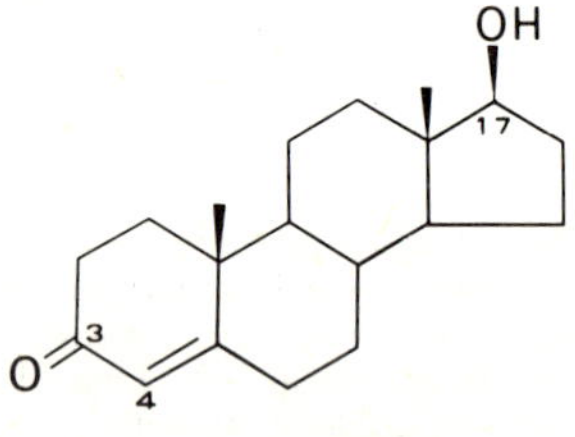

Testosterone

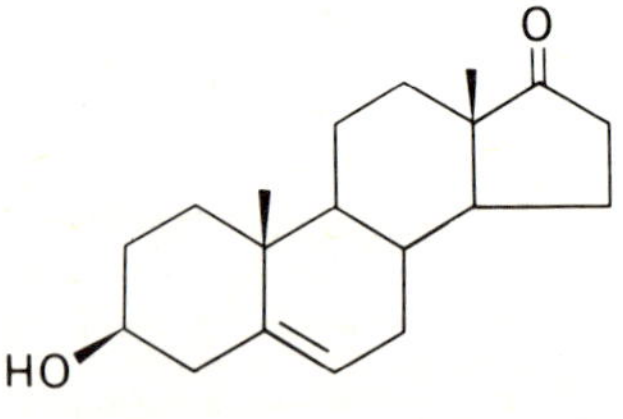

Dehydroepiandrosterone

with some androgenic activity are found in urine, such as *androsterone* (3α-hydroxy-5α-androstan-17-one) and *dehydroepiandrosterone*. These compounds are probably metabolic products of testosterone. They can also be found in the urine of castrates and females, since they are elaborated by the ovaries, placenta, and adrenals, as well as by testes.

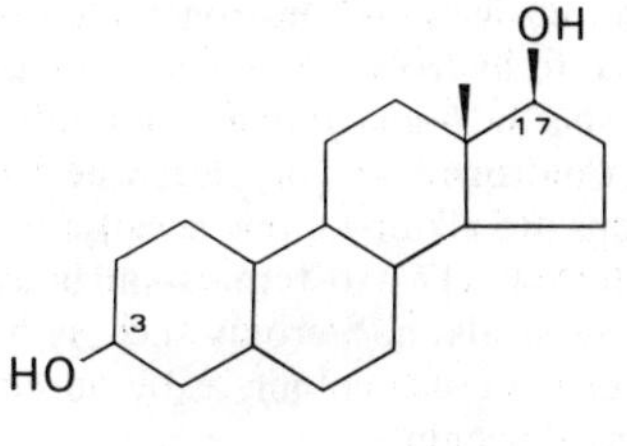

Oestradiol—17β

There are two groups of *female* sex hormones, the *oestrogenic hormones* and the *progestational hormones*. It is difficult to say whether there is one true oestrogenic hormone, but *oestradiol-17β* is found in greatest quantity in the ovaries. The 17α isomer is only weakly oestrogenic. It and *oestrone* (3-hydroxy,17-one) and *oestriol* (3,16α,17β-trihydroxy) are found in urine, both male and female. Oestrogenic compounds are secreted by the testis and adrenal, as well as by the ovary and placenta. Stallion urine is in fact the richest known source of oestrogen. Chemically, the oestrogens are unusual in that there is no methyl group at C–10, and ring A has 3 double bonds and is benzenoid. This means that the —OH at C–3 is *phenolic*, not alcoholic.

Progesterone, the progestational hormone, has a short side-chain at C–17, and one double bond in ring A. The chief derivative found in the urine is pregnanediol (3α,20α-dihydroxypregnane).

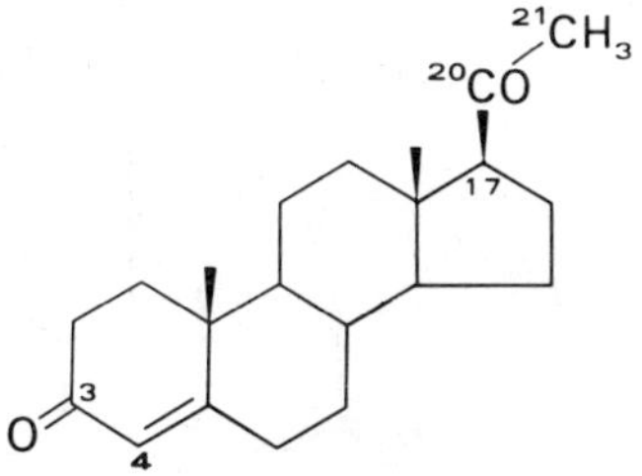

Progesterone

Excretion of sex hormones. The liver probably metabolizes most of the sex steroids, once they have been secreted into the blood. Their biological activity is destroyed, or very much reduced, by various oxidations and reductions described in more detail in Chapter 18. Some of these steroid derivatives may be excreted in the bile and later reabsorbed. They are often conjugated with glucuronic acid, which makes them water-soluble.

Paper chromatography, using suitable lipid solvents, is increasingly used for investigating the androgen and oestrogen picture in the urine. At the present time it is usual to hydrolyse any glucuronides by boiling the urine samples with acid. A specific *β-glucuronidase* may also be used. Oestrogens can be separated from androgens and progesterone derivatives by extracting with alkali. Oestrogens are alkali-soluble because of their phenolic —OH group. The '17-ketosteroids' (17-oxosteroids) can be estimated by treatment with *m*-dinitrobenzene in alkali. Steroids such as androsterone, with an oxo group at C–17, give a green colour, as do adrenal steroid derivatives which have lost their side-chain.

Adrenal cortical hormones. The three steroids which have been detected in human adrenal vein blood are the following:

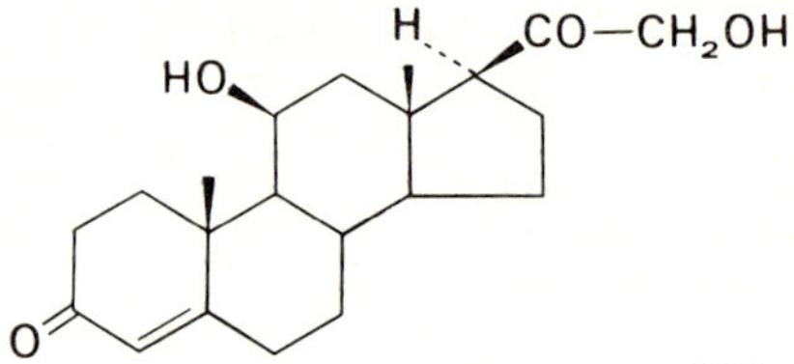

Corticosterone

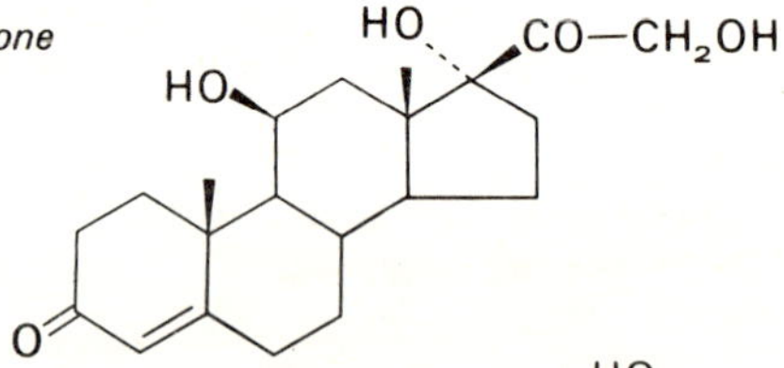

Cortisol

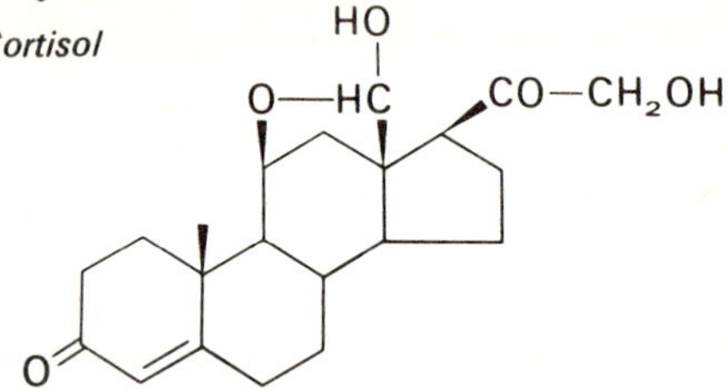

Aldosterone

The first two are chiefly concerned with regulation of carbohydrate and protein metabolism, the last with salt metabolism. Besides these compounds, some 25–30 other related steroids have been identified in extracts of adrenal cortex, besides quantities of cholesterol. Many of these compounds also possess hormonal activity. The groups required for this are an oxo group at C–3, double bond at C–4–5, the side-chain at C–17, and a hydroxyl group at C–11 (although there are compounds such as deoxycorticosterone, DOC, which have no oxygen function at C–11).

There is an obvious resemblance between these compounds and progesterone. A major route of formation of adrenal steroid hormones is via progesterone, which is found in small quantities in the gland. The androgens, too, can easily be formed from adrenal cortical steroid precursors, before the introduction of the hydroxyl at C–11. This may be the explanation of the secretion of sex hormones by the adrenal, which continually occurs.

Excretion of cortical steroid hormones. The number of compounds which are found in the urine, in larger or smaller amounts, is very large. As with the sex hormones, diminution in biological activity by reduction in ring A, and conjugation with sulphate or glucuronic acids, takes place in the liver. There are two main groups of excretion products : those in which the —CO—CH$_2$OH side-chain at C–17 has been removed—these fall in the group of 17-oxosteroids; and those in which it still exists—these are the 17-oxogenic steroids, since the side-chain can easily be oxidized away by chemical means (see also Chapter 18).

Other Steroids

Cardiac glycosides. Several closely related C$_{23}$ steroids have a stimulatory action, in small doses, on cardiac muscle. They are all poisonous if administered in excess. The drugs consist of a steroid part, known as a *genin*, conjugated with one or more sugars at C–3. This carbohydrate in the molecule makes the cardiac glycosides water-soluble (they could be, in fact, quite good detergents). Several glycosides come from extracts of foxgloves (the digitalins) and others from a tropical plant (strophanthins).

Other steroids with pharmacological activity are the toad poisons and the saponins. The *vitamins D* are dealt with in Chapter 20. They are not strictly steroids, since ring B is open, but they resemble steroids very much in physical properties.

Tests for Lipids

There is no general test for lipid material, other than its insolubility in neutral or acid aqueous solutions, and its solubility in fat solvents. It has already been mentioned in this chapter that a number of 'derived lipids' are insoluble in solvents such as acetone, alcohol, or ether. *Natural* fats, oils, and phospholipids almost always contain an unsaturated fatty acid radical, and will decolorize Br$_2$ or I$_2$ solutions, osmic acid or KMnO$_4$. Many synthetic fatty acid esters or derivatives, however, are completely saturated. Most fats will concentrate the so-called 'lipophilic' dyes, such as Sudan III.

The Liebermann–Burchardt and Salkowski tests for cholesterol also give a positive result with other sterols having the same arrangement in space of rings and hydroxyl group. Sterols with β-hydroxyl at C–3 give a precipitate with digitonin; this is sometimes used for estimating cholesterol. Other tests for steroids, such as Gmelin's test for bile acids, are not particularly specific.

Table 3.5
Relative Abundance of Lipids in Some Tissues
(% of dry weight)

	Total lipid	Fat	Cholesterol		Phospholipids		Glyco-lipid
			Free	Esters	Lecithin	Others	
Fat depot	93·0	93·0	about 0·1		trace		—
Brain	51·6	3·0	10·0	0·25	7·0	19·3	12·0
Liver	23·0*	5·8*	0·4	0·5	8·9	7·3	0
Heart	16·4*	4·0*	0·3	0·2	4·0	5·9	2·0
Skeletal muscle	14·2*	10·0*	0·2	0·03	1·7	1·4	0·9
Plasma	7·4*	2·5*	0·7*	1·8*	1·5	0·4	—

* These values are very variable.

Table 3.6
Relative Abundance of Lipids in Some Membranes

	Human erythrocytes	Human myelin	Rat liver mitochondria (inner membrane)
Lipid : protein	1 : 3	4 : 1	1 : 3
Phosphatidyl-choline	15 % of lipid	11	42
Phosphatidyl-ethanolamine	15	14	23
Phosphatidyl-serine	7	7⎫	12
Phosphatidyl-inositol	—	1⎭	
Sphingomyelin	14	6	—
Cardiolipin	—	—	12
Glycolipids	trace	26*	—
Cholesterol	46	25	2

* Mostly cerebrosides (p. 41).

4   Amino Acids, Peptides, and Proteins

Amino Acids

Amino acids are carboxylic acids which also contain an amino ($-NH_2$) group in the molecule. In the majority the amino group is in the α-position with respect to the carboxyl; they have the general formula

$$
\begin{array}{c}
R \\
| \\
CH \cdot NH_2 \\
| \\
COOH
\end{array}
$$

The structure of the individual amino acids and the usual abbreviations for them are given on the following pages. A few are known to occur in which the amino group is attached to a position β- or γ- to the carboxyl; proline is not strictly an amino acid but an imino acid.

Optical activity. In the α-amino acids the α-carbon atom is asymmetric (except in glycine) and two optically active isomers therefore exist. All those which occur in proteins are of the L-series (they have the same configuration as L-glyceraldehyde); some substances, particularly antibiotics and the capsular substance of *Bacillus anthracis*, contain D-amino acids.

Classification

The amino acids may be classified into three groups: the *aliphatic*, *aromatic*, and *heterocyclic* amino acids. Under these heads they may be further subdivided into *neutral*, containing one $-NH_2$ and one $-COOH$ in the molecule; *acidic*, containing one $-NH_2$ and two $-COOH$ groups; and *basic*, containing two $-NH_2$ groups (or one $-NH_2$ and some other basic group such as a guanidino or imidazolyl) and one $-COOH$. The structures of the common amino acids, the recognized abbreviations of their names, and some of their distinctive features are as follows:

Aliphatic

(a) *Neutral monoaminomonocarboxylic*

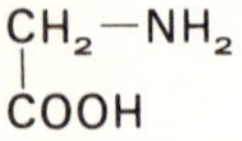

$$CH_2-NH_2$$
$$COOH$$
Glycine
Gly

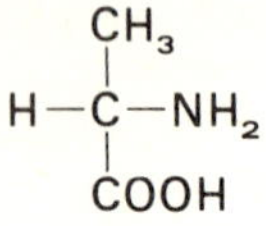

$$CH_3$$
$$H-C-NH_2$$
$$COOH$$
L-*Alanine*
Ala

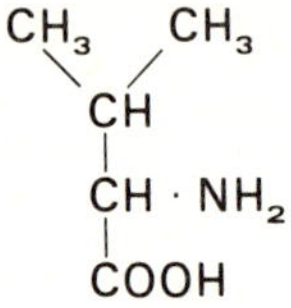

$$CH_3 \quad CH_3$$
$$CH$$
$$CH \cdot NH_2$$
$$COOH$$
L-*Valine*
Val

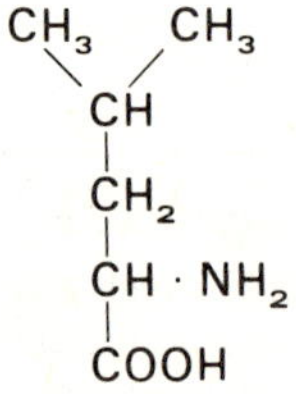

$$CH_3 \quad CH_3$$
$$CH$$
$$CH_2$$
$$CH \cdot NH_2$$
$$COOH$$
L-*Leucine*
Leu

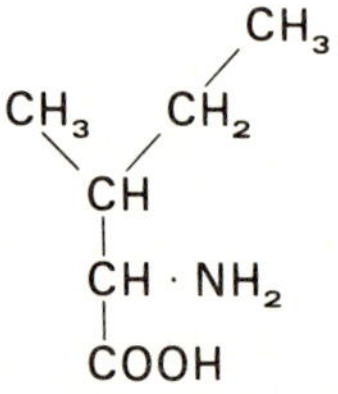

$$CH_3 \quad CH_2 \quad CH_3$$
$$CH$$
$$CH \cdot NH_2$$
$$COOH$$
L-*Isoleucine*
Ile

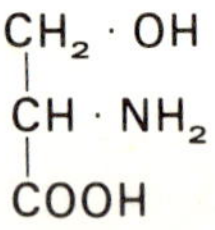

$$CH_2 \cdot OH$$
$$CH \cdot NH_2$$
$$COOH$$
L-*Serine*
Ser

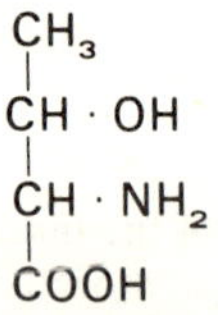

$$CH_3$$
$$CH \cdot OH$$
$$CH \cdot NH_2$$
$$COOH$$
L-*Threonine*
Thr

L-Isoleucine and L-threonine, together with L-hydroxylysine and L-hydroxyproline (see below), contain a second asymmetric carbon atom. There are therefore four possible isomers of these amino acids of which only one isomer occurs in proteins.

(b) *Neutral sulphur-containing amino acids*

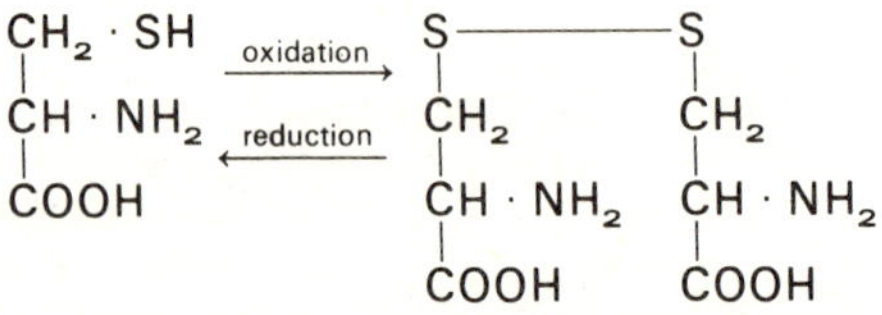

$$CH_2 \cdot SH \xrightarrow[\text{reduction}]{\text{oxidation}}$$
$$CH \cdot NH_2$$
$$COOH$$
L-*Cysteine*
CySH

$$S \text{———} S$$
$$CH_2 \quad CH_2$$
$$CH \cdot NH_2 \quad CH \cdot NH_2$$
$$COOH \quad COOH$$
L-*Cystine*

L-Cysteine occurs in proteins but is very easily oxidized when free to L-cystine, which is the disulphide of cysteine. L-Cystine also occurs in some proteins and is of importance in maintaining their structures. Cystine can be reduced in appropriate conditions to two molecules of cysteine.

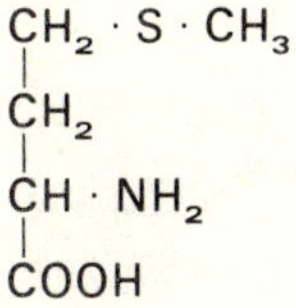

$$CH_2 \cdot S \cdot CH_3$$
$$|$$
$$CH_2$$
$$|$$
$$CH \cdot NH_2$$
$$|$$
$$COOH$$

L-*Methionine*
Met

(c) *Monoaminodicarboxylic* (*and their amides*)

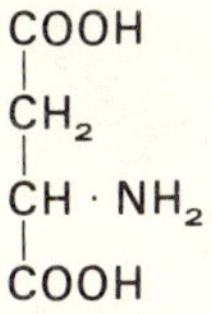

$COOH$	$CONH_2$	$COOH$	$CONH_2$
CH_2	CH_2	$(CH_2)_2$	$(CH_2)_2$
$CH \cdot NH_2$	$CH \cdot NH_2$	$CH \cdot NH_2$	$CH \cdot NH_2$
$COOH$	$COOH$	$COOH$	$COOH$
L-*Aspartic acid*	L-*Asparagine*	L-*Glutamic*	L-*Glutamine*
Asp	*Asn*	*Glu*	*Gln*

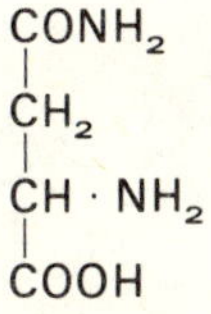

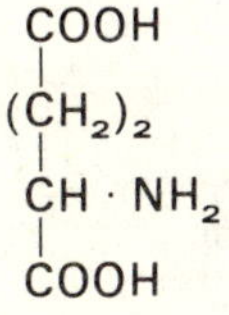

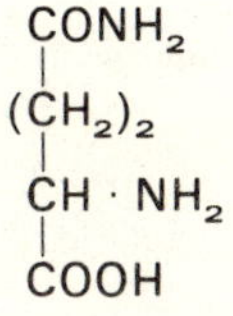

Aspartic and glutamic acids contain two carboxyl groups and only one amino group; their aqueous solutions are therefore acidic. The amides, asparagine and glutamine, are neutral.

(d) *Basic*

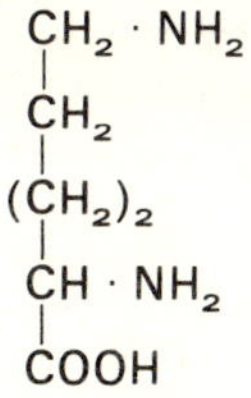

		NH_2
		$|$
$CH_2 \cdot NH_2$	$CH_2 \cdot NH_2$	$C{=}NH$
CH_2	$CH \cdot OH$	NH
$(CH_2)_2$	$(CH_2)_2$	$(CH_2)_3$
$CH \cdot NH_2$	$CH \cdot NH_2$	$CH \cdot NH_2$
$COOH$	$COOH$	$COOH$
L-*Lysine*	L-*Hydroxylysine*	L-*Arginine*
Lys	*Hylys*	*Arg*

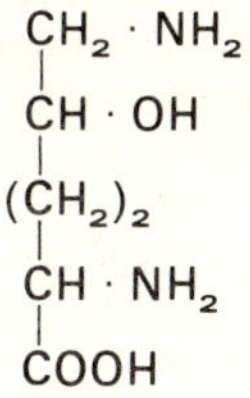

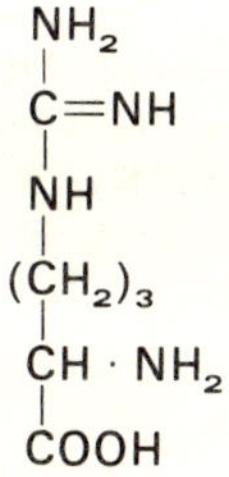

Lysine and hydroxylysine contain two amino groups and only one carboxyl group; their aqueous solutions are therefore alkaline. Arginine is very strongly basic because of the guanidino group in the molecule.

(e) *Others*

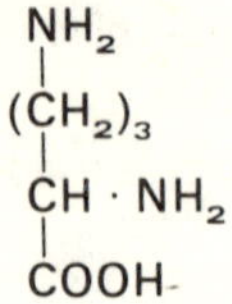

$$
\begin{array}{ccc}
 & \text{CO} \cdot \text{NH}_2 & \\
 & | & \\
\text{NH}_2 & \text{NH} & \\
| & | & \\
(\text{CH}_2)_3 & (\text{CH}_2)_3 & \text{CH}_2 \cdot \text{NH}_2 \\
| & | & | \\
\text{CH} \cdot \text{NH}_2 & \text{CH} \cdot \text{NH}_2 & \text{CH}_2 \\
| & | & | \\
\text{COOH} & \text{COOH} & \text{COOH}
\end{array}
$$

| L-*Ornithine* | L-*Citrulline* | β-*Alanine* |

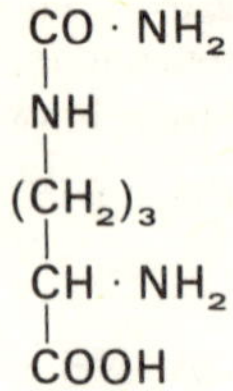

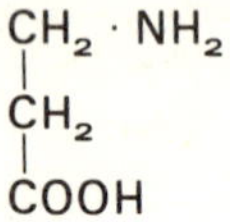

These three amino acids do not occur in proteins. L-Ornithine and L-citrulline are important in the metabolic formation of urea while β-alanine occurs in the vitamin pantothenic acid, in coenzyme A, and in the muscle peptides carnosine and anserine.

Aromatic

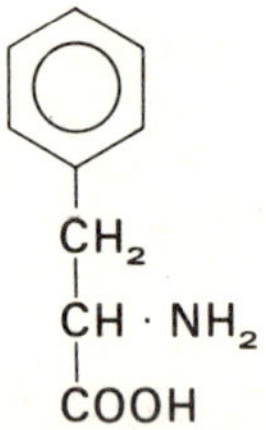

L-*Phenylalanine*
Phe

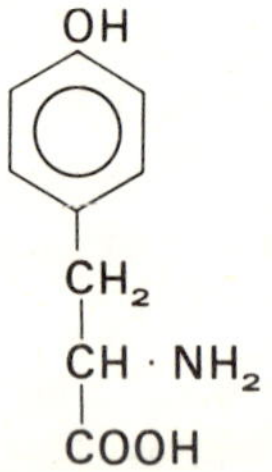

L-*Tyrosine*
Tyr

Tyrosine, together with tryptophan, is responsible for most of the intense absorption of ultra-violet light in the region 260–290 nm which is such a characteristic feature of many proteins.

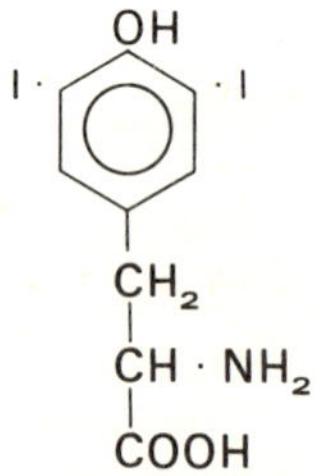

3,5-L-*Diiodotyrosine*

3,5-Diiodotyrosine occurs as a constituent of the protein thyroglobulin which is present in the thyroid gland.

Heterocyclic

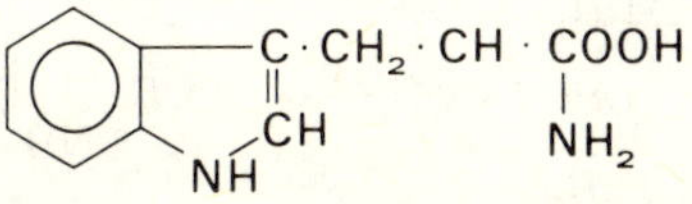

L-*Tryptophan*
Trp

Tryptophan gives a positive *xanthoproteic test* and *glyoxylic acid reaction*, it also absorbs ultra-violet light very strongly.

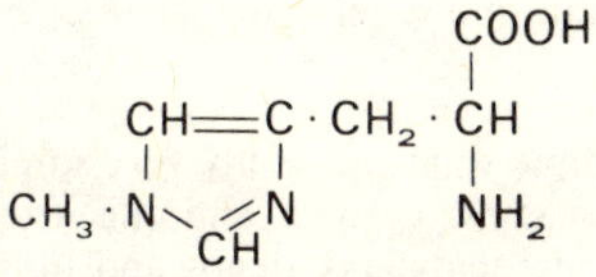

3-*Me*-L-*Histidine*

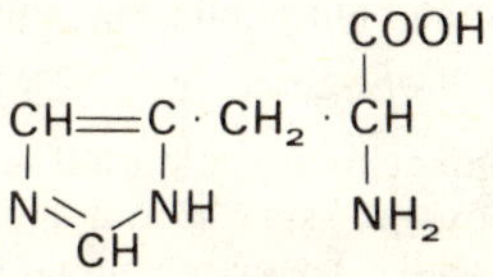

L-*Histidine*
His

Histidine is weakly basic because of its imidazolyl group; it is a constituent of most proteins. 3-Methylhistidine has been identified as a normal constituent of myosin.

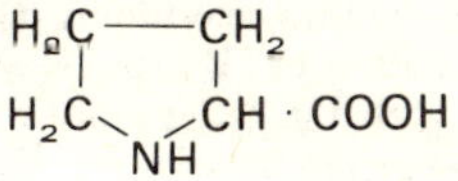

L-*Proline*
Pro

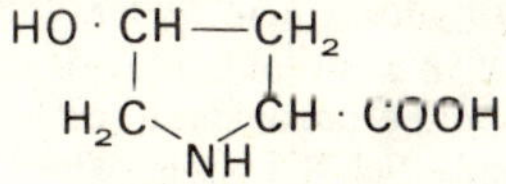

L-*Hydroxyproline*
Hypro

Proline and hydroxyproline do not contain primary amino groups but secondary ones; they are therefore imino acids. Proline gives a lemon yellow ninhydrin reaction and hydroxyproline a brownish yellow one. Hydroxyproline is not very widely distributed in proteins but constitutes a very large part of the structural protein collagen.

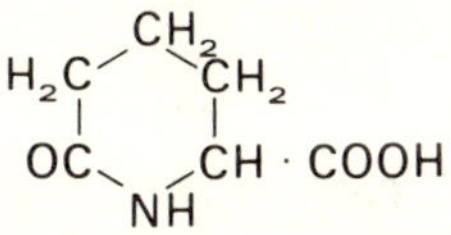

L-*Pyrrolidonecarboxylic acid*
Pca

Pyrrolidonecarboxylic acid is also an imino acid. It is readily formed when glutamate or glutamine solutions are heated to boiling:

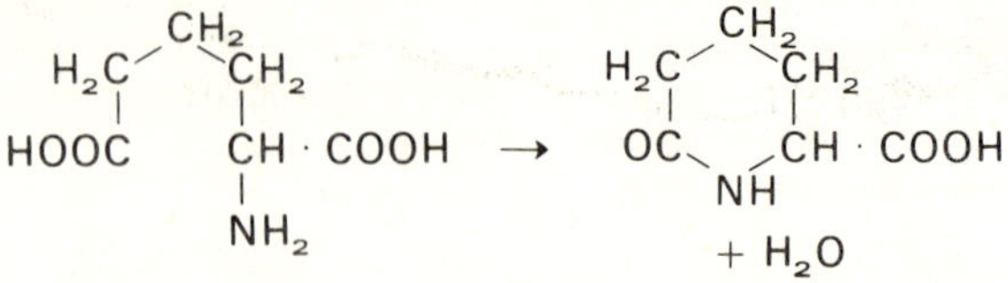

The ring closure occurs to a small extent even at 37°. It is not easily reversible, and enzymes which act on it are very rare. Although pyrrolidone-carboxylic acid is present only in traces in the body, it is the N-terminal amino acid of some proteins and peptides, notably the heavy chain of the immuno-globulins and some pituitary hormone-releasing factors.

General Properties

All amino acids are colourless, crystalline solids generally very soluble in water (except aspartic and glutamic acids, cystine, and the aromatic amino acids). They are soluble in dilute acids and alkalis and insoluble in organic solvents, being precipitated from aqueous solution by alcohol. Their melting points are rather indefinite and very high ($> 200°$).

Chemical Properties

They will undergo the reactions which are separately expected of both the amino and carboxyl groups. Thus the $-NH_2$ group reacts with HNO_2 to give a hydroxy acid, water, and N_2; in alkaline solution it reacts with CO_2 to give a carbamido acid; it forms salts with HCl, picric, phospho-tungstic, and other acids; it reacts readily with aldehydes to form addition compounds (Schiff's bases). The $-COOH$ group forms esters and it may be reduced, forming from an amino acid an amino alcohol.

Ninhydrin Reaction

α-Amino acids are oxidized by ninhydrin as shown below.

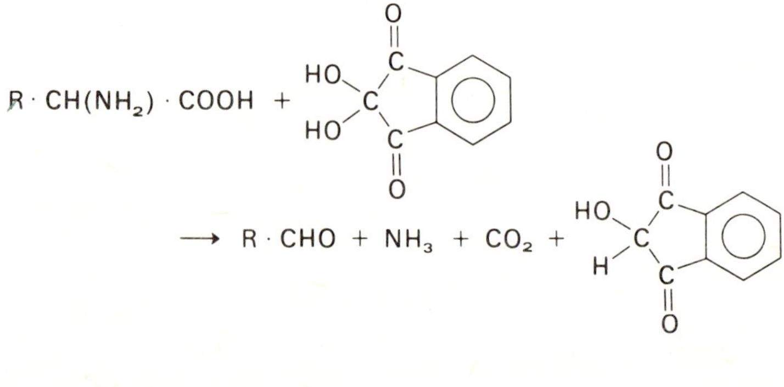

The reduced ninhydrin product then reacts with ammonia and a molecule of oxidized ninhydrin to give an intense blue product. This reaction is used for the qualitative detection and quantitative estimation of amino acids in very small concentrations. The imino acids proline and hydroxyproline give yellow colours, and asparagine a brown colour. In certain circumstances primary amines may react, but amino acids in peptide linkage (below) do not give a positive reaction.

Peptide Bonds

The most important property of amino acids is the ability to eliminate a molecule of water between the $-COOH$ of one and the $-NH_2$ of another to form a peptide bond, $-CO-NH-$. This condensation produces a dipeptide containing two amino acid residues. Further reactions can take place with more amino acids to give larger peptides. Very large molecules can be built up in this way containing hundreds of amino acid residues; these are the proteins. Energy is required to bring about the polymerization of amino acids to form proteins. In the body complex mechanisms exist to synthesize the numerous specific proteins it contains: these are dealt with in Chapter 16.

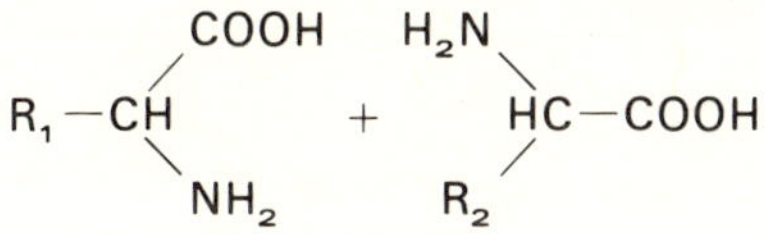

$$R_1-CH \overset{\displaystyle COOH}{\underset{\displaystyle NH_2}{}} \quad + \quad \overset{\displaystyle H_2N}{\underset{\displaystyle R_2}{}} HC-COOH$$

Amino acids

$$R_1-CH \overset{\displaystyle CO-NH}{\underset{\displaystyle NH_2}{}} \overset{}{\underset{\displaystyle R_2}{}} HC-COOH$$

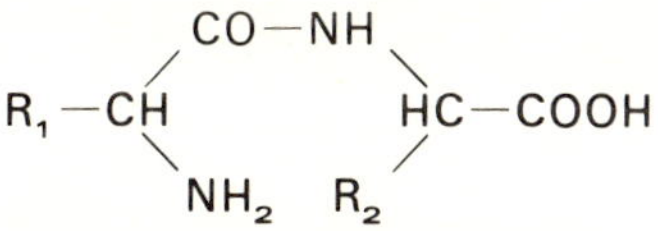

Dipeptide

Ionization of Amino Acids

The amino acids exist both in solution and in crystals as *dipolar ions* or *zwitter-ions* and not the *un-ionized* state. Evidence for this structure is the very high melting points, insolubility in organic (non-polar) solvents

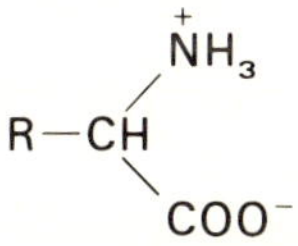

$$R-CH \overset{\displaystyle \overset{+}{N}H_3}{\underset{\displaystyle COO^-}{}}$$

compared with simple carboxylic acids and amines, the Raman and infra-red spectra, the pH titration curves with HCl and NaOH, and the behaviour on ionophoresis.

The pH titration curve of the monoaminomonocarboxylic amino acid glycine with HCl and NaOH is shown in Fig. 4.1. The pH of glycine dipolar ion solution is 6·0 (A), and on titration with NaOH the pH rises along the curve A, B, C. In this region a hydrogen ion is being removed from the dipolar ion to form an anion thus:

$$\begin{array}{ccc}
\begin{array}{c} COO^- \\ / \\ CH_2 \\ \backslash + \\ NH_3 \end{array}
& \underset{+H^+}{\overset{-H^+}{\rightleftharpoons}}
& \begin{array}{c} COO^- \\ / \\ CH_2 \\ \backslash \\ NH_2 \end{array}
\end{array} \qquad (1)$$

Net charge: 0 −1

The pK for this dissociation, glycine p$K_2 = 9\cdot6$.

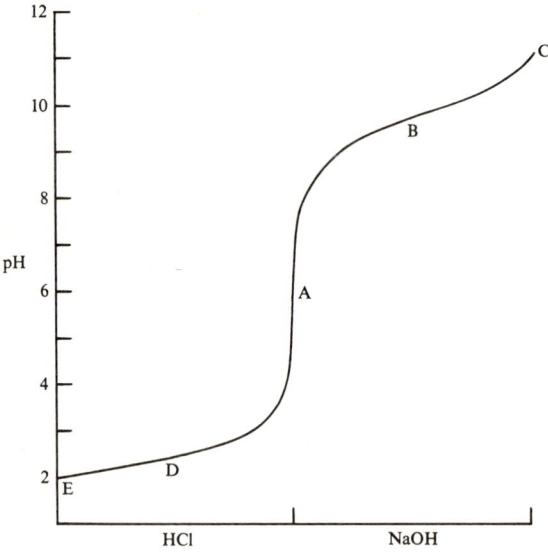

Fig. 4.1. Titration curve of glycine.

Conversely, on titration with HCl the pH falls along A, D, E, due to the addition of a hydrogen ion to the dipolar ion to form a cation, thus:

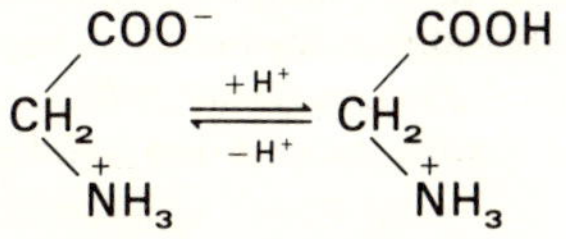

$$
\begin{array}{ccc}
\text{COO}^- & & \text{COOH} \\
| & \xrightarrow{+\text{H}^+} & | \\
\text{CH}_2 & \xleftarrow{-\text{H}^+} & \text{CH}_2 \\
| & & | \\
\overset{+}{\text{N}}\text{H}_3 & & \overset{+}{\text{N}}\text{H}_3
\end{array}
\qquad (2)
$$

Net charge: 0 -1

The pK for this dissociation, glycine $pK_1 = 2\cdot3$.

For the monoaminodicarboxylic aspartic and glutamic acids the following dissociation equilibria occur; the most acidic species is on the left and protons are progressively lost on going to the most basic species on the right. $n = 1$ for aspartic and 2 for glutamic acid.

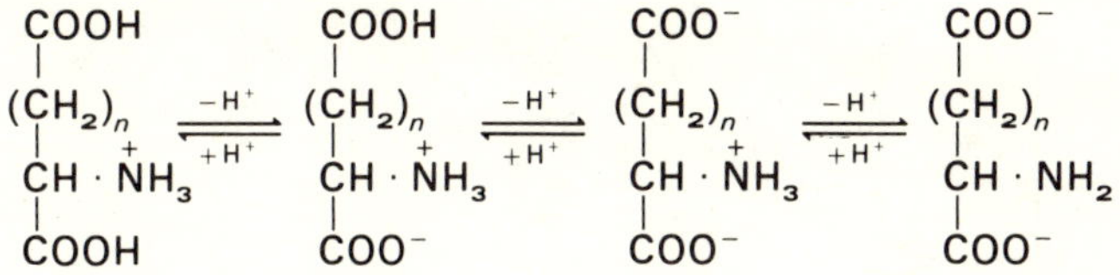

$$
\begin{array}{cccc}
\text{COOH} & \text{COOH} & \text{COO}^- & \text{COO}^- \\
| & | & | & | \\
(\text{CH}_2)_n & (\text{CH}_2)_n & (\text{CH}_2)_n & (\text{CH}_2)_n \\
| & | & | & | \\
\text{CH}\cdot\overset{+}{\text{N}}\text{H}_3 & \text{CH}\cdot\overset{+}{\text{N}}\text{H}_3 & \text{CH}\cdot\overset{+}{\text{N}}\text{H}_3 & \text{CH}\cdot\text{NH}_2 \\
| & | & | & | \\
\text{COOH} & \text{COO}^- & \text{COO}^- & \text{COO}^-
\end{array}
\qquad (3)
$$

$+1$ 0 -1 -2 Net charge

$pK_1 \sim 2$ $pK_2 \sim 4$ $pK_3 \sim 10$

Similar schemes may be drawn up for the basic amino acids. Thus for histidine we have:

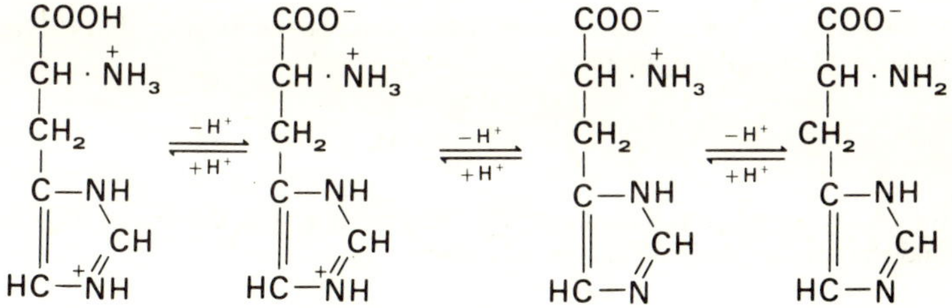

$$
(4)
$$

Net charge

$+2$ $+1$ 0 -1

$pK_1 \sim 2$ $pK_2 \sim 6$ $pK_3 \sim 9$

For lysine $pK_1(COOH) \sim 2$, $pK_2(-NH_3) \sim 9\cdot2$, and $pK(-NH_3) \sim 10\cdot8$ and for arginine $pK_1(COOH) \sim 1\cdot8$, $pK_2(-NH_3) \sim 9$, and pK_3(guanidinium) $\sim 12\cdot3$. The change in net charge on lysine and arginine molecules, as the pH changes, is the same as for histidine, becoming one less positive as the pH rises through the region of each pK.

The titration curves of glutamic acid and histidine are shown in Fig. 4.2.

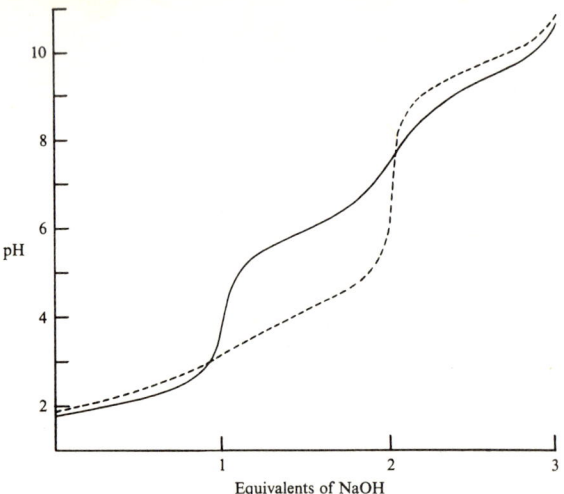

FIG. 4.2. Titration curves of glutamic acid and histidine. (Titrations were with NaOH and started from the most protonated forms of the amino acids). – – – – Glutamic acid hydrochloride, ———— histidine dihydrochloride.

Two other dissociations of amino acids must be mentioned: that of the sulphydryl group of cysteine

$$\underset{R}{\overset{SH}{|}} \rightleftharpoons \underset{R}{\overset{S^-}{|}} + H^+ \tag{5}$$

for which the pK when the cysteine is in a protein is in the region of 7–8; and that of the phenolic group of tyrosine

$$\underset{R}{\overset{OH}{\bigcirc}} \rightleftharpoons \underset{R}{\overset{O^-}{\bigcirc}} + H^+ \tag{6}$$

for which the pK, when the tyrosine is in a protein, is in the region 8·5–10·9.

Formol titration. The reaction of the $-NH_2$ group of a simple amino acid with excess formaldehyde shifts its pH titration curve with NaOH about 3 pH units towards the acid side. This effect is due to the reaction between the anionic species of the amino acid with formaldehyde thus:

$$\text{HCHO} + \text{H}_2\text{N} \cdot \overset{\overset{\textstyle R}{\displaystyle |}}{\text{CH}} \cdot \text{COO}^- \rightleftharpoons \text{HO} \cdot \text{CH}_2 \cdot \text{NH} \cdot \overset{\overset{\textstyle R}{\displaystyle |}}{\text{CH}} \cdot \text{COO}^- \quad (7)$$

$$\text{HCHO} + \text{HO} \cdot \text{CH}_2 \cdot \text{NH} \cdot \overset{\overset{\textstyle R}{\displaystyle |}}{\text{CH}} \cdot \text{COO}^- \rightleftharpoons$$

$$(\text{HO} \cdot \text{CH}_2)_2\text{N} \cdot \overset{\overset{\textstyle R}{\displaystyle |}}{\text{CH}} \cdot \text{COO}^- \quad (8)$$

These reactions displace equilibrium 1 (p. 58) to the right which means that the hydrogen ion is more easily liberated, i.e. in effect the amino group has become a stronger acid, the concentration of $-\overset{+}{\text{N}}\text{H}_3$ groups can be estimated by titrating to pH 8·5 with NaOH (phenolphalein).

Analytical Separation of Amino Acids

Ionophoresis. The direction in which an amino acid in solution will move in an electric field depends upon the pH. At the *isoionic point* (IIP) the molecule is electrically neutral, i.e. it has the same number of positive and negative charges and it does not move in the field. The pH of the IIP for a monoaminomonocarboxylic amino acid is given by:

$$\text{pH}_{\text{IIP}} = \frac{\text{p}K_{\text{amino}} + \text{p}K_{\text{carboxyl}}}{2} \quad (9)$$

For amino acids containing more than one basic (amino or other) group or more than one acidic (carboxyl) group in the molecule, the expression for pH_{IIP} is more complex.

At pHs *above the IIP* the amino acid will have a negative charge and will move to the anode, and at pHs *below the IIP* it will have a positive charge and will move to the cathode. Because of the different pKs of various amino acids and differences in their IIPs they may be separated from one another by differential migration in an electric field applied to some suitable supporting medium (paper, agar or starch gel, etc.). The pH of the supporting medium may be varied to effect the desired separations.

Ion-exchange chromatography. Ion-exchange resins are insoluble polymers containing either acidic ($-SO_3H$ or $-COOH$) or basic (quaternary amine hydroxide) groups. The former behave as cation exchangers because they can exchange H^+ ions for Na^+, K^+, Ca^{2+}, etc., the latter are anion exchangers because they can exchange OH^- ions for Cl^-, $R \cdot COO^-$, $H_2PO_4^-$, HPO_4^{2-}, etc. Because amino acids contain both anionic and cationic groups, mixtures of them, under suitable conditions, may be adsorbed on to either anion or cation exchangers and then eluted separately by washing the resin through with suitable solutions of acids, alkalis, or buffers. Often the temperature of the column in which the exchangers are held is varied in order to improve the separation. The concentration of each amino acid as it arrives in the effluent may be estimated with ninhydrin (p. 56), and by this means the amount of amino acids in the hydrolyate of a few milligrams of protein, or in a few millilitres of a biological fluid, may be estimated. The process is often completely automated. The separation of glutamine and asparagine is usually not so well studied as that of the other amino acids, as these two compounds are not found in acid protein hydrolysates.

Paper chromatography. Amino acids may be separated by repeatedly partitioning them between water and some other immiscible solvent in an apparatus which allows the fractions to be maintained in separate

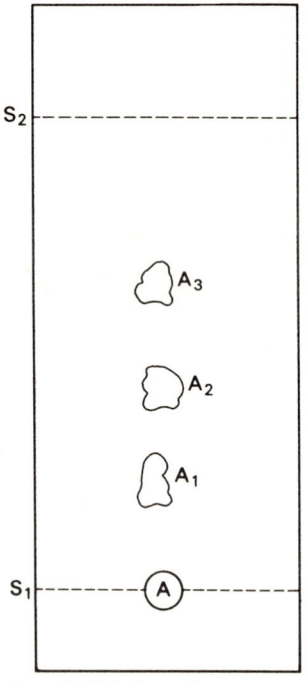

FIG. 4.3. Thin-layer chromatogram.

compartments. In paper chromatography separation is effected by the continuous partition of the amino acids between a stationary water phase adsorbed on the cellulose of filter paper and a moving organic solvent flowing over the paper. Chromatography on a thin layer of support (usually silica gel) spread on a glass plate is much faster than on paper, and the limit of detection is also much lower (*thin-layer chromatography, TLC*).

A small quantity (micrograms) of the mixture to be separated is dried on a thin-layer plate at A (Fig. 4.3). The bottom of the plate is immersed to below the level S, in a closed vessel containing solvent (phenol, butanol, etc.). The solvent travels slowly up the plate; when it reaches S_2 the solvent is dried off and the amino acids detected by spraying the plate with a solution of ninhydrin and heating in an oven. A number of blue-purple spots appear at A_1, A_2, A_3 etc., corresponding to the various amino acids present in the mixture.

The position reached by any particular amino acid is characteristic in a given solvent and at a given temperature. The relative flow rate, R_F, of a substance is defined as the distance moved by the substance divided by the distance moved by the solvent. Thus in Fig. 4.3 the R_F of substance A_3 is given by $(A_3-S_1)/(S_2-S_1)$. This is *ascending* chromatography; with paper strips a similar method may be used, or the paper may be reversed, folded over above S_1 and allowed to hang in a trough of solvent in a closed jar. The solvent then travels down the paper (*descending* chromatography).

If the resolution of a mixture is inadequate in any particular solvent, a two-dimensional chromatogram may be used (see p. 452).

Naturally Occurring Peptides

The term peptide is difficult to define precisely; the smallest peptides are dipeptides—2 amino acids joined by 1 peptide bond. The upper limit is less exact; peptides were originally thought of as amino acid polymers which could diffuse through a dialysis sac, but this depends to some extent on the shape, as well as the mol. wt., of the molecule. Some authors quote an upper limit of mol. wt. 6 000, others 10 000.

Peptide sequences conventionally start from the free amino end. The appearance of the grouping NH_2 at the right-hand end of some of the formulae below indicates that the terminal carboxyl group carries an amide residue.

A number of peptides occur in all living organisms; several hormones and other substances with powerful pharmacological actions are known to be comparatively simple peptides. The sequences of a number of such substances are given below, although it is not reasonable to expect to remember them in detail.

It should be borne in mind that all the peptides enumerated, and a number of more complex polypeptides besides, have been synthesized completely *de novo* in the laboratory after their sequences were elucidated.

Indeed, many peptides in clinical use are almost entirely of synthetic origin. It is relatively easy to introduce different amino acids into the chain, and to test such 'unnatural' products for biological activity. By this means new peptides with very different properties (e.g. potency, rate of destruction) from the naturally occurring prototypes may be prepared. Textbooks of basic medical science must continue to focus their attention on the natural products, but students should realize that the differences may sometimes be very great, as Table 4.1 illustrates.

Table 4.1

The Biological Activities of Some Vasopressin and Oxytocin Analogues

	Rat uterus	Pressor effect	Rat antidiuresis
Vasopressins			
1 2 3 4 5 6 7 8 9			
Cys · Tyr · Phe · Gln · Asn · Cys · Pro · Arg · GlyNH₂	20	400	430
* β-MP———————————————————		370	1 300
Cys——————————————Lys————	4	280	250
——————————————————Orn————		360	88
————————Ile——————————Orn——		103	2·5
Vasotocin			
————————Ile——————————Arg——	115	245	250
Oxytocins			
————————Ile——————————Leu——	500	5	5
——————Phe. Ile——————————Leu——	30	0·4	0·5

All values are in International Units/mg
* β-MP is β-mercaptopropionic acid: $HS \cdot CH_2 \cdot CH_2 \cdot COOH$
Human vasopressin and oxytocin are the first structures in their respective subsections

Similar remarks may be made about the synthetic analogues of the biologically active steroids, as has already been implied on p. 43.

The posterior pituitary hormones are touched on again in Chapter 16, as is the anterior pituitary *adrenocorticotrophic hormone* (ACTH), a larger peptide having at least 24 amino acid residues, and also the hypothalamic releasing factors. *Luteotrophin releasing factor* has the composition: PCA · His · Trp · Ser · Tyr · Gly · Leu · Arg · Pro · GlyNH₂, while *thyrotrophin releasing factor* is PCA · His · ProNH₂.

Insulin, the islet cell hormone, is a large peptide of molecular weight 6 000 which consists of 2 polypeptide chains held together by —S—S— linkages. The other islet cell hormone, *glucagon*, has 29 amino acids and a mol. wt. of 3 500. All these hormones show species differences in structure.

The *gastrins* stimulate the secretion of HCl by the stomach.

Human gastrin I has the composition: Glu · Glu · Pro · Trp · Leu · Glu · Glu · Glu · Glu · Glu · Ala · Tyr · Gly · Trp · Met · Asp · PheNH$_2$. The *Gastrins II* have a sulphate group esterifying the Tyr residue at position 12. The entire physiological activity resides in the C-terminal tetrapeptide, Trp · Met · Asp · PheNH$_2$.

Experimental renal hypertension, caused by constriction of the main renal arteries, is due to the release of a polypeptide—hypertensin—or angiotensin into the blood stream by the kidneys. The polypeptide is released by the kidney enzyme, renin, from a substrate in the α_2-globulin fraction of plasma (Chapter 14). *Hypertensin I*, which is produced by renin, is a decapeptide; it is itself inactive and is converted into the active octapeptide, *hypertensin II*, by an enzyme in blood plasma. This is the most powerful pressor substance known.

Plasma α_2-globulin

↓ *Renin*

Asp · Arg · Val · Tyr · Ile · His · Pro · Phe · His · Leu

Hypertensin I

↓ *Plasma enzyme*

Asp · Arg · Val · Tyr · Ile · His · Pro · Phe

Hypertensin II

Hypertensin II stimulates the secretion of aldosterone. It may not be of primary importance in the maintenance of chronic hypertension.

Among other pharmacologically active agents affecting smooth muscles are *kallidin*: Lys · Arg · Pro · Pro · Gly · Phe · Ser · Pro · Phe · Arg (kallidin II) and *bradykinin* (sometimes called kallidin I): Arg · Pro · Pro · Gly · Phe · Ser · Pro · Phe · Arg.

The dipeptides *carnosine* and *anserine*, β-Ala-L-His and β-Ala-L-methylhistidine respectively, occur in considerable amounts in voluntary muscle; their function is unknown.

The tripeptide *glutathione*, γ-L-Glu-L-CySH-Gly, is widely distributed in the body and is believed to play some part in oxidation–reduction systems. The related tripeptide *ophthalmic acid*, γ-L-Glu-L-α-amino-*n*-butyryl-Gly, occurs in the lens.

About 50 mg of peptides containing *hydroxyproline*, coming from the incomplete enzymic digestion of collagen, are excreted each day in normal urine.

Proteins

Proteins are large, nitrogen-containing, organic molecules built up of amino acid residues linked together by peptide bonds. They have molecular weights from about 6 000 up to several millions.

General Properties

Osmotic pressure. Because of their large size, proteins in solution will not pass through such semipermeable membranes as cellophane or the walls of the capillary blood vessels. When protein solutions are contained within such membranes they will exert an osmotic pressure.

Ionization. When the polypeptide chain is built up only of simple monoaminomonocarboxylic acids, the only groups which can be charged are the terminal amino and carboxyl residues:

$$\overset{+}{N}H_3 \cdot \underset{|}{\overset{\overset{\textstyle R}{|}}{C}}H \cdot CO \cdot NH \cdot \underset{|}{\overset{\overset{\textstyle R}{|}}{C}}H \ldots CO \cdot NH \cdot \underset{|}{\overset{\overset{\textstyle R}{|}}{C}}H \cdot COO^-$$

Usually, however, there are in the chain acidic amino acids such as aspartic and glutamic acids, and basic ones such as lysine, histidine, or arginine; as we have seen these residues may carry negative or positive charges. In addition, cysteine and tyrosine residues may be ionized. Such charged groups are water-attracting and their presence tends to make the molecule water-soluble.

Whether or not any particular group carries a charge depends on the pH of the solution. At low pHs there will be more positively than negatively charged groups and the protein will migrate to the cathode in an applied electric field. At high pHs the negatively charged groups will predominate and the molecule will migrate to the anode. The *iso-ionic point* of the protein is the pH at which the molecule carries the same number of positive and negative charges and it remains stationary in an electric field.

Solubility. Proteins vary greatly in their solubility. They may be soluble in water, dilute salt solutions, or dilute acids and alkalis; they are least soluble at their IIPs. They are insoluble in non-polar organic solvents such as benzene, carbon tetrachloride, and anhydrous acetone and alcohols, though they may dissolve in aqueous alcohols. They are precipitated from solution by concentrated salt solutions, strong mineral acids and bulky ions, e.g. tungstate.

Hydrolysis. Proteins may be hydrolysed by heating with acids or by the action of proteolytic enzymes, firstly to peptides and ultimately to the constituent amino acids.

Structure

Since proteins contain hundreds of amino acid residues in their molecules the problem of determining their detailed structure is very complex. In recent years great advances have been made in this field with a number of individual proteins. The problem may conveniently be divided into the elucidation of their primary, secondary, tertiary and quaternary structures.

By *primary structure* is meant the sequence of amino acids along the peptide chain, together with the position of interchain or intra-chain covalent bonds between residues such as —S—S— or desmosine. This sequence is unique for each protein or peptide and frequently determines completely the tertiary structure (see below). Semi-automatic methods of sequencing short stretches of peptide chain are now highly developed, and the primary structure of several hundred proteins is now known.

The *secondary structure* is the arrangement in space of atoms and groups belonging to neighbouring amino acid residues. This may conform to a pattern that extends over hundreds of residues, or it may change sharply within a few Ångström units. Not many configurations are possible, because the peptide grouping itself is planar (Fig. 4.4). This constraint means that for many polypeptides, only one of two secondary structures is energetically favourable, the *helical coil* (Fig. 4.5) or the more extended *pleated sheet* (Fig. 4.6). The former is the more common, and it has been shown (chiefly by optical means and by X-ray crystallography) that many proteins contain short regions of a right-handed α-helix. The stability of this configuration is ensured by *hydrogen bonds* between two nitrogen atoms or between a nitrogen and an oxygen atom. By a hydrogen bond is meant a hydrogen atom shared between two suitable atoms not too far away from one another. This is possible because the —OH and —NH bonds are very polar. The O and N atoms are very electronegative, and the hydrogen atom has some of the character of a naked proton. Fig. 4.5 shows that the hydrogen bonds tend to form between peptide bonds separated by 3–4 residues.

Proline and hydroxyproline have the wrong shape to take part in an α-helical coil, and several other amino acids, especially charged ones, do not favour this particular secondary structure. Thus it is rare for a protein molecule to be entirely helical. In addition, specialized proteins with extended repeating secondary structures are important, and are described later (p. 73).

In *denatured* proteins the complete chain, with its helical regions often almost intact, tends to be stretched out, or at best only haphazardly coiled,

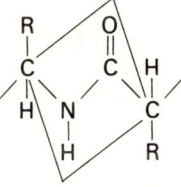

Fig. 4.4. Spatial conformation of the peptide bond. The bonds within the rectangle all lie within one plane.

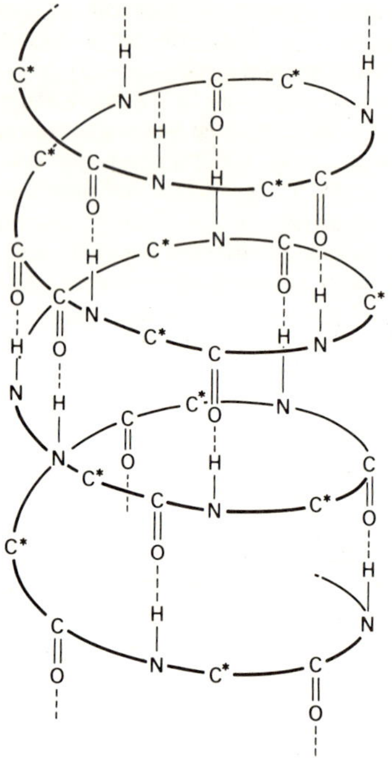

FIG. 4.5. The α-helix. Side chains are attached at C*.

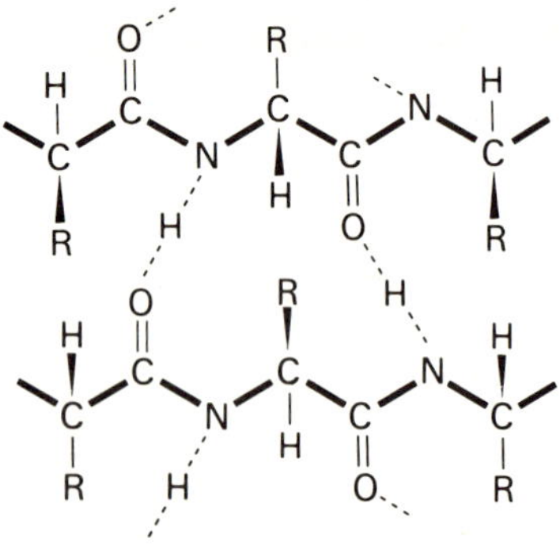

FIG. 4.6. Inter-chain hydrogen bonds in the β-form (parallel pleated sheet) of fibrous proteins.

like a discarded piece of string (a *random coil* structure). Treatment with detergents, especially sodium dodecyl sulphate (SDS) will produce this effect (see p. 75). Such a spatial arrangement is not characteristic of biologically active globular proteins, where the short helical regions of the chain are folded in among one another in a way that is absolutely specific for each protein. This is the *tertiary structure*, which may be further stabilized by intra-chain —S—S— bonds, but these usually form *after* the optimum conformation has been taken up. Frequently the folding takes place in such a way that almost all the polar residues are on the outside, while the non-polar residues (e.g. leucine, valine, and the aromatic amino acids) are tucked into the interior of the molecule. This conformation favours the maximum orderly arrangement—and thus the lowest entropy— of water molecules around the polar groups, and the maximum disposition of weak van der Waals forces (hydrophobic bonds) among the non-polar residues. It can be computed that the sum of such forces stabilizes the optimum tertiary structure considerably. Presumably in non-aqueous environments, as for example in membranes, the opposite configuration would be more stable, but methods for studying lipophilic proteins are not yet sufficiently advanced for this to be confirmed.

In many instances the polypeptide chain adopts its tertiary conformation spontaneously on release from the ribosome (p. 347). However, many polypeptides are known to change their conformation slightly, and some quite drastically, when quite small molecules (e.g. enzymic substrates or effectors) become bound (non-covalently) to them, and the change of shape with change of pH in the surrounding medium is often quite striking. The specific conformational folding of globular proteins brings into juxtaposition amino acid residues from quite different parts of the primary sequence, (see Fig. 4.7), and this is important, for instance, in the three-dimensional architecture of the active centre of enzymes (for examples see p.133 and p.135).

Somewhat different considerations apply to the tertiary structure of fibrous proteins, and these are discussed later in this chapter.

Many, perhaps most, large globular proteins have sub-units, themselves polypeptides with complex secondary and tertiary structures which are associated by non-covalent binding. The number is characteristic for each protein, but the minimum is 2 sub-units, and the maximum so far established is 15. The sub-units are sometimes called *protomers*, and the assembled proteins *oligomers*. This is called *quaternary structure*. Most is known about the quaternary structure of haemoglobin, but many enzymes such as lactic and glutamic dehydrogenases and phosphorylase have also been studied. Very often each sub-unit has a centre of biological activity, but this need not be the case. Interactions between the sub-units appear to affect the overall activity of the molecule. There is little absolute distinction between an assemblage of non-identical sub-units and a *multi-enzyme complex*, e.g. pyruvate oxidase (p. 165) or fatty acid synthetase (p. 190). The molecular weights of such complexes may be 10^6 daltons or more. Between these and ribosomes, which have a finite number of polypeptide components and a mol. wt. of about 4.5×10^6 daltons, the distinction is also blurred. Fibrous proteins, e.g. myosin, collagen, may also have a structure of repeating sub-units, and the mol. wt. of such polymers is very hard to define.

Classification

Simple Proteins

These contain only amino acid residues and no consistent amount of other intimately bound material; they may be classified partly according to their solubility.

1. *Protamines* are very basic proteins of low molecular weight (6 000 to 10 000); they contain a very high proportion of arginine and are not heat-coagulable. They occur in conjunction with nucleic acid in ripe spermatozoa, especially of fish.

2. *Histones* are basic proteins with higher molecular weights than the protamines; they have a high content of arginine, lysine, and histidine. They are soluble in dilute acid and in neutral solution, from which they can be precipitated by dilute ammonia, and are heat-coagulable. They are widely distributed in the body, being the usual protein moiety of nucleoproteins.

3. *Albumins* and *globulins* are typical globular proteins. The terms are now almost entirely restricted to the proteins of serum and of milk. They are neutral proteins with IIPs in the range pH 5–7. The chief solubility differences between albumins and globulins are shown in Table 4.2.

Table 4.2
Solubilities of Serum Proteins

	Albumins	Globulins
Water	soluble	insoluble
Dilute salt solutions	soluble	soluble
Half-saturated $(NH_4)_2SO_4$ solutions	soluble	insoluble
Saturated $(NH_4)_2SO_4$ solutions	insoluble	insoluble

4. *Contractile proteins* include the muscle proteins myosin, which belongs to the Kmef sub-group (p. 72), tropomyosin A and B, and troponin. Actin is in a different category, since the monomer is globular, and fibrous actin (F-actin) is a linear polymer of these units.

Contractile proteins are insoluble in water, but soluble in 0·6M–KCl.

5. *Epidermal proteins* also belong to the Kmef group and are the proteins of the skin, hair and nails. They are insoluble in most solvents including concentrated acids, but are fairly readily hydrolysed by concentrated alkalis. They have a high cystine content and there is considerable cross-linking of the polypeptide chains by —S—S— bridges. These bridges are broken by alkaline thiol solutions which dissolve these proteins.

6. *Scleroproteins* are connective tissue proteins such as collagen, ossein, reticulin, and elastin.

Conjugated Proteins

These contain in addition to the polypeptide chains defined amounts of other substances or groups including, in some instances, metal ions which impart characteristic properties. The following list is not exhaustive.

1. *Nucleoproteins* are found in all cells, they are formed by association of a histone or protamine with ribonucleic acid or by deoxyribonucleic acid.

2. *Chromoproteins* are generally soluble proteins combined with a chromophoric, or coloured, group. The chromophoric group is copper in ceruloproteins, haem in haemoglobin and the cytochromes, and riboflavin in the flavoproteins.

3. *Metalloproteins* contain one or more atoms of tightly bound metal per molecule. Examples are zinc (carbonic anhydrase), copper (superoxide dismutase, cytochrome oxidase), molybdenum (xanthine oxidase) and non-haem iron (ferredoxin).

4. *Phosphoproteins* contain either ortho- or pyro-phosphate or phospho-diesters. Milk casein is a globulin containing about 1 per cent phosphorus, ovalbumin of egg-white can contain 0, 1, or 2 phosphate residues per molecule of protein, and egg-yolk phosphoprotein contains 10 per cent phosphorus. Several enzymes (e.g. pepsin, phosphorylase) contain one molecule of covalently bound phosphate per molecule of polypeptide.

5. *Glycoproteins* are proteins combined with amino sugars, sugar acids, and sulphate. Two glycoproteins occur in plasma. Other glycoproteins are important constituents of cell walls and membranes, with functions of maintaining intercellular contact and intercell inhibition, and perhaps also some membrane transport. Cell-wall glycoproteins are strongly antigenic. For the carbohydrate moiety see p. 27.

Mucoids, the firm rubber-like proteins of ground substance, cartilage and bone (osseomucin), contain protein and chondroitin sulphuric acid which, on hydrolysis, yields N-acetylgalactosamine, glucuronic acid, and sulphate. *Mucins*, the thick slimy lubricants of the respiratory and alimentary systems, contain protein and mucoitin sulphuric acid which, on hydrolysis, yields N-acetylglycosamine, glucuronic acid, and sulphate.

6. *Lipoproteins* are combinations of proteins with lipids, which are found in plasma and in brain. In the plasma they may act as lipid transporters (see Chapter 10).

Proteins can also be defined in terms of physical properties, e.g. solubility, and especially their shape.

Globular Proteins

These are the generally soluble proteins of the body, the blood proteins and the many intra- and extra-cellular hormones and enzymes. Their primary and secondary structures are well defined, probably unique, and their individual physiological functions depend on the tertiary structure being intact. Many can be crystallized.

The molecules are compact and generally spherical or ovoid, either because the polypeptide chains bend back on themselves several times, or because several short chains are joined together as in the immuno-globulins. The chains are often partly in the α-helical form. Fig. 4.7 illustrates the complex folding of the peptide chain that occurs in globular proteins.

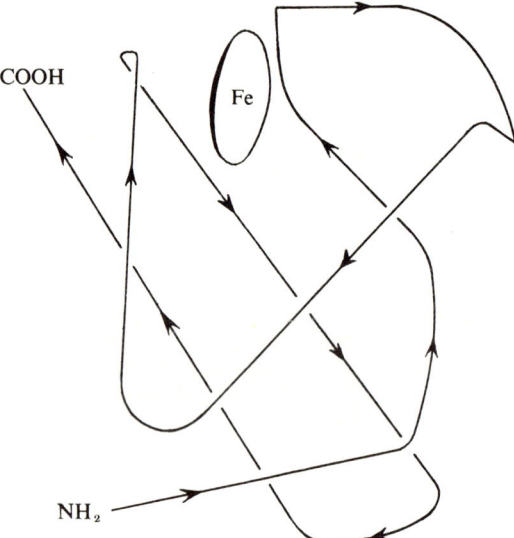

Fig. 4.7. Diagram of the tertiary structure of myoglobin.

Fibrous Proteins

The tertiary structure of fibrous proteins is such that the molecules are long and thin, and when assembled in bundles, they make up fibrils. The structural material of the body is largely made up of this group of insoluble proteins. *The Kmef sub-group* includes the keratin of the skin and hair, the epidermin of the skin, the myosin of the muscles and the blood-clotting protein fibrinogen. All these proteins are elastic, i.e., they can be stretched and will revert to their original length when released. When contracted the secondary structure is of the α-helical form, but when stretched it changes over into a β-form, which is known as the 'pleated sheet' conformation (Fig. 4.6). In this, the main peptide chain lies approximately in one plane, while the R-groups lie alternately above and below this plane. In the Kmef sub-group all the peptide backbones run in the same direction, and the hydrogen bonds holding the chains together are disposed as shown (in silk fibroin the chains are anti-parallel). With all fibrous proteins, bulky R-groups hinder the close packing of the chains in the β-form. In the third dimension, the chains are held together by intra-chain hydrogen bonds, or by intra-chain —S—S— bonds, which are frequent in keratin.

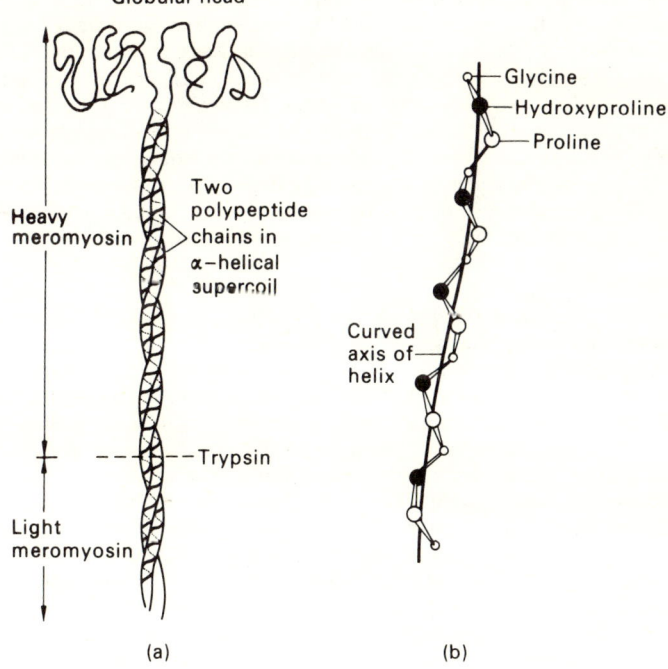

FIG. 4.8. (a) Schematic representation of the myosin molecule showing the globular head and long helical tail.

(b) A typical helix of a single collagen chain. In collagen fibrils three of these chains are coiled round each other.

Myosin is rather a special case. One end of each polypeptide chain, about 1 800 residues long, is globular. The rest is α-helical, and two such chains are super-coiled around each other like a rope (Fig. 4.8a). Trypsin cleaves the polypeptide into two unequal fragments: *light meromyosin*, and *heavy meromyosin*, which contains the head, and almost certainly the ATP-ase. In striated muscle the myosin filaments are stacked very regularly, in the A-band region, in a hexagonal pattern interspersed with F-actin polymers (see p. 70), and minor components such as troponin. The I-band region contains only actin filaments. The globular heads of the myosin molecules make and break contact with the actin as the two kinds of filaments slide past each other; thus normal contraction and relaxation of the myofibrils does not involve conformational change from the α-helical form.

Collagen. This inextensible supporting material makes up one-third of the total protein in the body. The primary structure contains a large number of repeats of the sequence -Gly-Pro-Hypro- (more than half of the residues are made up of these three amino acids). The chain can therefore not form an α-helix (see p. 67), but is arranged in a special helix, shown in Fig. 4.8b. Three of these helices form a supercoil of tropocollagen, which is arranged in staggered sections in the collagen fibrils. Hydroxyproline is formed by hydroxylation of proline *after* the primary sequence is complete (see p. 347 and p. 350), and this seems to be necessary for the extrusion of complete fibrils from the fibrocytes. It is also likely that the hydroxyproline —OH groups are important in binding the polysaccharides found in connective tissue, either by hydrogen bonding or by covalent links.

Denaturation

Globular proteins and some fibrous proteins such as myosin can lose their native state, or be denatured, when subjected to extremes of pH, heat, or violent shaking and foaming. Many organic solvents or high concentrations of heavy metal ions cause denaturation at room temperature. The characteristics of denaturation are a decrease in solubility, loss of crystallizability, and loss of specific biological activity. In the process of denaturation the tertiary structure of globular proteins changes into the random coil form (p. 69) and, if the process is prolonged, into the non-structure of insoluble protein, with the individual molecules not associated in bundles, but randomly oriented in space. Denaturation is not, however, a precise term, and mildly denatured proteins may re-assume their original conformation if the agent (heat, chemical) is removed slowly. Trypsin, for example, regains its activity rapidly after heating. This process is known as *annealing* or renaturing. Other polymers with precise structure, e.g. nucleic acids, can be denatured and annealed.

Some proteins, e.g. pyruvate kinase, are more stable at room temperature than in the cold. This shows that it is conformation, and not chemical energy *per se*, that is important in denaturation.

If the extended random coils of denatured globular proteins can be arranged parallel to one another, as by extruding through a fine orifice, the resulting bundles can be given the texture and elasticity of muscle fragments. This is the basis of *spun vegetable protein* as a meat substitute.

Methods of Separation (see also Chapter 21)

Fractional precipitation. Proteins may be separated into groups by taking advantage of their differing solubilities in ammonium sulphate and other salt solutions, or in various strengths of aqueous alcohol. The protein is subsequently freed from salt or alcohol by dialysis.

Electrophoresis. Since the charge carried by a protein molecule depends on the pH, it is possible to obtain separation by migration in an electric field under controlled conditions of pH, concentration, etc. Such separations are now rarely effected in the liquid phase in a U-tube, but much more often in aqueous solutions supported in filter papers, starch blocks, or agar or silica gels. Polyacrylamide gel is a very convenient support.

Chromatography. Proteins may be separated on columns of silica gel, ion exchange resin, or modified polysaccharides. Some of these methods have a very high resolution, especially with proteins of low molecular weight.

Molecular sieves. The polysaccharides mentioned above may be modified in such a way that the particles have a controlled pore size. Then proteins too bulky to enter the pores will be 'excluded', and will travel rapidly down the column. Smaller proteins will be retarded as they diffuse in and out of the pores. As some adsorption to the support also takes place, molecular sieve chromatography can be very discriminating. Materials with larger pores, e.g. glass beads, can separate larger macromolecules, for example, viruses.

Ultracentrifugation. Because of their high molecular weight, proteins can be made to sediment in very high speed centrifuges. The sedimentation rate depends, amongst other things, on the molecular weight and also on the shape of the molecules.

Molecular Weight Determinations

The classical methods depend on physical attributes related to size and shape—sedimentation velocity in the ultracentrifuge, velocity of free electrophoresis, rate of diffusion. More approximate but faster methods like molecular sieve chromatography or polyacrylamide gel electrophoresis are found to work well if the system is first calibrated with polypeptides of known molecular weight. Globular proteins may first be denatured by various agents and then subjected to electrophoresis in the presence of SDS (p. 69). This binds intensively to the extended polypeptide chain, and gives the molecule a uniform charge density. The effect of shape and charge on migration velocity then becomes minimal. This variant of calibrated gel electrophoresis provides a very reliable estimate of the molecular weights of single polypeptides or sub-units.

Detection of Proteins

Dye adsorption. Protein bands on paper or in gels are often detected by flooding the support with a dye that binds to protein (wool), e.g. naphthalene black or Coomassie blue, and washing off the excess.

Indicator papers. The 'protein error' in pH indicators dried on paper, caused by the selection adsorption of one species of the indicator molecule to the protein, is used as a qualitative test for protein.

Precipitation tests

(a) *Denaturation.* Most soluble proteins will denature sufficiently on heating with acid to precipitate. Dilute HNO_3 or acetic acid are often used.

(b) *'Alkaloidal' reagents.* Many polyvalent acids will precipitate proteins in the cold. Examples are salicylsulphonic (sulphosalicylic), picric, tannic, tungstic, molybdic, and trichloroacetic acids.

Biuret test. This is a very sensitive test for soluble proteins, depending on the formation of a purple copper complex with peptides in alkaline solution.

Immuno-assay. An antibody to a specific protein will co-precipitate with it even if both antibody and antigen are present in a mixture of other proteins (e.g. antiserum and tissue extract). In *radioimmuno-assay* the method is made more sensitive by making a sample of the antigen labelled with radioisotope. Labelled and unlabelled antigen are then made to compete for a fixed amount of antibody, and the amount of radioactivity in the precipitate is *inversely* proportional to the amount of antibody protein. The method has been extended to non-protein antigens, e.g. steroids. The method will detect nanograms of material. In the *fluorescent antibody* technique the antibody is covalently coupled to a fluorescent dye, and the precipitate may be visualized on a tissue section.

5 Nucleotides, Nucleic Acids, and Related Compounds

Nucleic acids occur in all living cells in both the nucleus and the cytoplasm. They are polymers and may have very high molecular weights; the building units are mononucleotides. In the nucleus the nucleic acids are frequently associated with protein, rich in basic amino acids, which is positively charged at neutral pHs, while the nucleic acids will themselves be negatively charged. The nucleic acids may be separated from protein by a number of methods, but only rather mild procedures, such as extraction with phenol and/or detergents, yield undegraded, protein-free, high molecular-weight nucleic acids. The material thus obtained contains, besides carbon, hydrogen, and oxygen, 15–16 per cent nitrogen (about the same as in the proteins) and, strikingly, 7–10 per cent phosphorus. On hydrolysis of the nucleic acids one obtains, successively, the substances shown in Table 5.1.

Table 5.1
Successive Hydrolysis Products of Nucleic Acids

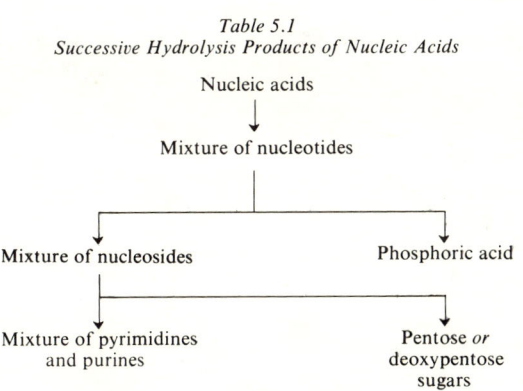

Nucleic acids
↓
Mixture of nucleotides

Mixture of nucleosides   Phosphoric acid

Mixture of pyrimidines   Pentose *or*
and purines   deoxypentose
                                sugars

It will be convenient to consider the hydrolysis products shown in Table 5.1 in the reverse order to their formation, so that we can see how the nucleic acids are built up.

Pyrimidines

Pyrimidines are derivatives of one parent substance, a six-membered ring with aromatic character, containing two nitrogen atoms. The numbering of the atoms given here is in accord with modern practice; it differs from that found in some older works. The structures of these bases are as follows:

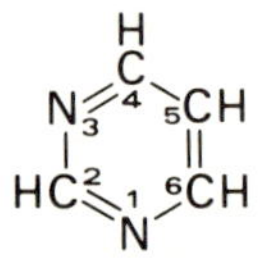

Pyrimidine

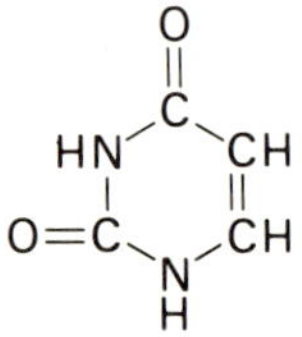

Uracil
(2,4-dioxy-
(pyrimidine)

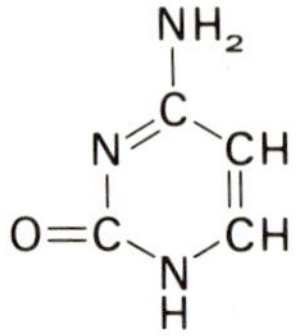

Cytosine
(2-oxy-4-amino-
pyrimidine)

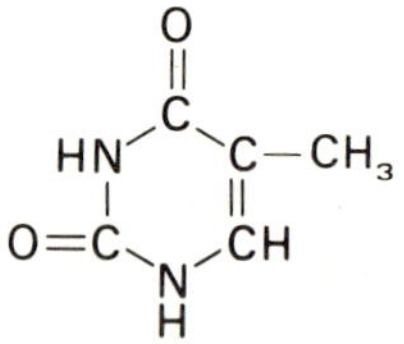

Thymine
(5-methyl uracil)

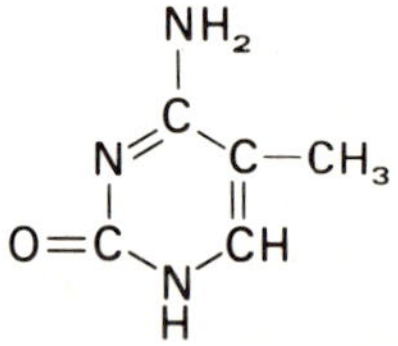

5-Methyl cytosine

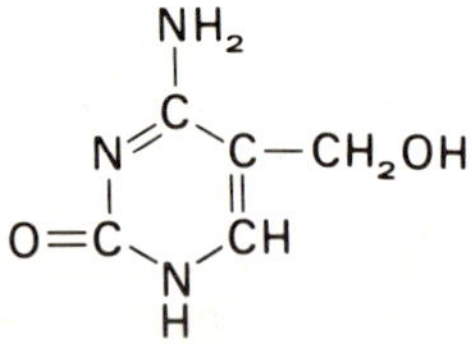

5-Hydroxymethyl cytosine

There are two pyrimidine derivatives which do not occur in the nucleic acids but which are of importance in the biosynthesis of the pyrimidines, they are:

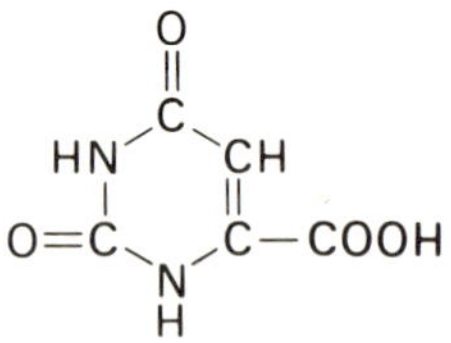

Orotic acid
(uracil-6-carboxylic
acid)

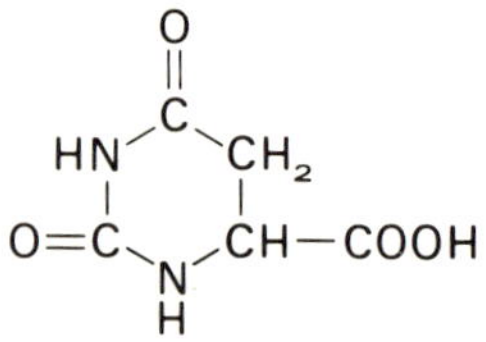

Dihydro-orotic acid
5,6-dihydrouracil-6-carboxylic
acid)

The pyrimidines are moderately soluble in water, particularly cytosine; they are less soluble in alcohol and ether. In the structures above, the oxygen-containing pyrimidines are shown in the keto (or lactam) form they have when linked by a glycosidic bond to ribose or 2-deoxyribose; it is this form which predominates at physiological pHs. An equilibrium, however, exists between keto (lactam) and enol (lactim) forms of these compounds, both of which behave as Bronsted acids at neutral pHs.

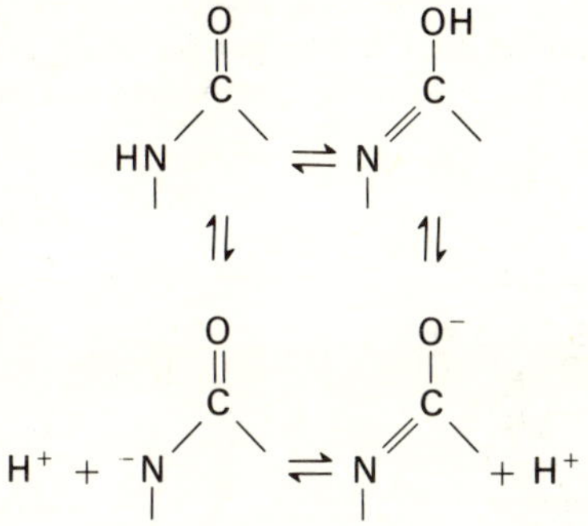

Lactam   Lactim

The nature of the predominant form will, therefore, depend on the pH. The conjugated double bond system in the pyrimidines causes them to have strong absorption bands in the ultra-violet.

Purines

These bases are derivatives of a parent substance which consists of a pyrimidine ring fused to an imidazole ring; the molecule contains four nitrogen atoms.

The structures of these bases are as follows:

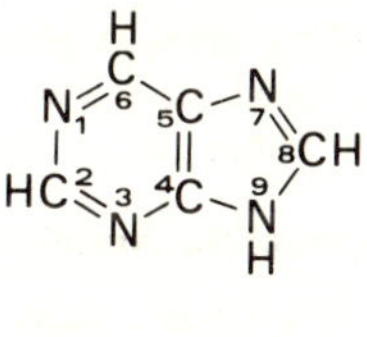

Purine

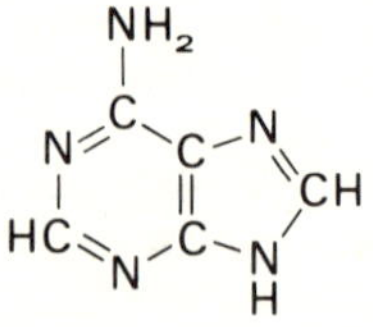

Adenine
(6-*aminopurine*)

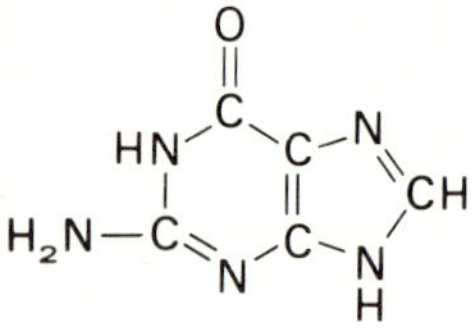

Guanine
(2-*amino*-6-*oxypurine*)

Adenine and guanine are the only purines which commonly occur in the nucleic acids. Others which are of importance in metabolism are:

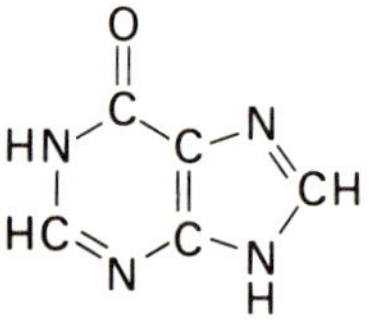

Hypoxanthine
(6-*oxypurine*)

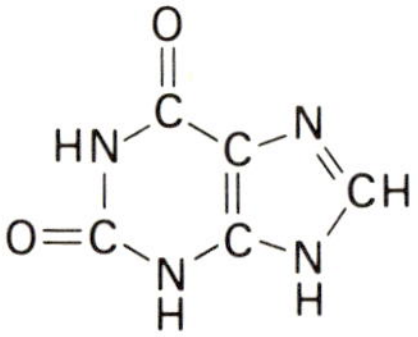

Xanthine
(2,6-*dioxypurine*)

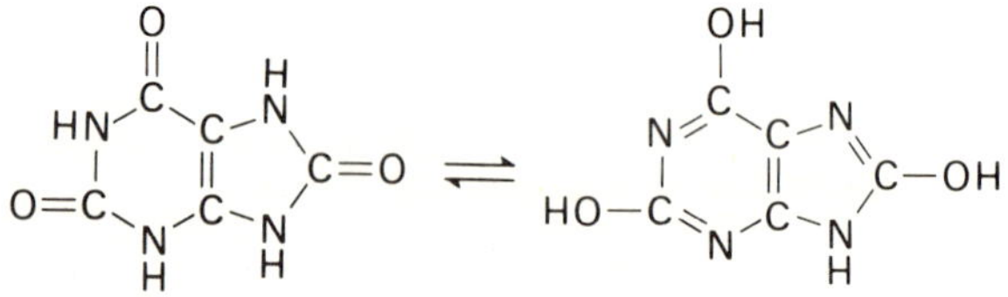

Lactam   *Lactim*

Uric acid
(2,6,8-*trioxypurine*)

The purines are relatively insoluble in water, particularly xanthine and uric acid. As in the pyrimidines there is ample opportunity for tautomerism, and a number of different forms may ocur depending on the pH; they also have strong absorption bands in the ultra-violet.

Sugars of the Nucleic Acids

Two sugars, *ribose* and 2-*deoxyribose*, occur in the two classes of nucleic acids. Only one kind of sugar occurs in any one nucleic acid, hence the

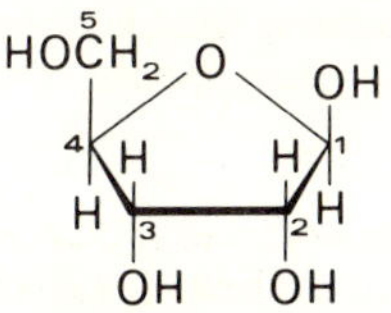

β-D-*Ribose*

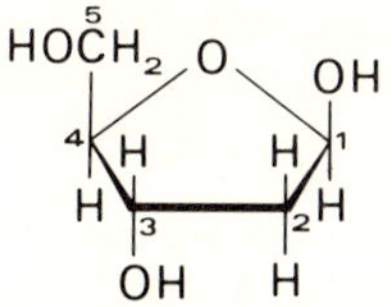

β-D-2-*Deoxyribose*

nucleic acids may be divided into *ribonucleic acid* or RNA and *deoxyribonucleic acid* or DNA according to whether they contain ribose or deoxyribose.

Both these sugars exist in the nucleic acids in the furanose form and the configuration about C–1 is β.

Nucleosides

The nucleosides are compounds of a pyrimidine or a purine and either D-ribose or D-2-deoxyribose. In these compounds the sugar is linked by its C–1 to N–1 of the pyrimidines or N–9 of the purines, except for *pseudouridine*, an unvarying minor constituent of transfer RNA (p. 84). This is uracil with the ribose attached not at N–1, but at C–5:

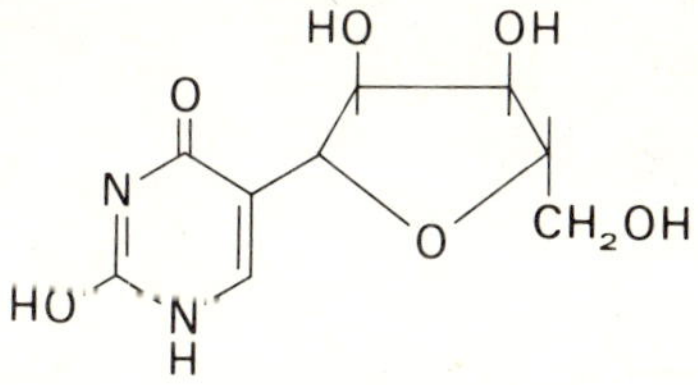

Pseudouridine

Ignoring the nature of the pentose moiety the nucleosides are designated as follows:

uridine ⎫
pseudouridine ⎭ are uracil nucleosides,
cytidine is cytosine nucleoside,
thymidine is thymine nucleoside,
orotidine is orotic acid nucleoside,
adenosine is adenine nucleoside,
guanosine is guanine nucleoside,
inosine is hypoxanthine nucleoside,
xanthosine is xanthine nucleoside.

In the nucleosides the atoms are numbered in the ordinary way, 1, 2, 3, etc., in the base, but primed numbers, 1′, 2′, 3′, etc., are used to designate the carbon atoms of the sugar. They are moderately soluble or soluble in water and have characteristic ultra-violet absorption spectra.

Nucleotides

The nucleotides are nucleosides with a phosphoric acid radical esterified to one of the pentose hydroxyls. They include the repeating units found in the nucleic acids, which contain base, sugar and phosphate in the ratio $1:1:1$, and also a number of other similar compounds, some containing more than one phosphate per sugar residue.

In the ribonucleotides it is possible for the phosphate to esterify a hydroxyl on either carbon atom 2′, 3′, or 5′, while in the deoxyribonucleotides the only free hydroxyls are on carbon atoms 3′ and 5′. According to the conditions of hydrolysis of the nucleic acids one obtains from DNA either 3′ or 5′ phosphates, though sometimes a mixture of the two types. From RNA it is also possible to get 2′-phosphates and 2′,3′-cyclic phosphates.

The structure of the nucleotides is illustrated in the following:

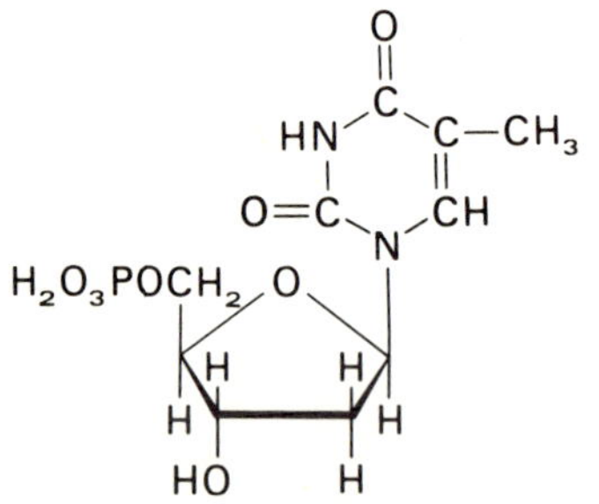

Thymidylic acid, or thymidine 5′-phosphoric acid

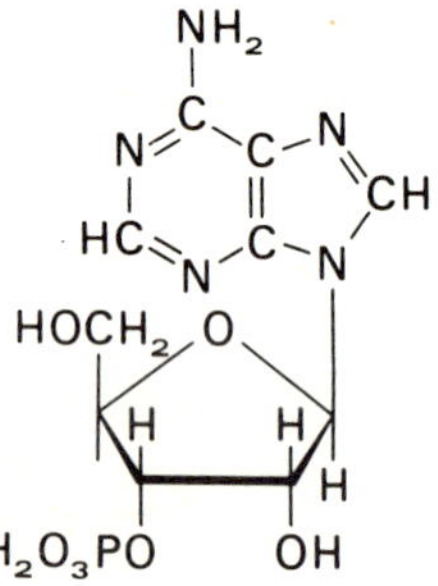

Adenosine 3′-phosphoric acid

The 5′-phosphate nucleotides are known by the following names:

adenylic acid (AMP) is adenosine 5′-phosphate,
guanylic acid (GMP) is guanosine 5′-phosphate,
inosinic acid (IMP) is inosine 5′-phosphate,
xanthylic acid (XMP) is xanthosine 5′-phosphate,
orotidylic acid is orotidine 5′-phosphate,
uridylic acid (UMP) is uridine 5′-phosphate,
cytidylic acid (CMP) is cytidine 5′-phosphate,
thymidylic acid (dTMP) is thymidine 5′-phosphate.

Note that because thymine occurs only in DNA, thymidylic acid is assumed to contain deoxyribose. With the other nucleotides found in nucleic acids, both possibilities exist, so that dATP (adenine-deoxyribose-5'-phosphate), and so on, are valid names where appropriate.

Coenzyme Nucleotides

Many important coenzymes have nucleotide structure; they include the nicotinamide–adenine dinucleotides (NAD and NADP), flavin–adenine dinucleotide (FAD), and coenzyme A, all of which are complex derivatives of 5'-adenylic acid. There is also a group of uridine coenzymes concerned in many sugar transfer reactions, others containing cytidine and guanosine are also known. The structures of several nucleotide coenzymes are given in Chapter 7.

When the phosphate is in the 5' position, the nucleotides are able to add on further molecules of phosphoric acid to form, for example, *adenosine diphosphate* (ADP) and *adenosine triphosphate* (ATP); these may be considered as esters of pyro- and tri-phosphoric acid with adenosine. In addition an important mediator in the control of a number of enzyme systems is 3',5'-cyclic adenosine monophosphate (cyclic AMP). This compound is present in tissues in very much lower concentrations than the other mononucleotides. It is formed from ATP by *adenyl cyclase*:

$$ATP \longrightarrow 3'5'\text{-AMP} + P-P_i$$

Cyclic AMP is hydrolysed back to 5'-AMP by a *phosphodiesterase*.
The structures of these important compounds are as follows:

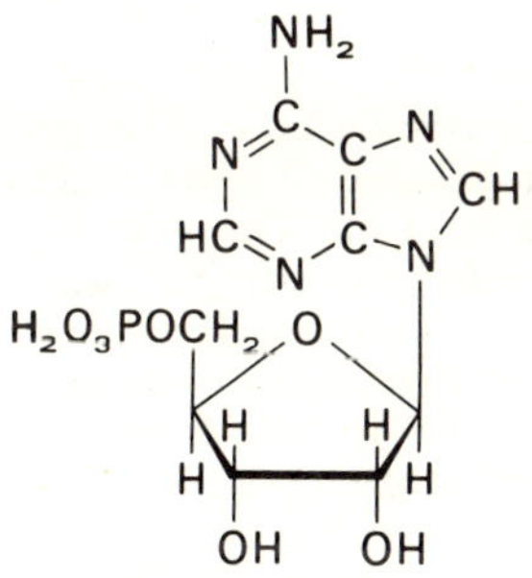

AMP
(*5'-adenylic acid*)

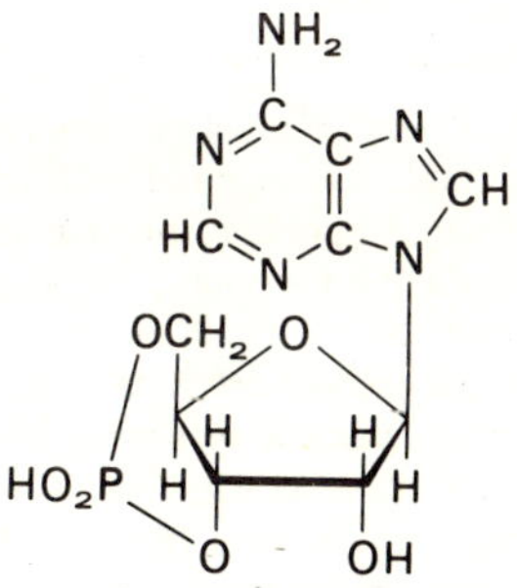

Cyclic AMP
(*3',5'-cyclic adenosine monophosphate*)

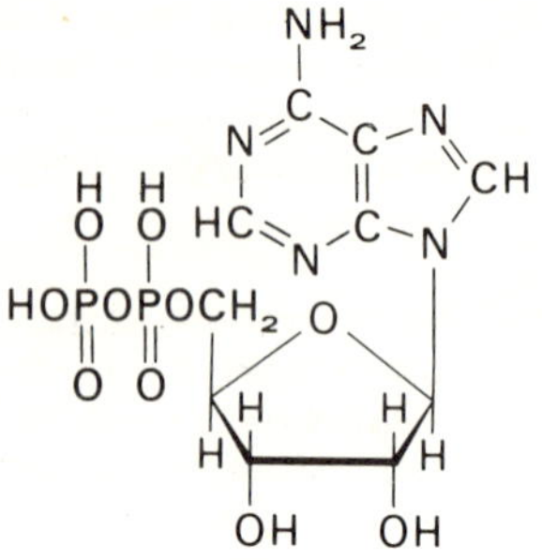

ADP

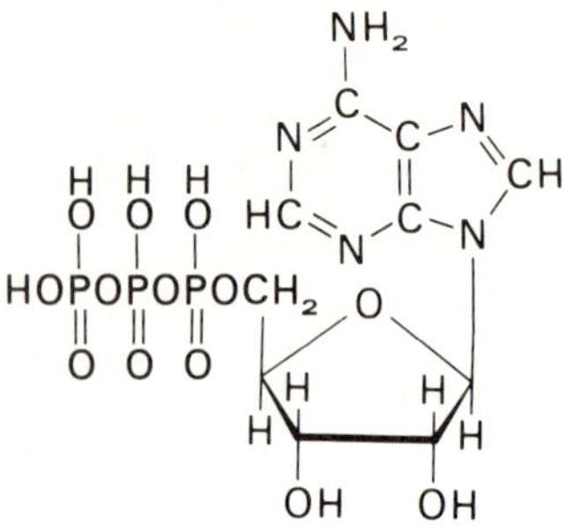

ATP

Analogous di- and tri-phosphates derived from the other nucleotides also exist. They are also designated by abbreviations, for example GTP (guanosine 5′-triphosphate) and UDP (uridine 5′-diphosphate). The deoxynucleoside phosphates are abbreviated similarly but are distinguished by the prefix 'd'.

The nucleotides are very acidic substances and are soluble in water; they have strong absorption bands in the ultra-violet. The nucleotide polyphosphates, as a class, form strong complexes with Mg^{2+} ions, e.g., at pH 7, $MgATP^{2-}$.

Nucleic Acids

Analyses of nucleic acids have shown that there are differences in both the kinds and the proportions of bases they contain. Both RNA and DNA contain adenine and guanine among the purines, but while the principal pyrimidines in RNA are cytosine and uracil those in DNA are mainly cytosine and thymine.

Ribonucleic Acid

RNA is found in most parts of the cell. The largest fraction, 50–80 per cent of the total, is associated with small particles called *ribosomes* which contain about 50 per cent protein and 50 per cent RNA. The ribosomes may be free in the cytoplasm arranged in groups or clusters (*polysomes*), or they may be attached to the surfaces of the endothelial reticulum. The ribosomal RNA (*rRNA*) consists of single-stranded molecules with molecular weights of about 0·7 and $1·7 \times 10^6$ in animals and 0·6 and $1·1 \times 10^6$ in bacteria. For a more complete description of ribosomal structure see Chapter 16.

The next fraction, known as soluble or transfer RNA (*tRNA*), comprises 10–20 per cent of the total. It consists of small polyribonucleotides containing about 75 nucleotides, with molecular weights near 23 000 (see Fig. 5.1). The third group consists of molecules of messenger RNA

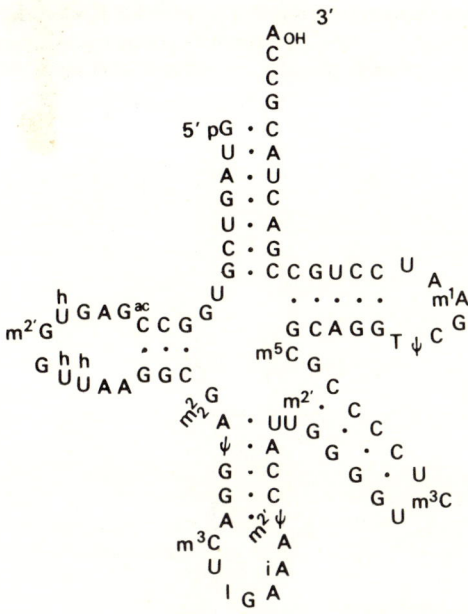

Fig. 5.1. Primary sequence and probable secondary structure of seryl-tRNA. Characteristic features are the 'clover-leaf' conformation with a 'hump' at the lower right (terminating in 3-Me-cytosine), and the high proportion of unusual bases including pseudouridylic acid (ψ). The anti-codon is at the bottom (AGI), while the amino acid is attached to the topmost adenylyl residue.

(*mRNA*) and accounts for 5–10 per cent of the cellular RNA. These molecules are single-stranded and have molecular weights up to 2×10^6. They are concerned with transcribing the genetic message from the DNA and taking it to the sites of protein synthesis on the ribosomes (see p. 333). In addition there is a little RNA in the mitochondria and some is associated with the nucleolus. As much as 20 per cent of the total nuclear nucleic acid may be RNA (*nRNA*).

The RNA from different sources contains different proportions of the various bases; nevertheless the data show that the sum of the 6-amino bases (adenine + cytosine) is often approximately equal to the sum of the 6-hydroxy bases (guanine + uracil).

RNA is labile to alkali and is difficult to obtain free from the hydrolytic enzyme ribonuclease (RNAase) which is widely distributed in cells. For these and other reasons the examination of intact RNA is difficult. Nevertheless it is certain that RNA is a polymer consisting of chains built up of a -ribose-3′-phosphate-5′-ribose-3′-phosphate- backbone, with the bases

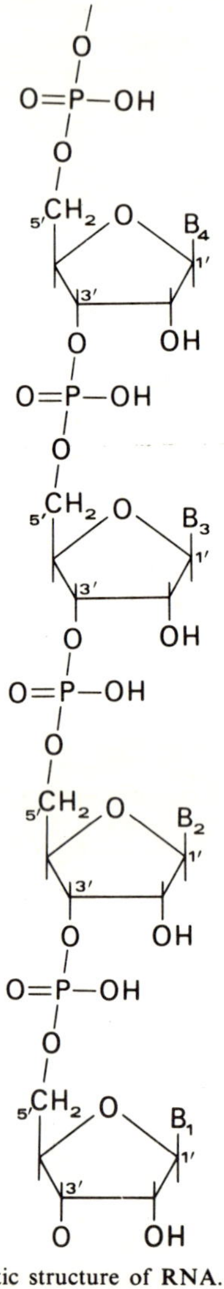

FIG. 5.2. Diagrammatic structure of RNA. B_1, B_2, B_3 and B_4 are pyrimidine or purine bases.

attached at C′–1 of the ribose units. This is shown diagrammatically in Fig. 5.2. At some Mg^{2+} ion concentrations, various RNAs have well-defined secondary and tertiary structures.

Deoxyribonucleic Acid

Analyses show that DNA from different sources has varying proportions of the bases mentioned above. As with RNA, the sum of the 6-amino bases (adenine + cytosine + 5-methylcytosine) is equal to the sum of the 6-hydroxy bases (in DNA, guanine + thymine). Other systematic trends are that the sum of the purines is equal to the sum of the pyrimidines and that the ratios adenine/thymine and guanine/(cytosine + 5-methylcytosine) are unity. DNAs tend to fall into two main types, the 'AT' type in which (adenine + thymine) are in excess and the 'GC' type in which (guanine + cytosine) are in excess.

In DNA there are only two positions for the internucleotide link to occur, these are the 3′ and 5′ carbons of the deoxyribofuranose. Thus it is certain that DNA is built up of deoxyribonucleotides linked 3′—5′ by phosphate bridges, the structure basically being the same as in RNA (above). DNA can be obtained as crystalline fibres which may be examined by X-ray crystallography and other physical methods. It can have a very high molecular weight, up to about 10^9 daltons.

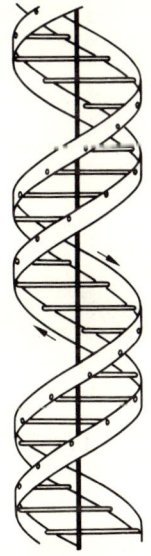

Fig. 5.3. Diagrammatic representation of the helical structure of DNA (see text).

In all but a few primitive organisms, DNA is built up of two right-handed helical chains, coiled around the same axis, and held together by hydrogen bonds in a double helix (Fig. 5.3). In this structure the phosphate-sugar chains are on the outside of the helix, represented by ribbons, with the bases, at right-angles to the long axis, lying towards the centre. The latter are shown as horizontal rods. The hydrogen bonds exist between a single base from one chain and another single base from the other chain, so that the two bases lie side by side. From molecular models it is concluded that the formation of these hydrogen bonds is highly specific; only certain pairs will fit the dimensions deduced from X-ray crystallography. These pairs are *adenine with thymine* and *guanine with cytosine*, which fits in excellently with the chemical analyses mentioned above. The arrangement of these base pairs is shown in Fig. 5.4. In the adenine-thymine pair there are two hydrogen bonds, while when the base pair consists of guanine and cytosine three hydrogen bonds are possible, as shown in Fig. 5.4.

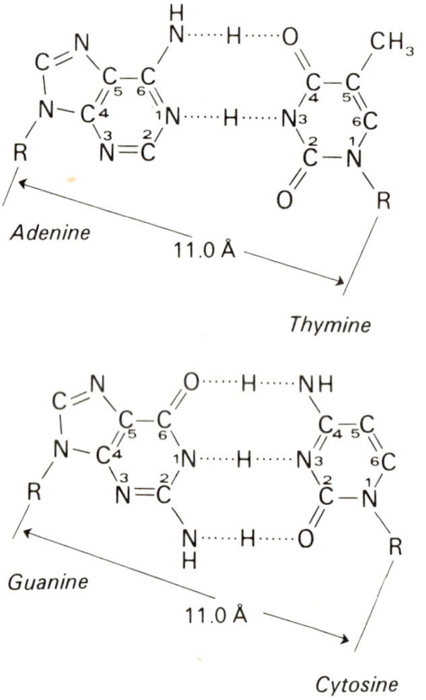

Fig. 5.4. Hydrogen bonding between purine and pyrimidine bases in DNA.

It follows from the specificity of the base-pairing that the two strands in a DNA molecule will have complementary base sequences. Further, it is known that the two strands are of *opposite polarity* in the sense that the 3′ and 5′ positions of the deoxyribose residues on one strand run in the opposite direction to those on the complementary strand. This is shown in Fig. 5.5. (See also Fig. 15.3, Chapter 15.)

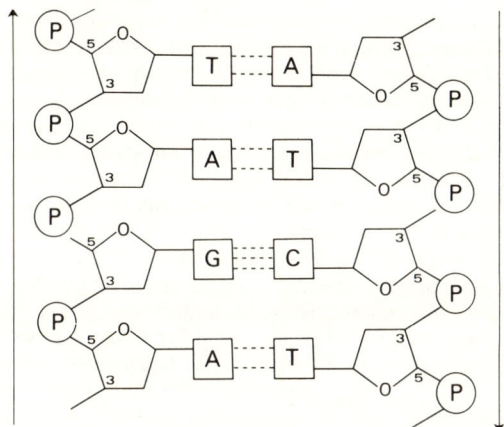

Fig. 5.5. Diagrammatic representation of part of the nucleotide sequences in two complementary strands of DNA of opposite polarity. The arrows indicate that the 3′ and 5′ positions of the deoxyribose residues in one strand run in the opposite direction to those in the other strand.

Distribution of Nucleic Acids and Their Biological Function

DNA is always present in the nuclei of all cells and its amount in the resting nucleus remains constant. The chromosomes are largely DNA-protein and the quantity of DNA in a nucleus increases before mitosis so that each daughter nucleus has its full complement. There is little doubt that the genetic material of the cell is DNA and that the DNA molecules carry the genetic information in the sequence of the nucleotides in the DNA molecule, although not all the nuclear DNA may be genetically active. The expression of the genetic information contained in DNA is achieved in the first place by the synthesis of RNA molecules, the nucleotide sequences of which are determined by the DNA (transcription, see Chapter 15). Some of this RNA, messenger RNA, directs protein synthesis (see Chapter 16). It is probably the vulnerability of DNA to such agents as

X-rays, ultra-violet light, nitrogen mustards, etc., which accounts for the mutagenic actions of these agents.

Isolated DNA from a capsulated strain of pneumococci can transform an uncapsulated strain into a capsulated one. Further, the newly capsulated strain is of the same serological type as that supplying the DNA, and it will now propagate indefinitely producing unlimited amounts of DNA. Many similar examples of transformations in bacteria brought about by DNA, such as the transfer of resistance to penicillin, streptomycin, etc., are known. There is good evidence that stretches of DNA are incorporated into the bacterial chromosome when this happens, and it is possible that similar insertions take place in animal chromosomes.

RNA occurs in highest concentration in organs with a high synthetic activity such as liver, pancreas, small intestine, brain, and the reticuloendothelial system; it is lowest in the active 'mechanical' organs, namely, kidney, heart, and skeletal muscle. Increased energy intake and treatment with cortisone, thyroxine, oestradiol, and insulin will increase the proportion of RNA to protein; high fat diets and thiamine and vitamin B_{12} deficiencies have the opposite effect. The distribution of RNA among the various subcellular components has been mentioned above.

Viruses consist mainly of protein and either RNA or DNA. In general the smaller viruses contain RNA while the larger ones contain DNA; a further differentiation may be made since the plant viruses contain only RNA while the animal viruses may contain either form of nucleic acid; bacteriophages (see Table 5.2) contain over 40 per cent DNA.

Table 5.1
Characteristics of Some Viruses

		Weight daltons	Per cent Nucleic acid	Size (nm)	Shape
E. coli phage T₇	DNA	38×10^6	41	6 (long)	Tadpole
Polyoma	DNA	21×10^6	13	45	Spherical
Adenovirus	DNA	20×10^7	5	70 (long)	Spherical
Vaccinia	DNA	20×10^8	7	230 (long)	Rectangular
Poliomyelitis	RNA	6.7×10^6	28	30	Spherical

Several small animal viruses have been crystallized. They are small spheres less than 50 nm in diameter, containing RNA but no DNA and the nucleic acid is not infective in the absence of the protein. The influenza viruses are larger, 70–80 nm in diameter and contain only 1 per cent RNA. Some DNA-containing animal viruses are about this size, while others, e.g. vaccinia (cowpox), Shope (rabbit) papilloma, psittacosis and rabies are much larger (Table 5.2).

Bacteriophages are parasites of bacteria; the T group, which parasitize

Escherichia coli, have been extensively studied and purified. One of the smallest of these, the T2 coliphage, is a tadpole-like particle with an elongated hexagonal head attached to a tail about 100 nm long. The proximal part of the tail contains a contractile protein. The phage attaches itself to a host cell tail first, enzymes in the tail attack the host cell-wall, and the tail then contracts forcing the tail cone into the bacterium. The way is now open for the phage DNA to enter the host cell. Once this has happened there is a latent period during which the metabolic activity of the host is reorientated by the phage DNA so as to produce a large number of new phage particles. Ultimately the host cell bursts (lyses) and the new phage particles are liberated into the medium where they can attack new bacterial cells.

Conformation of Nucleic Acid Molecules

Nucleotide polymers (especially DNA) denature on heating. Double-stranded helical DNA has a very sharp and characteristic temperature at which it unwinds, with the two strands subsequently assuming random coil configuration. This is known as the *melting temperature* (T_m), and it can be detected by various changes in physical properties. It is linearly related to the proportions of G/C base pairs in the polymer.

If the double helix has not completely unwound, renaturation is very rapid when the temperature is lowered, but it is much slower if the two chains have completely separated ('annealing', p. 74). It is possible, during the annealing process, for single DNA strands to form a double helix with single DNA strands from another organism, or with strands of RNA. This will only happen if stretches of the two strands have *complementary* base sequences, i.e., G coming opposite C and A opposite U or T, as in Fig. 5.5. The minimum number of base pairs for a segment of a double helix to form is about 12. The experimental use of this phenomenon, known as *hybridization*, is of great value in demonstrating relationships between DNAs and RNAs from different sources. The ease and completeness with which double-stranded DNA re-forms a helix on cooling demonstrates the almost complete complementarity between the two chains.

RNA, although it can contain up to 50 per cent of the bases G and C, shows practically no tendency to form a double helix, presumably because the individual strands are not usually complementary in the *anti-parallel* direction (p. 89). However, it is not unusual for short stretches of base sequence on the same RNA chain to be complementary, giving the possibility of forming a segment of double helix terminating in a loop. The physical and X-ray crystallographic evidence from relatively small RNA molecules shows that this possibility is frequently realized. The 'clover-leaf' pattern of tRNA in Fig. 5.1 is a symbolic representation of intra-chain helix formation in a specific molecule. Fig. 15.2 (Chapter 15) is a speculative representation of the same process as it might occur in a longer molecule.

the long-range folding of the strands or helices—will prove to be as important in their functional behaviour as it is for proteins, but physical methods of studying the tertiary structure of these very long molecules have not yet been worked out. It is, however, already known that the circular DNA of bacteria must be greatly folded, and there must also be great differences between the folding of the chromatid material of animal cell nuclei in interphase and in meiosis. Indirect evidence, e.g., the resistance of sections of the strand to attack by ribonuclease, also points to the importance of tertiary structure in RNA polynucleotides. It is not known at present how important is the role of cations (e.g., Mg^{2+}, spermidine, histones) in maintaining the conformation of nucleic acids. The probable tertiary conformation of a tRNA molecule, deduced from X-ray crystallographic evidence, is shown in Fig. 5.6.

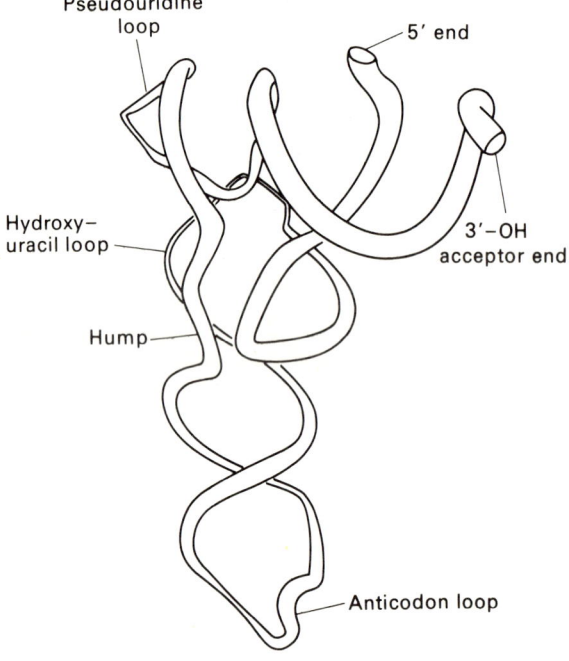

FIG. 5.6. Probable tertiary structure of yeast phenylalanine-tRNA.

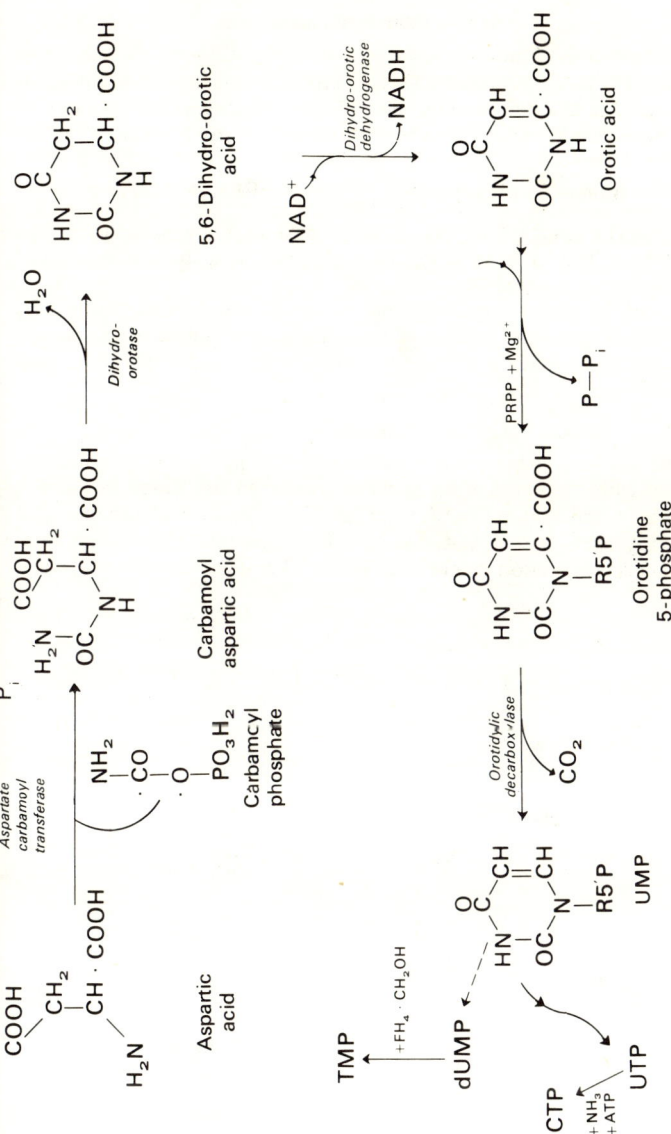

Fig. 5.7. The biosynthesis of pyrimidine nucleotides.

The Biosynthesis of Nucleotides

The biosynthesis of nucleic acids, and its control, forms the subject of Chapter 15. The *ribofuranose 5'-phosphate* component of nucleotides is synthesized in the hexose shunt (see Chapter 9) after which it is further phosphorylated by ATP to 5'-phosphoribosyl-1'-pyrophosphate (PRPP):

$$\text{Ribose 5'-phosphate} + \text{ATP} \;\rightarrow\; \text{PRPP} + \text{AMP}$$

The ribosyl-phosphate moiety is introduced into the molecule at the beginning of the synthesis of purine nucleotides, and towards the end of the synthesis of pyrimidine nucleotides.

Nucleotides containing *deoxyribofuranose-5'-phosphate* are formed by reduction of the sugar residue in the completed ribonucleotides, except that thymidylic acid is formed by methylation of deoxyuridylic acid (p. 247).

The Biosynthesis of Pyrimidine Nucleotides

The steps of the biosynthetic pathway to the pyrimidine nucleotides, together with the names of the enzymes concerned, are shown in Fig. 5.7. It' will be noted that N–3 is derived from ammonia (or glutamine), C–2 from CO_2 and C–4, C–5, C–6, and N–1 from aspartate. The conversion of uracil nucleotides into cytosine nucleotides takes place at the triphosphate

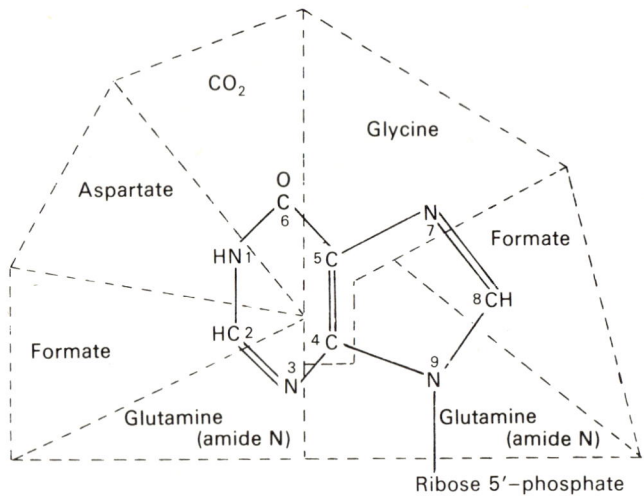

FIG. 5.8. The precursors of inosinic acid.

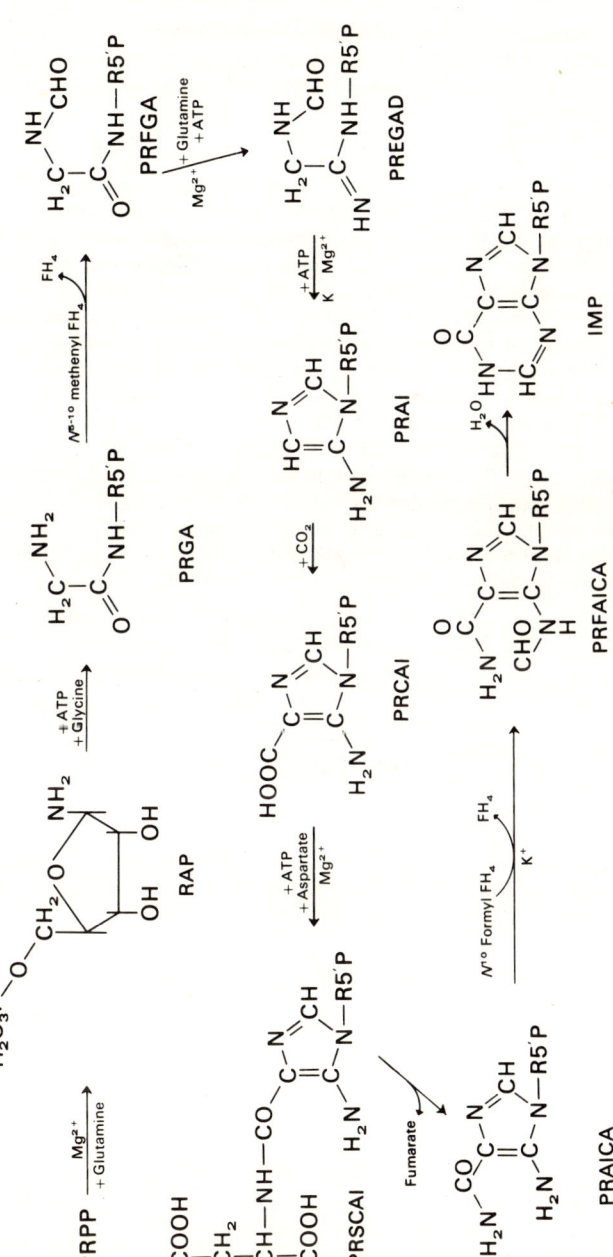

FIG. 5.9. The biosynthesis of purines.

PRPP = 5-phosphoribosyl pyrophosphate; RAP = 5-phosphoribosylamine-5-phosphate; R5'P = β-D-ribosyl-5-phosphate; PRGA = 5'-phosphoribosyl-glycinamide; PRFGA = 5'-phosphoribosyl-N-formylglycinamide; PRFGAD = 5'-phosphoribosyl-n-formyl-glycinamidine; PRAI = 5'-phosphoribosyl-5-aminoimidazole; PRCAI = 5'-phosphoribosyl-4-(N-succinocarboxamide)-5-amino-imidazole; PRCAI = 5'-phosphoribosyl-5-amino-4-imidazole carboxamide; PRFAICA = 5'-phosphoribosyl-5-formamamide-4-imidazole carboxamide; IMP = inosine monophosphate (inosinic acid); FH₄ = tetrahydrofolic acid.

level. The biosynthesis of thymidine derivatives is more complex since it involves both the reduction of the ribose and the introduction of the methyl group. The former is discussed on p. 98, while the latter is outlined in Chapter 11.

The Biosynthesis of Purine Nucleotides

From isotopic tracer evidence it has been possible to find the origin of all the ring atoms in the purines of the nucleic acids; the various precursors are shown in Fig. 5.8. The biosynthetic routes to both adenylic and guanylic acids pass through 5'-inosinic acid; they are shown in Figs 5.9 and 5.10. The main points to be noted are that, except for the addition of a whole glycine molecule to give C–4, C–5, and N–7 of the purine, all the atoms are added singly and that tetrahydrofolic acid is involved in two places in the addition of one-carbon units (see Chapter 11). The product, 5'-inosinic acid is aminated by condensation with aspartate to give adenylosuccinic acid which is split to fumarate and AMP. GMP is formed in a two-stage process: the NAD-linked oxidation of IMP to form 5'-xanthylic acid (XMP), followed by the amination of the latter from glutamine.

Most tissues appear to be able to synthesize all the purine and pyrimidine bases they require, but there is evidence that some peripheral tissues utilize bases absorbed from the gut or synthesized in other tissues for some of their requirements (see p. 100).

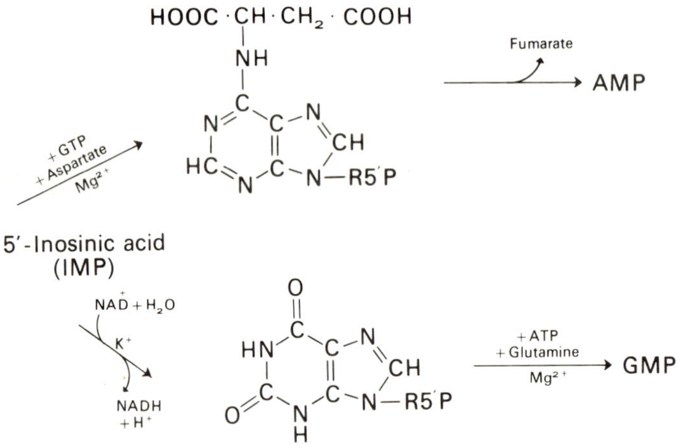

FIG. 5.10. The biosynthesis of adenylic and guanylic acids from inosinic acid.

Chemical Inhibition of Nucleic Acid Biosynthesis

Many substances inhibit the synthesis of nucleic acids (see Chapter 15), and others inhibit the synthesis of nucleotides. A number of them have been used as bacteriostatic or antitumour agents. The effective compounds are often unnatural analogues of purines and pyrimidines.

Purine analogues include:

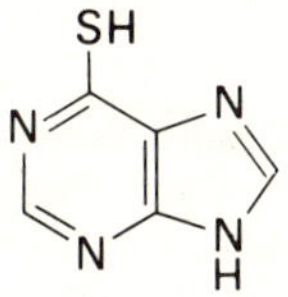

6-*Mercaptopurine*
(6*MP*)

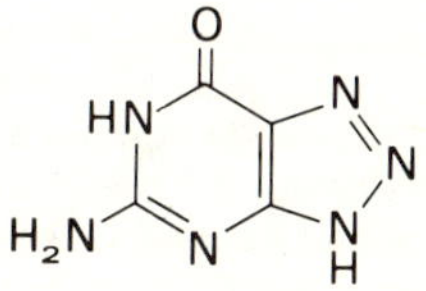

8-*Azaguanine*
(8*AG*)

Pyrimidine analogues include:

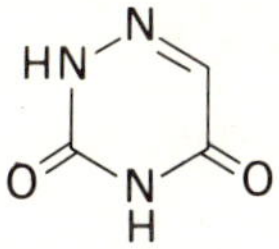

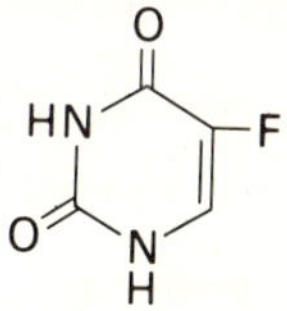

5-*Fluorouracil*
(5*FU*)

Glutamine analogues include the *Streptomyces* antibiotic azaserine which inhibits aminations where glutamine is the amino group donor, e.g., the amination of PRPP (p. 95).

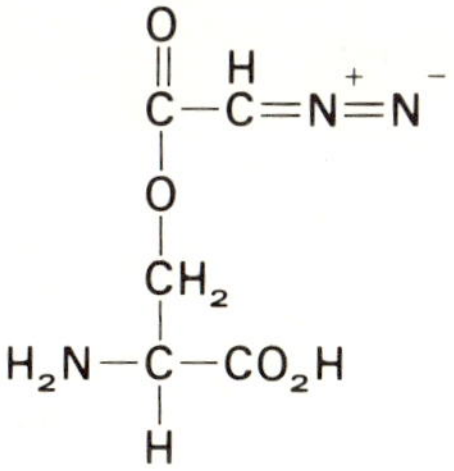

Azaserine
(*O*-*diazoacetyl*-L-*serine*)
(*AZS*)

Formation of Deoxyribonucleotides

In *E. coli*, the replacement of the 2'–OH in ribonucleotides by –H requires the nucleoside diphosphates, QDP, as substrates, and *thioredoxin*, an iron-sulphur protein of the type described in Chapter 16, as proximate reducing agent. The system which obtains in animals has not been elucidated, but it is thought to be similar.

There is no evidence for the re-use of deoxyribose-5'-phosphate.

Catabolism

The *a* and *b* type specificities of nucleases and phosphodiesterases are described in Chapter 8. Endo- and exo-nucleases of both categories have been isolated from various tissues, as well as the pancreas. It is difficult to reconcile the activity of *b* type nucleases with the observation that the individual nucleotides of animal tissues are overwhelmingly of the 5'-phosphate type.

1. *Mononucleotides*. These are acted on by two types of enzyme:

a. *Phosphatases* (*nucleotidases*), which catalyse the reaction (where Q stands for purine or pyrimidine):

$$Q-ribose-P \rightarrow Q-ribose + P_i$$

b. *Pyrophosphorylases*, which catalyse the reaction:

$$Q-ribose-P + \quad P-P \quad \rightleftharpoons \quad Q \quad + \quad PRPP$$

| | pyro-phosphate | free base | phosphoribose-pyrophosphate |

The fact that this reaction is reversible is of importance in the 'scavenger' pathways of nucleotide synthesis described below.

2. *Nucleosides*. These are also acted on by two types of enzyme:

a. *Nucleosidases*:

$$Q-ribose \rightarrow Q + ribose$$

b. *Nucleoside phosphorylases*:

$$Q-ribose + P_i \rightleftharpoons Q + ribose-5-P$$

There is reason to suspect that intracellularly the latter enzyme is more important than the former. This must especially be true for deoxyribonucleotides, as deoxyribose is only catabolized as its 5'-phosphate (to glyceraldehyde-3-phosphate and acetaldehyde), and is not re-utilized.

Pseudouridine is not attacked by either enzyme, and is excreted as such. No enzyme attacking guanosine has yet been described, so that

GMP is presumably degraded only by the pyrophosphorylase (see below). The final stages of catabolism require the free base.

The *pyrimidines* are degraded according to the scheme set out in Fig. 5.11. They are usually completely broken down in the body and do not appear as such in the urine, except when large quantities are ingested. The catabolic pathways of uridine and thymine are shown in Fig. 5.11.

The *purines* are broken down according to the scheme illustrated in Fig. 5.12.

The oxidation of hypoxanthine and xanthine to uric acid is catalysed

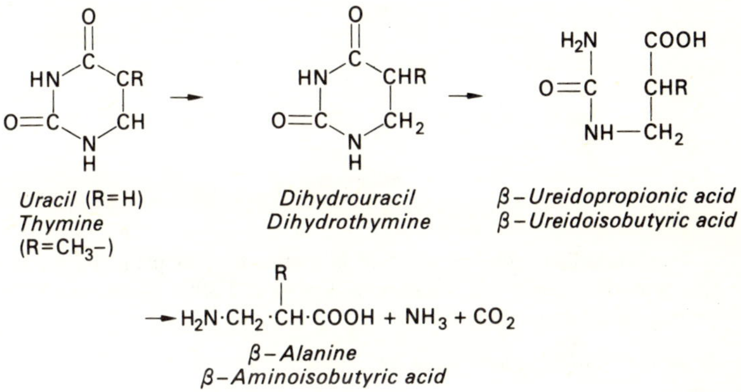

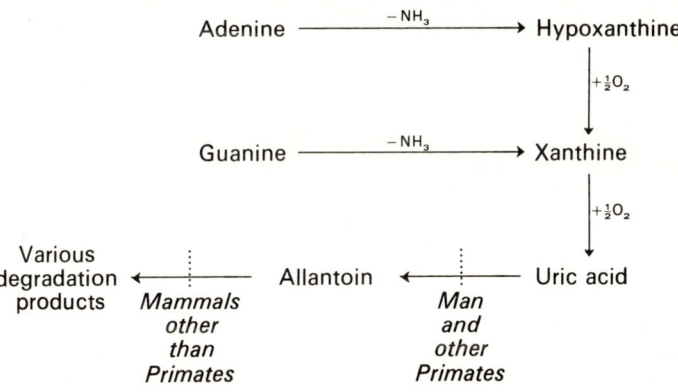

FIG. 5.11. The catabolism of pyrimidines.

FIG. 5.12. The catabolism of purines.

by *xanthine oxidase*, a complex dehydrogenase containing a riboflavin prosthetic group, non-haem iron, and Mo^V. It uses O_2 as a direct hydrogen acceptor, and is capable of oxidizing in addition several aldehydes, and also $NADH_2$. It occurs in liver and in milk.

Hyperuricacidaemia, whether due to raised exogenous intake or endogenous overproduction, leads to the formulation of painful nodules of uric acid and monosodium urate in joints (gout). The symptoms may be repressed by the chronic ingestion of a xanthine oxidase inhibitor such as *allopurinol*.

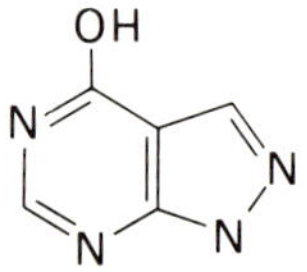

Allopurinol

The catabolism stops at uric acid in man and other primates, but continues to allantoin in other mammals. Fishes and other lower animals can further break down allantoin.

Scavenger Pathways for Purine Nucleotides

The existence of the pathways shown in Fig. 5.13 has been known for some time, but for long they were thought to have minimal importance, except in relation to oxidations by xanthine oxidase. Recent observations, however, suggest that they are more important than this. Among others, the *Lesch–Nyhan syndrome*, a disorder of children characterized by neurological abnormalities, self-mutilation and uric acid production up to 20 times the normal rate, has been found to be associated with the congenital absence of *guanine-hypoxanthine phosphoribosyl transferase*, GPRT (enzyme 2 in Fig. 5.13 and identical with 1(b) on p. 98). Adenine has its own specific enzyme, APRT (enzyme 1 in Fig. 5.13). The concentration of GPRT is normally highest in brain, especially in the basal ganglia. In normal tissues, the purine bases strongly inhibit one of the early stages in purine synthesis. There is some evidence that in the Lesch–Nyhan sufferer, purines *activate* the same reaction, thus explaining the overproduction of uric acid.

Taken together, these findings suggest that free purine bases are important in controlling the rate of synthesis of the purine nucleotides, and that APRT and GPRT play an important role in determining the concentrations of the free purines. Apart from this, there is some reason to think that peripheral tissues may not be completely autonomous with regard to

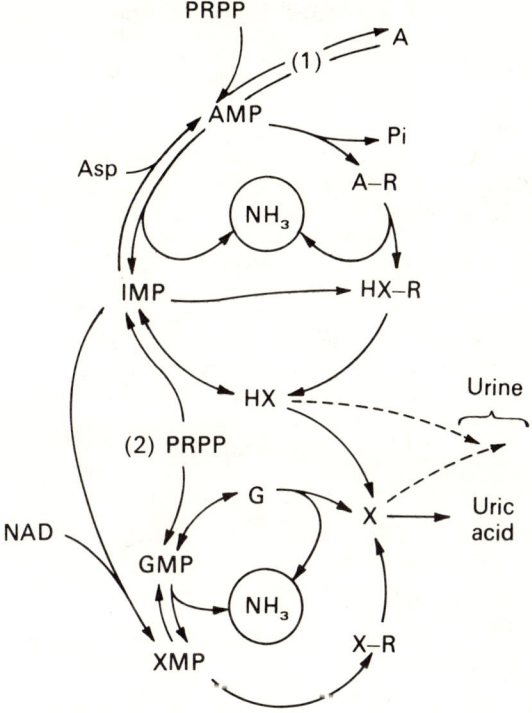

FIG. 5.13. Interrelations between the purines: 'scavenger' pathways. AMP, IMP, GMP and XMP represent the nucleotides; A–R, HX–R and X–R the nucleosides; and A, G, HX and X the bases (see p. 80). PRPP is phosphoribose pyrophosphate.

purine nucleotides (and perhaps also pyrimidines), but may take up at least some of their requirements from blood, and doubtless ultimately from the liver.

The deamination of AMP, shown in Fig. 5.13, is a very rapid process in many tissues, especially in skeletal muscle, where the product is IMP (and ammonia). This may be an emergency mechanism for maintaining ATP levels in anaerobic muscle, by preventing the attainment of the equilibrium catalysed by adenylate kinase (p. 37). IMP is normally re-aminated during recovery. In other tissues, e.g., cardiac muscle, adenosine deaminase is more important.

Control of Nucleotide Biosynthesis

Apart from the control mechanisms described above, *purine nucleotide* synthesis is controlled in two ways. All purine nucleotides inhibit the first step in their own synthesis, the amination of PRPP. Apart from this the nucleotides control each others' synthesis, since GTP is required for the amination of IMP, and ATP is required for the amination of XMP, the immediate precursor of GMP (see Fig. 5.14).

Pyrimidine nucleotide synthesis is controlled in several ways. The condensation of aspartate with carbamoyl phosphate is inhibited by several nucleotide products, chiefly CTP, although it is difficult to observe this in cell homogenates. The feedback control of aspartate transcarbamoylase is treated in more detail in Chapter 17. A series of interlocking inhibitions operate at the interconversions between the pyrimidine monophosphates and diphosphates. Finally, the conversion of thymidine monophosphate, TMP, to TDP is catalysed by an enzyme, *thymidylate kinase*, which has a very short half-life and whose synthesis may be one of the limiting factors in the formation of TTP for DNA synthesis.

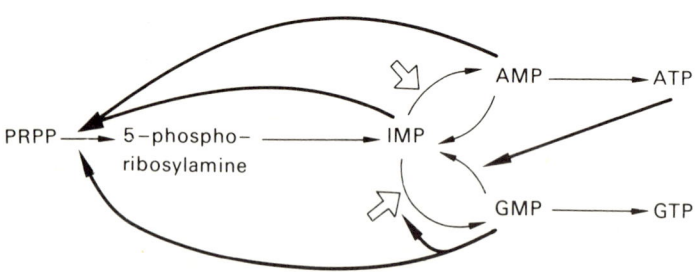

Fig. 5.14. Regulatory mechanisms in the synthesis of purine nucleo-tides. The heavy arrows indicate enzyme inhibitors. The open arrows indicate coenzyme requirements.

6 Haem Pigments and Bile Pigments

The important haem-containing compounds in the body are haemoglobin, myoglobin, the enzymes catalase and peroxidase, and the cytochromes. They are conjugated proteins in which the prosthetic group (haem) is a porphyrin ring containing an iron atom.

Porphyrins are all derivatives of *porphin*. They are large, flat, hetero-cyclic ring structures made up of four *pyrrole* rings linked together by methine (—CH=) bridges. The structures of pyrrole and porphin are shown in Fig. 6.1. In all the naturally occurring porphyrins there are eight side-chains on the ring in the positions marked 1–8. This allows a large number of different isomers to exist, since all the side-chains are not the same. The relationships between these isomers will be clearer after the biosynthesis of the porphyrins has been described; they are dealt with below.

When the side-chains contain carboxyl radicals, the prophyrins are amphoteric with iso-electric points in the range pH 3–4·5. They have four

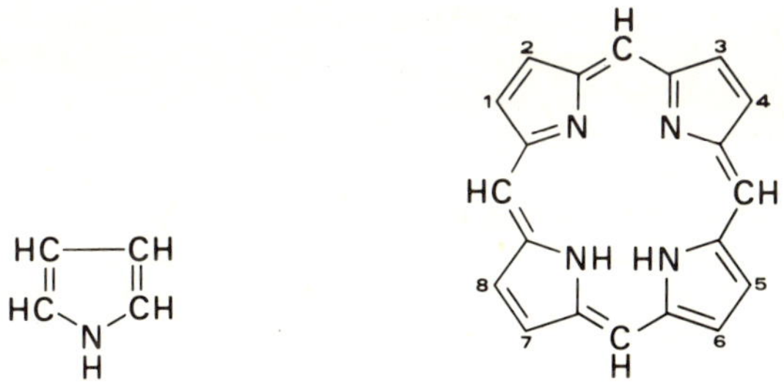

Pyrrole   *Porphin*

Fig. 6.1.

absorption bands in the visible spectrum and show ultra-violet fluorescence in solution in mineral acids. When the tetrapyrrole ring is broken the resulting compounds are the bile pigments which are described later in this chapter.

Porphyrinogens are compounds similar to the porphyrins but the pyrrole rings are linked together by $-CH_2-$ (methylene) bridges; they contain six more hydrogen atoms than do the corresponding porphyrins.

Biosynthesis of Porphyrins

The pathways by which the porphyrins are made in vivo are shown in Figs 6.2 and 6.3. These figures also show the way in which the nitrogen atom and the methylene carbon atom of glycine are incorporated into the porphyrin structure; all the other carbon atoms come from succinyl coenzyme A.

Reaction *I* (Fig. 6.2), the condensation of succinyl coenzyme A and glycine, requires pyridoxal phosphate. The hypothetical reaction product, α-amino-β-oxoadipic acid, decarboxylates spontaneously at neutral pH to form δ-aminolaevulinic acid, thereby eliminating the glycine carboxyl carbon atom. The enzyme that catalyses the reaction, *δ-aminolaevulinic acid synthetase*, is found to be present in very elevated amounts in the liver of patients suffering from two forms of congenital porphyria. Acute attacks of the disease, which are frequently precipitated by barbiturates, are accompanied by excretion of large amounts of δ-aminolaevulinic acid and of porphobilinogen. The increased concentration of the synthetase is of particular interest because inborn errors of metabolism are usually associated with a *deficiency* of an enzyme.

Reaction *II* (Fig. 6.2) involves the condensation of two molecules of δ-aminolaevulinic acid, with the elimination of water, to form the pyrrole derivative porphobilinogen. The enzyme catalysing this reaction is *δ-aminolaevulinic acid dehydrase*. Both this and the preceding enzyme are inhibited by haem and some haemoproteins, end-products of the process.

The next step, Reaction *III* (Fig. 6.3), the deamination and simultaneous polymerization of four molecules of porphobilinogen to form a cyclic tetrapyrrole is imperfectly understood. Because the two side-chains of porphobilinogen are different there are different ways of putting the molecules together, leading to four isomeric porphyrinogens. The crude enzyme preparation which catalyses this reaction brings about the formation of only *two* isomers, uroporphyrinogens I and III. Only porphyrin-(ogen)s derived from these two isomers, by alterations in the side-chains, occur in the body.

The four acetic acid side-chains of the uroporphyrinogens can be decarboxylated enzymically (Reaction *IV*, Fig. 6.3) to the corresponding coproporphyrinogens.

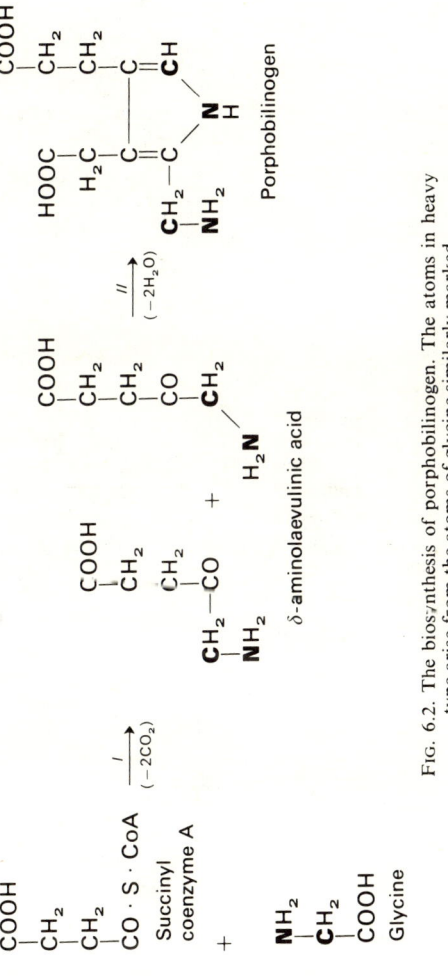

Fig. 6.2. The biosynthesis of porphobilinogen. The atoms in heavy type arise from the atoms of glycine similarly marked.

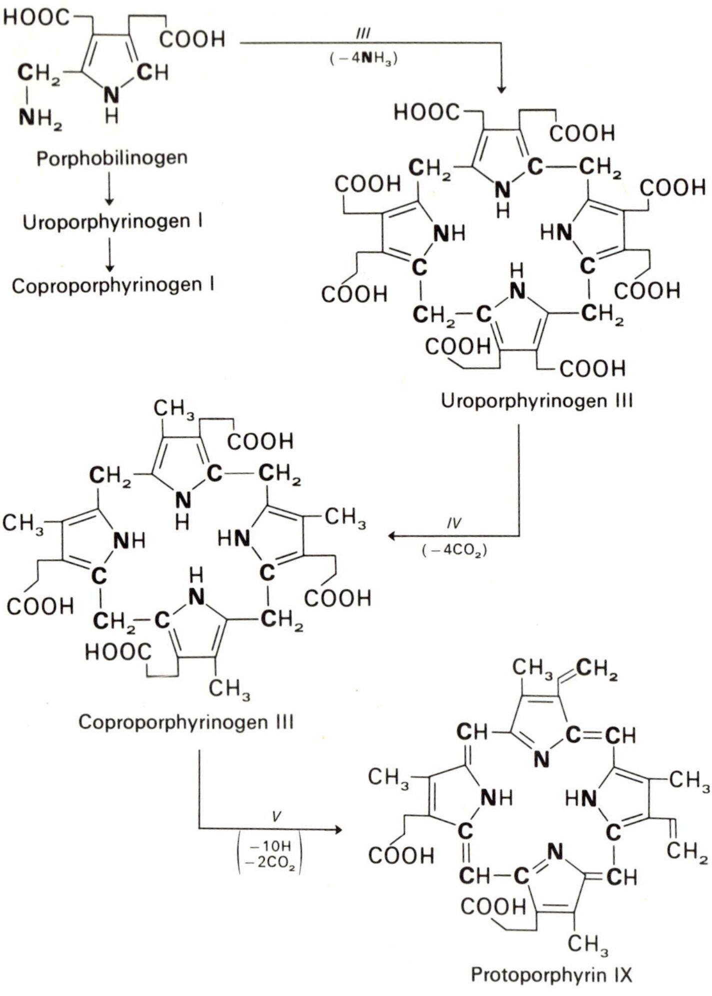

FIG. 6.3. The biosynthesis of porphyrins. All the atoms in heavy type arise from the atoms of glycine (see Fig. 6.2), the others from acetate. The uroporphyrins and coproporphyrins I and III arise from the corresponding porphyrinogens.

Coproporphyrinogen III is converted into a protoporphyrin (Reaction *V*, Fig. 6.3) by an enzyme which removes two hydrogen atoms and a carboxyl group from each of two of the four propionic acid side-chains, turning them into vinyl groups. At the same time the four methylene bridges are dehydrogenated to methine bridges, the porphyrinogen thereby being converted into a porphyrin. Protoporphyrin, therefore, has four methyl, two vinyl, and two propionic acid side-chains. There are 15 possible isomers of such a porphyrin but only one, type IX, is formed.

Table 6.1
The Arrangement of the Side-chains of the Porphyrins

Compound	Side-chains at the various positions of porphin (*Fig. 6.1*)							
	1	2	3	4	5	6	7	8
Uroporphyrin I	A	P	A	P	A	P	A	P
Coproporphyrin I	Me	P	Me	P	Me	P	Me	P
Uroporphyrin III	A	P	A	P	A	P	P	A
Coproporphyrin III	Me	P	Me	P	Me	P	P	Me
Protoporphyrin IX	Me	V	Me	V	Me	P	P	Me
Deuteroporphyrin IX	Me	H	Me	H	Me	P	P	Me
Mesoporphyrin IX	Me	Et	Me	Et	Me	P	P	Me
Haematoporphyrin IX	Me	HE	Me	HE	Me	P	P	Me

$$
\begin{aligned}
&\text{A} = -\text{CH}_2 \cdot \text{COOH} & &\text{Et} = -\text{CH}_2 \cdot \text{CH}_3 \\
&\text{H} = \text{hydrogen} & &\text{HE} = -\text{CH(OH)} \cdot \text{CH}_3 \\
&\text{Me} = -\text{CH}_3 & &\text{P} = -\text{CH}_2 \cdot \text{CH}_2 \cdot \text{COOH} \\
&\text{V} = -\text{CH}{=}\text{CH}_2 & &
\end{aligned}
$$

The arrangement of the side-chains in the naturally occurring porphyrins are shown in Table 6.1.

Occurrence of Porphyrins in the Body

Porphyrins are present in the excreta, bile, bone marrow, and the blood. In *blood* most of the porphyrins are in the red cells but their concentration is very small, about 20 μg protoporphyrin and less than 1 μg coproporphyrin per 100 ml blood. The *bile* contains 40–60 μg porphyrin per 100 ml and a considerable amount of this undergoes enterohepatic circulation. The *faeces* contain porphyrins of both endogenous and exogenous origin. The former enter the gut in the bile and are mainly protoporphyrin and coproporphyrin. The protoporphyrin can be altered by the bacteria in the gut to mesoporphyrin and deuteroporphyrin. Between 150 and 400 μg total porphyrin per day is normally excreted in the faeces, more than half being protoporphyrin and most of the rest coproporphyrin.

In normal human males the *urine* contains 166 $\pm$ 45 μg and in females 134 $\pm$ 42 μg total coproporphyrin per day. It is believed that it is excreted by the kidney as coproporphyrinogen and is oxidized spontaneously to the porphyrin. Normal urine also contains small amounts of uroporphyrin. The coproporphyrin occurs as a mixture of both the I and III isomers. Isotope tracer experiments have shown that urinary porphyrins are by-products of haem or porphyrin synthesis and are not derived from degradation of haemoglobin.

Haematin Compounds

The porphyrins can form coordination complexes with heavy metals and the iron complexes of the porphyrins are the haematin compounds. The insertion of an iron atom into protoporphyrin IX, in the in vivo synthesis of haem, is catalysed by a specific ferrochelatase. The structure is shown in Fig. 6.4, and the chief members of the group are shown in Table 6.2.

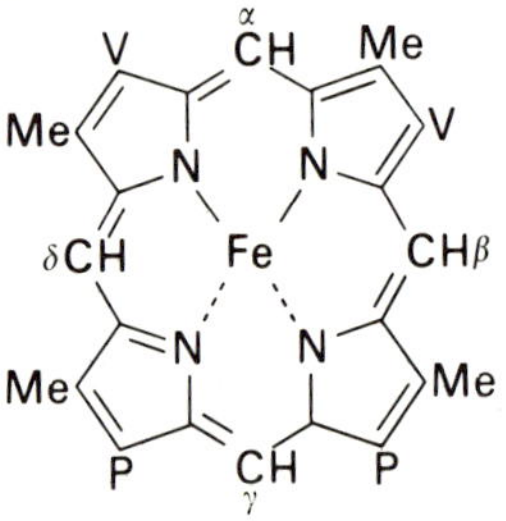

Haem

$Me = -CH_3$
$P = -CH_2 \cdot CH_2 \cdot COOH$
$V = -CH{=}CH_2$

Fig. 6.4.

Haem is the prosthetic group of haemoglobin. It contains ferrous iron and can easily be oxidized to haemin or haematin in which the iron is ferric. The iron atom is quite strongly held in the porphyrin. Haem alone cannot combine reversibly with oxygen as can haemoglobin.

Haemin and haematin. These are both protoporphyrin IX containing a ferric iron atom, coordinated only to N atoms in the planar tetrapyrrole ring, i.e., the iron is not coordinated to N atoms in protein nor to chromogenic nitrogenous compounds. The name haemin is reserved for the compound with a net positive charge (Table 6.2).

Haemochromes (these contain ferrous iron). Haem is able to combine with a large number of nitrogenous bases to form haemochromes. Among these bases are denatured protein, ammonia, amino acids, pyridine, etc.;

Table 6.2
Structure of Haematin Compounds

Compound	Structure	Charge	Iron atom	
			Valency	Ferrous or ferric
Haem	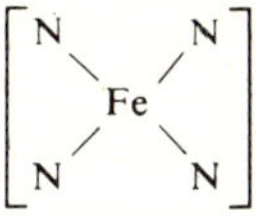 [N N \ / Fe / \ N N]	0	2	Ferrous
Haemochrome	[N' N | N \ / Fe N / | N N']	0	2	Ferrous
Haematin (in alkaline solution)	[HOH N | N \ / Fe N / \ N OH⁻]	0	3	Ferric
Haemin (chloride, etc.) (in acid solution)	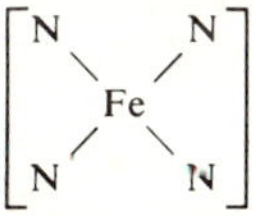 [N N \ Fe / N N]	+ 1	3	Ferric
Haemichrome (chloride, etc.)	[N' N | N \ / Fe N / \ N N']	+ 1	3	Ferric

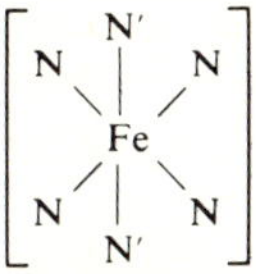

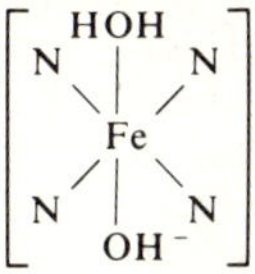

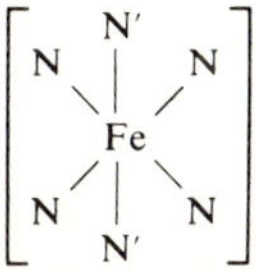

N = porphyrin nitrogens which contribute 2 negative charges to the complex. N' = non-porphyrin nitrogens.

combination occurs in the proportion of two molecules of base per ferrous iron atom. It is important when making haemochromes to add a reducing agent to keep the iron in the ferrous state. For example, when haemoglobin is treated with alkali the globin is denatured but remains attached to the haem to give *denatured globin haemochrome*. Pyridine haemochrome is formed by treating haemoglobin with pyridine, sodium hydroxide and a little dithionite as a reducing agent. The haemochromes have absorption spectra with a very sharp band near 550 nm; they do *not* form compounds reversibly with molecular oxygen.

Haemichromes (these contain ferric iron). They correspond to the haemochromes apart from the oxidation state of the iron. However, negatively charged ligands (e.g., CN^-, OH^-) bind more strongly to the haemichromes, which have a net positive charge, than they do to the haemochromes (cf. cyanmethaemoglobin, p. 114). The association constant for the hydroxyl ion is in fact so large that compounds with neutral ligands such as pyridine can only be formed in acid conditions.

Haemoglobin

Haemoglobin is the pigment of the red cells which is responsible for the transport of oxygen from the lungs to the tissues. It consists of a basic protein, globin, linked with haem (Type IX ferro-protoporphyrin). Its molecular weight is 64 500 and it contains 4 haems, and hence 4 iron atoms per molecule.

The haemoglobin molecule is built up of four polypeptide chains, consisting of two each of two different kinds known as the α and β chains in normal adult haemoglobin. Between 40 and 50 per cent of the amino acid residues are identical in the two chains, but the β chain is 5 residues longer. Each polypeptide chain is attached to a haem group through the coordination of a histidine residue in one of the polypeptide chains to the haem iron atom, as is shown in Fig. 6.5. It is possible to dissociate the two α chains from the two β chains by treatment with mildly acid solutions; the molecule may be reconstituted by raising the pH to near neutrality.

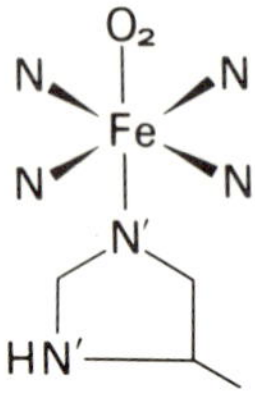

N = Nitrogens of porphyrin
N' = Nitrogens of histidine

Fig. 6.5. Linkage between haem and an imidazole ring of globin.

From X-ray diffraction studies, it has been shown that the oxyhaemoglobin molecule is, to a first approximation, a spheroid 64 Å long, 55 Å wide, and 50 Å high. The haem groups are buried in pockets formed by the polypeptide chains. It is striking that the polypeptide chain-fold (tertiary structure) of both pairs of chains is very similar and that it is also very similar to that found in the single polypeptide chain of myoglobin (Fig. 4.7, p. 72).

The most important property of haemoglobin is its ability to combine reversibly with one molecule of O_2 per iron atom to form oxyhaemoglobin, *without* the concomitant oxidation of the ferrous iron. This is a property conferred on the haem moiety by the nature of the protein ligand, cf. the haemochromes.

If we represent the molecular weight unit, containing one iron atom, as Hb, then we may represent the reversible combination with oxygen as follows:

$$Hb + O_2 \rightleftharpoons HbO_2$$

the amount of oxygen combined with haemoglobin being dependent on the oxygen tension (partial pressure). Thus at high oxygen tensions, in the lungs, the reaction goes to the right and oxygen is taken up; at low oxygen tensions, in the tissues, it goes to the left and oxygen is given up. The oxygen dissociation curves of haemoglobin and myoglobin are shown in Fig. 6.6. The haemoglobin curve is characteristically S-shaped, unlike that of myoglobin which is hyperbolic. The shape of the haemoglobin curve is due to the four haem groups not being independent; the equilibrium constant for the combination of oxygen with any one of them is influenced by the state of oxygenation of the others. Also, near the pH of blood (7·4), for each oxygen molecule given up one acid group 'disappears'. This is

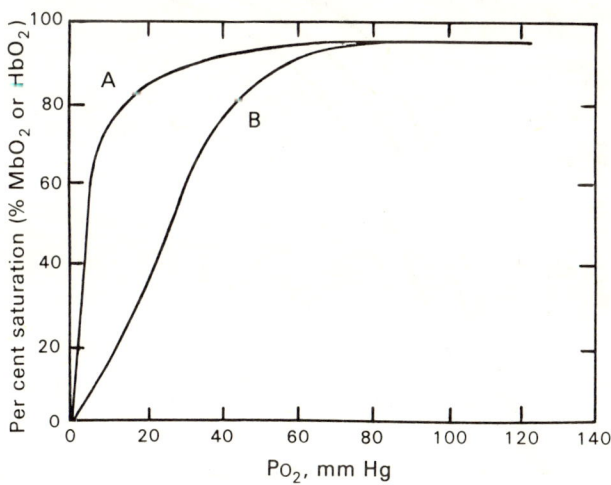

FIG. 6.6. Oxygen dissociation curves of myoglobin (A) and of haemoglobin (B).

of great importance in buffering the CO_2 produced in the tissues (see Chapter 14). Another manifestation of these haem-linked acid groups is the effect of pH on the affinity of haemoglobin for oxygen; the affinity increases at higher pHs and diminishes as the pH is decreased. This pH dependence is known as the *Bohr* effect (Fig. 6.7).

Oxyhaemoglobin has one binding site for 2,3-diphosphoglycerate, which shifts the dissociation curve to the left, making unloading of O_2 more difficult. This effector does not seem to come into play, however, during the cycle of erythrocytes between tissues and lungs.

Detailed X-ray studies have shown that when oxygen is removed from oxyhaemoglobin the two β chains move apart a distance of about 7 Å, which is about 10 per cent of the diameter of the whole molecule. It is by this means that the effect of oxygen binding to the haem groups is transmitted from the separate polypeptide chains throughout the whole molecule. The process is interactive in that these changes in the quaternary structure (sub-unit distance) affect the environment of haem groups which have not yet bound an oxygen molecule.

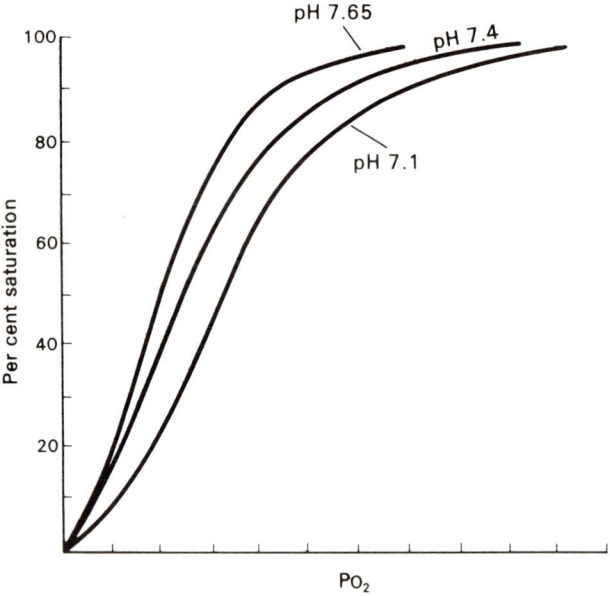

FIG. 6.7. The effect of pH on the oxyhaemoglobin dissociation curve.

Haemoglobin can also combine with other neutral molecules such as carbon monoxide (to form carboxyhaemoglobin) and nitric oxide.

Haemoglobin Variants

Fetal haemoglobin contains two α-chains and two so-called γ-chains. These are of the same length as β-chains, but 39 of the residues differ. Most of the haemoglobin in the infant's blood at birth is HbF, but production of the γ-chains then ceases and after 4–6 months the changeover to adult (A or $\alpha_2\beta_2$) haemoglobin is complete, except in a few individuals who have haemoglobin F in their blood throughout life. Fetal red cells have a higher affinity for oxygen than adult cells, which is an advantage for placental oxygen transfer. The difference is not necessarily due to the difference between the γ- and the β-chains.

A small fraction of adult haemoglobin, A_2, has two δ-chains, which differ in many residues from both β- and γ-chains.

Besides these variants, many inherited polymorphisms are known in which only one amino acid residue has changed, presumably because of a point mutation. About 150 have so far been identified, but there are likely to be many more. Some variant haemoglobins are harmless, but others are less stable than the normal or cause more rapid lysis of the red cells; yet others favour oxidation of the iron, giving methaemoglobin (see p. 114). The best known is haemoglobin S (sickle cell haemoglobin), in which the glutamic acid residue at position 6 on the β-chain is replaced by a valine. Deoxygenated HbS is less soluble than normal HbA, and on precipitating out causes a characteristic sickle-shaped distortion of the erythrocytes, lysis and consequent anaemia. It does however protect against infection by malarial parasites.

For imbalances in the rates of synthesis of the two haemoglobin chains (*thalassaemia*) see Chapter 13.

A number of pigments related to haemoglobin are of importance in physiology and pathology; the chief ones are as follows:

Oxyhaemoglobin (HbO_2) is bright red in colour with two sharp absorption bands in the green. On extreme dilution it gives a yellowish-green colour (cf. carboxyhaemoglobin). The spectrum is insensitive to pH between 5·5 and 10.

Haemoglobin (reduced haemoglobin, Hb) is prepared by the evacuation of HbO_2 or its treatment with sodium dithionite ($Na_2S_2O_4$). It has a broad absorption band in the yellow-green which is insensitive to pH in the range 5·5 to 9·5.

Carboxyhaemoglobin (HbCO). The affinity of haemoglobin for carbon monoxide is 200 times that for oxygen, thus CO will displace O_2 from HbO_2. Its absorption spectrum is very like that of oxyhaemoglobin but the two bands in the green are shifted to slightly shorter wave-lengths. It is cherry red in colour and remains the same on extreme dilution. It is made by passing coal gas through a solution of haemoglobin or oxy-haemoglobin.

Methaemoglobin is made by the oxidation of the iron of haemoglobin to the ferric state, by potassium ferricyanide or by the auto-oxidation of oxyhaemoglobin. It is found in blood in small quantities; when present in abnormally large amounts the condition is called methaemoglobinaemia. Methaemoglobin does not combine reversibly with O_2. Its solution is brown and it has four bands in its absorption spectrum; it may be reduced to haemoglobin by sodium dithionite. MetHb combines with cyanide; cyanmethaemoglobin has a sharp absorption band at 542 nm, which can be used for estimating Hb and its derivatives.

Tests for Blood by Means of Haemoglobin

The most sensitive tests for haem depend on the catalytic oxidation of the aromatic compounds *guaiacol*, *benzidine* or *o-toluidine* by hydrogen peroxide, to deeply coloured (green-blue) compounds. The oxidation is catalysed by haem, but it is non-enzymic, i.e. it survives boiling.

The presence of haemoglobin or its derivatives may be detected by spectroscopy, or by the preparation of haemin crystals.

Other Haem-containing Proteins

Myoglobin has been referred to above (see also Chapter 14).

The *cytochromes* contain ferriprotoporphyrin linked to protein; they are of the utmost importance in cellular oxidation-reduction reactions (see Chapter 12). One of them, cytochrome *c*, has a molecular weight of 13 000 and contains one haem group (and hence one iron atom) per molecule. It is very stable to heat and to acids, and the haem group can only with difficulty be separated from the protein moiety, because the two are covalently linked. The haem group is derived from protoporphyrin IX,

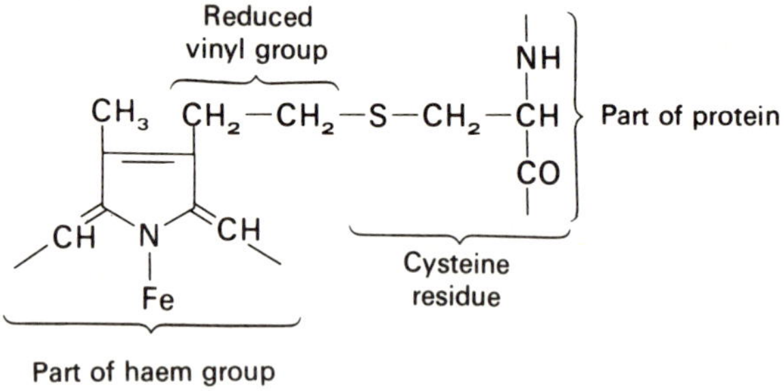

Fig. 6.8. The thioether bond between haem and protein in cytochrome *c*.

but the two vinyl side-chains are reduced and are linked by thioether bonds to cysteine residues of the protein, as is shown in Fig. 6.8.

Unlike haemoglobin, cytochrome c does not combine reversibly with oxygen but the iron atom can change reversibly from the ferrous to the ferric state.

Several enzymes such as *catalase* contain a haem prosthetic group related to protoporphyrin IX. In *cytochrome a* the prosthetic group is of a different type called haem A. The major differences from the more common haems are a formyl group at position 8 and a 15-carbon farnesyl chain at position 2. Haem A is not covalently linked to the enzyme protein.

Bile Pigments

The molecules of the bile pigments consist of an open chain of four pyrrole rings joined together by either methylene ($-CH_2-$) or methine ($=CH-$) groups. They are derived from protoporphyrin IX by the oxidative scission of its α-methine link (see Fig. 6.4), to form biliverdin, and subsequent reduction to other members of the group.

The bile pigments may be classified into bilanes, bilenes, biladienes, and bilatrienes, which contain 0, 1, 2, or 3 methine bridges respectively. The structures of representative members of these groups are shown below. The 'ogen' compounds (bilanes) are colourless, the other compounds are highly coloured.

Bilanes

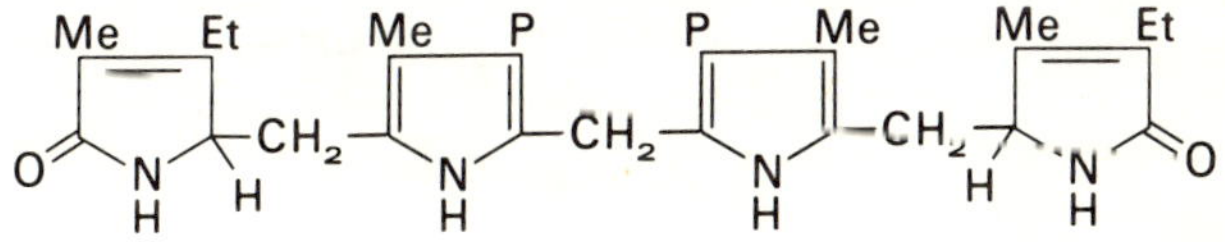

Mesobilirubinogen

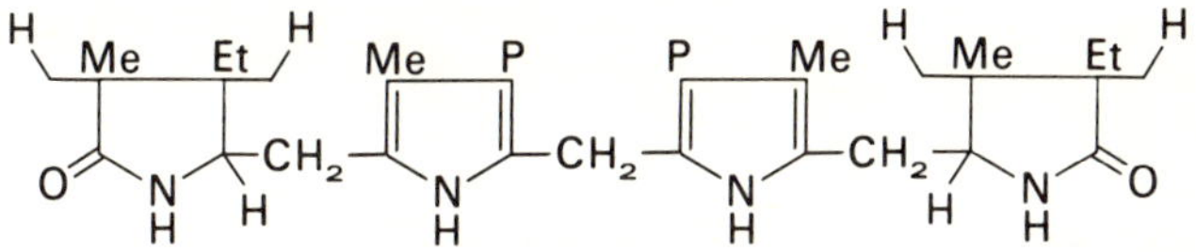

Stercobilinogen

This is the tetrahydro derivative of mesobilirubinogen.

Also: **d**-*urobilinogen* in which an Et group of mesobilirubinogen is replaced by a V group.

Bilenes

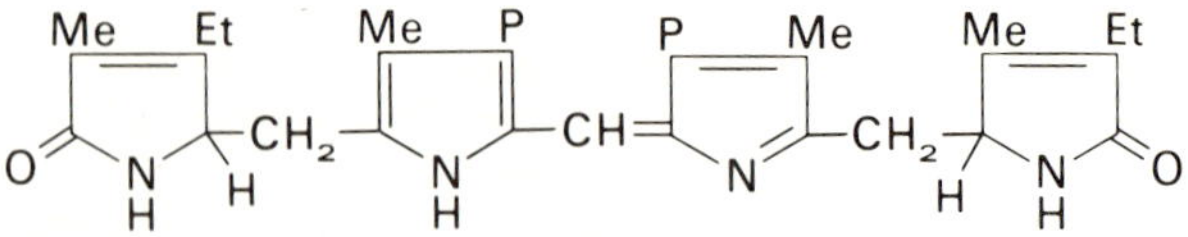

Urobilin IXα

Also: **d**-*urobilin* and *stercobilin* with side-chains corresponding to *d*-urobilinogen and stercobilinogen respectively.

Biladienes

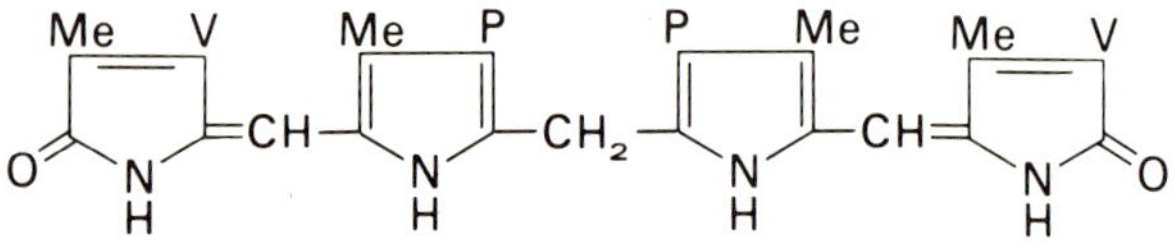

Bilirubin

Also: *mesobilirubin, mesobiliviolin,* and *mesobilirhodin,* in which the two V groups are replaced by two Et groups; the methine groups are in different places in the last two.

Bilatrienes

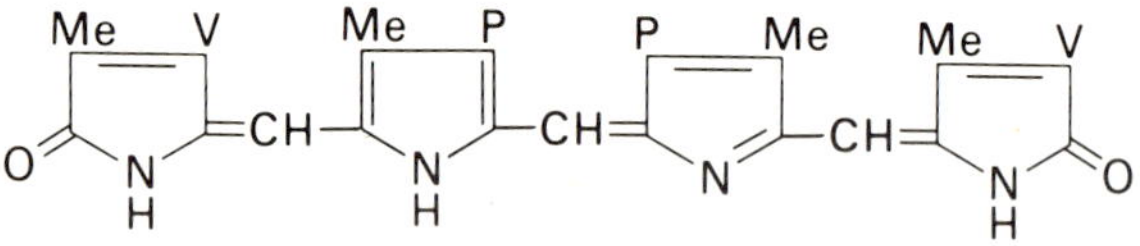

Biliverdin

$Me = -CH_3$
$Et = -CH_2-CH_3$
$V = -CH=CH_2$
$P = -CH_2-CH_2-COOH.$

Origin of Bile Pigments—The Degradation of Haemoglobin

After a life-span of about 120 days, the red cells are taken up by the cells of the reticulo-endothelial system and are destroyed. The breakdown of haemoglobin starts with the oxidative scission of the α-methine bridge of the haems to form a green biliverdin-iron-globin complex known as verdohaemoglobin (or choleglobin). About 15 per cent of the haem, however, is excreted a few days after synthesis, apparently never reaching the general circulation. The proportion of this 'haemolytic bile pigment'

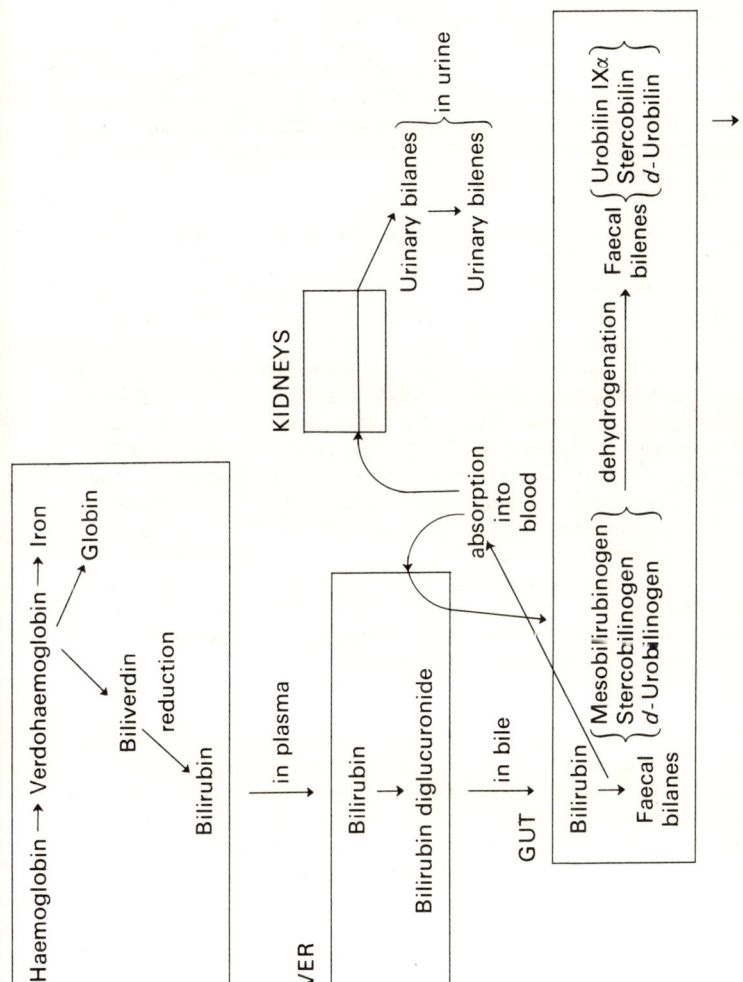

Fig. 6.9. The formation and metabolism of bile pigments.

is greatly increased in several diseases of differing aetiology. Iron is removed and attached to transferrin for transport to storage sites, globin is liberated and enters the general pool of protein metabolism, while the biliverdin is reduced at the γ-methine bridge to bilirubin.

The bilirubin is transported to the liver as a complex with plasma proteins (both albumins and α-globulins). In the liver each of the two propionic acid side-chains of bilirubin are esterified to the hydroxyl group at C–1 of glucuronic acid to form the water-soluble bilirubin diglucuronide. The glucuronic acid is transferred to bilirubin from uridine diphosphate glucuronate.

$$\text{Bilirubin} + 2\,\text{UDP glucuronate} \rightarrow \text{Bilirubin diglucuronide}$$

The soluble diglucuronide is secreted by the liver into the bile. The ability to conjugate bilirubin is almost lacking in the liver of the new born, and some degree of jaundice is usual. If there is severe haemolysis, e.g. because of incompatible maternal antibodies, the blood bilirubin level rises tremendously, and the pigment is selectively absorbed by brain tissues, causing irreversible damage (*kernicterus*). This can usually be prevented by blood analysis, followed by remedial measures, in the first few days of life. In the gut, probably largely by the action of the bacterial flora, bilirubin is further reduced to various bilanes, frequently collectively called 'urobilinogen'; some of these are dehydrogenated to bilenes, also known as 'urobilin'. Some of the bilanes are absorbed from the gut into the blood, from which they are removed by the liver and kidneys, which excrete them into the bile and the urine respectively. In the urine, the bilanes ('urobilinogen') are spontaneously dehydrogenated to bilenes ('urobilin'). All these pathways are summarized in Fig. 6.9.

The Van den Bergh Reaction

Diazotized sulphanilic acid reacts with bilirubin or mesobilirubin to give a red colour. It has long been known that in normal serum this reaction will not take place without the addition of ethanol ('indirect' reaction), while in some cases of jaundice the colour develops without the ethanol ('direct' reaction). Direct reacting bilirubin in sera is a mixture of the mono- and di-glucuronides of bilirubin, while the indirect reacting pigment is bilirubin itself.

7   Enzymes

Most of the chemical reactions which occur readily in the body will not proceed in vitro except under the influence of high temperatures, strong reagents, or other unphysiological conditions. In living systems reactions proceed rapidly at low temperatures and in dilute solutions under the influence of biological catalysts called enzymes. The name enzyme, meaning literally 'in yeast', was coined following the demonstration of the catalytic properties of yeast and yeast juices.

General Properties

The essential property of both enzymes and other catalysts is that, while they enter into chemical reactions, they remain intact in the process; ideally they remain quite unchanged indefinitely and do not lose their activity, though this does not often happen. Enzymes are frequently much more effective, mole for mole, than many inorganic catalysts such as H^+ and OH^- ions, platinum black, etc. Most enzyme-catalysed reactions studied in the laboratory do not proceed to any appreciable extent in the absence of the enzyme so, within limits, the rate of reaction is proportional to the enzyme concentration.

In common with inorganic catalysts enzymes catalyse both forward and reverse reactions, that is they assist the attainment of the equilibrium state which is governed by energetic considerations alone. Consider the reaction:

$$A + B \underset{v_2}{\overset{v_1}{\rightleftharpoons}} C + D \tag{1}$$

where v_1 and v_2 are the forward and reverse reaction velocities respectively. Then

$$v_1 = k_1[A][B] \tag{2}$$

and

$$v_2 = k_2[C][D] \tag{3}$$

where k_1 and k_2 are the forward and reverse rate constants. Now at equilibrium the forward and reverse velocities are equal, that is:

$$v_1 = v_2 \tag{4}$$

whence

$$k_1[A][B] = k_2[C][D] \tag{5}$$

and

$$\frac{k_1}{k_2} = \frac{[C][D]}{[A][B]} = K \tag{6}$$

where K is the equilibrium constant of the reaction. Since K depends only on thermodynamic factors, it cannot be altered by an enzyme or catalyst. Thus if an enzyme increases the forward reaction velocity v_1, the reverse velocity v_2, and the rate constants k_1 and k_2 must be increased in the same proportion.

Enzymes are proteins, and their conformation is determined by the general influences that were discussed in Chapter 4. The conformation of enzymes is important not only as it affects their behaviour as a class, but because there is reason to suspect that strain induced in the substrate on binding to a particular steric conformation at the active site of the enzyme is often partly responsible for the effectiveness of the protein as a catalyst. The great influence exerted by the protein environment on chemical reactivity has already been pointed out in connection with the binding of oxygen to haemoglobin (Chapter 6, p. 112).

The activity of enzymes is markedly dependent on the pH of the medium and often on the ionic strength as well, and there is little doubt that this reflects the changing charge distribution on the enzyme molecule and the packing of the hydrophobic residues. If the substrate is also charged, effects of pH on enzyme-substrate binding are likely to be even more marked.

A number of enzymes are not active unless the $-SH$ groups of one or more cysteine residues are free.

Prosthetic Groups and Coenzymes

Some enzymes, for example pepsin and ribonuclease, have been shown to be pure proteins while others have some non-protein part in the molecule. In the latter, the protein moiety is called the *apoenzyme* and the non-protein moiety the *prosthetic group*; the whole enzyme being known as the *holoenzyme*. When the holoenzyme can be easily separated (e.g. by dialysis) into an apoenzyme and a diffusible part, the latter is usually called a *coenzyme*. There is no essential difference so far as the enzymic activity is concerned between prosthetic groups firmly bound to the apoenzyme and loosely bound coenzymes. When the enzymic activity depends on the attachment of ions to the enzyme protein, such ions are known as *activators*; they can usually be removed by dialysis.

Coenzymes may, however, couple separate enzymic reactions, as for example those catalysed by triose phosphate dehydrogenase and lactic dehydrogenase. It is also true that coenzymes which are co-substrates (see also p. 127) are required in amounts stoichiometrically equal to the substrate as the reaction proceeds; this is not so for firmly attached prosthetic groups.

The structures and functions of the principal coenzymes are dealt with at the end of this chapter.

Enzyme Specificity

One of the principal differences between inorganic catalysts and enzymes is the much greater specificity of the latter. Inorganic catalysts frequently bring about a wide range of differing reactions while enzymes are more or less specific as to the substrate upon which they act or the nature of the reactions they catalyse.

Some enzymes are very specific, like urease, which hydrolyses only urea:

$$CO \begin{cases} NH_2 \\ NH_2 \end{cases} + H_2O \rightleftharpoons 2NH_3 + CO_2$$

Others, like the proteolytic enzyme pepsin, act on a range of substrates all of which possess certain definite structural features. Pepsin will hydrolyse peptide bonds:

$$-CO-NH-\underset{\underset{R}{|}}{CH}-CO- + H_2O$$

$$\rightleftharpoons -COOH + NH_2-\underset{\underset{R}{|}}{CH}-CO-$$

but only between certain amino acid residues.

The carboxyl esterases are specific in their esterase action, they will only hydrolyse carboxylic acid esters:

$$R-CO-O-R' + H_2O \rightleftharpoons R-COOH + HO-R'$$

the nature of R and R′ being immaterial. Differences are, however, observed in the rates at which different substrates are acted on, thus liver carboxyl esterases usually act more rapidly on esters containing short-chain fatty acids.

Most enzymes show a high degree of spatial specificity. Arginase will only catalyse the hydrolysis of the guanidino group of L-arginine:

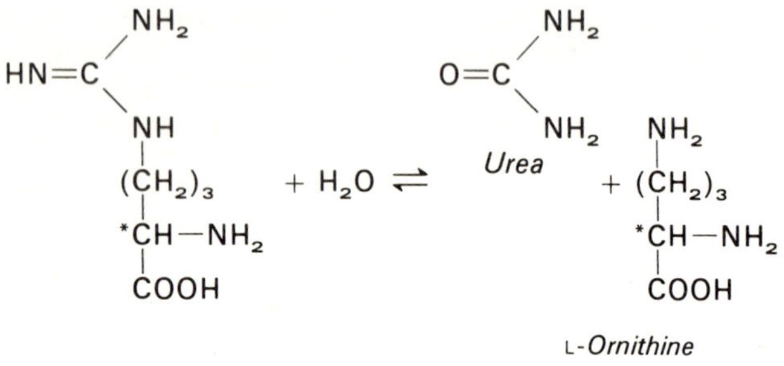

$$HN{=}C \begin{array}{l} NH_2 \\ \\ NH \\ | \\ (CH_2)_3 \\ | \\ {}^*CH{-}NH_2 \\ | \\ COOH \end{array} + H_2O \rightleftharpoons \overset{Urea}{} O{=}C \begin{array}{l} NH_2 \\ \\ NH_2 \end{array} + \begin{array}{l} NH_2 \\ | \\ (CH_2)_3 \\ | \\ {}^*CH{-}NH_2 \\ | \\ COOH \end{array}$$

L-*Ornithine*

The enzyme has no action on D-arginine even though the asymmetric carbon atom (*C) is 3 carbon atoms removed from the bond hydrolysed.

In some enzyme-catalysed reactions a substrate which is structurally symmetrical from a chemical point of view behaves, in the enzyme reaction, as if it were asymmetrical. Thus in the following reaction sequence, which occurs in the tricarboxylic acid cycle (see Chapter 12), the citric acid,

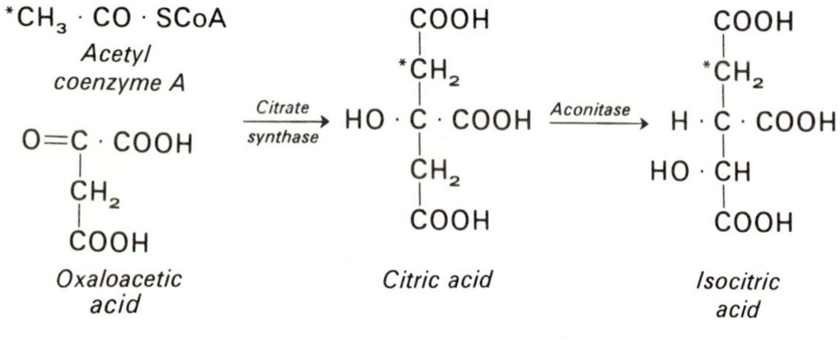

$$\begin{array}{lll} {}^*CH_3 \cdot CO \cdot SCoA & COOH & COOH \\ \textit{Acetyl} & | & | \\ \textit{coenzyme A} & {}^*CH_2 & {}^*CH_2 \\ & | & | \\ & \xrightarrow[synthase]{Citrate} HO \cdot C \cdot COOH \xrightarrow{Aconitase} H \cdot C \cdot COOH \\ O{=}C \cdot COOH & | & | \\ | & CH_2 & HO \cdot CH \\ CH_2 & | & | \\ | & COOH & COOH \\ COOH & & \\ \textit{Oxaloacetic} & \textit{Citric acid} & \textit{Isocitric} \\ \textit{acid} & & \textit{acid} \end{array}$$

formed from the condensation of the acetyl moiety of acetyl coenzyme A and oxaloacetic acid, is apparently a symmetrical molecule, but in the next step in the sequence, the isomerization of citric acid to isocitric acid, the hydroxyl group moves to the carbon atom initially derived from oxaloacetate and not to the carbon atom derived from the methyl group of the acetyl radical. This means that the enzyme–substrate relationship is asymmetrical and that there are at least three points of specific interaction between the enzyme and substrate, and the affinities of the three enzymic sites must be different or asymmetric.

The Kinetics of Enzyme-catalysed Reactions

The Measurement of Reaction Rate

The progress of an enzyme-catalysed reaction may be represented by a graph of the concentration of one of the products (or of the diminution in the concentration of the substrate) against time; such a plot is shown in Fig. 7.1. The rate of the reaction in moles per litre per second at any particular time is given by the slope of the curve at that time. The fall-off of the rate with time is due to a number of factors such as fall in substrate concentration, inhibition of the reaction by the accumulating products, poisoning or denaturation of the enzyme, changes in the pH of the medium due to one of the products, etc. Thus in any simple study of the effects of changes in enzyme or substrate concentration, or of different substrates or conditions, on the reaction rate, comparisons must be made only of the *initial velocities* (given by the tangent to the progress curve at zero time, as shown in Fig. 7.1).

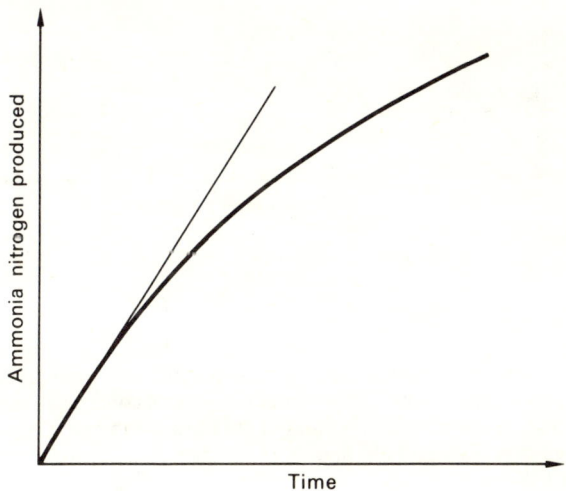

FIG. 7.1. The time course of the hydrolysis of urea in the presence of urease.

Factors Influencing the Rate of Reaction

Enzyme Concentration

In vitro the enzyme is usually present at a very much lower molar concentration than is the substrate and the initial velocity is found to be

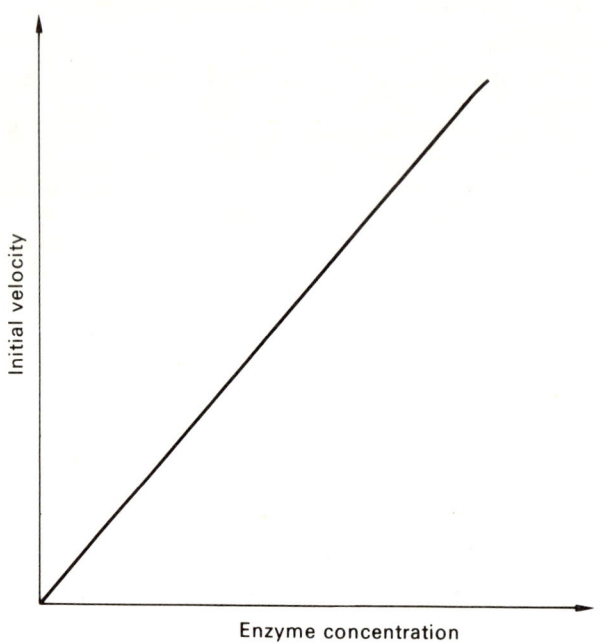

Fɪɢ. 7.2. The effect of enzyme concentration on the initial
reaction velocity.

directly proportional to the enzyme concentration (Fig. 7.2). In these circumstances the reaction is said to be of the *first order* in enzyme as its rate is proportional to the first power of the enzyme concentration. In vivo the concentrations of substrates and of enzymes are in many instances of the same order of magnitude, and more complex considerations apply (see Chapter 17).

Substrate Concentration

When the initial reaction velocity is investigated at different substrate concentrations various types of behaviour are found, the commonest being illustrated in Fig. 7.3. At *low substrate* concentrations, the rate is directly proportional to the substrate concentration, that is the reaction is *first order* in substrate. At *high substrate* concentrations the rate becomes independent of the substrate concentration, i.e. the reaction is *zero order* in substrate. When an enzyme reaction behaves like this an equation of

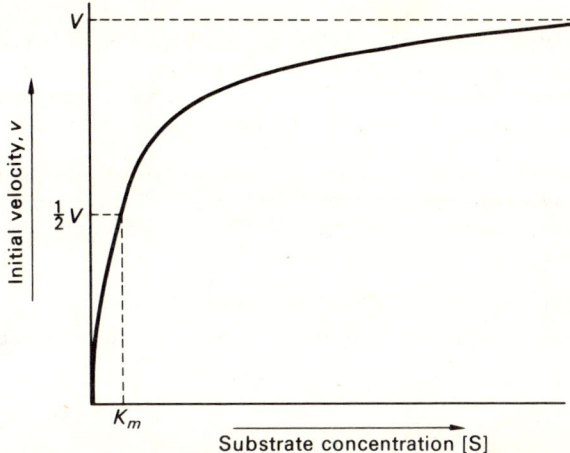

Fɪɢ. 7.3. The effect of substrate concentration on the initial
reaction velocity.

the type

$$v = \frac{a[S]}{1 + b[S]} \tag{7}$$

can be applied where [S] is the substrate concentration and a and b are
constants which will fit the results. Equations of this general type describe
a hyperbola (see Fig. 7.3):

$$v \rightarrow a \quad \text{as} \quad [S] \rightarrow \infty$$

These findings were first interpreted theoretically by Michaelis and
Menten, and equation (7) is the *Michaelis–Menten Law*. Their interpreta-
tion postulates a two-step process; the enzyme E first combining with the
substrate S to form an enzyme-substrate complex ES, which then breaks
down into the products P and the enzyme, which is thus regenerated. The
two steps can be represented by the following reactions:

$$[E] + [S] \underset{k_2}{\overset{k_1}{\rightleftharpoons}} [ES] \tag{8}$$

$$[ES] \overset{k_3}{\rightarrow} [E] + [P] \tag{9}$$

At *low substrate* concentrations much of the enzyme is free and the amount of the complex formed is proportional to the substrate concentration: equilibrium (8) is over to the left. Since the rate of formation of products is proportional to the concentration of enzyme-substrate complex [ES] reaction, the overall rate of conversion of substrate into products is proportional to the substrate concentration, i.e. the reaction is *first order* in *substrate*.

At *high substrate* concentrations all the enzyme is as ES, equilibrium (8) is over to the right and since a further increase in [S] cannot increase [ES] the overall reaction rate is independent of [S], i.e. the reaction is *zero order* in *substrate*.

Reaction Velocity and the Michaelis Constant

The velocity at which the products are formed is seen from (9) to be:

$$v = k_3[ES] \tag{10}$$

and the maximum velocity, V, when all the enzyme is in the complex ES, is given by:

$$V = k_3[E_t] \tag{11}$$

where $[E_t]$ is the *total* concentration of enzyme. At substrate concentrations below those giving maximum velocity, $[E_t]$ is the sum of $[E]$ and $[ES]$.

If the binding of product, P, to the enzyme may be neglected, and if [S] may be taken to be constant during the period in which v is measured, we have, from (8) and (9):

$$k_1([E_t] - [ES])[S] = (k_2 + k_3)[ES] \tag{12}$$

Rearranging we get:

$$\frac{([E_t] - [ES])[S]}{[ES]} = \frac{k_2 + k_3}{k_1} = K_m \tag{13}$$

where K_m is the *Michaelis constant*; it has the dimensions of a concentration and contains the three velocity constants k_1, k_2, and k_3.

From (13) the following expression for the concentration of the enzyme-substrate complex may be obtained:

$$[ES] = \frac{[E_t][S]}{K_m + [S]} \tag{14}$$

Combining (10), (11), and (14) an expression for the velocity is obtained:

$$v = \frac{k_3[E_t][S]}{K_m + [S]} = \frac{V[S]}{K_m + [S]} \tag{15}$$

From (15) it is evident that when $v = \frac{1}{2}V$, $K_m = [S]$. That is to say the Michaelis constant is equal to the substrate concentration which gives half the maximum velocity, as is shown in Fig. 7.3.

This method of obtaining K_m is not always satisfactory since it is often difficult to determine the exact value of V. A common method is to make use of an equation due to Lineweaver and Burk which is the reciprocal form of (15):

$$\frac{1}{v} = \frac{K_m}{V} \cdot \frac{1}{[S]} + \frac{1}{V} \tag{16}$$

If $1/v$ is plotted against $1/[S]$, as in Fig. 7.4(a), a straight line is obtained.

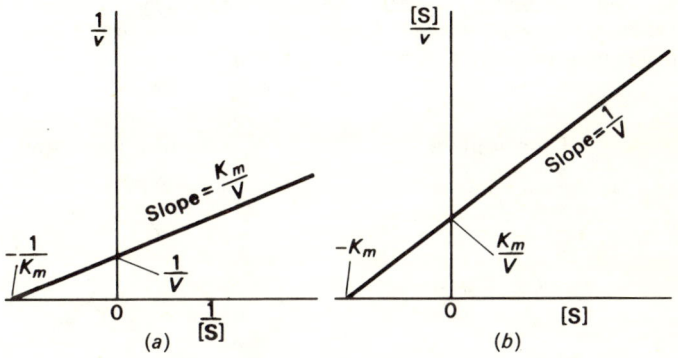

Fig. 7.4. Reciprocal plots of the effect of substrate concentration on the initial reaction velocity, (a) according to Equation (16) (Lineweaver and Burk) and (b) according to Equation (17) (Wilkinson).

This line has a slope of K_m/V, it cuts the $1/v$ axis at $1/V$ and the $1/[S]$ axis at $-1/K_m$. This method has the practical disadvantage that the slope of the line depends strongly on values of v corresponding to low $[S]$. A better variant, due to Wilkinson, uses the related equation:

$$\frac{[S]}{v} = \frac{K_m}{V} + \frac{1}{V} \cdot [S] \tag{17}$$

See Fig. 7.4(b).

When an enzyme has a coenzyme associated with it, as for example with many dehydrogenases, the initial velocity (at constant $[S]$) varies with the coenzyme concentration according to equation (7). Thus the coenzyme may be considered as a second substrate forming an active complex with the enzyme which saturates at high coenzyme concentrations.

The Effect of Inhibitors

Many instances are known in which the addition of a substance to an enzyme system reduces the rate of the reaction. Such substances are known as inhibitors. Two types of inhibitors may be distinguished, *reversible* and *non-reversible*. The latter cause more or less permanent inactivation of a fraction of the enzyme molecules, and their effect will be proportional to the concentration of the inhibitor. As an example, the fact that some enzymes need thiol ($-SH$) groups for activity may be recalled. A number of substances such as iodoacetamide ($CH_2I \cdot CONH_2$) and mercurials react with thiols to form covalent derivatives that do not break down when the inhibitors are removed. Such modified thiol enzyme molecules will be inactive, and the effective concentration of enzyme $[E_t]$ will be reduced, which will decrease the maximum velocity of the system [equation (15)]. The effect on the inverse plot cannot readily be distinguished from reversible non-competitive inhibition (Fig. 7.5(b)).

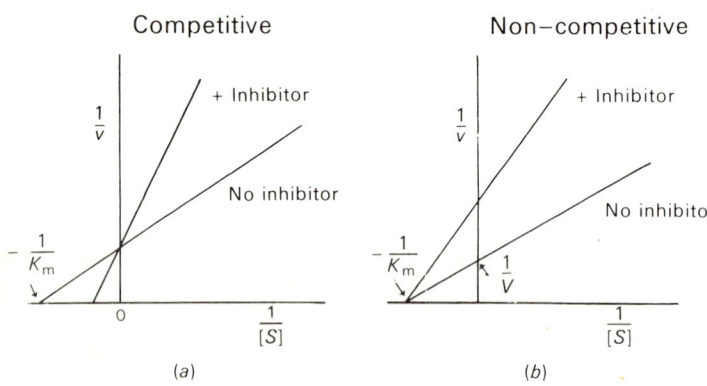

FIG. 7.5. Reciprocal plot of the effect of substrate concentration on the initial reaction velocity in the presence of (a) a competive inhibitor [Equation (21)] and (b) a non-competitive inhibitor.

Reversible inhibitors may also be divided into two main categories: *competitive* in which the inhibition may be nullified by raising the substrate concentration sufficiently, and *non-competitive* where there is always some inhibition, even at very high substrate concentrations. With competititive inhibition we assume that there are two equilibria:

$$[E] + [S] \underset{k_2}{\overset{k_1}{\rightleftharpoons}} [ES] \tag{8}$$

and

$$[E] + [I] \underset{k_5}{\overset{k_4}{\rightleftharpoons}} [EI] \tag{18}$$

It is further assumed that EI cannot break down to give any products but that equilibrium (18) is reversible. ES of course breaks down to the products as follows:

$$[ES] \overset{k_3}{\rightarrow} [E] + [P] \tag{9}$$

The rate at which the products are formed is again given by:

$$v = k_3[ES] \tag{10}$$

During the initial part of the reaction period we have, from (8), (18), and (9), remembering that $[E_t]$ is now the sum of $[E]$, $[ES]$, and $[EI]$:

$$k_1([E_t] - [ES] - [EI])[S] = (k_2 + k_3)[ES] \tag{12a}$$

and

$$k_4([E_t] - [ES] - [EI])[I] = k_5[EI] \tag{19}$$

From (12a) and (19) we obtain an expression analogous to (14) for the complex ES:

$$[ES] = \frac{[E_t][S]}{[S] + \left(\dfrac{k_2 + k_3}{k_1}\right)\left(1 + \dfrac{k_4[I]}{k_5}\right)} \tag{20}$$

and the velocity of the inhibited reaction is then given by:

$$v = \frac{V[S]}{[S] + K_m\left(1 + \dfrac{[I]}{K_i}\right)} \tag{21}$$

where $K_i = k_5/k_4$, i.e. the dissociation constant of the complex EI. Equation (21) is analogous to (15).

The reciprocal form of equation (21) is:

$$\frac{1}{v} = \frac{K_m}{V} \cdot \left(1 + \frac{[I]}{K_i}\right) \cdot \frac{1}{[S]} + \frac{1}{V} \tag{22}$$

If $1/v$ is plotted against $1/[S]$, at a *fixed inhibitor concentration*, $[I]$, a graph such as Fig. 7.5(a) is obtained.

This kind of plot allows the distinction between a competitive and a non-competitive inhibitor to be made. When the inhibitor is competitive, the slope of the line is greater than when no inhibitor is present by the factor $(1 + [I]/K_i)$, but the line cuts the $1/v$ axis at the same value of $1/V$. This is because the greater [S] the more equilibrium (8) is over to the right. Consequently, the free enzyme concentration, [E], is diminished and equilibrium (18) moves to the left and the concentration of inhibited enzyme, [EI], is diminished. At infinite substrate concentration $(1/[S] = 0)$ there will be no EI and consequently the maximum velocity $1/V$ will be the same as in the absence of a competitive inhibitor. On the other hand, the apparent Michaelis constant will be greater than the true K_m by the factor $(1 + [I]/K_i)$.

When the inhibitor is non-competitive the equation corresponding to (22) is:

$$\frac{1}{v} = \left[\frac{K_m}{V} \cdot \frac{1}{[S]} + \frac{1}{V} \right] \left(1 + \frac{[I]}{K_i} \right) \qquad (23)$$

Thus the slope of the reciprocal plot (Fig. 7.5(b)) is greater, and it cuts the $1/v$ axis at a higher point than when no inhibitor is present. The inhibitor reacts with ES, and even though the reaction is reversible, the effective concentration of ES is lowered, even if [S] is infinitely large. The line usually cuts the $1/[S]$ axis at the same point and the K_m for the substrate therefore remains the same. In another form of inhibition, *uncompetitive inhibition*, the line of the inverse plot is parallel to that without inhibitor, but cuts the ordinate at a higher value of I/V. Both this and non-competitive inhibition are relatively infrequent with one-substrate enzymes. With two-substrate enzymes, such as kinases and dehydrogenases, they are more often seen, because an inhibitor which is competitive for one substrate will automatically be non-competitive for the other.

The Influence of Temperature

The rate of enzyme-catalysed reactions passes through a maximum with increasing temperature as is shown in Fig. 7.6.

The temperature at which the maximum velocity occurs is known as the optimum temperature; it varies with the substrate concentration and the pH, and the time interval over which the initial rate is measured.

The explanation of the behaviour shown in Fig. 7.6 is to be found in two opposing effects. Firstly the rate of an enzyme-catalysed reaction increases with increasing temperature as do the rates of chemical reactions generally. Secondly enzymes, being proteins, are denatured and hence inactivated at high temperatures. Thus with increasing temperature the first effect is tending to increase the reaction velocity while the second is tending to reduce it by removing the enzyme. At the optimum temperature the two effects are balanced to the best advantage.

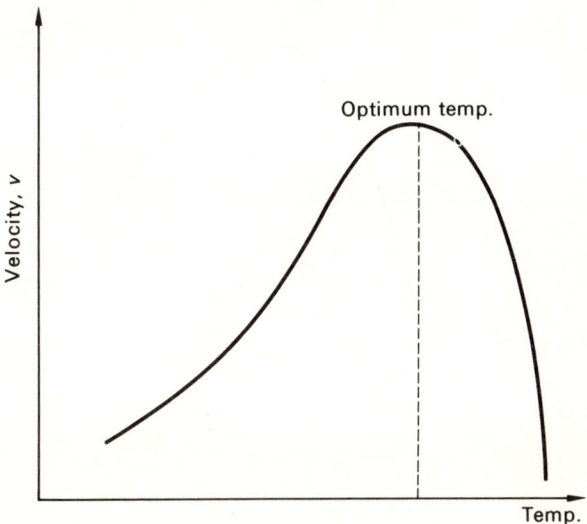

FIG. 7.6. The effect of temperature on an enzyme-catalysed reaction.

The Effect of pH

The effect of the pH of the medium on the rate of enzyme-catalysed reactions is very complex, different effects being found with different enzymes
and even with different substrates and the same enzyme. Generally the
velocity is found to pass through a maximum as the pH is varied as is
shown in Fig. 7.7.

The pH at which the velocity is maximal is known as the optimum pH
of the enzyme; its value depends on the substrate, its concentration and
the precise conditions in which the velocity is measured. Near the optimum
value the effects of changing the pH are usually reversible, i.e. the maximum
activity can be restored by bringing the pH back to the optimum value. If
the pH is taken too far to the acid or alkaline side, the changes may become irreversible. This is because these large changes in pH bring about
the denaturation of the enzyme protein. As mentioned at the beginning of
this chapter, the effects of changes of pH on the velocity are no doubt to
be explained in terms of differences in the charge distribution on the enzyme
and substrate molecules.

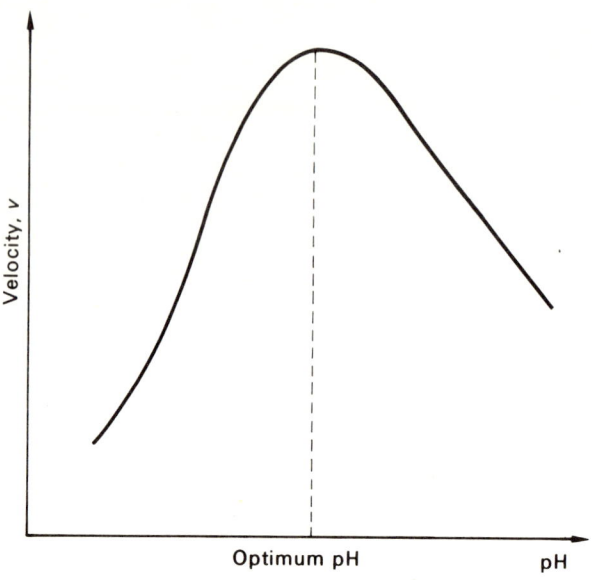

Fig. 7.7. The effect of pH on the velocity of an enzyme-catalysed reaction.

The Mechanism of Enzyme Action

Whether we can be said to understand the mechanisms of enzyme action depends on the rigour with which the question is asked. Certainly we can show that enzymes, like other catalysts, reduce the activation energy of reactions, so that the fraction of reactants at a given temperature possessing sufficient energy to make and break covalent bonds is significantly increased. Putting this in another way, enzymes stabilize transition state compounds—intermediates between reactants and products—and, as discussed below, evidence for the existence of such intermediates can sometimes be found. The search for precise reaction mechanisms has been most successful with enzymes that have prosthetic groups, such as phospho-pantotheine or pyridoxal phosphate, whose role can be mimicked with model compounds. In the special class of 'ping-pong' enzymes where a stable covalently modified form of the enzyme is an intermediate, the reaction pathway is also fairly clear.

Even in these instances it is at present impossible to give a convincing and detailed explanation for the observed fact that the rate of a reaction when catalysed by an enzyme is typically 10^6–10^9 times faster than in model reactions. With those enzymes that are pure proteins we have even less grasp of the reasons for the remarkable effectiveness of enzymes. Notwithstanding these provisos, many important observations have been made over the last half-century, and have been strengthened by recent progress in elucidating the primary sequence and the conformation of many enzymes. In the following paragraphs some of our present ideas are outlined.

The Enzyme-substrate Complex

The Michaelis theory requires the substrate to be bound temporarily to the enzyme to form an enzyme-substrate complex which then reacts to give the products and regenerate the enzyme. This reaction between the substrate and the enzyme necessarily takes place at a limited number of *active sites* on the enzyme protein; the saturation characteristics of the catalysis, and the very strong restrictions on the steric conformation of many substrates (p. 122) dictate that this must be so. There may be several active sites per enzyme molecule, but not usually more than one per polypeptide chain. As pointed out in Chapter 4, the conformation of globular proteins is such that the amino acid residues making up the active site, although in proximity to one another, are not usually in adjacent positions in the polypeptide chain (see also Fig. 7.8).

In some instances, as when binding alters the optical properties of a substrate or of a prosthetic group, it is possible to obtain fairly direct evidence for the existence of an enzyme-substrate complex (e.g., cytochrome oxidase). It may sometimes be possible by careful hydrolysis to isolate, bound to a residue at the active site, an irreversible inhibitor (e.g., di-isopropyl-phosphonate bound to a seryl residue of cholinesterase), or even a stable intermediate. An example is phosphoglucomutase (p. 162), where the mechanism is:

$$\text{G-6-P} + \text{P-Enz} \rightleftharpoons \text{G-1:6-di-P} + \text{Enz}$$
$$\text{G-1:6-di-P} + \text{Enz} \rightleftharpoons \text{G-1-P} + \text{P-Enz}$$

On incubating the enzyme with the intermediate glucose-1:6-diphosphate it is completely converted to the phospho-enzyme. On hydrolysing the latter, all the phosphate is recovered esterified to the hydroxyl of one serine residue.

Convincing evidence is also provided by the so-called 'transition state inhibitors', which resemble *both* substrates of a two-substrate enzyme, and are bound to it far more strongly than analogues of either substrate alone.

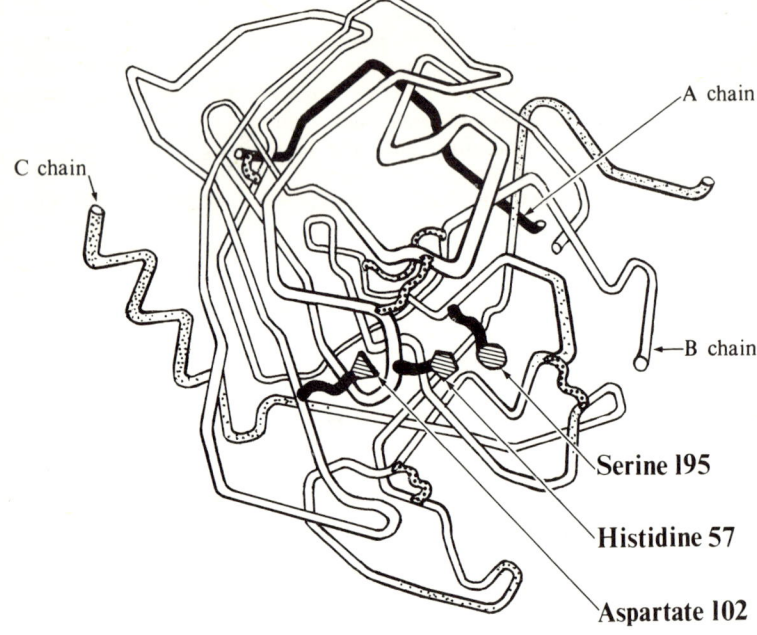

C chain

A chain

B chain

Serine 195

Histidine 57

Aspartate 102

FIG. 7.8. Schematized diagram of the secondary and tertiary structures of α-chymotrypsin. There are three peptide chains, whose —COOH terminal ends are indicated by the arrows. The serine and histidine residues at the active site are shown.

Carnitine-acyl transferase catalyses the reaction

$$\text{Acetyl-carnitine} + \text{CoA} . \text{SH} \rightleftharpoons \text{Carnitine} + \text{acetyl-CoA}$$

On incubating the enzyme with CoA·SH and the analogue bromoacetyl-carnitine, the transition state inhibitor

$$\text{Carnitine} \cdot \text{O} \cdot \text{CO} \cdot \text{CH}_2 \cdot \text{S} \cdot \text{CoA}$$

is formed in situ. It is barely possible to remove it from the enzyme, although it is not covalently bound.

The Mechanism of Action of Proteinases

Many of the active sites that have been studied contain residues with reactive groups in their side-chains, including the hydroxyl group (serine), the sulphydryl group (cysteine) and the imidazolyl group (histidine). At the present time interest is concentrated on enzymes, chiefly proteinases and esterases, which contain both serine and histidine in the active centre and often also aspartic acid in proximity to the histidine (Fig. 7.9). The evidence that an intermediate acylated serine is formed in these enzymes is

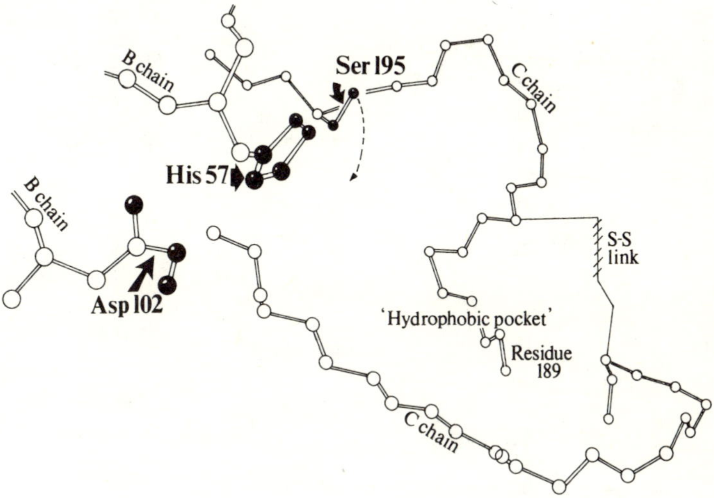

FIG. 7.9. Active centre of α-chymotrypsin, from X ray studies of the crystalline enzyme.

The amino acid on the −NH− side of the peptide bond to be cleaved sits in the 'hydrophobic pocket', which is lined with non-polar residues. This helps to determine the specificity of the enzyme. In trypsin, residue 189 is aspartyl, which binds a basic amino acid residue in the peptide substrate, rather than the aromatic residue favoured by chymotrypsin.

When the −OH group on Ser 195 is acylated, during the hydrolysis procedure, it moves down towards His 57 as indicated by the dotted arrow.

now very strong; more indirect evidence suggests that the aspartate-histidine pair provides a mechanism for the transfer of a proton, essential in the reaction (p. 136). This view is strengthened by the remarkable fact that several enzymes, e.g., chymotrypsin, trypsin, thrombin, all possess the same spatial arrangement of these three residues. The limitation, characteristic for each enzyme, on the type of residue adjacent to the bond to be hydrolysed (see Chapter 8) no doubt arises from the nature of the residues adjacent to the active site.

The precursor of chymotrypsin, chymotrypsinogen, is secreted as a single polypeptide chain, and it is of interest that after unmasking the three residues mentioned above are found on two peptide chains, held together only by $-S-S-$ bonds. A description of the postulated mechanism follows.

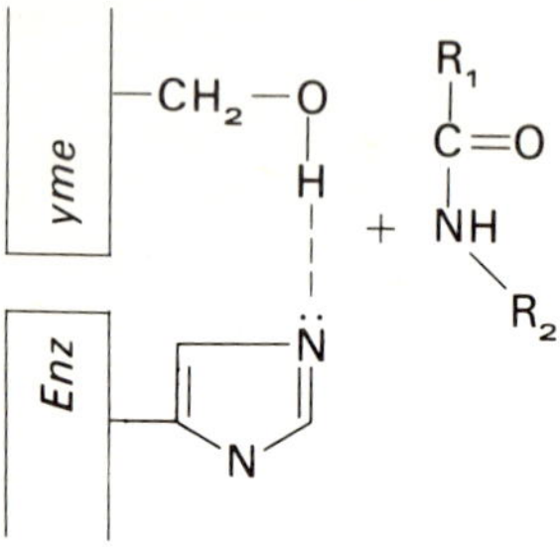

I

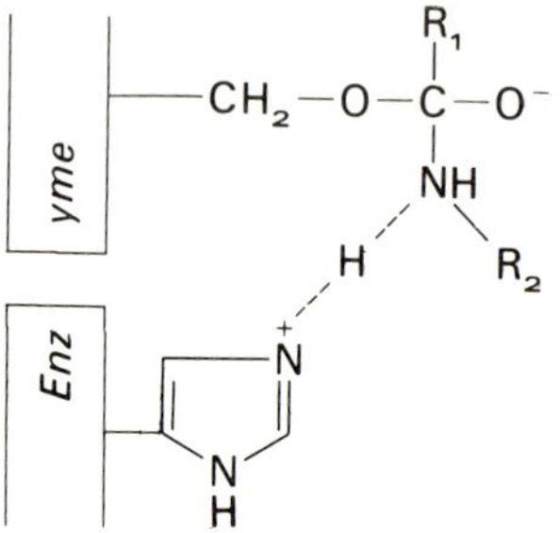

II

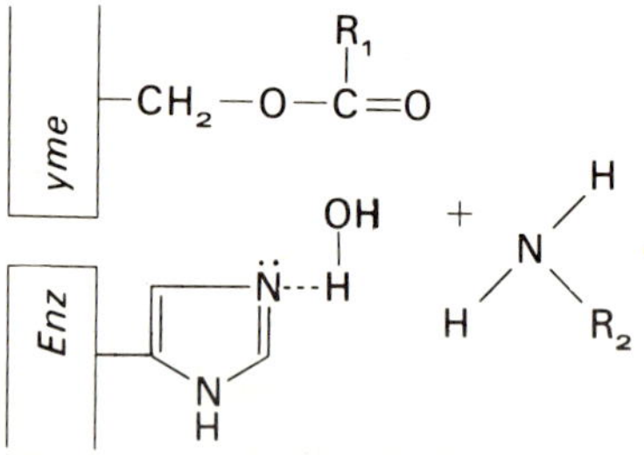

III

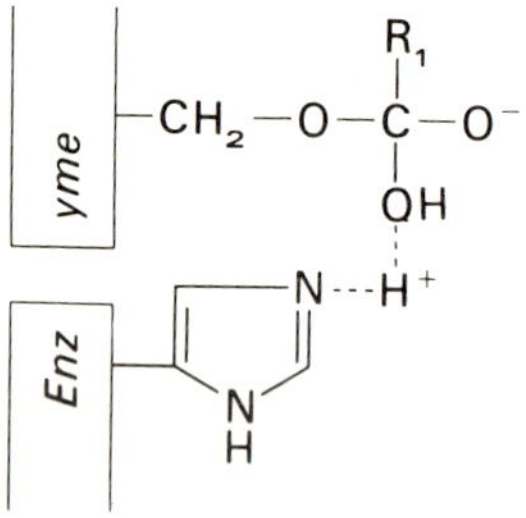

IV

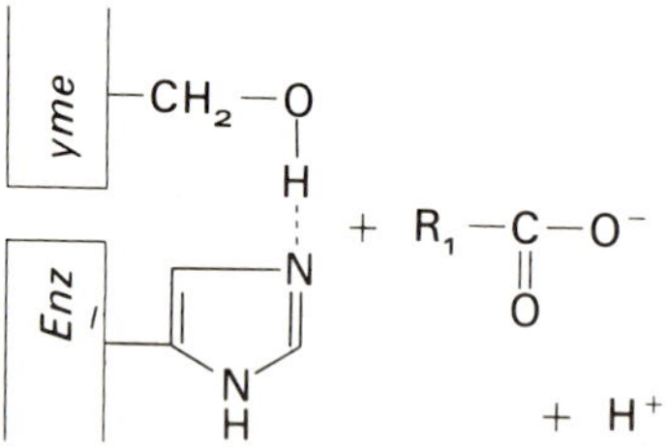

V

As the peptide substrate approaches the active centre (I), the hydroxyl hydrogen of the seryl residue moves as a proton to the nitrogen of the histidine. At the same time an adduct to the carbonyl carbon of the peptide is formed (II). The rupture of the carbon-nitrogen bond releases a free amine, and its place at the histidine nitrogen is taken by a hydrogen-bonded water molecule. A fully acylated seryl residue is coincidentally formed (III). The next two stages involve the hydrolysis of the acyl-enzyme; first an attack by the water molecule on the carboxyl carbon (IV), and then release of the free acid (V) and of the enzyme, ready to accept another substrate molecule.

Classification of Enzymes

The most generally useful classification of enzymes is in terms of the reactions they catalyse, individual enzymes being named by adding the suffix *-ase* to the name of the substrate acted on or to the reaction brought about. The following list is not exhaustive, but the numbering of the classes is that recommended by the Enzyme Commission; thus all transferases have a code number which begins EC 2.—.

1. *Oxidoreductases*

Oxidoreductases catalyse oxidation–reduction reactions. They are subdivided into:

(*a*) *Dehydrogenases* which catalyse the removal of two atoms of hydrogen from the substrate and their transference to a coenzyme acceptor.

(*b*) *Oxidases* which catalyse the direct reduction of oxygen.

(*c*) *Oxygenases* which catalyse the incorporation of O_2 into substrate molecules.

(*d*) *Mixed-function oxidases* which catalyse the incorporation of one atom of oxygen from O_2 into the substrate, the other going to a second acceptor.

(*e*) *Oxidative deaminases* which catalyse the oxidation of amino compounds with the elimination of a molecule of ammonia.

2. *Transferases*

These bring about an exchange of groups between two substrates

$$AB + CD \rightleftharpoons AC + BD$$

They are subdivided into:

(*a*) *Aminotransferases* which bring about the exchange of amino and keto groups between an amino- and a keto-acid.

(*b*) *Phosphotransferases or kinases* which bring about the transfer of a phosphate radical.

(*c*) *Acyltransferases* catalyse the transfer of an acyl (including acetyl) group, often from an acyl-CoA molecule, to a suitable acceptor.

(*d*) *Glycosyltransferases*: these enzymes bring about the transfer of a glycosyl residue from a donor to an acceptor other than water. The group includes the phosphorylases, where the acceptor is phosphate; the reaction is effectively that of splitting the substrate by reaction with a molecule of phosphoric acid.

3. Hydrolases

This large group includes all the digestive enzymes. They catalyse the reaction

$$AB + H_2O \rightleftharpoons A \cdot OH + HB$$

and may be subdivided according to the type of substrate acted on.

(a) *Proteolytic enzymes* (*peptide hydrolases*) which catalyse the hydrolysis of the peptide bond.

(b) *Carbohydrases* (*glycoside hydrolases*) which catalyse the hydrolysis of the glycosidic bond.

(c) *Carboxyl esterases* which catalyse the hydrolysis of carboxylic acid esters.

(d) The *phosphatases* which catalyse the hydrolysis of phosphoric acid esters.

(f) *Phosphodiesterases*. The substrates for this group of enzymes contain a phosphate radical linked by ester bonds to two separate alcohols; only one of the ester links is hydrolysed. The phospholipases and the nucleases are among the enzymes in this group.

(g) *Deaminases* (*aminohydrolases*) which catalyse the hydrolysis of amines liberating ammonia and a hydroxy compound.

(h) *Deamidases* (*amidohydrolases*) catalyse the hydrolysis of amides.

4. Lyases

These enzymes usually remove groups from substrates non-hydrolytically, leaving double bonds (or adding groups to double bonds). The group, however, also includes the decarboxylases. The reactions catalysed may be represented by:

$$AB \rightleftharpoons A + B$$

5. Isomerases

These catalyse the isomerization of substrates.

6. Ligases

These bring about the formation of $C-O$, $C-S$, $C-N$, or $C-C$ bonds. The ligases bring about reactions which require the expenditure of energy provided by the simultaneous breakdown of ATP or another triphosphate. They are often known as *synthetases* (cf. group 4).

Isoenzymes

This term has been used to describe groups of enzymes which have the same main enzymic activity but which show different physical properties. Although the main enzymic activity of members of a family of isoenzymes is the same, there may be differences in the details of their enzymic action such as their K_m values and their rates of reaction with different substrates or coenzymes. These differences are often due to alterations in quaternary structure (see p. 69).

Lactate dehydrogenases (LDH) from different tissues can be classified into five distinct types, or isoenzymes, by electrophoresis on starch gel. These types also differ in their behaviour in DEAE cellulose chromatography and show differences in amino acid composition, immunochemical structure, relative rates of reaction with analogues of NAD, and other properties. One of these enzymes, known as the H enzyme, predominates in heart muscle; it is more negatively charged than the other LDH isoenzymes at pH 7·0 and consists of four identical sub-units each having a molecular weight of about 30 000. The other main LDH type predominates in skeletal muscle and is called the M enzyme. The M enzyme also consists of four identical sub-units which are different from those of the H enzyme.

When these two different kinds of sub-unit are produced in the same cell they combine in groups of four in a random fashion. There are, therefore, five possible different isoenzymes of LDH with the following sub-unit composition: H_4, H_3M_1, H_2M_2, H_1M_3, and M_4. It is these five different combinations of H and M sub-units which are responsible for the five electrophoretically distinguishable lactate dehydrogenases.

Isoenzymes of a number of other enzymes are known, such as alkaline phosphatase, glutamate-oxaloacetate transaminase, and creatine phospho-kinase. It should, however, be emphasized that the appearance of several bands of enzymic activity on starch-gel electrophoresis (or in some other method of separation) does not necessarily imply that one is dealing with distinct isoenzymes since the results may be artifacts. Such artifacts may arise from reaction between some of the enzyme with substances (such as metal ions) in the separation medium.

Coenzymes

Pyridine Nucleotides

These two coenzymes are known as *nicotinamide-adenine dinucleotide* (NAD) and *nicotinamide-adenine dinucleotide phosphate* (NADP). They are also frequently called diphosphopyridine nucleotide (DPN), or coenzyme I, and triphosphopyridine nucleotide (TPN), or coenzyme II, respectively. Both of them contain the nucleotide adenosine monophosphate (AMP) linked through the phosphate to the phosphate of another nucleotide containing nicotinamide as the base.

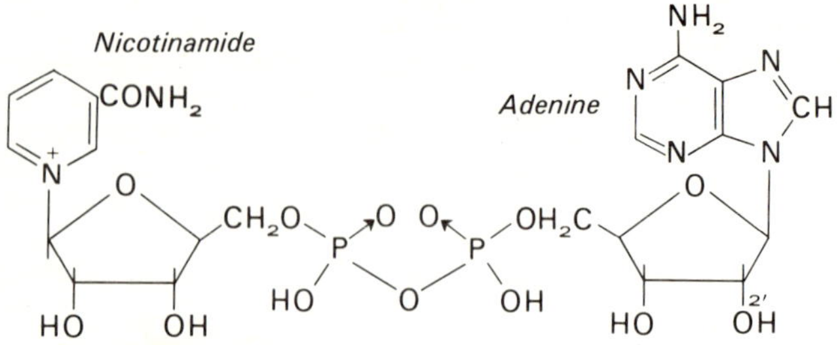

Nicotinamide-adenine dinucleotide (NAD)

NADP (*nicotinamide-adenine dinucleotide phosphate*) has the same structure as *NAD* plus a phosphate on position 2′ of adenosine.

The pyridine nucleotides are coenzymes for a large number of dehydrogenase enzymes; a list of these is given in Chapter 12. In these reactions two hydrogen atoms are removed from the substrate and transferred to the coenzyme, as for example in the dehydrogenation of lactate:

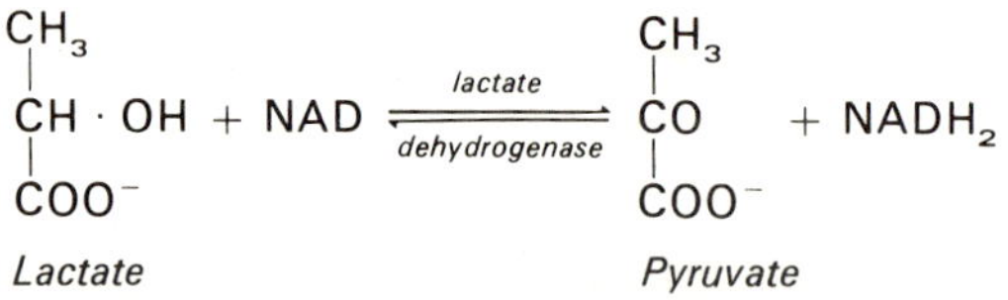

$$
\begin{array}{cc}
\text{CH}_3 & \text{CH}_3 \\
| & | \\
\text{CH} \cdot \text{OH} + \text{NAD} \xrightleftharpoons[dehydrogenase]{lactate} & \text{CO} \quad + \text{NADH}_2 \qquad (23) \\
| & | \\
\text{COO}^- & \text{COO}^-
\end{array}
$$

Lactate   Pyruvate

Equilibrium (23), however, is not a strictly accurate representation of the reaction. Examination of the formula of NAD above shows the nitrogen atom in the pyridine ring to be quaternary and to be positively charged. The reaction shown in (23) is better represented by:

$$\text{Lactate} + \text{NAD}^+ \rightleftharpoons \text{Pyruvate} + \text{NADH} + \text{H}^+ \qquad (24)$$

because the nitrogen atom of the pyridine ring is uncharged in the reduced (quinonoid) form. For details of this relationship see Chapter 12 (p. 262).

There is a marked difference in the ultra-violet absorption spectrum of oxidized and reduced pyridine nucleotides (Fig. 7.10) and changes from one form to the other are easily followed by measuring the optical density of a solution at 340 nm.

Flavin Coenzymes

There are two flavin coenzymes, *flavin mononucleotide* (FMN) and *flavin-adenine dinucleotide* (FAD). FMN is simply the phosphate of the

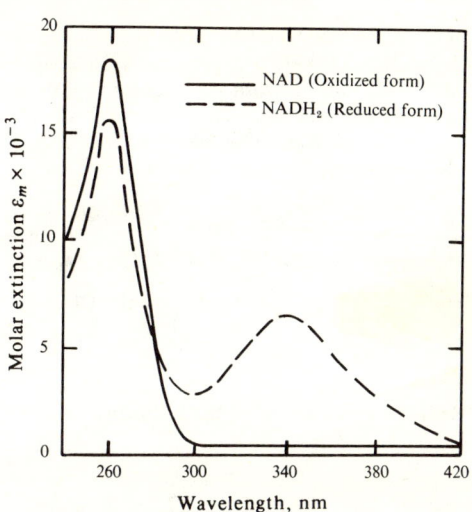

FIG. 7.10. Absorption spectra of oxidized and reduced nicotinamide-
adenine dinucleotides.

vitamin riboflavin (see Chapter 20), while FAD contains riboflavin
phosphate linked to AMP through the phosphate groups. Flavin mononucleotide is not strictly speaking a nucleotide, since the C_5 chain attached
to the flavin is the sugar alcohol ribitol and not the aldose ribose.

Flavin-adenine dinucleotide (FAD)
(FMN consists of riboflavin, ribitol, and one phosphate)

These coenzymes are important in dehydrogenation reactions; often they are tightly bound to the enzyme protein. A list of a number of them is given in Chapter 12, together with the structures of the reduced forms.

Electron transporting quinones. These are 2,3-dimethoxy-5-methyl-benzoquinones substituted at position 6 by a homologous polyisoprenoid side-chain. In beef heart mitochondria $n = 10$.

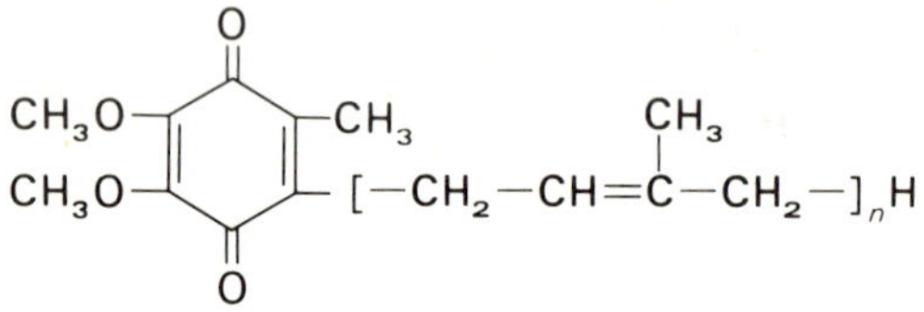

Reduced ubiquinone has the following structure

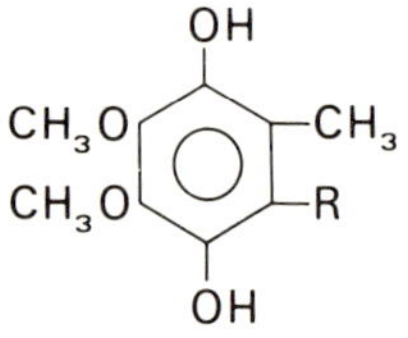

Ubiquinone (Coenzyme Q) [UQ]

Cytochromes. These electron-transporting haem compounds have some of the attributes of coenzymes. Their structure is discussed in Chapter 6 and their functioning in Chapter 12.

Coenzyme A. This compound contains adenosine which carries a phosphate group on the C'–3 carbon atom and a pyrophosphate group on the C'–5 carbon atom. The pyrophosphate group is esterified, distally from the adenosine, to the vitamin pantothenic acid (p. 427), the carboxyl of which forms an amide with 2-mercaptoethylamine.

Coenzyme A functions as a very important carrier of acyl groups by

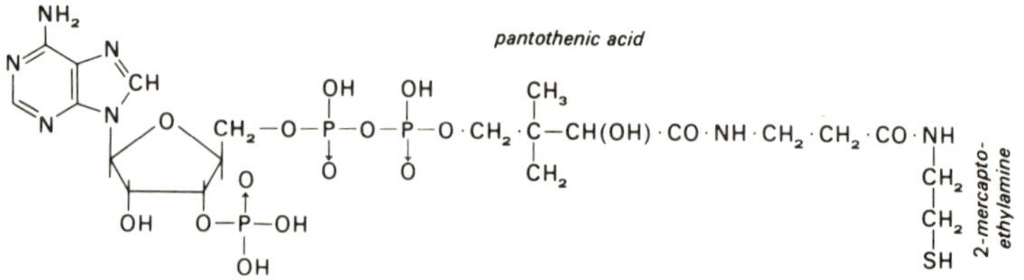

Coenzyme A (CoA · SH)

forming thioesters with carboxylic acids. Thioesters such as acyl coenzyme A compounds are high energy compounds and for their hydrolysis:

$$R \cdot CO \cdot S \cdot CoA + H_2O \rightleftharpoons R \cdot COOH + HS \cdot CoA$$

ΔG is about $-33\,500$ joules per mole (J/mol) at pH 7·0. This free energy change is much greater than for oxygen esters because the resonating form

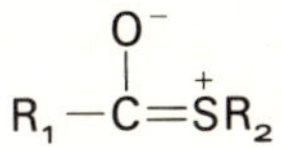

$$R_1 - \overset{\overset{\displaystyle O^-}{\displaystyle |}}{C} = \overset{+}{S} R_2$$

is not possible for thioesters, as sulphur has little tendency to form double bonds. Numerous examples of the carrier functions of coenzyme A are given in the chapters on metabolism.

The *acyl carrier protein* (p. 190) has pantotheine (2-mercaptoethylamine + pantothenic acid) esterified through a phosphodiester bond to a serine hydroxyl of the protein.

Thiamine pyrophosphate (*TPP*). This coenzyme is the pyrophosphate of thiamine or vitamin B_1 (Chapter 20). The structure is

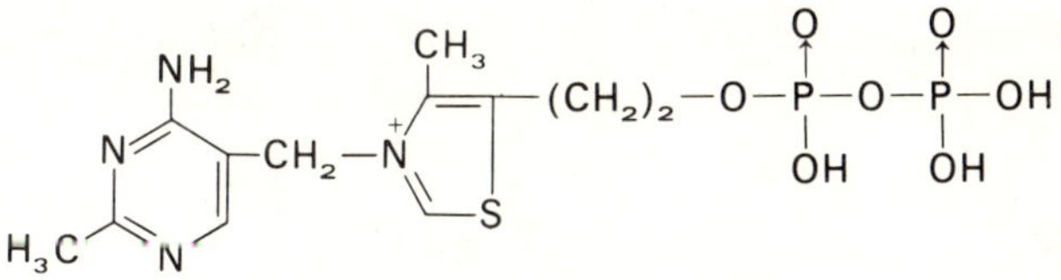

Thiamine pyrophosphate (*TPP*)

TPP is the coenzyme in the following enzyme systems: the pyruvate decarboxylase of yeast, the α-oxo-acid decarboxylases (pyruvate, α-oxoglutarate, and the α-oxo-acids corresponding to the branched chain amino acids), transketolase, and phosphoketolase. In all these reactions the substrate becomes attached to C–2 of the thiazole of TPP which easily dissociates to form a carbanion

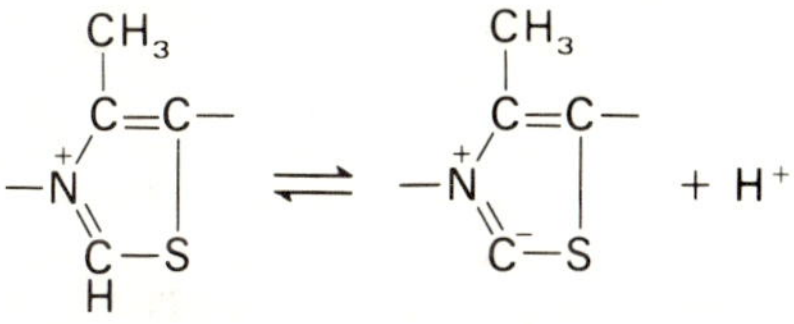

Thiazole ring
of TPP   *Carbanion*

The carbanion is able to form an adduct with the carbonyl carbon atom of α-oxo-acids. The α-oxo-acid after appropriate rearrangement of the electrons undergoes decarboxylation

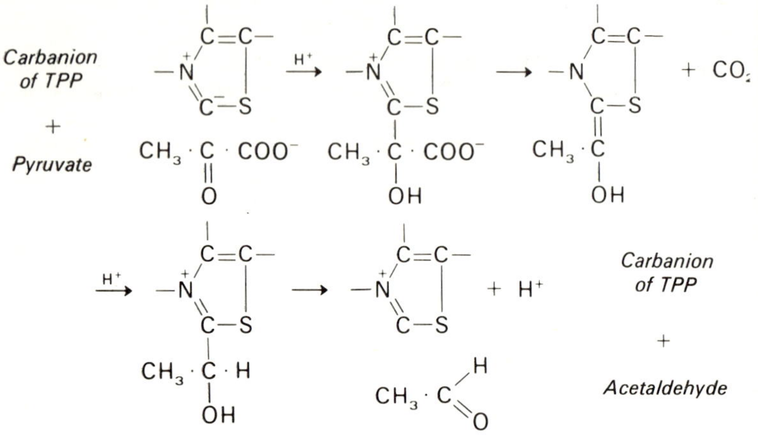

In the participation of TPP in the oxidative decarboxylation of α-oxo-acids the

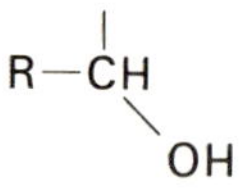

$$R-\overset{|}{\underset{\diagdown OH}{CH}}$$

group is transferred to a receptor discussed under lipoic acid below.

In the transketolase reaction the group $-CO \cdot CH_2OH$ is transferred by means of TPP in an analogous manner.

Lipoc acid. Lipoic acid is a disulphide derivative of octanoic acid.

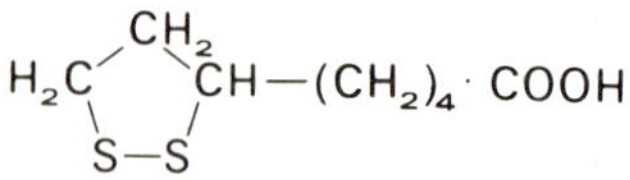

It functions as a coenzyme in enzyme reactions which generate acyl groups and in acyl transfer reactions. In the oxidative decarboxylation of α-oxo-acids it is intimately associated with thiamine pyrophosphate (TPP) (see above) and NAD (Chapter 9). An acyl–TPP complex reacts with lipoic acid to form an addition complex which, on addition of a proton, splits into an acyl–lipoic acid complex and TPP. Lipoic acid is usually covalently bound to the enzyme through an amide link to the ε-amino group of a lysyl residue.

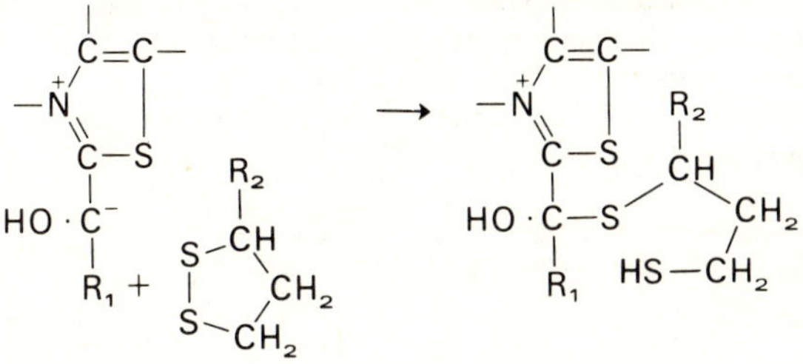

Acylol–TPP   Lipoic acid   Addition complex
complex

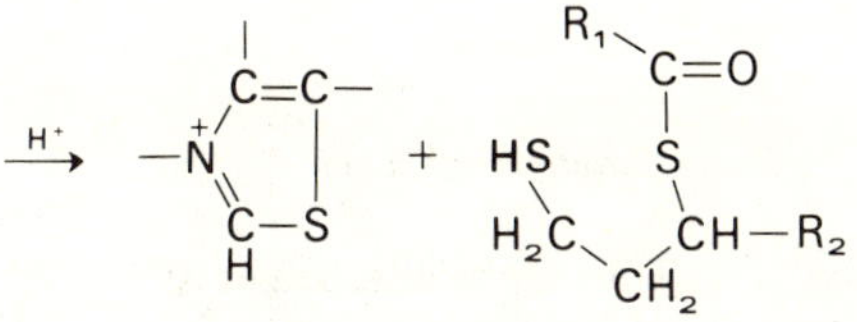

TPP   Acyl–lipoic acid complex

Biotin is the prosthetic group in a number of carboxylation reactions where it functions as the actual carrier of CO_2. One ring nitrogen of free

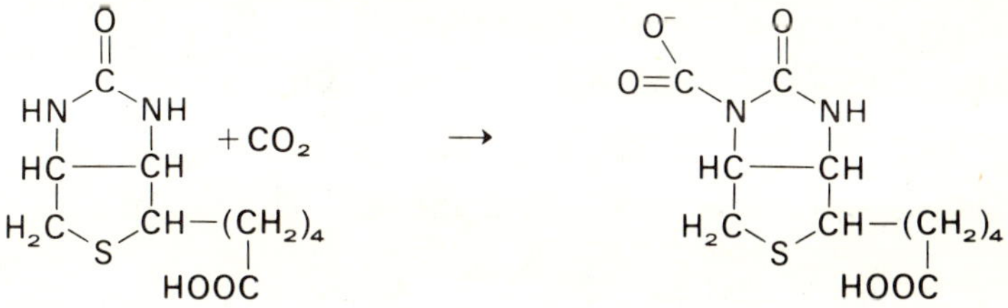

biotin, or of enzyme-bound biotin (see below), is carboxylated in a reaction requiring ATP. The carboxyl group is subsequently transferred at another site on the enzyme to an acceptor. E.g.:

$$HCO_3^- + biotin + ATP \rightleftharpoons biotin - COO^- + ADP + P_i$$

$$biotin - COO^- + CH_3 \cdot CO \cdot S \cdot CoA \rightleftharpoons biotin +$$

$$^-OOC \cdot CH_2 \cdot CO \cdot S \cdot CoA$$
Malonyl-CoA

Biotin is normally bound to the enzyme by an amide linkage between the $-COOH$ group and an $\varepsilon-NH_2$ group on a lysine residue of the protein. This provides a flexible 'arm' which can swing the carboxylated biotin from the carboxylating site to the carboxytransferase site. A similar role has been proposed for lysyl-bound lipoic acid.

Tetrahydrofolic acid. This is the reduced form of the vitamin folic acid (see Chapter 20)

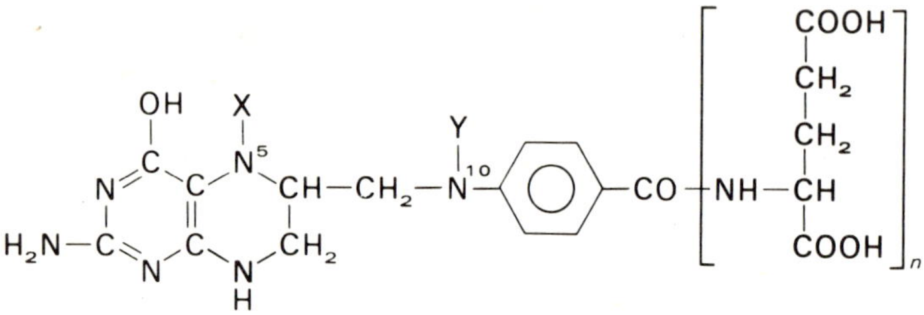

Tetrahydrofolic acid (FH₄)

A 'one-carbon' fragment may be attached at $X(N^5)$, $Y(N^{10})$, or at both X and Y. It is transferred from Y. *n* is usually 3.

It is of importance in the metabolism of one-carbon fragments; the details of its participation in a number of reactions are given in Chapter 11.

Pyridoxal phosphate. This coenzyme is a derivative of pyridoxine (vitamin B_6) (Chapter 20). It also occurs as pyridoxamine phosphate

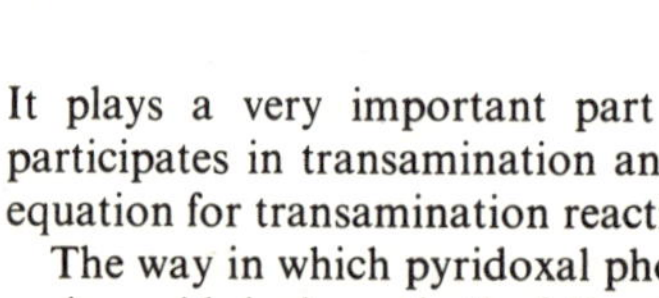

Pyridoxal phosphate

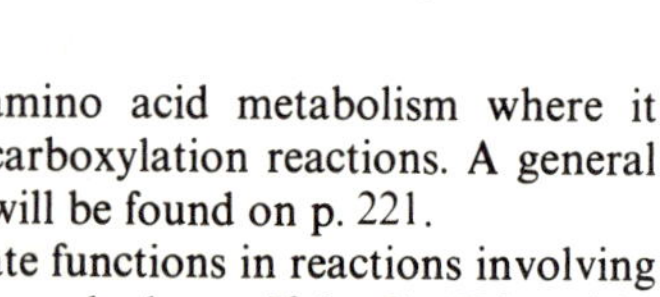

Pyridoxamine phosphate

It plays a very important part in amino acid metabolism where it participates in transamination and decarboxylation reactions. A general equation for transamination reactions will be found on p. 221.

The way in which pyridoxal phosphate functions in reactions involving amino acids is shown in the following general scheme. If the $C-H$ bond (*a*) is broken and reformed the enzyme is a *racemase* (i.e. D-amino-acid → L-amino acid): if the $C-C$ bond (*b*) is broken, the enzyme is a *decarboxylase* (product a primary amine).

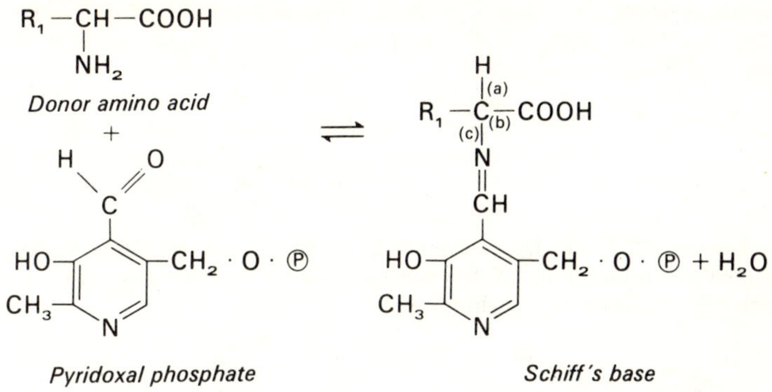

$R_1-CH-COOH$
    $|$
    NH_2

Donor amino acid

    +

Pyridoxal phosphate

$\rightleftharpoons$

Schiff's base

Finally, if the C—N bond (*c*) is broken, an oxo-acid together with pyridoxamine phosphate (above) is formed. For this to happen, the double bond of the Schiff's base has to migrate to the C—N link (*c*). This is promoted by electron shifts elsewhere in the molecule especially in the pyridine ring (quinonoid intermediate).

In transamination, the amino group of pyridoxamine phosphate is then transferred to the oxo-acid acceptor by the reversal of this reaction. In the free enzyme, the pyridoxal phosphate is frequently bound to the $\varepsilon-NH_2$ of a lysyl residue through a Schiff's base.

Adenosine phosphates. Adenosine triphosphate (ATP)

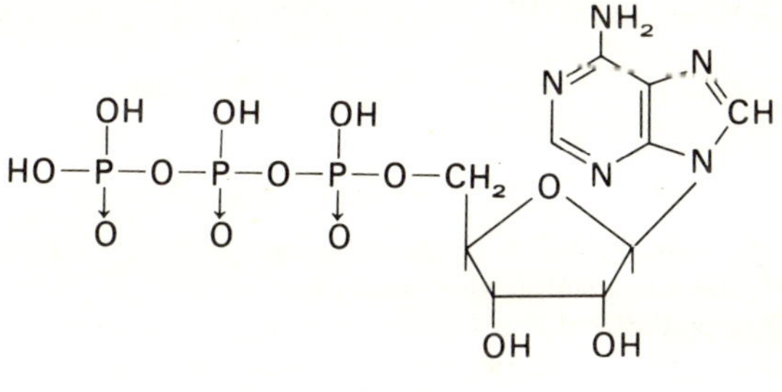

Adenosine triphosphate (ATP)

has a most important place in numerous metabolic reactions as a link between exergonic reactions, in which it is generated, and endergonic reactions, where its hydrolysis provides the energy for driving the reaction. ATP is the prime example of a group of so-called *energy-rich* compounds. The energy-rich nature of ATP is shown by the much greater change in free energy on the hydrolysis of the terminal phosphate group of ATP than when a simple phosphate ester such as glucose-6-phosphate is hydrolysed.

$$\text{Adenosine} - \circledP - \circledP - \circledP + H_2O \rightleftharpoons \text{Adenosine} - \circledP - \circledP + P_i$$

ATP ADP

$$\Delta G = -30\,500 \text{ J (pH 7·0)}$$

$$\text{Glucose-6-} \circledP + H_2O \rightleftharpoons \text{Glucose} + P_i$$

$$\Delta G = -13\,800 \text{ J (pH 7·0)}$$

The reasons for the large negative ΔG for the hydrolysis of ATP include the electrostatic repulsion between the induced positive charges on the phosphorus atoms of the triphosphate

$$\text{Adenosine} - O - \overset{\overset{\displaystyle O^{\delta-}}{|}}{\underset{\underset{\displaystyle OH}{|}}{P^{\delta+}}} - O - \overset{\overset{\displaystyle O^{\delta-}}{|}}{\underset{\underset{\displaystyle OH}{|}}{P^{\delta+}}} - O - \overset{\overset{\displaystyle O^{\delta-}}{|}}{\underset{\underset{\displaystyle OH}{|}}{P^{\delta+}}} - OH$$

which results from the attraction of electrons to the double-bonded oxygen atoms. The energy necessary to maintain such a structure is released on hydrolysis. Such a situation does not occur in G-6-P, where there is no atom bearing a positive charge near the phosphorus atom. Also the number of possible resonance forms for each of the phosphate groups is less in ATP than in ADP plus inorganic phosphate. This effect is less marked with simple phosphate esters than with pyro- and tri-phosphate compounds.

The hydrolysis of ADP is similarly a strongly exergonic reaction

$$\text{Adenosine} - \circledP - \circledP + H_2O \rightleftharpoons \text{Adenosine} - \circledP + \circledP$$

ADP AMP

$$\Delta G = -27\,000 \text{ J (pH 7·0)}$$

The hydrolysis of AMP is much less exergonic, having a ΔG near that for the hydrolysis of a simple phosphate ester.

ATP is synthesized in a number of exergonic reactions of metabolism, particularly in the reoxidation of reduced coenzymes (see oxidative phosphorylation, Chapter 12). The energy of these reactions is trapped in the energy-rich compound ATP. The energy made available by the exergonic hydrolysis of ATP is used in a multitude of synthetic reactions which include amide synthesis, amino acid activation, carboxylation reactions, fatty acid activation, and the formation of phosphate esters and amides. These and other reactions are described in the chapters on metabolism (see also Chapter 12). It has been established that the true coenzyme in the majority of these reactions is a magnesium-ATP complex ($MgATP^{2-}$ at physiological pHs). Most of the ATP in the cell is in this form, while ADP and AMP bind Mg^{2+} less strongly, and mostly exist as the free nucleotides.

In some of these synthetic reactions ATP is hydrolysed to AMP and pyrophosphate instead of to ADP and phosphate. The free energy of hydrolysis is even larger for this reaction ($-36\,000$ J/mol).

ATP sometimes acts as a carrier as in the activation of amino acids for protein synthesis. Here the amino acid reacts with the enzyme and ATP to form an aminoacyl-AMP-enzyme complex. In other reactions the adenosyl moiety, or part of it, is transferred (see pp. 230 and 238).

Uridine phosphates. Uridine diphosphate is the chief transferring coenzyme for carbohydrates. UDP-glucose is used in the synthesis of

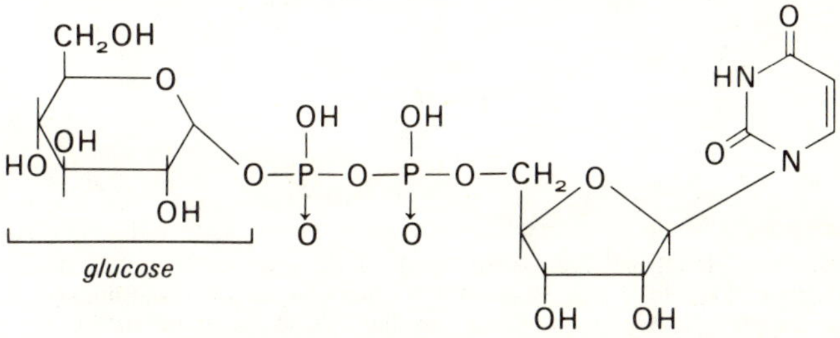

Uridine diphosphate glucose (UDPG)
(UDP-galactose, UDP-glucuronate, and UDP-glucosamine
have corresponding structures)

glycogen, it can be epimerized to UDP-galactose for the synthesis of lactose and chondroitin, or oxidized to UDP-glucuronate for the synthesis of glucuronides and chondroitin (see Chapter 9).

Cytidine phosphates.

Cytidine diphosphate is a coenzyme involved in the transfer reactions of choline and ethanolamine in the synthesis of phospholipids (Chapter 10).

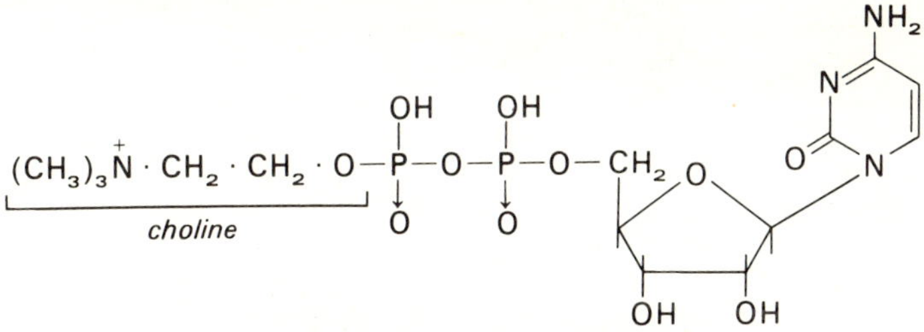

Cytidine diphosphate choline (CDP-choline)
(CDP-ethanolamine, CDP-glycerol, and CDP-ribitol
have corresponding structures)

B_{12} *Coenzymes*

Deoxyadenosyl B_{12} (shown schematically in Fig. 7.11) is the coenzyme for methylmalonyl CoA isomerase.

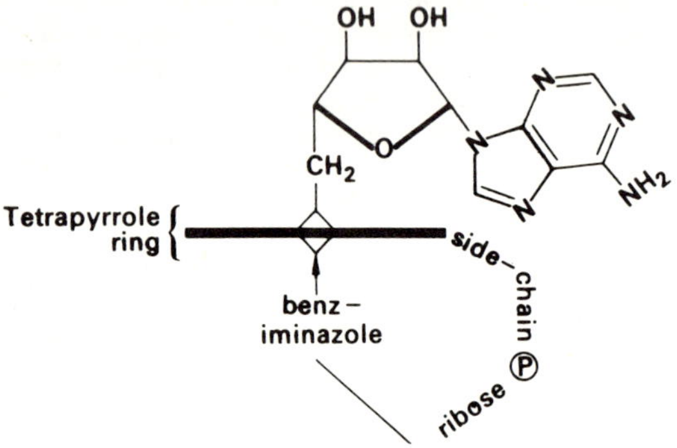

FIG. 7.11. Schematic formula of deoxyadenosyl cobalamin, a B_{12} coenzyme. In vitamin B_{12} itself CN^- occupies the sixth coordination position of Co^{3+} (shown by the diamond shape in the centre of the ring).

8 Digestion

In general the main classes of foodstuffs, carbohydrates, proteins, and fats, have to be broken down to small molecules before they can be absorbed. From the chemical point of view, this breakdown is effected by the enzymes secreted in the various digestive juices. Further enzymic action on absorbed substances occurs in the mucosal cells of the small intestine. The chief characteristics of the various digestive juices are shown in Table 8.1.

Table 8.1
Chief Characteristics of Digestive Juices

Saliva	Daily volume: 1–1·5 l. pH: 5·6–7·6, mean 6·7. Enzymes: α-amylase. Na$^+$ and Cl$^-$ 14·8 and 10 mM resting, 44·6 and 43 mM stimulated, respectively.
Gastric juice	Daily volume: 2–3 l. pH: 1·1–2·6. Enzymes: Pepsin secreted as precursor. Cl$^-$ 75–160 mM. Free acid 8–66 mM.
Pancreatic juice	Daily volume: 1·2–1·4 l. pH: 7·0–8·5. Enzymes: Trypsin, chymotrypsin, and carboxypeptidases, secreted as precursors. Elastase, α-amylase, lipase, esterase, phospholipase A, RNAase, DNAase. Na$^+$ 91–142 mM, HCO$_3^-$ 30–82 mM, Cl$^-$ 35–97 mM.
Bile: *Hepatic* *Gall bladder*	Daily volume: 0·5–1·0 l. pH: 7·15. Water 97–98 per cent w/v. pH: 6·5–9·0. Water 82 per cent w/v. Na$^+$ 122–275, mean 191 mM. Bile pigments 0·003–1·78 per cent w/v. Bile acids 5·4–12·9 per cent w/v. Phospholipids (lecithin) 1·7–5·1 per cent w/v. Fatty acids 2·4 per cent w/v. Cholesterol 0·23 per cent w/v.
Intestinal juice	pH: 6·0 in duodenum, rising to 8·0 in lower ileum. Enzymes: numerous enzymes present in desquamated cells but of low activity in solution.

Carbohydrates

Digestion

The main carbohydrates in food are starch, glycogen, and the disaccharides sucrose and lactose; cellulose is also present but cannot be utilized by man, because there is no β-glucosidase activity in the intestine. *Starch* contains two different polysaccharides, *amylose* which is a coiled unbranched chain of α-D-glucose units joined by 1,4 links, and *amylopectin* which is a branched polysaccharide made up of chains of 1,4 linked α-D-glucose units, the chains being joined at one end to the next by α-1,6 links; *glycogen* resembles amylopectin but is more highly branched. Starch and glycogen are rapidly hydrolysed by the *salivary* and *pancreatic α-amylases*. Both these enzymes act in the lumen of the gut and the amylase activity of the intestinal contents is very high in the duodenum and jejunum. Since α-amylases bring about the random hydrolysis of non-terminal α-1,4 glucoside links in such polysaccharides, the final products of their action on starch and glycogen are principally α-maltose, α-maltotriose, and a branched pentasaccharide:

Glucose-1,4-glucose-1
|
6
|
Glucose-1,4-glucose-1,4-glucose

Further hydrolysis of these substances to glucose is brought about by two enzymes which occur in the epithelial cells of the small intestine: *maltase* (α-glucosidase), which hydrolyses maltose, maltotriose, and other α-1,4-oligosaccharides, and an enzyme *oligo-1,6-glucosidase*, which hydrolyses α-1,6 links in oligosaccharides. Also in the epithelium of the small intestine, there are the enzymes *sucrase* (β-fructofuranosidase) and *lactase* (β-galactosidase), which will hydrolyse sucrose and lactose respectively.

The activity of the latter two enzymes falls off dramatically in Protein–Calorie malnutrition (see Chapter 19). It should be emphasized that these enzymes are most active in the epithelial cells and that their activity in the so-called intestinal juice is due to the presence therein of desquamated epithelial cells. The enzyme activity in the lumen is very low and is insufficient to account for much hydrolysis before absorption. This contrasts with the α-amylase activity.

Absorption

Monosaccharides such as D-glucose and D-galactose, which have a hydroxyl group on C–2 and a methyl or hydroxymethyl group on C–5, both in the D-glucose configuration, are actively absorbed from the lumen of the small intestine. By this is meant that they can be absorbed against a concentration gradient and that their active absorption is dependent on the intact aerobic metabolism of the mucosa. Further, it is known that Na^+ ions are necessary for the active absorption process, and it is postulated that these sugars are transported into the cells as some form of complex with Na^+ ions and that in order to keep the concentration of these ions low in the cell, they are excreted back into the lumen by an energy-requiring ion-pump mechanism. Sugars without the structural characteristics given above, e.g., fructose, may be absorbed by simple diffusion. Active absorption of sugars can be inhibited specifically by low concentrations of *phlorrhizin*, probably because it combines with the sugar-transport process and blocks it. Inhibitors of metabolism, such as dinitrophenol, inhibit active absorption as does *ouabain* which is an inhibitor of Na^+ ion transport. It is noteworthy that the Na^+-linked active transport of sugars is not found in tissues other than the intestine (cf. amino acid transport, p. 157).

In man, glucose, galactose, lactose, and maltose are mainly absorbed in the middle of the small intestine (jejunum) and there is evidence that sucrose absorption occurs mainly in the upper jejunum. It should be remembered that though much carbohydrate is absorbed into the intestinal mucosa as di- and oligo-saccharides, it passes into the blood stream only as monosaccharides. When disaccharides are injected into the blood they are rapidly excreted in the urine.

Digestion and Absorption of Fats

Triglycerides

There is no lipase in the saliva nor in the gastric juice, though some regurgitated pancreatic lipase may be found in the gastric contents. It is in the stomach that ingested fats are warmed to body temperature which softens them and assists their subsequent emulsification.

In the duodenum the bile salts, which act as detergents because of their lipid-soluble (steroid) unit at one end and polar tail (glycine or taurine), and the bile phospholipids, assist in the emulsification of triglycerides and stabilize the emulsion formed. The fat particles in the emulsion have a diameter of 0·5 to 1·0 μm. This emulsification exposes a large surface area of glyceride to the action of *pancreatic lipase*, an enzyme which accumulates at any oil–water interface. Lipase hydrolyses triglyceride by acting specifically on the outer (1-) ester bonds with the production of free fatty acids (FFA) and a mixture of 1,2-diglycerides and 2-monoglycerides. The fatty acid esterified in the 2-position is only split off slowly and it is known that migration of the fatty acid residue from the 2- to the 1-position must take place before hydrolysis can occur.

As hydrolysis proceeds the principal products, FFA and 2-mono-glycerides, together with bile salts and bile phospholipids, are dispersed into smaller particles known as micelles, which have a diameter of 3–8 nm.

These micelles may also contain small amounts of di- and tri-glycerides and it is probable that fat enters the mucosal cells in this form. Once inside the cells this very finely dispersed lipid aggregates just below the terminal web, and further within the cells the droplets become larger and are surrounded by endoplasmic reticulum. During their passage through the mucosal cells, the products of lipase action are largely resynthesized into triglycerides. This resynthesis proceeds by two pathways, (i) the reacylation of 2-monoglycerides, and (ii) the acylation of *sn*-3-glycerophosphate (by the mechanism described on p. 193). The 'new' glycerol used in the second pathway is derived largely from glucose but may also come from absorbed glycerol.

Whatever the nature of the ingested fat, the triglycerides leaving the mucosal cells contain only long-chain fatty acids, because the short-chain acids are absorbed directly into the portal blood. A certain proportion of the long-chain fatty acids also escape esterification; these free fatty acids become bound to plasma proteins.

Before the droplets of resynthesized triglycerides can leave through the base of the mucosal cell they have to be coated with a layer of protein and phospholipid. These protein-coated triglyceride droplets are known as *chylomicrons*; they have a diameter of about $0.5\,\mu$m. Three specific polypeptides, probably synthesized in the intestinal wall, are found in chylomicrons, but the total protein content does not exceed 2 per cent of the total weight. The formation of chylomicrons, and hence the absorption of fat, may be reduced by inhibitors of protein synthesis such as puromycin. After passing through the base of the mucosal cell, the chylomicrons enter the extracellular fluid and then proceed through pores into the lymphatics.

Steroids and Fat-soluble Vitamins

The major fraction of secreted bile salts are readily re-absorbed, although in the jejunum rather than higher up in the intestine. The fraction not absorbed forms a large part of the total excretion of cholesterol from the body. Cholesterol itself is absorbed only to a limited extent and plant sterols, which are chemically very similar to cholesterol are about 95 per cent not absorbed. Vitamins A and D are absorbed with fat as is vitamin K. Carotenes are also well absorbed by man. It is probable that the absorption of these three fat-soluble vitamins as well as cholesterol and carotene is completely dependent on the fine dispersion of fat into micelles in the intestine. The absorption of all these substances is markedly reduced when micelle formation is depressed. Thus patients with obstructive jaundice, when the concentration of bile salts in the gut lumen is decreased, frequently suffer from deficiencies of vitamins A and D and particularly of vitamin K (the latter deficiency is more important as vitamin K is stored in the body only to a very limited extent).

While in the absence of bile the absorption of fat-soluble vitamins is negligible, some absorption of triglycerides may still occur. On the other hand, impaired splitting of dietary fat due to lipase deficiency, causes a failure in triglyceride absorption (pancreatic steatorrhoea) but not in the absorption of the fat-soluble vitamins. Presumably in these conditions sufficient fat-soluble vitamins can be leached from the oily into the micellar phase to allow adequate absorption.

Digestion of Proteins

The proteolytic enzymes of the digestive tract are of two kinds; the *endopeptidases* which attack polypeptide chains at peptide bonds away from the ends and the *exopeptidases* which attack the terminal peptide bonds. From experiments with synthetic substrates it was believed that these proteolytic enzymes have fairly rigid specificities, but recent work has shown that they can attack a wide variety of peptide bonds. It seems, however, that proteins are much more readily attacked if they have first been denatured by cooking or the action of gastric HCl.

Endopeptidases

Pepsin is secreted in the gastric juice as an inactive precursor, *pepsinogen*. Pepsinogen (mol. wt. 42 500) is converted into a pepsin-inhibitor complex and various small peptides spontaneously at pHs below 6·0. The reaction is very slow at pH 6·0 but almost instantaneous at pH 2·0. The reaction is autocatalysed by pepsin. At pHs below 5·4 the inhibitor (mol. wt. 3 100) dissociates from pepsin (mol. wt. 34 500), but the complex will re-form at pH s above 5·4. Both pepsin and the inhibitor are hydrolysed to peptides by pepsin itself.

Pepsin is a very acidic protein with an isoelectric point less than pH 1 and an optimum pH of 1·5–2·5 depending on the substrate. It is stable in acid solution but is rapidly inactivated in neutral or alkaline solutions; it has no prosthetic group. When acting on synthetic substrates it attacks peptide bonds between an acidic amino acid and an aromatic one ((Asp or Glu)-(Tyr or Phe)). From its action on insulin, however, it is known to attack, among others, links between two aromatic amino acids (Phe-Phe or Phe-Tyr) and links adjacent to leucine (Leu-Val and Tyr-Leu). Pepsin, like rennin, will coagulate milk by converting the phospho-protein caseinogen to casein which forms an insoluble complex with calcium.

The *gastric HCl* is secreted by the oxyntic cells of the gastric mucosa; at the same time the blood coming from the mucosa is made more alkaline. The cells contain carbonic anhydrase whose presence appears necessary for HCl secretion; about 80 per cent of the H^+ secretion into gastric juice may fail to appear in the presence of the carbonic anhydrase inhibitor acetazolimide (diamox). The juice then contains much more Na^+ ion than usual. It is probable that the hydrogen ion, secreted together with chloride, is derived from water and the remaining hydroxide ion reacts with CO_2 under the influence of carbonic anhydrase to form bicarbonate.

$$H_2O$$
$$OH^- + H^+ \longrightarrow$$
$$HCO_3^- \xleftarrow[\text{anhydrase}]{\text{Carbonic}}$$
$$CO_2$$
$$Na^+ \qquad\qquad Cl^- \longrightarrow$$

Oxyntic cell   *Gastric lumen*

It is very doubtful whether non-ruminant gastric juice contains *rennin*, an endopeptidase with a particularly powerful milk-clotting action. It has been reported that human milk is not clotted by calf rennin.

Trypsin is secreted by the pancreas as an inactive precursor, *trypsinogen*, which is activated by the enzyme *enterokinase*, secreted by the intestinal mucosa, and then autocatalytically by trypsin itself. A hexapeptide, $Val \cdot (Asp)_4 \cdot Lys$, is split off the *N*-terminal end of trypsinogen during activation by trypsin, leaving an *N*-terminal isoleucine.

Trypsin has no prosthetic group, it is relatively stable to heat in acid solution but less so in alkaline solution. The optimum pH is in the range 7–9 and it has a low Michaelis constant indicating that the substrate is firmly bound to the enzyme. It catalyses the hydrolysis of peptides, amides, and esters where a diaminomonocarboxylic amino acid (Lys or Arg) provides the carboxyl group. When acting on a natural substrate it also splits other bonds, e.g. Arg-Gly, Lys-Ala, Phe-Tyr, Lys-Tyr, Arg-Arg, Arg-Ala, and Tyr-Leu.

The *chymotrypsin* group of enzymes is all derived from a common precursor *chymotrypsinogen* secreted by the pancreas. The activation is initially brought about by trypsin to give an active chymotrypsin which may be converted to other chymotrypsins by autolysis; there is no change in the molecular weight in the first step, but there are differences in the subsequent products. The optimum pH is 7–8, the Michaelis constant is high, and the enzyme has no prosthetic group. These enzymes attack peptides or esters of a number of amino acids, but particularly non-polar ones (Leu-, Tyr-, Phe-, Met-, Trp-). See also Chapter 7, p. 135.

Elastase (pancreatopeptidase E), from the pancreas, solubilizes elastin by hydrolysis of peptide bonds, especially those adjacent to neutral amino acid residues.

These endopeptidases, pepsin, trypsin, the chymotrypsins, and elastase bring about the hydrolysis of large protein molecules to smaller peptide fragments. Their further hydrolysis then depends on the action of a number of exopeptidases and dipeptidases either secreted by the pancreas or in the intestinal mucosa.

Exopeptidases

A number of these enzymes, in contrast to the endopeptidases, require a metal ion as activator.

The two *carboxypeptidases* are secreted as precursor *procarboxypeptidases* which are activated by trypsin. Carboxypeptidase A contains firmly bound Zn^{2+} and hydrolyses off the carboxyl-terminal amino acid unless this is lysine or arginine. Carboxypeptidase B hydrolyses peptides with carboxy-terminal lysine or arginine. Neither will attack dipeptides.

Leucineaminopeptidase brings about the hydrolysis of amino-terminal residues from peptides, but not from dipeptides. In spite of the name, it is rather unspecific.

Dipeptidases. There are many specific dipeptidases.

Prolidase catalyses the hydrolysis of proline peptides which are mainly derived from the breakdown of collagen.

Carnosinase hydrolyses the dipeptides carnosine and anserine which occur in muscle.

Absorption

From experiments involving the feeding of isotopically labelled proteins, it is known that absorption is very rapid in man. The extent of hydrolysis of the food protein is low in the stomach, 10–15 per cent, but quickly reaches values of 50–60 per cent in the duodenum. In the duodenal contents the enzymes trypsin and chymotrypsin are present at concentrations of 200–800 μg per ml of fluid, within a short time of stimulation. These concentrations of enzymes are capable of rapidly hydrolysing food proteins to small peptides. Absorption of fed protein fragments takes place in the duodenum and the jejunum, most of it absorbed as di- and oligo-peptides. Although it is generally believed that all ingested protein is completely split to amino acids (either in the lumen or in the mucosa) before passing into the body fluids, it must be pointed out that this has never been rigorously demonstrated by quantitative experiments.

In *coeliac disease* a peptide of mol. wt. about 1 500 formed by enzymic hydrolysis from the gliadin fraction of cereal gluten has a toxic effect on the mucosal cells of the jejunum. Although it is far from established that this is the primary cause of the disease, it indicates that quite large peptides can penetrate the brush border.

The absorption of L-amino acids and some L-peptides is an active process and requires the metabolism of the mucosa to be intact. This absorption is interfered with by the presence of the D-isomers. The active absorption of amino acids in the intestine, as in other types of cell, takes place down a Na^+ gradient (see also p. 153).

In the new-born infant considerable amounts of colostrum proteins can be absorbed, particularly in the first 48 hours of life. This process is, however, very selective, greatly favouring globulins at the expense of albumins. It is believed that absorption of whole proteins ceases after the first 2 weeks of life.

Digestion of Nucleic Acids

Enzymes hydrolysing nucleic acids have two types of specificity, as shown in Fig. 8.1.

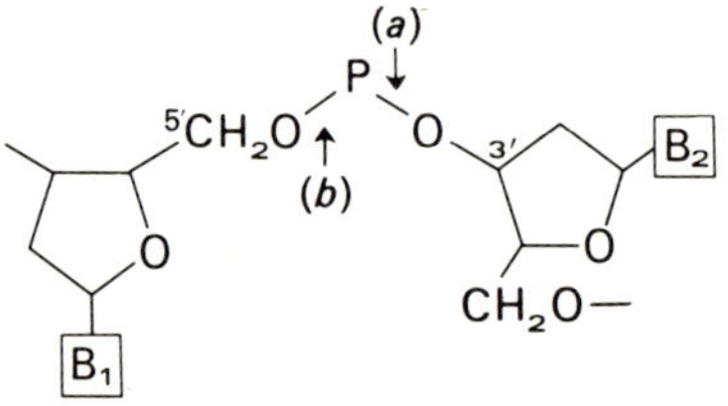

FIG. 8.1. Specificity of polynucleotidases. The a-type enzymes give rise to a free 5'-phosphate and the b-type to a free 3'-phosphate.

The enzymes can also be divided into endonucleases, which hydrolyse in the middle of a polynucleotide chain, and exonucleases which cut off the terminal nucleotide. The best-known enzymes in pancreatic juice are: *deoxyribonuclease I*, which is a type a endonuclease, hydrolysing between all pyrimidine–purine pairs; pancreatic *ribonuclease*, a type b enzyme, requiring that the a-linkage is to a pyrimidine (i.e. B_2 in Fig. 8.1); and *phosphodiesterase*, a b-type enzyme hydrolysing nucleotides successively from the 3' end of both DNA and RNA oligonucleotides. It requires a free 3' hydroxyl. Rather unspecific *nucleotidases* can account for breakdown to nucleosides, which are probably absorbed intact.

9 Metabolism of Carbohydrate

If a single dose of glucose, labelled with ^{14}C, is given to almost any living organism, it will be found that within a few hours, up to 90 per cent of the labelled carbon will have been excreted again in the form of $^{14}CO_2$. The remaining few per cent of ^{14}C will, however, be found in greater or lesser amounts, distributed among almost all the carbon-containing compounds of the organism, with a few known exceptions. If the organism could be re-examined a few weeks or months later it would be found that, except for traces of isotope in a few structural materials, all the ^{14}C had now completely disappeared.

This simple experiment, which could be repeated to a certain extent with a labelled fat, is a dramatic demonstration of two things: first, that glucose is very rapidly used as a source of energy for living things (since oxidation of complex organic compounds to CO_2 and H_2O is the way in which organisms obtain their energy); second, that there is a complex network of enzyme-catalysed chemical reactions which relates almost every biological compound to glucose. In fact, man could, in principle, build up every organic compound in his body, with the exception of some of the amino acids, some unsaturated fatty acids and all the vitamins, from this single compound. In practice, some of these transformations would be rather slow, and normal food is often an easier source of many of the compounds. Conversely, glucose can be formed from many, although not all, of the organic compounds in the body.

Our knowledge of the pathways by which these transformations occur grows more detailed each year, as the metabolic maps of intermediary metabolism indicate. It would be unreasonable to expect students to be familiar with such an increasing mass of detail. It is more important to understand the general principles, and the rather limited number of mechanisms that recur again and again. In this text an attempt is made to stress these general ideas and mechanisms, while retaining enough detail for the background of individual diseases of metabolism to be appreciated.

Only a few of the compounds classified as carbohydrates, i.e. compounds approximating to the empirical formula $(CH_2O)_n$, can be metabolized by man. In vertebrate biochemistry glucose, and a few of the polymers containing glucose, are the carbohydrates of main interest. D-glucose is the only sugar normally present in any quantity in blood. In man, but not in all animals, it is distributed almost equally between red cells and plasma. It is freely diffusible into extracellular fluid. The true blood glucose of normal subjects (determined with glucose oxidase) is 50–95 mg per 100 ml blood. There is a wide variation between individuals. The value obtained is rather more than 20 mg per 100 ml lower than that given by chemical methods which are not specific for glucose. During digestion the blood sugar level rises to 120–150 mg per 100 ml or even higher, and remains at this concentration for a length of time depending on the speed of digestion and on the carbohydrate content of the meal.

Two factors govern the metabolism of carbohydrate in the human body. The first is that except in unusual circumstances, further discussed in Chapter 10, the brain and nervous system oxidize *only* carbohydrate, i.e. glucose, for energy purposes. If the true glucose concentration in the blood falls to 25 mg per 100 ml for any length of time, the subject falls into a stupor. This is followed in 10–30 minutes by a coma which, if prolonged for more than 30 minutes, may be irreversible. It is therefore essential that glucose should always be present in the blood in concentrations of more than 25 mg per 100 ml, whatever the dietary régime. Very efficient physiological and hormonal mechanisms in fact keep the true blood sugar at 50–60 mg per 100 ml even in prolonged starvation (see also Chapter 18).

The second factor is the efficiency of the kidney in reabsorbing glucose from the glomerular filtrate. No glucose appears in urine, in the normal person, until the blood sugar reaches a threshold concentration of 150 mg per 100 ml (there are again wide individual variations). Above this concentration, increasing amounts are excreted; with a blood sugar level of 500 mg per 100 ml, up to 100 g may be lost in a day. The metabolism of the normal body is very carefully adjusted to prevent this loss of a substance which is a source of readily available energy. It can be calculated that the absorption of glucose from the alimentary canal during the digestion of the carbohydrate in an ordinary meal might raise the blood sugar level to 300 mg per 100 ml, if there were no possibility of storing some of it. This is after allowing for the diffusion of the absorbed glucose from the blood to the extracellular fluid space, and for the oxidation of glucose to the exclusion of all other energy-supplying compounds, up to the limit set by the body's energy needs. Such a calculation suggests that an efficient mechanism of carbohydrate storage must come into play after digestion in order to prevent loss of sugar in the urine. This is indeed the case.

The reactions by which the blood sugar is maintained at a normal level both in starvation and during absorption, will be discussed in outline in this chapter. Since both these mechanisms are subject to very precise hormonal control, the functioning of the normal complete organism will be dealt with in Chapter 18.

Glucose Catabolism

Complete oxidation of glucose requires 6 molecules of oxygen for each molecule of glucose:

$$C_6H_{12}O_6 + 6O_2 \rightarrow 6CO_2 + 6H_2O$$

The respiratory quotient for this process is 1·0. A complex series of enzymic reactions occurs before most of the oxidation takes place. These enzymic reactions lead to the production of *pyruvic acid*. The overall equation is:

$$C_6H_{12}O_6 + O_2 \rightarrow 2CH_3 \cdot CO \cdot COOH + 2H_2O$$

The reactions are shown in detail in Fig. 9.1, frequently known as the Embden–Meyerhof scheme.

It must be stressed that the production of pyruvic acid is *not* an anaerobic process. Except in special circumstances, pyruvic acid is decarboxylated and the resulting '2-carbon-fragment'—or acetylcoenzyme A—is oxidized by the reactions of the tricarboxylic acid cycle, described in detail in Chapter 12. The following notes about the reaction sequence shown in Fig. 9.1 refer to the main reactions of interest.

1. The intermediates are all *phosphate esters*, not simple sugars or sugar derivatives.

2. Phosphoryl transfer from ATP (the only effective donor) to glucose is catalysed by the enzyme *hexokinase. Phosphofructokinase* catalyses phosphoryl transfer from ATP to fructose 6-phosphate.

There are several hexokinases in animal tissues. They resemble one another in being active towards several hexoses, and in having very low K_ms towards both $MgATP^{2-}$ and sugars. In vivo the enzymes are always effectively saturated with the former substrate, but in many tissues the concentration of free glucose is low, so that the rate of phosphorylation of glucose is first order with respect to this substrate. In effect, in these tissues the rate of glucose phosphorylation is determined by the rate of membrane transport. In liver, however, the intracellular glucose concentration approximates to that in plasma, so that hexokinase activity must be zero order with respect to this substrate as well as to MgATP. Hepatocytes contain rather little hexokinase, but considerable quantities of a more specific *glucokinase* with a much higher K_m for glucose (approximately 10 mM), so that the overall rate of glucose phosphorylation in liver does vary with the plasma glucose concentration.

3. The splitting of the hexose molecule into two trioses, catalysed by the enzyme *aldolase*, is perfectly reversible and indeed, except at very low

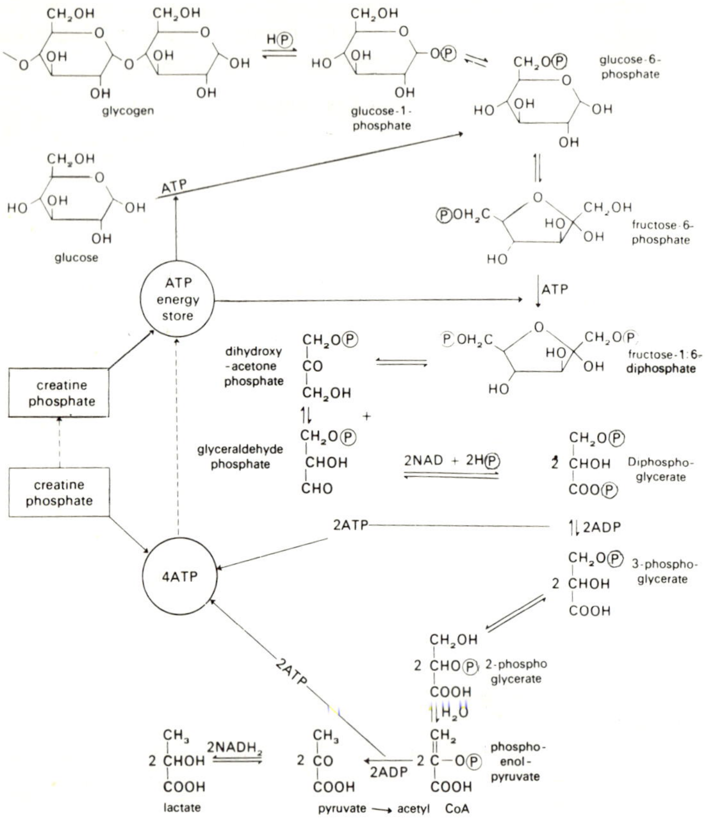

FIG. 9.1. The Embden–Meyerhof scheme of glucose metabolism.

triose phosphate concentrations, the equilibrium lies towards fructose diphosphate. There are several aldolases in different tissues with different affinities for the three substrates, but in accordance with the principles discussed on p. 120, these differences cannot affect the position of equilibrium. There are two products of this reaction, dihydroxyacetone phosphate and glyceraldehyde phosphate. Only the latter is metabolized in this scheme, but dihydroxyacetone phosphate does not accumulate because it is converted into the other triose by the action of the enzyme *triose phosphate isomerase*.

4. Phospho-glyceraldehyde is oxidized to phosphoglyceric acid in two stages. The reaction catalysed by the first enzyme, *glyceraldehyde phosphate dehydrogenase*, may be represented as follows:

$$CH_2O \textcircled{P} \cdot CHOH \cdot CHO + HS \cdot Enz \cdot NAD$$

$$\underset{|}{\overset{OH}{}}$$

$$\rightleftharpoons CH_2O \textcircled{P} \cdot CHOH \cdot \overset{OH}{\underset{|}{CH}} \cdot S \cdot Enz \cdot NAD \tag{i}$$

$$\rightleftharpoons CH_2O \textcircled{P} \cdot CHOH \cdot CO \cdot S \cdot Enz \cdot NADH_2 \tag{ii}$$

$$+ NAD \rightleftharpoons CH_2O \textcircled{P} \cdot CHOH \cdot CO \cdot S \cdot Enz \cdot NAD + NADH_2 \tag{iii}$$

$$+ P_i \rightleftharpoons CH_2O \textcircled{P} \cdot CHOH \cdot COO \textcircled{P} + HS \cdot Enz \cdot NAD \tag{iv}$$

The symbol $\textcircled{P}$ refers to the phosphoryl group i.e. $-PO_3H_2$.

The oxidation of the adduct (i) to the thio-ester (ii) takes place with reduction of NAD which is so firmly bound to the enzyme that its reoxidation is thought to be at the expense of free NAD (iii). The thio-ester is cleaved by phosphorolysis (iv), leading to the production of the mixed anhydride 1:3-diphosphoglyceric acid.

$\Delta G_0'$ for the straightforward oxidation of the aldehyde to the acid is -43 kilojoules per mole (kJ/mol); for the reaction outlined above it is $+6.3$ kJ/mol. This is the classic example of *energy trapping* by *substrate level phosphorylation*.

In a second reaction catalysed by *phosphoglycerate kinase* the following reaction takes place:

$$CH_2O \textcircled{P} \cdot CHOH \cdot COO \textcircled{P} + ADP$$
$$\rightleftharpoons CH_2O \textcircled{P} \cdot CHOH \cdot COOH + ATP$$

Thus the trapped energy is transferred to the more general energy link, ATP. $\Delta G_0'$ for the overall reaction is -12.5 kJ/mol.

The reduced NAD formed in this complex reaction can be reoxidized by two mechanisms. It can be assumed that the oxygen supply in the normal organism is adequate for reoxidation through a hydrogen transport system (see Chapter 12).

$$NADH_2 + \ldots \tfrac{1}{2}O_2 \rightarrow NAD + \ldots H_2O$$

If the oxygen supply is temporarily inadequate, $NADH_2$ is reoxidized, using pyruvate as hydrogen acceptor:

$$NADH_2 + CH_3 \cdot CO \cdot COOH \rightleftharpoons NAD + CH_3 \cdot CHOH \cdot COOH$$

The whole system of glucose breakdown can then work anaerobically, and is known in these conditions as *anaerobic glycolysis*. Note that *lactate*, not *pyruvate*, is the final product.

5. *Phospho-enol pyruvic acid* is another so-called energy-rich compound, that is to say that the phosphate ester of enol pyruvic acid is very unstable, and although the transfer of the phosphate group to ADP is theoretically reversible, the equilibrium of the reaction

$$ADP + PEP \rightleftharpoons Pyr + ATP$$

is so far to the right as to make the synthesis of PEP by this reaction very unlikely. This fact is important in considering the synthesis of glucose from small molecules (see later in this chapter).

6. For each molecule of glucose which is catabolized, two molecules of pyruvate are formed, and 4 molecules of ATP are synthesized, 2 in the oxidation of two molecules of glyceraldehyde phosphate and 2 in the transfer of two phosphate groups from phospho-enolpyruvate to ADP. Two molecules of ATP were used in the preliminary phosphorylation of glucose and of fructose-6-phosphate, so that there is a net gain of 2 ATP in the conversion of one glucose molecule to two molecules of pyruvate.

ATP, or adenosine triphosphate, appears to be the common 'energy currency' of most cells, and only that part of the total energy set free in the catabolism of organic compounds which is 'trapped' by the synthesis of ATP is available to the cell for energy-requiring processes, except for the 'reducing power' required in, for example, fatty acid synthesis. These topics are discussed more fully in Chapters 7 and 12.

In muscle cells, but only to a very small extent in other tissues, a substance called *creatine phosphate* is found. This can act as an immediate source of useful energy in the first few seconds of muscular contraction, before the whole process of glucose (or fatty acid) catabolism can be speeded up to produce the extra energy required. An active enzyme in muscle catalyses the reaction

$$\text{creatine}\circledP + ADP \rightleftharpoons \text{creatine} + ATP$$

going from left to right. In resting muscle, creatine phosphate is formed by reversal of this reaction, as ATP is resynthesized in reactions coupled to catabolic processes. The formula of creatine is given on p. 227.

7. In erythrocytes, the most abundant phosphate ester is 2,3-diphospho-glycerate. This is formed from 1,3-diphosphoglycerate by a mutase that in effect converts a high-energy phosphate compound into a low-energy one. 2,3-Diphosphoglycerate can be converted into 3-phosphoglycerate and P_i by a phosphatase. This is a shunt that reduces the net ATP yield of the glycolytic pathway to zero, and appears to come into operation when the ATP concentration in the cells is high. For diphosphoglycerate and oxyhaemoglobin see p. 112.

Pyruvic Acid

This is a very important compound in carbohydrate metabolism because it can undergo many reactions, and thus the carbon atoms of glucose can be directed into one of several channels.

1. It can be *decarboxylated*. In animals this is always an oxidative reaction, whose overall equation is:

$$CH_3 \cdot CO \cdot COOH + HS \cdot CoA + NAD \rightarrow$$
$$CH_3 \cdot CO \cdot S \cdot CoA + CO_2 + NADH_2$$

It is catalysed by a multi-enzyme system, *pyruvate oxidase*, partial mechanisms for which have been discussed on p. 144 and p. 145. The complete reaction may be represented as:

$$Enz_1 - TPP + CH_3 \cdot CO \cdot COOH \rightarrow$$
$$Enz_1 - TPP \cdot CHOH \cdot CH_3 + CO_2 \quad (i)$$

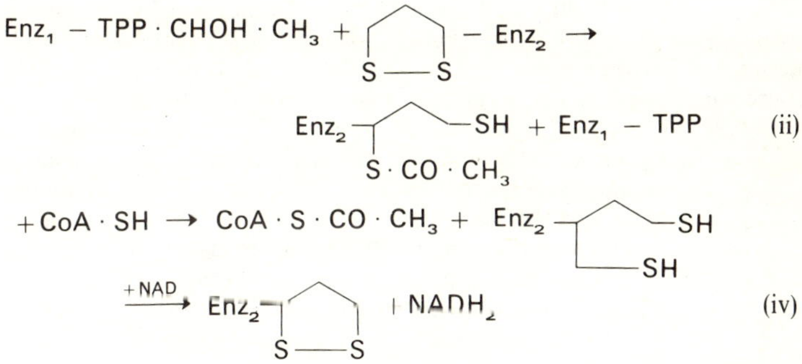

$$+ CoA \cdot SH \rightarrow CoA \cdot S \cdot CO \cdot CH_3 + Enz_2$$

The enzyme catalysing stage (iv) is a flavoprotein.

In yeasts, by contrast, the adduct formed after decarboxylation (i) breaks down to free enzyme and acetaldehyde:

$$CH_3 \cdot CO \cdot COOH \rightarrow CH_3 \cdot CHO + CO_2$$

which may be reduced to ethanol.

The acetyl-coenzyme A formed in this reaction can, with oxaloacetic acid, form citric acid and so be oxidized in the tricarboxylic acid cycle (see Chapter 12), or it can be synthesized into long-chain fatty acids.

2. Pyruvic acid can be carboxylated by two reactions:

(*a*)   pyruvate + ATP + CO_2 $\rightarrow$ oxaloacetate + ADP + P_i.

This reaction, catalysed by *pyruvate carboxylase*, a biotin-containing enzyme, is essentially irreversible. The enzyme has an absolute requirement for acetyl-CoA as a cofactor.

(b)   pyruvate + $NADPH_2$ + CO_2 $\rightleftharpoons$ malate + NADP

The *malic enzyme* is cytoplasmic, and may be used to oxidize malate in the cytoplasm, perhaps for the generation of reduced NADP.

In addition, the reactions leading to carboxylation of PEP, which are described below, cannot be left out of account in considering the disposition of pyruvate.

3. Pyruvate can be transaminated by glutamic acid to form alanine:

$$CH_3 \cdot CO \cdot COOH + HOOC \cdot (CH_2)_2 \cdot CH(NH_2) \cdot COOH \rightleftharpoons$$
$$CH_3 \cdot CH(NH_2) \cdot COOH + HOOC \cdot (CH_2)_2 \cdot CO \cdot COOH$$

In muscle, this reaction leads to a net production of alanine which leaves the tissue and is taken up by the liver (the *alanine cycle*, see Chapter 11).

4. It can be reduced to *lactic acid*, by an $NADH_2$— requiring dehydrogenase. The cytoplasm of most cells contains much lactic dehydrogenase activity, and the [lactate]/[pyruvate] ratio in tissues usually reflects the cytoplasmic $[NAD]/[NADH_2]$ ratio. However, as pyruvate is not freely permeable between many tissues and the blood, the blood [lactate]/[pyruvate] ratio does not necessarily reflect the tissue redox ratios.

The importance of glycolysis, i.e. the breakdown of carbohydrate (strictly glycogen) to lactic acid, can be exaggerated, although sudden muscular movement will sharply increase the lactic acid level of venous blood. It is often said that anaerobic glycolysis is inefficient as a means of producing useful energy (i.e. as ATP). This is not strictly true; the free energy released in the change glucose $\rightarrow$ 2 lactate is only 210 kJ per mole, and of this about 40 per cent is trapped as ATP (compared with the 65 per cent for the complete oxidation of glucose). The real inefficiency lies in the *weight* of glucose or glycogen which must be broken down to lactic acid to provide a given number of ATP molecules. The energy contained in the lactic acid (about 2 660 kJ for 2 moles of lactate oxidized to CO_2 and H_2O) is not usually lost to the organism. Small amounts of lactic acid may be found in urine, but it is rare for significant amounts of carbohydrate to be lost in this way.

Skeletal muscle does not appear to re-utilize lactate readily, but it is rapidly oxidized by cardiac muscle. Liver also rapidly removes lactate from the blood stream; in this organ most of the lactate is probably resynthesized to hexose units. There is no evidence that mammalian muscle, unlike frog muscle, can re-synthesize glycogen from lactate.

Several iso-enzymes of lactic dehydrogenase exist (p. 139) and are characteristic for various tissues. The 'heart' enzyme $LDH(H_4)$ is strongly inhibited by pyruvate, and it has been suggested that this form favours the oxidation of lactate. However, as already pointed out in connection with aldolase iso-enzymes (p. 162), catalysts cannot alter the position of equilibrium. An enzyme to which an inhibitor (in this case, pyruvate) is bound, is unavailable for catalysis in *either* direction.

The Hexose Shunt

There is an alternative metabolic pathway of glucose catabolism which does not involve the Embden–Meyerhof sequence. The overall importance of this path in mammals is probably small, but in certain tissues, notably liver, a considerable fraction of the glucose used may be metabolized in this way. One of the intermediates of the alternative pathway is ribose 5-phosphate, and it is very probable that the ribose required for the synthesis of NAD, NADP, coenzyme A, ATP, and all the nucleotides forming part of nucleic acids is made either from glucose or from glyceraldehyde phosphate, the terminal compound of the pathway. In most tissues, then, the 'hexose shunt' may work as a source of chemicals rather than of energy. This has not been definitely established; it is not easy to estimate the amount of glucose catabolized in this way. It is probably also important that operation of the hexose shunt provides quantities of reduced NADP, which are needed for fat synthesis (see Chapter 10).

Fig. 9.2 shows the main relationships of the reaction sequence. All the enzymic reactions are reversible, except for the decarboxylation of 6-phosphogluconate.

The first stage is the 'direct' oxidation of glucose-6-℗ to 6-phosphogluconate (the oxidation of the aldehyde group on C–1). Next, 6-phosphogluconate is oxidized and decarboxylated to the ketopentose, ribulose-5-℗. Both these reactions require NADP. The next series of reactions, whose validity has been established by following the movement of isotopic labels, depends in essence on the activities of two transferring enzymes. *Transketolase* (*TK*) transfers the group

$$-CO \cdot CH_2OH$$

from a suitable donor, usually a keto-sugar, to an aldose acceptor. This '2-carbon fragment' is not free during the transfer, but is attached to the prosthetic group of the enzyme (thiamine pyrophosphate). *Transaldolase* (*TA*) catalyses the transfer of $-CHOH \cdot CO \cdot CH_2OH$ from a donor to glyceraldehyde-3-℗ (Glyc-3-℗). The aldolase of the Embden–Meyerhof scheme is an enzyme of this type.

The following are the reactions which then ensue:

1. Ribulose-5-℗ is isomerized to a mixture of the ketopentose xylulose-5-℗, and ribose-5-℗. The latter is the starting-point of nucleotide synthesis (Chapter 5), so that some of the original glucose-6-℗ may be drained off here. Xylulose-5-℗ is the source of the '2-carbon fragment' in the transketolase reactions which follow.

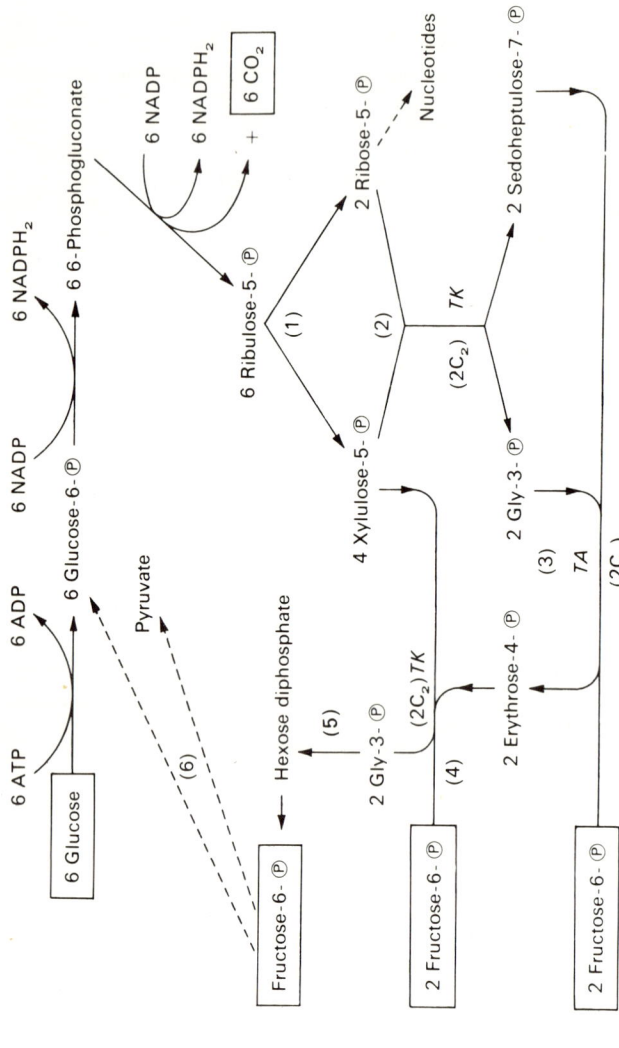

Fig. 9.2. The alternative metabolic pathway of glucose. Gly-3-℗—glyceraldehyde 3-phosphate. TA—Transaldoase. TK—Transketolase. C_2—Two-carbon fragment transferred by TK. C_3—Three-carbon fragment transferred by TA. (Starting materials and end-products are enclosed in panels.)

2. $-CO \cdot CH_2OH$ moves from xylulose-5-$\circledP$ to ribose-5-$\circledP$, forming the 7-carbon sugar sedoheptulose-7-$\circledP$. Glyc-3-$\circledP$ is left. (*TK*.)

3. $-CHOH \cdot CO \cdot CH_2OH$ moves from sedoheptulose-7-$\circledP$ to Glyc-3-$\circledP$, forming fructose-6-$\circledP$, and leaving the 4-carbon sugar erythrose-4-$\circledP$. (*TA*.)

4. $-CO \cdot CH_2OH$ moves from xylulose-5-$\circledP$ to erythrose-4-$\circledP$, forming fructose-6-$\circledP$, and leaving Glyc-3-$\circledP$. (*TK*.)

5. Glyc-3-$\circledP$ by partial isomerization to dihydroxyacetone phosphate can be converted to hexose diphosphate, and so to fructose-6-$\circledP$. Alternatively, since the Embden–Meyerhof enzymes are always present in the cell, Glyc-3-$\circledP$ can be oxidized and converted into pyruvic acid.

6. Fructose-6-$\circledP$ may be isomerized to glucose-6-$\circledP$, which can go through the cycle of reactions again. Alternatively, for the reasons given in (5), it may be degraded to pyruvic acid.

If the scheme of reactions is made to start with 6 molecules of glucose-6-$\circledP$, as in Fig. 9.2, the net result of the cyclical process is

$$6 \text{ G-6-}\circledP + 12 \text{ NADP} + 7 \text{ H}_2\text{O} \rightarrow$$

$$6 \text{ CO}_2 + 12 \text{ NADPH}_2 + 5 \text{ G-6-}\circledP + \text{H}_3\text{PO}_4$$

These enzymes also occur in green plants and are, with one to two more, thought to be very important in the trapping of atmospheric CO_2 which occurs in photosynthesis.

Note that in this formulation all the CO_2 has come from C–1 of 6 glucose molecules. Attempts to measure the quantitative importance of this pathway are based on this fact: the amount of $^{14}CO_2$ coming from glucose labelled at C–1 is measured. It is more difficult to measure the total amount of glucose being catabolized: it is usual to use glucose labelled at C–6, but not all the label will appear in CO_2 since there may be considerable synthesis of fat and other compounds.

Gluconeogenesis (Glucose Synthesis)

The reactions by which hexoses are built up from smaller molecules are mainly reversals of those in the Embden–Meyerhof sequence. As already pointed out, three of the reactions in this sequence are practically irreversible, and other enzyme-catalysed reactions exist to bypass them. These are:

1. The formation of phospho-enol pyruvic acid from oxaloacetic acid

$$HOOC \cdot CH_2 \cdot CO \cdot COOH + GTP \rightleftharpoons$$

$$CH_2{=}\underset{\underset{O\,\circledP}{|}}{C} \cdot COOH + CO_2 + GDP$$

where GTP stands for *guanosine* triphosphate.

By this reaction any compound which can be transferred into the tricarboxylic acid cycle can also be converted into phospho-enol pyruvic acid. This applies particularly to several amino acids: glutamic acid which can form α-oxoglutaric acid (proline and arginine, among others, can be converted into glutamic acid), aspartic acid, which forms oxaloacetic acid, and phenylalanine and tyrosine which are partly converted to fumaric acid (see Fig. 11.6).

Fatty acids, although they form acetyl residues which are incorporated into the tricarboxylic acid cycle, do not, as far as is known, give rise to any *net* gain of glucose. This is because in order to transform an acetyl residue into oxaloacetic acid, it must first be condensed to form citric acid, which is then oxidized with the loss of 2 carbon atoms as CO_2:

$$\text{Fat} \rightarrow C_2 + C_4 \rightarrow C_6 \rightarrow 2CO_2 + C_4$$

The glycerol of neutral fat can, however, give rise to a net gain of carbohydrate.

Pyruvic acid itself can be converted into oxaloacetic acid by the ATP-requiring carboxylase described above, or via the malic enzyme with subsequent re-oxidation. Thus this acid, or any compound which can be transformed into it, such as lactic acid, alanine, serine, or cysteine, can give rise to hexose molecules.

2. The conversion of fructose diphosphate and glucose-6-phosphate into fructose-6-phosphate and glucose respectively is carried out by two phosphatases, e.g.:

$$\text{Fructose 1,6-di} \, \textcircled{P} \rightarrow \text{Fructose 6-} \textcircled{P} + \textcircled{P}$$

These enzyme reactions are again irreversible. The phosphatases, unlike those of blood and bone, are specific, and only kidney besides liver contains a significant amount of a specific *glucose-6-phosphatase*. Liver can form G-6-P from many compounds by the pathways outlined above, and can thus secrete free glucose into the blood when necessary, but glucose is not secreted by any other organ (except for an unimportant amount from the kidney). Hexose phosphates formed from small molecules in tissues other than liver can be converted only into glycogen, as described below.

Gluconeogenesis is clearly an endergonic (energy-requiring) process.

Glycogen

This is a branched chain polysaccharide (see p. 25) which occurs in most mammalian cells, but particularly in liver and muscle. It functions as a store of glucose which is especially convenient from the osmotic point of view. Quantitatively, fat is more important than glycogen for the storage of those carbon atoms of ingested glucose which are not oxidized immediately after absorption. Nevertheless, glycogen is an effective store of immediately available energy. In muscles the amounts are rarely more than 2 per cent and often under 1 per cent of the wet weight and there is some evidence that it is not very readily broken down unless the muscle is worked to exhaustion. Liver glycogen is much more labile. Its concentration can vary from 5, or on special diets even 10 per cent, of the total liver weight immediately after a meal, down to 0·1 per cent after 24 hours fasting.

In the stomach, and in the intestinal lumen, starch and glycogen are broken down by hydrolytic amylases some of which attack internal glycosidic bonds. In cells on the other hand, glycogen is broken down by a *phosphorylase* (*not* a phosphatase) which catalyses the reaction:

$$\text{Glycogen} + n\,H_3PO_4 \rightleftharpoons n\,\text{Glucose 1-}Ⓟ$$

Only terminal residues attacked by α-1:4 bonds are attacked, and these only if they are not too near a branch point (see below). The isomerization of the product, glucose-1-phosphate, to glucose-6-phosphate is catalysed by a *phosphoglucomutase*, and by this means glycogen is brought into the main stream of glucose metabolism

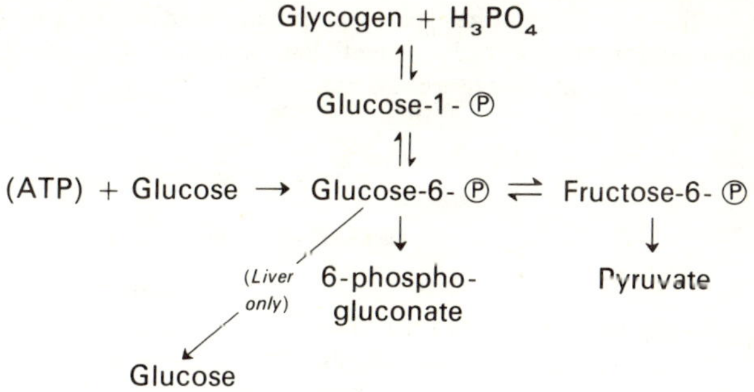

$$\text{Glycogen} + H_3PO_4$$
$$\updownarrow$$
$$\text{Glucose-1-}Ⓟ$$
$$\updownarrow$$

(ATP) + Glucose → Glucose-6-Ⓟ ⇌ Fructose-6-Ⓟ

(Liver only)	6-phospho-gluconate	Pyruvate

Glucose

This scheme indicates the importance of glucose-6-phosphate as the central compound from which all pathways of glucose metabolism begin.

The reaction catalysed by phosphorylase is reversible, but the cellular concentrations of the reactants glycogen, P_i and G-1-P always favour breakdown. It is difficult to see how net synthesis of glycogen can be forced to occur in this freely reversible system (cf. synthesis of fatty acids, Chapter 10). However, it can be shown that enzymes occur in cells, especially muscle and liver, which catalyse the following reactions.

$$\text{Glucose-1-}Ⓟ + UTP \rightarrow UDPG + Ⓟ\text{-}Ⓟ \qquad (1)$$

UDPG is the coenzyme *uridine diphosphate-glucose* (formula on p. 149)

$$\text{UDPG} + [n \text{ Glucose}] \rightarrow \text{UDP} + [(n + 1) \text{ Glucose}] \qquad (2)$$
$$\qquad\qquad \textit{glycogen} \qquad\qquad\qquad\qquad\qquad \textit{glycogen}$$

$$\text{UDP} + \text{ATP} \rightleftharpoons \text{UTP} + \text{ADP} \qquad (3)$$

Reactions (1) and (2) are not reversible partly because pyrophosphate, one product of reaction (1), is promptly hydrolysed to inorganic phosphate by a *pyrophosphatase*. The enzyme catalysing reaction (2) is called *glycogen synthetase*.

It is thought that glycogen synthesis takes place by this pathway, and breakdown via phosphorylase action. This belief has been strengthened by the discovery of rare inborn errors of metabolism in which the tissues contain glycogen but no detectable phosphorylase.

Both phosphorylase and the UDPG-transferring enzyme can synthesize glycogen only by adding glucose residues to pre-existing glycogen molecules, or primers, and both enzymes also catalyse only the synthesis of α-1,4 bonds between the glucose residues. Glycogen proper, however, has a large number of *branch points*, at which a chain of glucose residues is attached by the reducing end to C–6 of another residue (see p. 25). This α-1,6 link must be synthesized by a special *branching enzyme* which transfers a section of an already formed chain from a C–4 of one glucose residue to a C–6 on another (the α-1,6 link is shown by $\downarrow$):

$$\begin{array}{c} \text{G–G–G–G–G–G–G} \\ | \qquad\qquad\qquad\qquad \downarrow \\ \text{G–G–G–G–G–G–G–|–G–G–G–G–G} \rightarrow \text{G–G–G–G–G} \\ | \end{array}$$

The optimum length of chain for transferring is 7 glucose units long.

When breaking down glycogen, phosphorylase can only split α-1,4 links, furthermore it is unable to act between branch points so its action ends with the production of a *limit dextrin* in which short chains of α-1,4-linked glucose residues are left attached to the branch points. Debranching is brought about by two reactions; firstly the transference of a piece of α-1,4 chain from one side chain to another chain, leaving a single glucose residue linked α-1,6; secondly hydrolysis of the α-1,6-linked 'stump' to give a molecule of free glucose. (See p. 173). Phosphorylase can now attack the remaining α-1,4-linked chain. The two reactions can be separated by the choice of suitable substrates, but they both appear to be catalysed by the same protein (*debranching enzyme*), which is possibly a multi-enzyme complex.

Lysosomes contain an *acid glucosidase* (with optimum activity at pH 4) which appears to hydrolyse only glycogen which has in some unknown manner become absorbed into the lysosomes. It will hydrolyse both

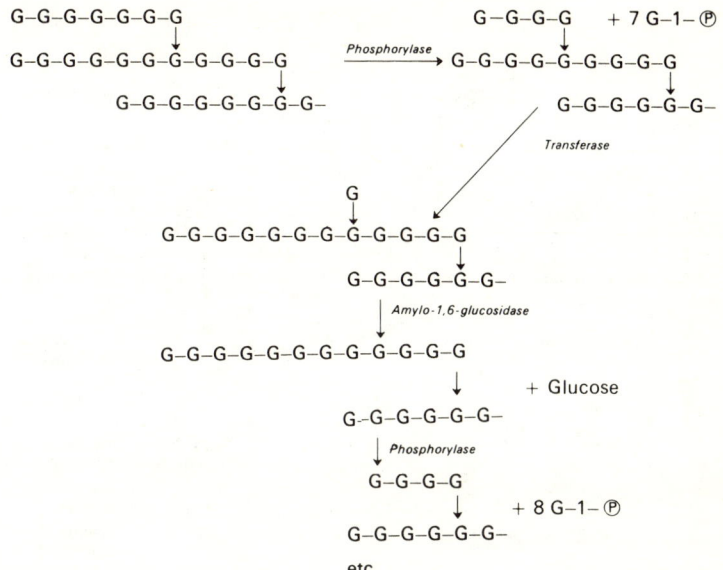

terminal α-1,4 and α-1,6 bonds, and is therefore capable of hydrolysing glycogen completely to free glucose. It is not known, however, how important this process is in normal glycogen degradation. The enzyme was only discovered as a result of studies on some of the glycogen storage diseases described in the next paragraph.

Glycogen Storage Diseases

These are inborn errors of metabolism in which one of the enzymes of glycogen metabolism is missing or inactive.

Table 9.1 shows some of the most common glycogenoses.

Diseases of glycogen metabolism are not common; there are a number of sub-types not shown in the table which are very rare indeed. Nevertheless, investigation of these diseases has advanced and modified our ideas about the importance of glucose metabolism and its non-hormonal controls very considerably. Inability to metabolize glycogen is not fatal in itself, although there may be physiological handicaps, particularly those due to organ enlargement. The most fatal of these diseases is the type II glycogenosis; in the most severe form death usually occurs before 1 year. For some reason the heart is most susceptible to acid glycosidase deficiency, and cardiac lysosomes become strikingly enlarged. On the other hand,

Table 9.1
Glycogenoses

Cori type	Organs affected	Hypo-glycaemia	Glycogen structure	Glycogen content per cent	Enzyme defect
I	Liver, kidney	+	Normal	5–15 (L)	G-6-P-ase absent
II	Generalized	–	Normal	5–15 (L) 5–15 (M)	Lysosomal acid glucosidase absent
III	Liver, muscle	+	Abnormal, excessive branching	10–20 (L) 2–6 (M)	Debranching enzyme absent
IV	Generalized	–	Abnormal, very little branching	3 (L)	No branching enzyme
V	Muscle	–	Normal	2–5 (M)	Phosphorylase absent
VI	Liver	+	Normal	5–20 (L)	Phosphorylase 50% of normal
VII	Muscle	–	Normal		Phospho-fructokinase absent
VIII	Liver, muscle	+ +	Normal	0·5 (L)	Glycogen syn-thetase absent
IX	Liver		Normal		Phosphory-lase kinase reduced

(L) = liver; (M) = muscle

in type V glycogenosis (McArdle's disease) it is established that muscle glycogen cannot break down during contraction and in adults there is pain and weakness in the muscles to an extent depending on the severity of the exercise. However, up to the age of puberty this distress is absent. In fact some children later found to have type V glycogenosis have been satisfactory members of football teams.

Especially in type I disease, the blood lactate and pyruvate levels may be very high, particularly when adrenaline secretion is likely to have occurred. In the hypoglycaemic states, which may occur with some forms of the disease, ketosis is common.

One family is now known in which several members have no phospho-fructokinase in their muscles. Since enzymes of the hexose shunt are almost absent in muscle, this presumably means that glucose cannot be catabolized at all in these muscles. The clinical picture is very similar to that of type V glycogenosis: marked weakness and stiffness after vigorous or prolonged exertion, but no other symptoms.

Uridine Diphosphate Glucose

UDPG is an important substrate. It can be transformed into UDP-Galactose (p. 177), and can also be dehydrogenated by 2 NAD to give UDP-Glucuronic acid, important in mucopolysaccharide synthesis and in detoxication (Chapter 14). Thus we have:

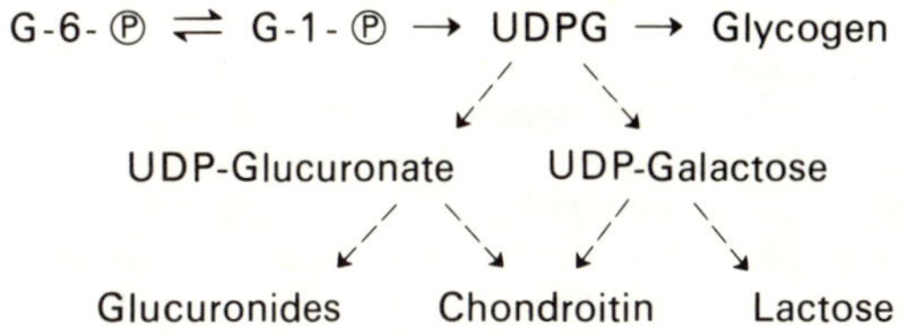

$$\text{G-6-} \textcircled{P} \rightleftharpoons \text{G-1-} \textcircled{P} \rightarrow \text{UDPG} \rightarrow \text{Glycogen}$$

UDP-Glucuronate   UDP-Galactose

Glucuronides   Chondroitin   Lactose

The Direct Oxidation of Glucose

An NAD-linked enzyme catalysing the direct oxidation of glucose to gluconic acid, $CH_2OH \cdot (CHOH)_4 \cdot COOH$, is found only in small amounts in liver, but several micro-organisms, notably *Penicillium notatum*, contain large quantities of an enzyme, *glucose oxidase*, which catalyses the reaction below:

$$CH_2OH \cdot (CHOH)_4 \cdot CHO + H_2O + O_2 \rightarrow$$
$$CH_2OH \cdot (CHOH)_4 \cdot COOH + H_2O_2$$

This enzyme is widely used for the qualitative detection and quantitative analysis of glucose in biological specimens. If peroxidase is added to the system, the hydrogen peroxide may be made to oxidize a suitable leuco-dye to a coloured product. A *galactose oxidase* from micro-organisms has been purified and is used for the detection of galactosaemia (p. 176).

The Metabolism of Other Sugars

Pentoses

Ribose derivatives, i.e. the nucleotides and nucleic acids, are present in all cells. They do not usually originate from ingested ribose, which is a variable constituent of foodstuffs and is rather slowly phosphorylated in vivo, but from glucose by way of the pentose phosphate pathway, or by synthesis from smaller molecules (see p. 167). The deoxyribose of DNA is formed by the reduction of ribose already in nucleosides or nucleotides.

Free ribose, and also xylose and arabinose, are found in small quantities in blood from time to time. They come from the intestine during the digestion of plant polysaccharides, particularly those of fruit. They are rather slowly metabolized, since pentosuria is quite common in people who eat a great deal of fruit. *Essential pentosuria* is a harmless inborn error of metabolism in which about 1 g of xylulose is found each day in the urine. Its stereochemical configuration indicates that it cannot have originated in the pentose phosphate pathway; it is probably excreted as the result of an enzymic defect in the metabolism of glucuronic acid.

Fructose

Considerable quantities of this sugar are ingested each day, chiefly combined in sucrose. It is normally converted into glucose or glycogen in the liver, and perhaps also to some extent in the small intestine. Fructose (like galactose) is phosphorylated in the C–1 position, and the resulting fructose 1-P is split by a specific aldolase to a mixture of dihydroxyacetone phosphate and glyceraldehyde. The latter can be reduced to glycerol; more usually, however, it appears to be oxidized to glycerate, which is then phosphorylated. If this latter compound is reduced to glyceraldehyde phosphate, the two trioses can form one molecule of hexose diphosphate.

$$F\text{-}1\text{-}\circled{P} \longrightarrow DHAP + Glyceraldehyde \longrightarrow Glycerate$$

$$GAP \longleftarrow 2\text{-}\circled{P}\text{-}Glycerate \longleftarrow$$

HDP

Pyruvate

Glycogen

Tissues such as brain and muscle also have a fructokinase catalysing the phosphorylation of fructose to F-6-P. This enzyme appears, however, to be competitively inhibited by glucose, so that fructose feeding is of no use when, as in diabetes, glucose metabolism is inhibited. Hexokinase, but not glucokinase, will slowly phosphorylate fructose in the absence of glucose.

A fructose tolerance test, like a galactose tolerance test (see below) is sometimes used as a test of liver function. *Fructosuria* is a condition in which a small part of any fructose eaten is found in the urine. In fetal blood, and in seminal fluid, fructose is normally found as well as glucose.

Galactose

This sugar is found in milk-containing diets. It is rather slowly transformed into glucose in the liver. It is made from glucose in large quantities in actively secreting mammary glands, and the blood and urine of pregnant and lactating women often contain both galactose and lactose.

Galactosuria or galactosaemia is an inborn error of metabolism which can have fatal results in infants if they are not quickly put on a lactose-free diet. For reasons largely unknown, galactose in excessive quantities is toxic.

This is a problem that does not arise in older children or adults but can be very serious in infants. The symptoms are failure to gain weight, lethargy, vomiting, liver enlargement, and often jaundice, and at a later stage, cataract and mental defect. Some of these symptoms have been duplicated in adult rats by feeding diets containing 35 per cent lactose.

The transformation of galactose into glucose has the following steps:

(1)   Galactose + ATP $\rightarrow$ Galactose-1-(P) + ADP

(2)   Uridine di-(P)-glucose + Galactose-1-(P) $\rightleftharpoons$
                         Uridine di-(P)-galactose + Glucose-1-(P)

(3)   Uridine di-(P)-galactose $\rightleftharpoons$ Uridine di-(P)-glucose

Reaction (3) is the isomerization of the —OH on C–6 of the galactose molecule. The uridine diphosphate glucose so formed can then react with more galactose-1-phosphate.

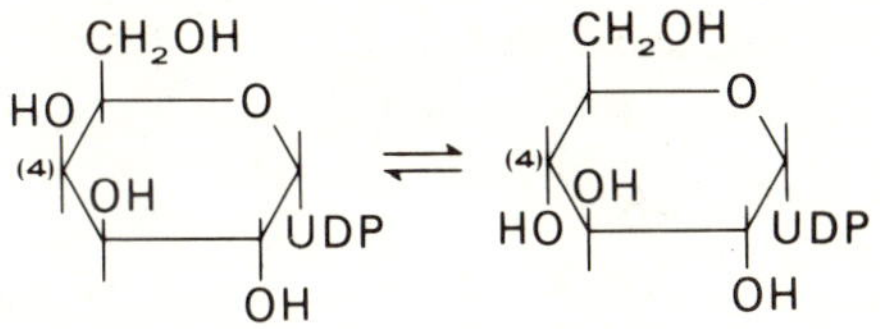

For the structure of uridine diphosphate glucose see p. 149.

It has been established that in galactosurics the enzyme for reaction (2) is missing, so that there arises an accumulation of galactose-1-phosphate, and eventually of galactose, in the cells, with toxic consequences. The galactosuria can be diagnosed soon after birth (p. 28), and the infant placed on a lactose-free diet.

Lactose is synthesized in lactating breast tissue by the following steps:

G-1-(P) + UTP $\rightarrow$ UDPG + (P)-(P)

UDPG $\rightarrow$ UDPGal

UDPGal + G-1-(P) $\rightarrow$ Lactose-1-(P) + UDP
    (*or* glucose)   (*or* lactose)

Amino Sugars

The two most important members of this class are 2-amino-2-deoxyglucose and 2-amino-2-deoxygalactose which are found (usually with the amino

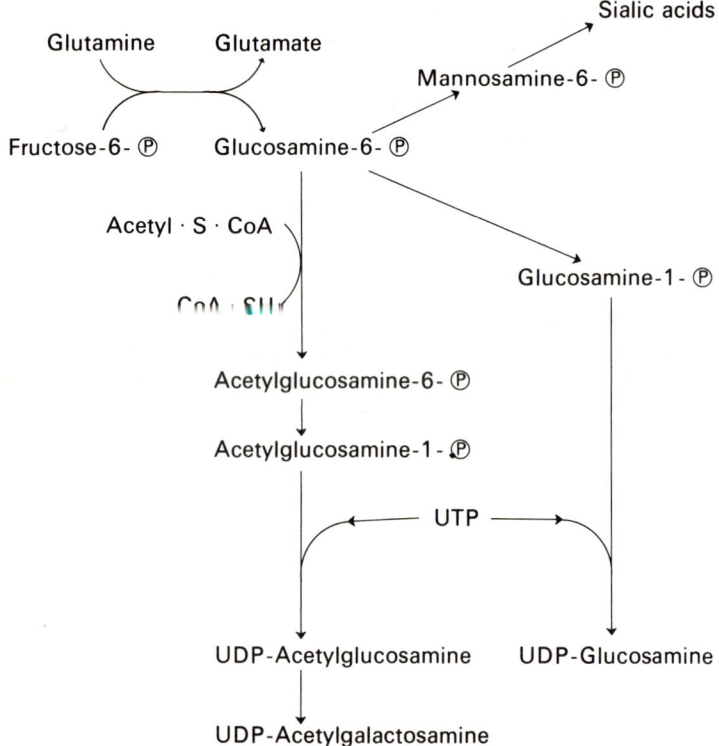

Glucosamine

Galactosamine

group acetylated) in the mucopolysaccharides. They are synthesized by the following pathways:

Enzymes have been found, in cartilage and elsewhere, which incorporate these sugars, together with glucuronic acid (from UDP-Glucuronate) into the characteristic mucopolysaccharide structures, and ingenious mechanisms have been proposed to account for the regular alternation of amino sugar and glucuronic acid residues which is characteristic of these polymers. One or more of the —OH groups on the hexose residues is frequently sulphated (see p. 26).

A General Consideration of Carbohydrate Metabolism

It has already been pointed out that the blood sugar level must always be maintained, whether or not carbohydrate is immediately available from food. The part that the liver plays in this process cannot be overestimated. It was shown many years ago that after the experimental removal of the liver from an animal, the blood sugar steadily decreases until death. Two separate processes in liver maintain the blood sugar level—*glycogenolysis* and *gluconeogenesis*.

Glycogenolysis, the breakdown of liver glycogen to free glucose, is the first process to come into play. It is under hormonal control by adrenaline and glucagon. The contribution which it can make to the body's glucose needs must depend on the previous dietary history, but it is usual to find that glycogen has almost completely disappeared from the liver after a 24-hour fast. Muscle glycogen does not usually disappear, however, even after a long fast and, as explained above, muscle glycogen can make no direct contribution to blood glucose.

Gluconeogenesis becomes vitally important when liver glycogen is exhausted. The extent of gluconeogenesis in a prolonged fast can be estimated with some accuracy. Urea-N excretion is maintained at about 6 g per day, even in the absence of all food. Since proteins, on the average, contain about 16 per cent N, this is equivalent to $6 \times 100/16$, or 37 g protein or about 40 g amino acids. It is usual to regard about half the amino acids of an average tissue protein as glucogenic. Thus of the 40 g amino acids formed from endogenous protein per day, about 15–20 g glucose can, as a maximum, be produced. The glycerol moiety, but not the fatty acids, of triglycerides can also be a source of glucose carbons. Now, if it is assumed that the metabolic rate of the subject is 8 400 kJ per day, about 210 g fat must be oxidized (460 kJ of the total are supplied by protein). Fat contains about $\frac{1}{9}$ of its weight as glycerol, i.e. about 20–25 g, and this can be converted to glucose. One may conclude, therefore, that in prolonged fasting of the order of 40 g glucose can be formed per day by gluconeogenesis, rather less than 10 per cent of the total energy requirements. In a less prolonged fast, the metabolic rate might be as high as 12 500 kJ per day, which would be equivalent to about 55 g glucose formed by gluconeogenesis. This accords fairly precisely with the estimated energy requirements of the brain, leaving nothing over for the rest of the body. Observations on sufferers from the glycogen storage diseases (especially types V and VII) suggest that the musculature can function without glucose as an energy source, but that stiffness and weakness after exercise are common. Since this does not occur during fasting, it may be suspected that the brain and the muscles compete to some extent for the available glucose. One must then reckon with the possibility that the brain can use another fuel during fasting. This point is discussed at the end of Chapter 10.

One should not think that gluconeogenesis is separated from glycogenolysis in time, during the adaptation of the subject to starvation. Gluconeogenesis is not an alternative source of energy which can be switched in, in the sense of a reserve fuel tank on a motor car. The two processes occur simultaneously; nevertheless, gluconeogenesis from protein can be prevented to a very considerable extent by the consumption of a small amount of carbohydrate, even if it is quite inadequate as an energy source. This is known as the *protein-sparing* action of carbohydrate; it can be easily understood in terms of gluconeogenesis, and the essential role of glucose and of pyruvate.

It is somewhat surprising after these considerations to recall the importance of fat as a storage form of carbohydrate as compared with glycogen. The facts revealed by isotopic tracer experiments are not, however, disputed; indeed obesity—the result of excessive carbohydrate storage as fat—is a major clinical problem today. The teleological answer to this paradox may be the superior efficiency of fat as an energy store when it is compared with carbohydrate on a weight basis. In a mobile organism, such as man, this is important; for example, the fetus in utero has a large store of glycogen but very little fat; the latter begins to be synthesized only after birth. Obesity is frequently treated by providing moderate calorie diets which are high in fat and protein. The effectiveness of these diets depends partly on gluconeogenesis, both from food protein and endogeneous protein. As soon as some essential amino acids have been transformed into carbohydrate, the other (non-essential) amino acids arising from the protein which has been hydrolysed must also be catabolized since there is no amino acid storage in the body.

The Control of Carbohydrate Metabolism

It is very noticeable that chronic accumulation of intermediates of glucose catabolism is very rare. Almost the only examples are the raised blood pyruvate of vitamin B_1 deficiency (when pyruvate decarboxylation is deficient), and the raised blood pyruvate and lactate of type I glycogen storage disease, when liver glycogen can be broken down to hexose phosphates which cannot be hydrolysed to free glucose, but are catabolized to pyruvate in excess. It is also remarkable that production of acetyl units from carbohydrate is so controlled that ketone body formation from glucose, although technically possible, is never found to take place in vivo.

The occurrence of a high blood and tissue lactate in ischaemia or after severe muscular exercise has not been ignored in formulating the above remarks. The rapid production of lactic acid in these circumstances is an example of a phenomenon of general occurrence in biology called the *Pasteur effect*, which makes possible the rapid catabolism of hexoses in anaerobic conditions by arranging that the hydrogen atoms removed to NAD in the oxidation of glyceraldehyde phosphate are transferred to a suitable acceptor. In vertebrates the acceptor is pyruvate, in some insects dihydroxyacetone phosphate, in yeasts acetaldehyde.

Pyruvate and lactate do not accumulate in aerobic conditions, in cells which have a cytochrome system. That this is so is good evidence of a very effective control mechanism, and there is a great deal of interest in discovering how such a mechanism works. In the short term, the coupled state of activity of phosphofructokinase and pyruvate kinase, together with the supply of cofactors to triose phosphate dehydrogenase, appears chiefly to determine the rate of flux through the Embden–Meyerhof pathway, but this in itself would only cause hexose monophosphates to accumulate. The relative activation and inactivation of phosphorylase and glycogen synthetase helps to regulate the disposal of these units, but this is of minor importance in brain and kidney, which normally contain little glycogen. In liver, glucose-6-phosphate may be hydrolysed to free glucose, thus setting up a 'futile cycle'. It seems that in muscle, at least, the rate of flux through the pathway must determine the rate of glucose uptake into the tissue. It is not at present known how this control is exerted; suggestions that it is by a direct inhibition of hexokinase by its product glucose-6-phosphate seem unlikely to be correct, since this compound is a rather weak inhibitor of all except brain hexokinase. These problems also make it difficult to explain fully the observations that glucose uptake from plasma by muscle is inhibited by ketone bodies and by free fatty acids.

It is over-simplistic to assume that tissue metabolism is controlled merely by the maximal activities of certain enzymes. For example, the neonate is amazingly more resistant to anoxia than the adult, although the tissues of both then depend on anaerobic glycolysis for survival, in spite of the fact that the concentrations of several glycolytic enzymes rise steeply in the weeks *after* birth. The resistance is partly due to an abnormally high glycogen level in the heart just before birth, which prolongs blood circulation, and partly to rapid hypothermia, which slows down the rate of energy dissipation. However, there is also evidence that the brain of the neonate has a resistance to anoxia that cannot be explained simply on metabolic grounds.

Although acetyl units not required for oxidation in the TCA cycle can in liver normally be synthesized into fatty acids, this pathway is of little importance in muscle. Moreover, the effective switch from carbohydrate to fat oxidation in the post-absorptive animal, even when carbohydrate remains in muscle glycogen, suggests a control at the pyruvate oxidase level which has not been completely worked out. The rate of gluconeogenesis in liver from substrates other than pyruvate and its precursors, must depend very much on the activity of phospho-enol pyruvate carboxykinase and the supply of substrates to this enzyme; the chief controlling factors have not yet become clear. It has been established that in adipose tissue and liver almost all the hydrogen atoms transferred to NADP during the oxidation of glucose by the hexose shunt are used for the synthesis of fatty acids. During fasting, the rate of fat synthesis falls sharply, but it is not clear whether the proportion of glucose metabolized in these tissues through the shunt is controlled by anything other than the supply of oxidized NADP.

10 Lipid Metabolism

It is more difficult to describe lipid metabolism than carbohydrate metabolism since, as outlined in Chapter 3, the term lipid covers a number of compounds of rather dissimilar function and chemical composition, with little similarity other than relative insolubility in water and solubility in many organic solvents. A large proportion of lipid substances contain esterified long-chain fatty acids. It is perhaps simplest to define lipids as substances whose major component is synthesized from acetyl units as building blocks.

The oxidation of long-chain fatty acids provides more than half the body's total energy needs in most circumstances, excepting in the few hours after ingesting carbohydrate foods. The normal body contains a very large store of these fatty acids—sufficient to supply its energy needs for several weeks. The fat (triglyceride) content of individuals is very variable, but a representative value for the mythical 70 kg man might be 5·5 kg in the adipose tissue alone. If one assumes a daily expenditure of energy of 8 500 kJ, this could be supplied by the oxidation of 225 g fat (producing 38 kJ/g, see Chapter 19). The fat stores alone in this individual would then be capable of supplying his energy needs for 25 days or more. Clearly, then, fat metabolism is a very important aspect of the total body economy, but there is one crucial proviso: glucose can, by way of pyruvate and acetyl CoA (Chapter 9), be converted into fatty acids, but the latter cannot, in man, be reconverted into glucose. The fact that this is so immediately brings protein metabolism (because of gluconeogenesis, see Chapter 9) into the picture.

In biochemistry the term *fat* is reserved for the triglycerides (triacyl-glycerols) discussed above. There are many other lipids, particularly the so-called structural lipids—cholesterol and its esters, the phospholipids and glycolipids—which are presumed to be equally important because they are found in almost all cells, particularly in cell structures, and do not disappear on fasting even when the fats are exhausted. It must be confessed that we know little, in precise chemical terms, of the function of these lipids. Fig. 10.1 shows some of the interrelationships of lipid metabolism.

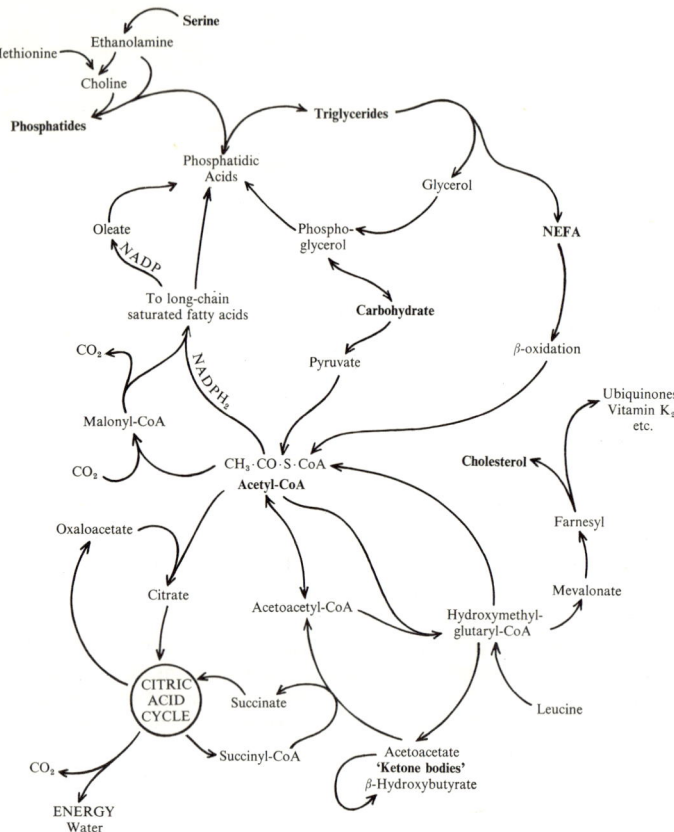

FIG. 10.1. Some of the interrelationships of lipid metabolism.

It is particularly important in lipid metabolism to grasp the dynamic nature of cellular reactions. To take one example, the turnover in a few hours of plasma non-esterified fatty acids may be sufficient to account for 200 per cent of tissue oxygen consumption. It is clear that there must be considerable recycling of these acids.

The digestion and absorption of fat has been dealt with in Chapter 8. The absorbed fat reaches the blood by way of the portal system (short-chain fatty acids) and the thoracic lymph duct (triglycerides of long-chain acids, in chylomicrons).

The Nature of Plasma Lipids

The concentration of lipids in normal blood is given in Table 14.3 (Chapter 14). It is necessary to analyse these figures further, since the complexity of lipid metabolism is reflected to some extent in the complexity of plasma lipids.

Chylomicrons. These appear to be formed only in the mucosal cells of the small intestine, although the protein component, which is small but specific, is similar to that of the plasma lipoprotein (Table 10.1). In normal feeding the chylomicron count of plasma is small and transient, but it may be very large in heavy fat feeding or in uncontrolled diabetes. The chylomicrons can be taken up and metabolized by most tissues; the process seems to involve adsorption of the globule to the plasma membrane of the cell and partial hydrolysis during the translocation. Resynthesis of triglyceride occurs rapidly as soon as the lipids are inside the cell. Although tissues such as muscle, but not brain, are able to utilize triglycerides directly in this way, it seems likely that in normal feeding the chylomicrons are mostly taken up by liver and adipose tissue. How much of the ingested fatty acids will be immediately oxidized in liver must depend on the percentage of carbohydrate in the diet and the immediate dietary history. No doubt in many circumstances the chylomicron triglycerides will be used in the formation of β-lipoproteins. There is, in any event, a good deal of replacement of acyl radicals, both in adipose tissue and liver, as well as in the intestine (Chapter 8), so that the composition of the depot fat will only slowly approach that of the food fat (if the latter has a composition characteristic of a single source, see Tables 3.2 and 3.3, p. 34).

A small fraction of chylomicrons is removed in the plasma by *lipoprotein lipase* (clearing factor), an enzyme which hydrolyses fats only when they are attached to protein. Heparin is in some way connected with clearing factor activity. Adipose tissue cells also contain a lipoprotein lipase, which seems to be essential for chylomicron assimilation by this tissue, but liver in vivo shows no lipoprotein lipase activity.

Lipoproteins. Apart from chylomicrons, all the lipids in plasma are bound. The chief 'vehicles' are two groups of lipoproteins, the high-density or α-lipoproteins and the low density or β-lipoproteins. The prefixes α and β refer to the occurrence of these entities in the α and β globulin fractions of plasma, respectively. The low density lipoproteins may even be lighter than water; a standard method of analysis is centrifugation in a medium of density $1\cdot063$ $[\equiv 1\cdot75\text{M NaCl}]$. See Table 10.1.

Table 10.1
Composition of Plasma Lipoproteins

	Chylomicrons	Very low density lipoproteins (VLDL)	Low density lipoproteins (LDL)	High density lipoproteins (HDL)
Plasma concentration, mg/100 ml	variable	150	350	300
Density, g/ml	<0.95	0.95–1.006	1.006–1.063	1.063–1.21
Paper electrophoretic migration	Chylomicrons	pre-β	β	α
Average composition:				
Phospholipid	7	19	22	24
Unesterified cholesterol	2	7	8	2
Cholesterol esters	5	13	37	20
Triglyceride	84	51	11	4
Protein	2	8	21	50
Major protein (apolipoprotein constituents)	Probably include: *apoLP-ser apoLP-glu apoLP-ala$_1$ apoLP-ala$_2$	apoLDL apoLP-ser apoLP-glu apoLP-ala$_1$ apoLP-ala$_2$	apoLDL	apoLP-gln I apoLP-gln II

* The apolipoproteins are identified by their carboxy terminal amino acids, e.g., ser $\equiv$ serine.

According to present views, the β-lipoproteins have two functions. The main one is to carry triglyceride from liver to fat depots and other tissues of the body. Although the proportion of protein in these molecules is small, the peptide is perfectly characteristic. Like the phospholipid, it is made in the liver, and the triglyceride-poor LDL may well be the 'boat' which carries triglyceride in amounts depending on the nutritional status of the organism. Because of the rapid turnover of the protein it is thought that each 'boat' only makes one journey; the half life of the laden vehicle is 8–12 hours.

The second function is to act as a store for cholesterol. The sterol content—particularly of cholesterol ester—in β-lipoproteins is very high, and the concentration of LDL in plasma rises in conditions in which the body's cholesterol balance becomes positive.

In familial absence of β-lipoproteins (abetalipoproteinaemia) no chylomicrons are found in the plasma; there is very little lipid in the plasma altogether, although on electrophoresis the bond in the α-globulins corresponding to the protein moiety of HDL can be detected. Fat absorption is deranged.

Much less is known about the function of the α-lipoproteins, which are also made in liver. They are relatively richer in phospholipid, and their concentration in plasma is very constant. Nevertheless it is found that in lecithin-cholesterol acyl-transferase deficiency (LCAT deficiency, see p. 201), the ultramicroscopic structure of HDL changes from spheroidal with hexagonal symmetry, to discs stacked one upon another. Less marked changes are seen in LDL, which are also spheroidal, while VLDL does not have such a high degree of reproducible morphology. Both the HDL and LDL fractions have a high free cholesterol and low cholesteryl-ester content in LCAT deficiency, and a high lecithin content, while the total amount of the HDL fraction is low. Thus it is clear that the proportion of free to esterified cholesterol is important in forming both HDL and LDL.

In *Tangier* disease the HDL are absent from plasma.

Both α- and β-lipoproteins carry also the trace lipids of the plasma— fat-soluble vitamins, steroid hormones, etc.

Non-esterified fatty acids. The concentration of long-chain non-esterified fatty acids (NEFA, also abbreviated UFA or FFA) in plasma is rather less than 1 μmole/ml—about 20–30 mg/100 ml. They come almost entirely from adipose tissue, and are bound to the albumin, except for a small fraction carried by HDL. The association constant for at least the first two molecules of fatty acid bound to each molecule of albumin is very large indeed, and there is no satisfactory explanation at present of how the NEFA dissociate from the protein as they enter the cells in which they are metabolized. Although the concentration of NEFA is small, the turnover rate is high (half life 2–3 minutes) and it is calculated that their rate of entry into and disappearance from plasma would be sufficient to account for all the fat oxidized by a fasting animal. They are the immediate substrate for fat oxidation in all organs (brain excepted). The rate of release of NEFA from fat depots is strongly influenced by many factors (see later in this chapter).

The Oxidation of Saturated Fatty Acids

In in vitro systems, the acids must first be activated by condensation with *coenzyme A*. This coenzyme, whose structure has been given on p. 142, is widely distributed in cells, and is necessary not only for the oxidation of the fatty acids, but for the transfer of acetyl and other acyl radicals from one compound to another. The activation reaction is:

(1) $R \cdot COOH + HS \cdot CoA + ATP \rightleftharpoons$

$$R \cdot CO \cdot S \cdot CoA + AMP + P—P_i$$

where $P—P_i$ represents pyrophosphate.

The activated fatty acid, as a *thio-ester* of coenzyme A, is then subjected to the following series of reactions which result in the production of the CoA thio-ester of an acid with 2 carbon atoms less in the chain. This is known as *β-oxidation*.

(2) $R \cdot CH_2 \cdot CH_2 \cdot CH_2 \cdot CO \cdot S \cdot CoA \rightarrow$

$$R \cdot CH_2 \cdot CH=CH \cdot CO \cdot S \cdot CoA + 2 H$$

The hydrogen acceptor here is flavin-adenine dinucleotide (FAD), the prosthetic group of the enzyme.

(3) $R \cdot CH_2 \cdot CH=CH \cdot CO \cdot S \cdot CoA + H_2O \rightleftharpoons$

$$R \cdot CH_2 \cdot \underset{\underset{OH}{|}}{CH} \cdot CH_2 \cdot CO \cdot S \cdot CoA$$

(4) $R \cdot CH_2 \cdot \underset{\underset{OH}{|}}{CH} \cdot CH_2 \cdot CO \cdot S \cdot CoA + NAD \rightleftharpoons$

$$R \cdot CH_2 \cdot CO \cdot CH_2 \cdot CO \cdot S \cdot CoA + NADH_2$$

(5) $R \cdot CH_2 \cdot CO \cdot CH_2 \cdot CO \cdot S \cdot CoA + HS \cdot CoA \rightleftharpoons$

$$R \cdot CH_2 \cdot CO \cdot S \cdot CoA + CH_3 \cdot CO \cdot S \cdot CoA$$

This last reaction deserves particular attention for two reasons. One of the products is acetyl-CoA, which can be readily oxidized (see Chapter 12), while the other product is also an acyl-CoA, which can undergo β-oxidation by reactions (2) to (5). This means that the process can be repeated again and again until the original long-chain acid has been completely split into acetyl groups. As almost all natural fatty acids have an even number of carbon atoms, the final product of reaction (5) will be two molecules of acetyl-CoA. When odd-numbered fatty acids are oxidized, the final products are acetyl-CoA and propionyl-CoA, which has to be oxidized by a special pathway (see p. 203). The normal fate of acetyl-CoA is condensation with oxaloacetic acid, with subsequent oxidation of the citric acid thus formed (Chapter 12).

Fatty acid oxidation by this sequence of reactions is the most important mechanism for all saturated straight-chain acids of whatever chain length. Enzymes which catalyse ω-oxidation (i.e. the introduction of an $-OH$ into the terminal methyl group) and α-oxidation are known to exist, but their activities, at least in man, are of little importance. The β-oxidation enzyme complex is located in the mitochondria, and the reduced FAD of the *acyl-CoA dehydrogenase* (which catalyses (2)) is re-oxidized by an electron transferring factor (ETF) which transfers reducing equivalents to the cytochrome c region of the hydrogen transport system (Chapter 12). The $NADH_2$ from reaction (4) is, of course, also produced inside the mitochondria, and the β-oxidation system will only remain operative if the reduced FAD and NAD are quickly re-oxidized. This means that the hydrogen transport system must be working; there is no equivalent, in fat metabolism, to anaerobic glycolysis of carbohydrates.

Each successive β-oxidation by this four-step cycle requires a fresh molecule of $CoA \cdot SH$ for the thiolytic splitting in step (5). The acetyl-CoA formed in this step must therefore be quickly metabolized, since the total concentration of coenzyme A molecules in any cell is rather small, and there can be no significant accumulation of any intermediate.

β-Oxidation is coupled with ATP regeneration; each turn of the cycle should result in the regeneration of 5 ATP molecules. The formation of 9 molecules of acetyl-CoA from a molecule of stearic acid can therefore produce 38 molecules of ATP from ADP and inorganic phosphate ($8 \times 5 - 2$ (for activation) = 38).

Carnitine, $(-)\gamma$-Trimethylamino-β-hydroxybutyric acid

$$(CH_3)_3 \overset{+}{N} \cdot CH_2 \cdot \overset{\overset{\displaystyle OH}{\displaystyle |}}{CH} \cdot CH_2 \cdot COOH$$

has been found to stimulate fatty acid oxidation in many types of tissue homogenate. Muscle mitochondria will not oxidize free fatty acids at all in vitro, but they will rapidly oxidize acyl-carnitine esters. It has been suggested that carnitine esters of fatty acids are a device for facilitating the transport of long-chain acyl radicals into the mitochondria, and an *acyl-carnitine transferase*, catalysing the reaction

$$R \cdot CO \cdot S \cdot CoA + Carn. \rightleftharpoons R \cdot CO \cdot Carn. + CoA \cdot SH$$

in which the acyl group is transferred to the secondary OH group of carnitine, has been found to be widely distributed.

Yet there remain some puzzles about the physiological role of carnitine. The fatty acid activating mechanisms are in the mitochondria, so that acyl carnitines could hardly be formed in the cytoplasm. Very often, indeed, mitochondria fail to oxidize fatty acids when these are presented together with carnitine, ATP and $CoA \cdot SH$. Finally, carnitine, as the name suggests, is found in high concentration in the cytoplasm of muscle, but there is little, and little of the transferase enzyme, in liver and almost none in brain. If there is a transfer mechanism it cannot therefore be of universal significance.

The Synthesis of Saturated Fatty Acids

The requirements for fat synthesis are acetyl-CoA units (from carbohydrate or the β-oxidation process), reducing equivalents (as reduced NADP produced in tissue oxidations, particularly in the hexose shunt, see Chapter 9), energy (as ATP), and glycerol phosphate (from carbohydrate). It is evident that a plentiful supply of carbohydrate is a prerequisite for net synthesis of fat.

The scheme of breakdown of fatty acids by β-oxidation, described on p. 188, is in principle reversible, although the equilibrium in reactions (2) and (5) lies far to the right. However, a fatty acid synthesizing system which is agreed to be much more important than the reversal of β-oxidation is described below. The complete system of enzymes is found as a synthetase particle, of mol. wt. approximately 10^6 daltons, in the cytoplasm. In this particle the acyl carrier is not CoA, but *acyl-carrier protein* (ACP), a polypeptide to which pantetheine is attached through a seryl link (see p. 143). Acyl radicals are transferred to it from CoA by a suitable transferase. In the synthetase of animal tissues, an easily dissociable ACP is not found, but the abbreviation is still convenient.

The key intermediate in this pathway is *malonyl-CoA*, formed by carboxylation of acetyl-CoA.

$$CH_3 \cdot CO \cdot S \cdot CoA + CO_2 + ATP \rightarrow$$

Acetyl-CoA
$$\underset{\underset{COOH}{|}}{CH_2} \cdot CO \cdot S \cdot CoA + ADP + P_i$$

The enzyme, acetyl-CoA carboxylase, catalysing this step contains biotin as a prosthetic group (see p. 145). It is not found in the multi-enzyme synthetase. The central $-CH_2-$ group of malonic acid and its esters is known to be very reactive, and the energy-requiring synthesis of malonyl-CoA has therefore produced an intermediate which will react with other small molecules very much in the direction of synthesis.

The reaction sequence proceeds as shown:

$$CH_3 \cdot CO \cdot S \cdot ACP + \underset{\underset{COOH}{|}}{CH_2} \cdot CO \cdot S \cdot ACP \rightarrow$$

Acetyl-ACP   *Malonyl-ACP*

$$CH_3 \cdot CO \cdot CH_2 \cdot CO \cdot S \cdot ACP + ACP \cdot SH + CO_2$$

Acetoacetyl-ACP

Acetoacetyl-ACP is then reduced to butyryl-ACP by a sequence of three reactions similar to the reversal of steps (4), (3), and (2) of the β-oxidation pathway, except that $NADPH_2$ is required in both the reduction steps. Finally, the long-chain fatty acid is hydrolysed off from the carrier protein. During the synthesis there are no free intermediates in the cytoplasm. It is a peculiarity of this system that it stops at palmityl-ACP. Stearyl-CoA must then be formed from palmitic acid by some variant of the pathway described on p. 188. Palmitic acid is, however, the most common saturated fatty acid in human fats.

Some mitochondria, notably those of the heart, can synthesize long-chain fatty acids by a mechanism which does not involve malonyl-CoA, and in which $NADH_2$ is the reducing agent. The condensation of two acetyl groups appears to be energy-dependent. A separate enzyme for elongating palmityl-CoA to stearyl-CoA is found on outer mitochondrial membranes, and elongation is found to be most active when the mitochondria are anaerobic, with acetyl release occurring when oxygen is readmitted. It is postulated that this is a storage device for reducing equivalents, and easily oxidizable substrate, within the mitochondrial domain.

The Metabolism of Unsaturated Fatty Acids

Table 3.3 (p. 34) shows that quantitatively the most important fatty acid constituent of lipids is the mono-unsaturated acid *oleic acid,*

$$CH_3 \cdot (CH_2)_7 \cdot CH = CH \cdot (CH_2)_7 \cdot COOH$$

There is almost more of this acid in depot fat triglycerides than of all the others put together. Isotopic tracer evidence shows that oleic acid is metabolized rather more slowly than the saturated acids, but nevertheless considerable quantities must be turned over each day.

An enzyme is known which will reduce oleic acid to stearic acid, using $NADPH_2$, but it is more likely that oleic is activated by conversion to the CoA thio-ester, and β-oxidized in the normal way. After 3 acetyl groups have been removed, the compound

$$CH_3 \cdot (CH_2)_7 \cdot CH{=}CH \cdot CH_2 \cdot CO \cdot S \cdot CoA$$

is left. This is a β,γ-unsaturated acyl-CoA and, moreover, the double bond is *cis*; whereas the configuration required for the hydratase reaction (Step (3) on p. 188) is *trans*. A special isomerase catalyses the conversion from β-γ-*cis* to α-β-*trans*.

The main desaturating enzyme system is found in microsomes, mainly in the liver, and acts on stearyl CoA. The reaction is catalysed by a mono-oxygenase (Chapter 12), but the 2 H atoms appear to be removed directly, without the formation of an intermediate hydroxy-compound. The mechanism and cofactors are not yet fully established.

The presence of so much oleic acid in triglycerides is fairly clearly connected with the lower melting point of fats containing unsaturated acids. Tristearin and tripalmitin are quite solid, even at 37°, and if they were present in high concentration in adipose tissue the flexibility of the organism would be disastrously reduced.

The poly-unsaturated acids (Table 3.1, p. 31) are relatively unimportant as constituents of depot fats (Table 3.3, p. 34). Linoleic, linolenic, arachidonic, and more complex acids, however, are found in high concentration in some of the structural lipids, particularly in some phospholipids and cholesterol esters. It appears that animals in general, probably including man, cannot insert a second double bond in a long-chain fatty acid on the *methyl* side of the Δ-9-10 double bond in oleate, although they can do so on the *carboxyl* side. Thus although 6,9 octadienoic acid can be formed in animals, only the 9,12-acid (linoleic) is of importance both in itself and as a precursor of other acids. Thus animals depend on plant sources for their linoleic acid; it is established that adults need about 10 g per day, but as more than this is usually present in the diet, it is not clear that in adults there exists a definite deficiency disease although it may exist in deprived children. The effects on fatty livers, hypercholesteraemia, and skin lesions have been most intensively studied.

γ-Linolenic acid

It will be appreciated that there are two isomers of linolenic acid, α-*linolenic* (9, 12, 15), which occurs in plants, and γ-*linolenic* (6, 9, 12), which can be formed in animals by desaturating linoleic. It is this isomer which is found in animal lipids, and it also appears to be the most important precursor of *arachidonic acid*, qualitatively the most important of the poly-unsaturated fatty acids in man. Arachidonic acid is formed by adding one acetyl residue on to γ-linolenyl-CoA followed by desaturation.

In the catabolism of these acids, hydrogenation to saturated acids appears to play little part. It is possible that the formation of peroxides (cf. p. 35), followed by cleavage to short-chain acids is important.

Triglycerides

A. *Glycerol*

The glycerol set free by the final destruction of a triglyceride is transferred to the pathway of carbohydrate oxidation by the following reactions:

(1) $CH_2OH \cdot CHOH \cdot CH_2OH + ATP \rightarrow$

$$CH_2OH \cdot CHOH \cdot CH_2O \, \circledP + ADP$$

(2) $CH_2OH \cdot CHOH \cdot CH_2O \, \circledP + NAD \leftrightarrows$
   Glycerol-3-sn-phosphate

$$CH_2OH \cdot CO \cdot CH_2O \, \circledP + NADH_2$$

*Dihydroxyacetone
phosphate*

The enzyme catalysing reaction (1) is not present in all tissues. Thus, in adipose tissue, triglycerides can only be synthesized if carbohydrate metabolism is working sufficiently well to provide enough dihydroxyacetone phosphate, a normal intermediate of carbohydrate metabolism, for reduction to glycerol phosphate.

B. *Synthesis of Triglycerides*

Free glycerol is not used for the synthesis of triglycerides, but glycerol-3-phosphate (see above). This is esterified to form a *phosphatidic acid*:

$$\begin{array}{ll}
CH_2OH & CH_2O \cdot CO \cdot R \\
| & | \\
CHOH + 2\,R \cdot CO \cdot S \cdot CoA \rightarrow & CHO \cdot CO \cdot R + 2\,CoA \cdot SH \\
| & | \\
CH_2O \, \circledP & CH_2O \, \circledP
\end{array}$$

The phosphate is removed to yield a diglyceride, which is then esterified by a third molecule of acyl-CoA. Only 1,2-*sn*-diacylglycerols are acylated by the transferase, thus phosphatidic acid is almost an obligatory precursor, except in intestinal mucosa. A relatively less important pathway recently discovered starts from dihydroxyacetone phosphate, which is first esterified, then reduced:

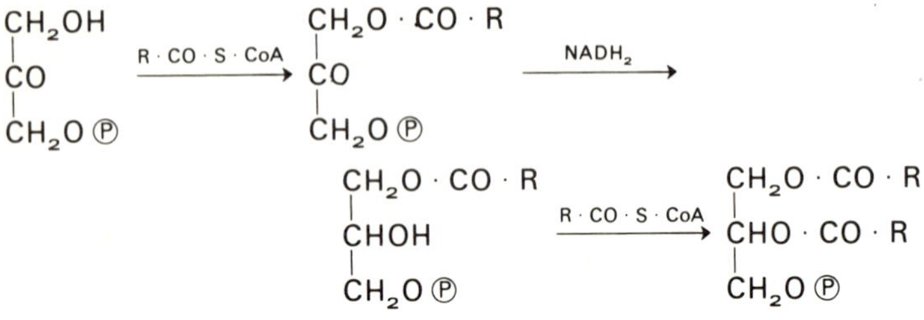

$$
\begin{array}{c}
\begin{array}{l}
CH_2OH \\
| \\
CO \\
| \\
CH_2O \; \circledP
\end{array}
\quad \xrightarrow{R \cdot CO \cdot S \cdot CoA} \quad
\begin{array}{l}
CH_2O \cdot CO \cdot R \\
| \\
CO \\
| \\
CH_2O \; \circledP
\end{array}
\quad \xrightarrow{NADH_2}
\end{array}
$$

$$
\begin{array}{l}
CH_2O \cdot CO \cdot R \\
| \\
CHOH \\
| \\
CH_2O \; \circledP
\end{array}
\quad \xrightarrow{R \cdot CO \cdot S \cdot CoA} \quad
\begin{array}{l}
CH_2O \cdot CO \cdot R \\
| \\
CHO \cdot CO \cdot R \\
| \\
CH_2O \; \circledP
\end{array}
$$

Although the acyl groups have been shown here as identical, triglycerides are mixed, and there is usually a tendency for unsaturated fatty acid residues to occupy the 2-position. The intermediates in both pathways appear to be bound to the endoplasmic reticulum, and are not found free in the cytoplasm.

In intestinal mucosa, both diglycerides and monoglycerides can be re-esterified by acyl-CoA to neutral fat. 2-Monoglycerides are transformed to diglycerides faster than are the 1-monoglycerides.

There are three main tissues in the body in which triglyceride synthesis takes place: the small intestine, particularly the lower half (resynthesis of fat digested in the intestinal lumen), the liver (together with the synthesis and secretion of β-lipoproteins), and above all adipose tissue. See also p. 214.

Phosphatides

Quantitatively the most important phosphatides are the *lecithins* (phosphatidyl choline, see p. 39) and the *cephalins* (phosphatidyl ethanolamine, serine, inositol, etc.). Another group of phosphatides, the sphingomyelins, contain phosphoryl choline, but it is esterified to the amino alcohol sphingosine instead of to diglyceride. (The cerebrosides, sulphatides, and gangliosides also contain sphingosine, but no phosphate, so they are not phosphatides.)

There are several ways in which the lecithin/cephalin groups of phosphatides can be synthesized. In one group of pathways the last step involves esterification with a 1,2-diglyceride; in the other, addition to a phosphatidic acid. The diglyceride may be formed from a phosphatidic acid, or (perhaps mainly in intestine) from a triglyceride. Lipase action on neutral fat may produce both 1,2- and 2-3-diacylglycerols; the latter are not substrates for phospholipid synthesis.

The most important relationships are shown in Fig. 10.2.

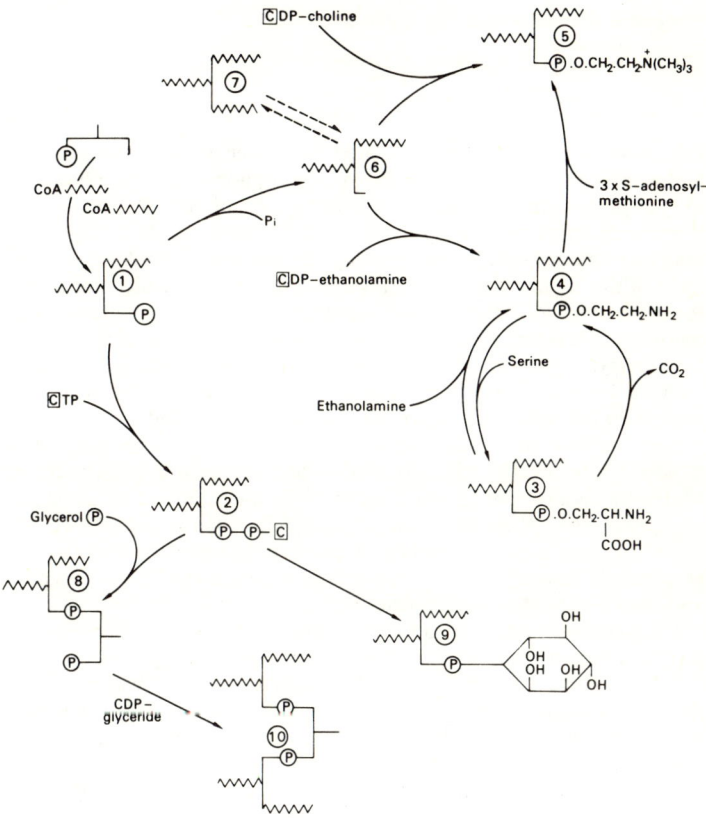

FIG. 10.2. The synthesis of phospholipids. [C] stands for cytosine. The numbered compounds are as follows: ① phosphatidic acid; ② cytidine diphosphate diglyceride; ③ phosphatidyl-serine; ④ phosphatidyl ethanolamine; ⑤ phosphatidyl-choline (lecithin); ⑥ diglyceride; ⑦ triglyceride; ⑧ phosphatidyl-glycerophosphate; ⑨ phosphatidyl-inositol; ⑩ cardiolipin.

CDP-choline and CDP-ethanolamine are formed in a transfer reaction similar to that used for the synthesis of UDP-glucose (Chapter 9).

$$\left.\begin{array}{l}\text{Choline phosphate}\\\text{ethanolamine phosphate}\end{array}\right\} + \text{CTP} \rightarrow \begin{array}{l}\text{CDP-choline}\\\text{CDP-ethanolamine}\end{array} + \text{P}-\text{P}_i$$

Choline is phosphorylated by a *choline kinase*:

$$(CH_3)_3\overset{+}{N} \cdot CH_2 \cdot CH_2OH + ATP \rightarrow$$

$$(CH_3)_3\overset{+}{N} \cdot CH_2 \cdot CH_2O\textcircled{P} + ADP$$

Ethanolamine is also phosphorylated by this enzyme, but free ethanolamine must occur rather rarely because phospholipase D (p. 39) is found only in plant tissues. A major immediate source of both phosphate esters is by the action of phospholipase C on phosphatides (recycling). In animals the rate of methylation of phosphatidyl-ethanolamine is not large, so that most of the choline that is not recycled comes from the diet; for choline as an essential food factor see Chapter 20. The source of ethanolamine (phosphate) is rather a puzzle; the only known mechanism of formation is by decarboxylation of the seryl residue in phosphatidyl-serine, but in animals the latter is only formed by the exchange reaction between serine and phosphatidyl-ethanolamine shown in Fig. 10.1, and isotopic tracer studies show clearly that the major part of the phosphatidyl-serine molecule turns over much more slowly than phosphatidyl-ethanolamine. Moreover, tissues contain much more of the latter, and of ethanolamine phosphate, than they do of the corresponding serine derivatives. This implies a great deal of recycling of phosphatidyl-ethanolamine without very much *de novo* synthesis of ethanolamine phosphate. The same is probably also true of phosphatidyl-choline.

It is clear that transfer from CDP-choline or CDP-ethanolamine directly to diacylglycerols is a major pathway in phospholipid synthesis, but this is not the whole story, because of the characteristic fatty acid patterns of the phospholipids. This is particularly important for liver (and hence plasma lipoprotein) phospholipids because they have a much higher content of poly-unsaturated fatty acids, particularly in the 2-position, than the usual run of glycerides. It is probable that special unsaturated phosphatidic acids are synthesized, and that the CDP-base transferases select diacylglycerols with poly-unsaturated acids at position 2 as substrates. Rearrangement of the acyl groups after synthesis is also possible because of the wide distribution of *phospholipase A_2*, which specifically hydrolyses the acyl residue esterified on the 2-position. This *lysophospholipid* can be re-esterified from an acyl-CoA.

The above remarks apply more specifically to phospholipids synthesized in the liver. Even in the kidney the turnover of phospholipids is not so rapid, and a proportion (but not all) of the phospholipid in all tissues examined (particularly brain) has a very long turnover time, that it is metabolically stable. When an animal is starved for a long period almost all the neutral fat may disappear, but a considerable proportion of the phospholipid remains in all tissues. These facts suggest that much of the phospholipid in the body is structural.

There are two fundamentally different types of membrane in animals. The myelin sheath of nerve is a double membrane wound spirally like a roll of paper, which has a high content of free cholesterol, sphingolipids (i.e. both sphingomyelin and cerebrosides), and much saturated fatty acid. The mitochondrial, nuclear and microsomal membranes, on the other hand, contain highly unsaturated phospholipids and much less cholesterol and sphingolipids. (See Table 3.6). Cardiolipin is a characteristic component of mitochondrial membranes. The inner mitochondrial membrane is not a simple lipid-protein leaflet, but is made up of a series of respiratory assemblies (Chapter 12), i.e. a set of proteins and co-factors which catalyse the complete oxidation of reducing equivalents from substrates. The lipids cannot be regarded merely as a 'glue' which holds these assemblies in place, since the enzymic activity of fragmented membranes can be much altered by addition or removal of phospholipids.

Little also is known about the functions of tissue phosphatides with rapid turnover. There is a large increase in turnover of phospholipids during secretion of protein by glands—pancreas, salivary glands, pituitary, thyroid. This may be related to the synthesis of the protein itself. In some tissues, notably brain and spermatozoa, phosphatides may be oxidized for energy. One or more phosphatides (not lecithin) are essential for the formation of thromboplastin in blood coagulation.

Phosphatidyl-inositol and sphingomyelin (p. 40) are phospholipids which are major components of brain lipids, but their turnover is very slow. Sphingomyelin is also found in membranes outside the brain; for example, it replaces lecithin in erythrocyte membranes in several species of mammals. The base sphingosine is formed by condensation of serine with palmityl-CoA.

Sphingosine is also the backbone of the large class of *glycolipids*. In classifying these it is now usual to define the sphingosine amido-acyl moiety as *ceramide*; the fatty acyl residues include a high proportion of branched-chain and odd-numbered fatty acids. The monoglycosyl-ceramides (formerly cerebrosides) are predominantly found in brain (myelin sheath), where the sugar residue is galactose. In other tissues it is glucose. Smaller amounts of di-, tri-, and tetraglycosyl-ceramides may be found in various tissues.

The *gangliosides* have an oligosaccharide chain attached to the ceramide moiety, thus:

cer-glu-galNAc-gal

with sialic acid (p. 22) attached to the innermost galactosyl residue. The fatty acyl is predominantly stearic.

There are several inborn errors of metabolism which involve these compounds. In *Niemann–Pick disease* there is a generalized accumulation of sphingomyelin, owing to absence of the enzyme hydrolysing it to sphingosine + phosphorylcholine. In *Gaucher's* disease there is progressive accumulation of glucosyl-ceramide in various tissues, because of the absence of the hydrolysing enzyme. In *Tay–Sachs disease* a characteristic ganglioside lacking the terminal galactose residue accumulates in brain, while in *generalized gangliosidosis* a similar lipid accumulates also in the visceral organs.

Cholesterol

Cholesterol is the parent 27-carbon atom steroid from which all other steroids in mammalian tissues are derived. The formula and conformation of this molecule are given on p. 44.

Synthesis. Cholesterol is synthesized by all nucleated mammalian cells. The simplest precursor in the biosynthetic pathway is acetyl-CoA. It is utilized as follows:

(1) $2 \text{ Ac} \cdot \text{CoA} \rightleftharpoons CH_3 \cdot CO \cdot CH_2 \cdot CO \cdot S \cdot CoA + CoA \cdot SH$

(2) $CH_3 \cdot CO \cdot S \cdot CoA + CH_3 \cdot CO \cdot CH_2 \cdot CO \cdot S \cdot CoA \rightarrow$

$$CH_3 \cdot \underset{\underset{OH}{|}}{\overset{\overset{CH_2 \cdot COOH}{|}}{C}} \cdot CH_2 \cdot CO \cdot S \cdot CoA + CoA \cdot SH$$

β-hydroxy-β-methylglutaryl-CoA

This synthesis is similar in principle to the synthesis of citric acid (Chapter 12).

$β$-Hydroxy-$β$-methylglutaryl-CoA (HMG-CoA) can be cleaved to form acetoacetate (p. 206), but in the synthesis of cholesterol the next step is the reduction of the carboxyl esterified with CoA to a $-CH_2OH$ group. The reaction takes place in two stages, each requiring $NADPH_2$, and the CoA is split off

(3) $HMG\text{-}CoA + 2 \text{ NADPH}_2 \rightarrow$

$$CH_3 \cdot \underset{\underset{OH}{|}}{\overset{\overset{CH_2 \cdot COOH}{|}}{C}} \cdot CH_2 \cdot CH_2OH + CoA \cdot SH$$

Mevalonic acid

There is very strong evidence that the limiting factor in cholesterol synthesis is the rate of formation of mevalonic acid, since this latter compound is very rapidly utilized by tissues which actively synthesize much cholesterol. The regulating stage appears to be reaction (3) (see also the section on ketosis, p. 206). Dietary cholesterol inhibits endogenous cholesterol synthesis strongly at this point, but the proximate agent is not cholesterol itself. However, the picture is not entirely clear, since the carbon skeleton of leucine, which yields HMG-CoA directly (p. 234), is rapidly utilized for cholesterol synthesis.

The next stage in the synthesis is the formation of mevalonic acid pyrophosphate

(4)

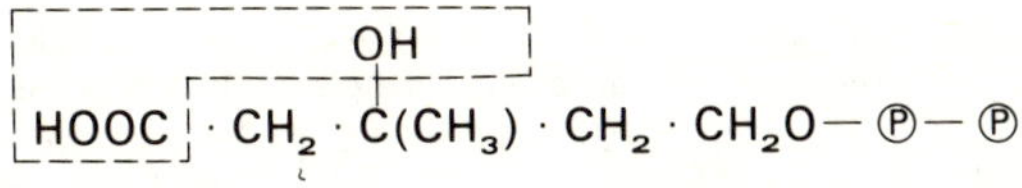

$$\boxed{HOOC} \cdot CH_2 \cdot \underset{|}{C}(CH_3) \cdot CH_2 \cdot CH_2O - \textcircled{P} - \textcircled{P}$$

with $\boxed{OH}$ above

a reaction requiring 2 molecules of ATP. A concerted reaction requiring ATP follows this, in which water and CO_2 are split off as indicated by the dotted line in the formula above to give *iso*pentenyl pyrophosphate.

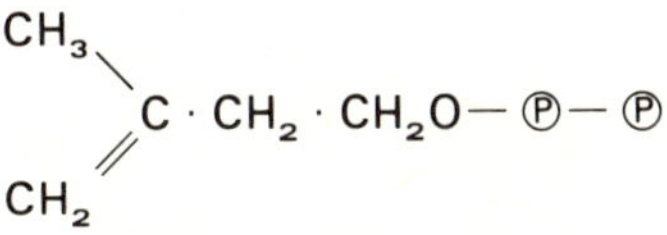

$$\underset{CH_2}{\overset{CH_3}{\diagdown}}C \cdot CH_2 \cdot CH_2O - \textcircled{P} - \textcircled{P}$$

$$\underset{CH_3}{\overset{CH_3}{\diagdown}}C = CH \cdot CH_2O - \textcircled{P} - \textcircled{P}$$

Isopentenyl pyrophosphate 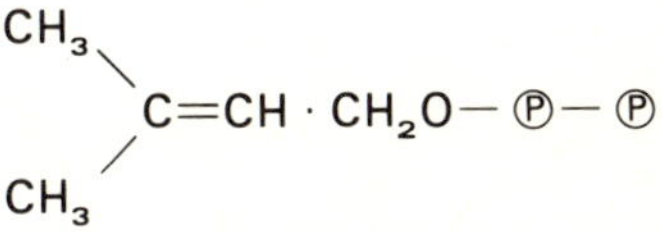   *Dimethylallyl pyrophosphate*

This is one form of the active '*isoprene unit*'; it tautomerizes readily to dimethylallyl pyrophosphate.

Hydrocarbons made up of repeating isoprene units:

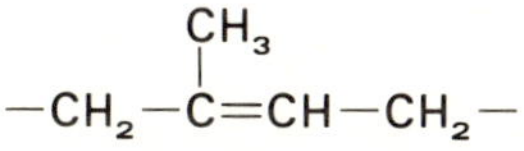

$$-CH_2-\underset{|}{\overset{CH_3}{C}}=CH-CH_2-$$

are widely distributed in nature. Active 'isoprene units' are key intermediates in the synthesis of, e.g. cholesterol and ubiquinone (p. 142) in animals, and carotenoids, essential oils, and rubber latex in plants.

(5) The next step in the synthesis of cholesterol is the condensation in two stages of isopentenyl and dimethylallyl residues to give *farnesyl* pyrophosphate, which contains 15 carbon atoms. (The stereochemistry of the double bonds is not shown, to save space.)

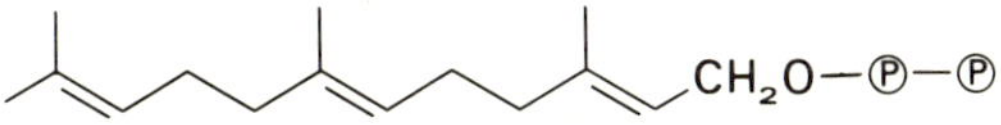

$$CH_2O - \textcircled{P} - \textcircled{P}$$

In this and subsequent reactions in the synthesis of cholesterol, the intermediates are enzyme-bound. From compounds of this kind the long hydrocarbon chains of many natural products are formed (compare the formulae of ubiquinone, p. 142, vitamins A, E, K_1, and K_2, Chapter 20).

(6) In the synthesis of cholesterol two farnesyl chains condense head to head to form the C_{30} compound *squalene*

This compound is ingeniously folded and simultaneous rearrangement of bonds takes place. The condensation to form the four rings of *lanosterol*, a C_{30} steroid precursor of cholesterol (not shown below), is initiated by the attack of a mixed function oxidase, (cf. p. 269), at the double bond indicated by the arrow. An epoxide is first formed (cf. Chapter 13), and a concerted electron rearrangement follows. From lanosterol, the C_{27} steroid cholesterol is formed.

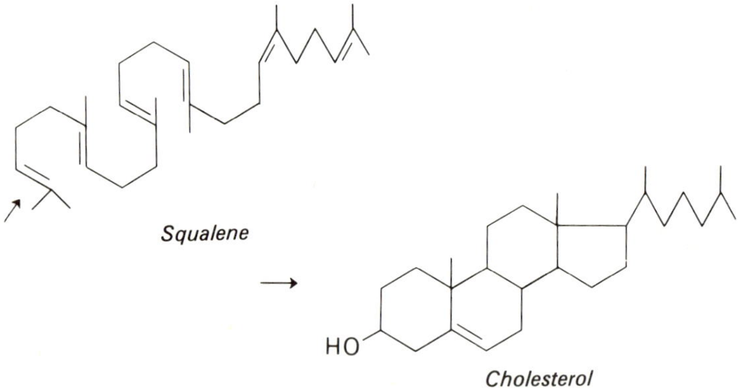

Squalene

$\rightarrow$

Cholesterol

The epoxide O atom remains as an —OH (or =O) group at C–3 in all steroids.

Cholesterol appears to be an essential component of all mammalian cells and may have a key role in maintaining the essential characteristics of living membranes. The cholesterol which is present in plasma originates largely from hepatic synthesis, although some is synthesized in the intestinal mucosa. It is present both as the free steroid and as long-chain fatty acid esters of cholesterol (about two-thirds of the total). As a rule the fatty acids which predominate in the cholesterol esters are linoleic and oleic acids. Both the free and esterified cholesterol molecules in plasma are concentrated in the lipoproteins (p. 185) but particularly in the LDL fraction. A good deal of cholesterol ester formation takes place in plasma, catalysed by the plasma enzyme *lecithin-cholesterol acyl transferase*, which catalyses the reaction:

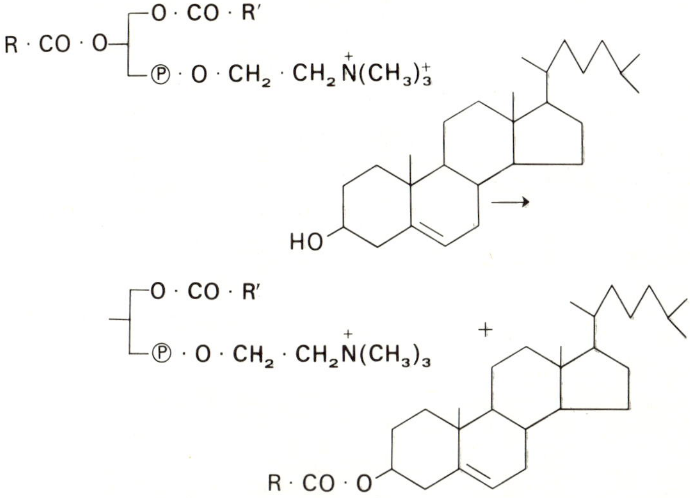

The changes in the structure and lipid content of plasma lipoproteins in LCAT deficiency have already been described (p. 187). A plausible view is that the free cholesterol which is found in the outer 'skin' of chylomicrons and VLDL, becomes concentrated in smaller particles as the triglyceride is removed by clearing factor. The function of LCAT is, by esterifying this cholesterol to facilitate the destruction of these particles.

There is reasonable evidence that unesterified cholesterol in plasma tends to replace cholesterol esters in cell membranes, with deleterious effects on their structures (cf. p. 35).

A high plasma cholesterol concentration related to type II hyperlipo-proteinaemia (Table 10.2) is associated with *atherosclerosis*, which in turn is associated with coronary thrombosis. The final composition of the atheromatous plaques that form on the walls of the aorta and other major blood vessels is complex, but their original composition is very high in esterified cholesterol. It is still not clear why the plaques form preferentially in blood vessels; in *familial hypercholesterolaemia* the steroid is deposited as xanthomata in many tissues, as well as in the cardiovascular system. The plasma cholesterol level tends to be higher in men than in women, and to rise with age. It is often high when diets contain much animal fat, and is strikingly but often not permanently reduced by the ingestion of some unsaturated fatty acids particularly linoleate. The level is also lower than normal in thyrotoxicosis. Isotope studies show that there is no

Table 10.2
Abnormal Lipoprotein Patterns in Familial Hyperlipoproteinaemia

Type	Plasma lipid peculiarity	Cholesterol/triglyceride ratio	Associated factors
I (rare)	Marked increase and persistence of chylomicrons	C↑TG↑ C/TG < 0·2	Deficiency in lipoprotein lipase
II (most common)	LDL increased	C↑TG↑ or normal C/TG > 1·5	Severe xanthomata. High percentage of ischaemic heart disease
III (uncommon)	'β-VLDL' present (LDL with very high percentage of TG)	C↑TG↑ C/TG ~ 1	With xanthomata. Responds well to restrictive diet
IV (common)	VLDL increased. Few chylomicrons	C↑ or normal, TG↑ C/TG variable	'Carbohydrate-induced' (excessive synthesis of TG from dietary CHO)
V (uncommon)	VLDL increased. Chylomicrons present	C↑TG↑ 0·15 < C/TG > 0·6	Genetically heterogeneous. 75 per cent have ketotic diabetes

These types have been distinguished on the basis of the appearance and composition of plasma 12–16 hours after the last (evening) meal, and on investigation of familial incidence of similar symptoms.

correlation between the plasma cholesterol level and the rate of synthesis but an inverse relationship between the plasma level and the excretion rate, suggesting that cholesterolaemia may be due to a defect in clearance or degradation.

Several inborn errors of lipid metabolism, manifesting particularly in abnormally high concentrations of one or other of the plasma lipoproteins, or of chylomicrons, have been established in recent years. The agreed classification is shown in Table 10.2. The underlying biochemical defects are still not clearly known, but they are being intensively investigated because of resemblances between their symptoms and those of hyperlipoproteinaemia in the population at large.

Degradation. While cholesterol is the parent substance from which oestrogens, androgens, corticoids and progesterone are derived, from a quantitative standpoint the most important degradative pathway is to the bile acids.

The reactions involved in the degradation of cholesterol to bile acids are incompletely understood, but it appears that one of the earliest reactions is an oxidative attack on the 'nucleus' of the molecule at C–7, followed by other modifications to the nucleus producing ultimately a hydroxylated coprostane structure (cf. p. 44) containing 27 carbon atoms. Probably one of the final stages is the cleavage of the side chain to produce molecules with 24 C atoms such as cholic acid, which are then conjugated with either glycine or taurine in the liver to produce bile salts.

About 1 g of cholesterol is excreted each day in the bile. It occasionally forms 'stones' in the gall-bladder. Much of it is found in faeces as coprostanol. About 2 g of bile salt are excreted each day, of which about 60 per cent is reabsorbed. The conversion of cholesterol to bile salts is inhibited by the accumulation of reabsorbed bile salt in the liver. A quite effective treatment for cholesterolaemia is the continuous ingestion of an ion-exchange resin of the type that binds anions of weak acids, but not Cl^-. As the resin with the bile acids bound to it is lost in faeces, the reabsorption of bile acids is necessarily reduced and the rate of conversion of cholesterol to bile acids is increased.

Prostaglandin Synthesis

The precursors of these substances are the C_{20} poly-unsaturated acids homolinolenic or arachidonic (cf. Fig. 3.6, p. 42). Ring formation follows attack at the 8- and 11-double bonds by an oxygenase, with the two atoms of oxygen from one molecule of O_2 appearing on either side of the prostaglandin ring. A mixed function oxidase inserts the hydroxyl at C–15.

Propionate Catabolism

It is clear that if an odd-numbered fatty acid is degraded by β-oxidation, the terminal fragment of the chain will not be acetyl-, but *propionyl*-CoA,

$CH_3 \cdot CH_2 \cdot CO \cdot S \cdot CoA$. Although fatty acids with odd numbers of carbon atoms are rare in triglycerides, propionyl-CoA is a product of catabolism of a number of amino acids (Chapter 11). In herbivorous animals considerable quantities of acetic and propionic acids are formed by the action of rumen bacteria on the carbohydrates of the fodder, and thus plasma propionate is an important energy-producing substrate in these animals. It has long been known to be glycogenic.

Propionyl-CoA is carboxylated by a biotin-containing enzyme (cf. p. 145) to methylmalonyl-CoA. This is converted to succinyl-CoA, and hence to succinate:

$$CH_3 \cdot CH_2 \cdot CO \cdot S \cdot CoA + CO_2 \rightarrow CH_3 \cdot \underset{\underset{COOH}{|}}{CH} \cdot CO \cdot S \cdot CoA$$

Methylmalonyl-CoA

$$\downarrow$$

$$\leftarrow HOOC \cdot CH_2 \cdot CH_2 \cdot CO \cdot S \cdot CoA$$

Succinyl-CoA

$$HOOC \cdot CH_2 \cdot CH_2 \cdot COOH + HS \cdot CoA$$
Succinic Acid

The unusual reaction in which the methyl group of methylmalonyl-CoA is inserted into the main carbon chain, is catalysed by an enzyme containing a vitamin B_{12} coenzyme (p. 150). Succinic acid is oxidized to malic acid, and can be converted to carbohydrate, by the reactions outlined in Chapter 9.

In rare inborn errors in which propionyl-CoA carboxylase or methylmalonyl-CoA mutase are absent, propionic or methylmalonic acids are excreted in the urine in amounts of 2 g per day. Presumably this chiefly represents the daily rate of formation of propionate from the catabolism of methionine and isoleucine (Chapter 11).

The Disposal of Acetyl-Coenzyme A

1. The most important reaction which acetyl-CoA undergoes is condensation with oxaloacetic acid to give citric acid and free $CoA \cdot SH$:

$$CH_3 \cdot CO \cdot S \cdot CoA + HOOC \cdot CO \cdot CH_2 \cdot COOH + H_2O \rightarrow$$
$$HOOC \cdot CH_2 \cdot \underset{\underset{COOH}{|}}{C(OH)} \cdot CH_2 \cdot COOH + HS \cdot CoA$$

The coenzyme A can then be used again to further the β-oxidation cycle.

This mechanism depends on the steady supply of oxaloacetic acid, and oxidation of the citric acid formed, in other words on the proper functioning of the tricarboxylic acid cycle.

2. In the cytoplasm, acetyl radicals can be used for fatty acid synthesis, via the malonyl-CoA pathway. This is of particular importance in liver and fat cells. Since fatty acyl-CoA molecules or free fatty acids do not accumulate in cells, and palmityl-CoA is known to inhibit acetyl-CoA carboxylase, it is tempting to suppose that the rate of fatty acid synthesis is effectively controlled by the rate at which α-glycerophosphate becomes available for triglyceride synthesis, thus removing the palmityl residues. One must however remember that triglycerides do not contain exclusively palmityl residues: the chain-lengthening and desaturating reactions which lead to the formation of oleyl-CoA must be very important in regulating triglyceride synthesis in vivo.

3. Another problem is also of importance. Acetyl-CoA does not readily cross the mitochondrial membrane (p. 272), and yet the acetyl residues which are the substrates for fat synthesis in the cytoplasm are produced, either by pyruvate oxidation or β-oxidation, in mitochondria. Acetyl-carnitine has been proposed as a transporting agent, but the highest concentrations of carnitine, acetyl carnitine and the transferase enzymes (p. 190), are found in tissues in which fat synthesis is unimportant. A citrate cleavage enzyme, *citrate lyase*, is found in the cytoplasm of many cells and catalyses the following reaction:

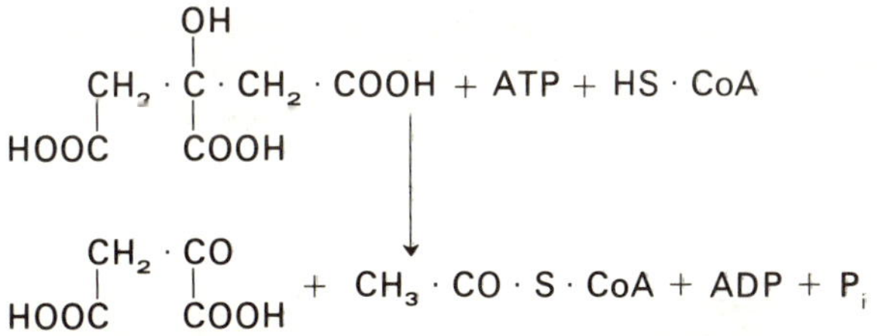

$$
\begin{array}{c}
\quad\quad OH \\
\quad\quad | \\
CH_2 \cdot C \cdot CH_2 \cdot COOH + ATP + HS \cdot CoA \\
| \quad\quad | \\
HOOC \quad COOH
\end{array}
$$

$$
\begin{array}{c}
CH_2 \cdot CO \\
| \quad\quad | \quad + \quad CH_3 \cdot CO \cdot S \cdot CoA + ADP + P_i \\
HOOC \quad COOH
\end{array}
$$

Oxaloacetate   Acetyl CoA

The oxaloacetate which is formed could either be reduced to malate or transaminated to aspartate. Both these substances enter mitochondria rather easily. The concentration of the citrate cleavage enzyme in many species is highest in those tissues which synthesize most fat, and a good deal of evidence has accumulated to suggest that the level of activity varies with the rate of fatty acid synthesis in various dietary and hormonal states. There is, however, still dispute about the importance of citrate. In some species citrate lyase is a mitochondrial enzyme, and not all workers accept that citrate readily leaves mitochondria, which will, in many circumstances, accumulate it against a concentration gradient. Finally, the distribution of labelled hydrogen atoms in synthesized fatty acids speaks against the role of citrate. The way in which acetyl units cross the mitochondrial membrane must still be regarded as unestablished.

4. Two acetyl radicals can condense to form acetoacetyl-CoA. HMG-CoA can be formed in a subsequent reaction as described on p. 198; this can be the beginning of steroid synthesis, or HMG-CoA can be cleaved to form free acetoacetate, as discussed in more detail under ketones, below.

5. Acetyl radicals are used in various detoxification reactions (Chapter 13), and for the synthesis of acetyl-choline, which is important in nervous tissue:

$$CH_3 \cdot CO \cdot S \cdot CoA + HO \cdot CH_2 \cdot CH_2 \cdot \overset{+}{N}(CH_3)_3 \rightarrow$$

$$CH_3 \cdot CO \cdot O \cdot CH_2 \cdot CH_2 \cdot \overset{+}{N}(CH_3)_3 + HS \cdot CoA$$

It is not, of course, quantitatively important in fat metabolism as a whole.

Ketones

Acetoacetic Acid

Acetoacetic acid and its reduction product *β-hydroxybutyric acid* accumulate in the blood in certain states, particularly of carbohydrate deficiency such as starvation or diabetes. Accumulation may also occur spontaneously, especially in children. In principle, free acetoacetate can be formed in most tissues by the splitting of HMG-CoA:

$$
\begin{array}{ccc}
& CH_2 \cdot CO \cdot S \cdot CoA & CH_3 \cdot CO \cdot S \cdot CoA \\
& | & \\
CH_3-\underset{|}{\overset{|}{C}}-CH_2 \cdot COOH \rightarrow & + \\
OH & CH_3 \cdot CO \cdot CH_2 \cdot COOH \\
\textit{Hydroxymethyl-} & \textit{Acetoacetic} \\
\textit{glutaryl CoA} & \textit{acid}
\end{array}
$$

A deacylase found only in liver can also produce free acetoacetate, by hydrolysis of acetoacetyl-CoA.

Acetoacetate is readily reduced to D(—)-β-hydroxybutyrate by a NADH$_2$-requiring enzyme found in most tissues, so that both acids accumulate together.

$$CH_3 \cdot CO \cdot CH_2 \cdot COOH + NADH_2 \rightleftharpoons$$

Acetoacetic acid

$$CH_3 \cdot CHOH \cdot CH_2 \cdot COOH + NAD$$

β-hydroxybutyric acid

The ratio of the two acids depends on the $NADH_2/NAD$ ratio *in mitochondria*.

Acetoacetate is a β-keto acid and spontaneously undergoes decarboxylation (cf. oxaloacetic acid, Chapter 12), but the process is accelerated by a decarboxylase

$$CH_3 \cdot CO \cdot CH_2 \cdot COOH \rightarrow CH_3 \cdot CO \cdot CH_3 + CO_2$$

Acetoacetic acid   *Acetone*

Traces of acetone are therefore found in the blood in ketotic states, hence the name *ketone bodies* for the three related compounds. As acetone is volatile it may be smelt on the breath.

Although ketone body production need not be a purely hepatic phenomenon, it is nevertheless certain that the ketone bodies found in blood in ketonaemia come from the liver. This was established many years ago, on the basis of the fall in ketone body concentration after experimental hepatectomy. The concentration of the two acids in blood at any time will be the resultant of the rates of production, excretion, and utilization (see Fig. 10.3). Acetoacetate is readily utilized by non-hepatic tissues, particularly muscles, in exercise as well as at rest. In fact, most tissues other than liver contain a transferase, which reactivates acetoacetate by CoA transfer from succinyl-CoA, a normal intermediate of the citric acid cycle (Chapter 12):

$$HOOC \cdot CH_2 \cdot CH_2 \cdot CO \cdot S \cdot CoA + CH_3 \cdot CO \cdot CH_2 \cdot COOH \rightarrow$$

Succinyl-CoA   *Acetoacetic acid*

$$HOOC \cdot CH_2 \cdot CH_2 \cdot COOH + CH_3 \cdot CO \cdot CH_2 \cdot CO \cdot S \cdot CoA$$

Succinic acid   *Acetoacetyl-CoA*

There is a rather low concentration of acetoacetyl-transferase in brain; the utilization of ketone bodies by brain is negligible at low blood ketone levels, but it has been shown, by A–V difference studies, that a significant fraction of the energy needs of brain can be met from acetoacetate in moderate or severe ketonaemia. The enzyme is absent from liver, so that any free acetoacetate which is formed in this tissue must necessarily diffuse out into the blood. β-Hydroxybutyrate is not activated in this way and indeed is very slowly metabolized, which is perhaps why it usually makes up most of the total ketone body concentration in blood or urine.

There are always traces of ketone bodies in blood and also in urine, since there is no renal threshold for the two acids. A fast of less than 24 hours may produce a significant ketonaemia, and after 3–4 days the blood ketone level may be around 20 mg/100 ml. After this the ketones only increase slowly, and an average excretion of 6 g β-hydroxybutyrate per day at the end of a 31-day fast has been recorded. In low carbohydrate-high fat diets the ketosis is proportional to the fat loading. Ketosis does not develop on high protein–high fat diets because the dietary amino acids can be used for gluconeogenesis.

The ketosis of starvation is not severe enough to cause any physiological disturbance, e.g. of acid–base status. In uncontrolled juvenile-type diabetes, however, the blood ketone body concentration may be 50–100 mg/100 ml with 50 g/day appearing in the urine. The glycosuria in itself creates problems, and the reduction in plasma bicarbonate and the loss of cations in urine in such severe ketosis are very dangerous (see Chapter 14).

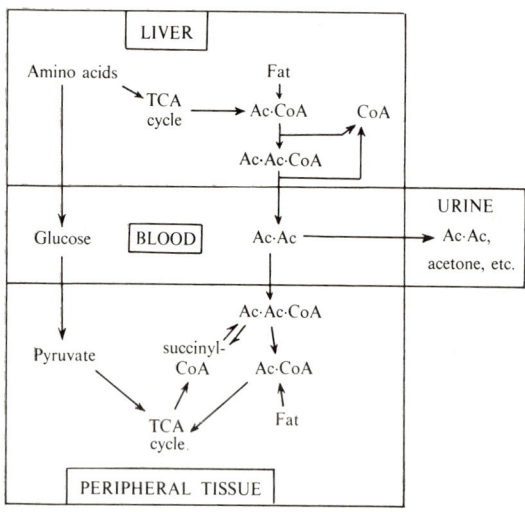

Fig. 10.3. Fat metabolism when carbohydrate is deficient. Ac represents CH_3CO and TCA represents tricarboxylic acid.

The overall picture is summed up in Fig. 10.3, which makes it clear that the amount of acetoacetic acid oxidized by the peripheral tissues depends both on the amount formed by the liver and secreted into the blood, and also on the rate of reactivation, which will depend on the activity of the TCA cycle.

The amount excreted represents fat which could have been converted to energy, and which has been lost to the organism, but the amount should not be exaggerated. There is no record of more than 50 g of ketone bodies per day being excreted, even in a severe diabetic not treated with insulin.

This is equivalent to about 25 g stearic acid or about 900 kilojoules. The clinical problem of ketonuria is not the loss of energy but the loss of Na^+ or K^+ which has to be excreted with the anions of these acids.

Ketosis

It is known that liver can produce more than one molecule of acetoacetate per molecule of fatty acid oxidized, so that at least some must be formed by the condensation of two acetyl units. The equilibrium constant for the reaction, i.e.

$$K_{eq} = \frac{[\text{Acetoacetyl-CoA}] [\text{CoA}]}{[\text{Acetyl-CoA}]^2}$$

is, however, 10^{-5}, which is very unfavourable to acetoacetyl-CoA formation, particularly as it has been found that the ratio of [free CoA] to [acetyl-CoA] is about unity in a variety of conditions. These facts probably explain why ketone body production is undetectable in tissues other than liver, but they do not explain how it can occur on a large scale in that organ. The mitochondrial synthesis of fatty acids (p. 191) starts with an energy-dependent condensation of acetyl-CoA residues, but there has never been any suggestion that this mechanism might operate in ketosis.

It is generally agreed that the major pathway of acetoacetate formation is via HMG-CoA. Hydroxymethylglutaryl-CoA synthetase is inhibited by cholesterol, but no effects of fasting or hormonal imbalance have been reported. On the other hand, HMG-CoA lyase is activated, perhaps tenfold, by fasting; the mechanism is unknown. This activation, and the consequent disturbance of the thiolase equilibrium (above), could account for the production of a few grammes per day of ketone bodies, but not for the severe ketonaemia and ketonuria, and intensive ketone body oxidation in tissues, that characterize uncontrolled diabetes.

Two theories have been developed to explain this massive synthesis of acetoacetate: underutilization and overproduction. The best-known form of the former hypothesis really starts from the observation that the oxidation of quite a small amount of glucose per day—perhaps 150 g—will suppress ketosis, and is an extension of the old adage 'fat burns in the flame of carbohydrates'. The suggestion is that the TCA cycle slows down in fasting or diabetes, because of a lack of oxaloacetate, which can only be replenished from pyruvate. This view, however, is untenable for the following reasons:

1. Liver oxaloacetate concentrations do not decrease markedly on fasting.

2. The flux from fumarate to oxaloacetate which is connected with the urea cycle (p. 224) does not decrease in fasting, and may even increase in diabetes.

3. Ketosis is usually associated with gluconeogenesis, and almost all the carbon atoms of glucose precursors must pass through oxaloacetate in order to bypass pyruvate kinase. These two points argue against any lack of oxaloacetate in ketotic states.

4. The TCA cycle can indeed be shown to slow down in extreme ketosis, but for a different reason : the rate of β-oxidation has become so rapid that a significant fraction of the ATP requirements of the liver can be met from this process alone (ATP synthesis associated with the reoxidation of co-enzymes reduced in the β-oxidation of a long-chain fatty acyl is one-third of that formed in the subsequent oxidation of the acetyl residues). How-ever, this cannot be regarded as a normal adjustment of metabolism, as one may see by imagining the consequences if a large part of the oxidation of glucose stopped at pyruvate during fasting.

Underutilization of acetyl residues does occur in ketotic states, in the sense that fatty acid synthesis is sharply reduced, even in fasting. There are several possible reasons for this discussed in the next section of this chapter. The rate of recycling of acetyl residues to and from long-chain fatty acids in liver tends to be under-estimated, so that the imbalance between production and utilization becomes large, if the catabolism of fatty acids is not curtailed to meet the changed circumstances. It is worth remarking that if the acetyl units produced by β-oxidation have to be returned to the cytoplasm through citrate, citrate synthase activity would be reduced in ketotic states, although not the activity of the whole TCA cycle. However, as discussed on p. 205, there are still objections to the citrate hypothesis.

The general conclusion from the foregoing is that even if the rate of fatty acid activation were to be unchanged in fasting and diabetic ketosis, this would lead to overproduction of acetyl residues, taking into account the reduced rate of fat synthesis. However, there is much evidence that the rate of activation is actually increased. It appears that in liver, activa-tion of fatty acids is proportional to the tissue non-esterified fatty acid concentration, which in turn is related to the plasma NEFA level. There is no feedback control of the initiation of catabolism.

Factors Affecting Fat Metabolism as a Whole

One cannot overemphasize the importance of fat as a source of energy, except during the period immediately after the absorption of carbohydrate from the gut. Even in muscle, the evidence is that the oxidation of glucose provides only a small proportion of the energy at rest while fasting, although the proportion may rise during exercise. Carbohydrate is the sole source of energy for brain, except in ketosis (p. 208).

A calculation for the basal state (i.e. resting, and after a 12-hour fast) gives the figures shown in Table 10.3.

Table 10.3

Substrate	Site of oxidation	Required O_2 consumption ml/min	Percentage of total
Glucose	Chiefly brain, but all tissues	45	18
Glycerides	All tissues	70	28
Amino acids ⎰NEFA	Chiefly liver ⎱ * Liver ⎰	58	23
⎱NEFA	Extrahepatic tissues	62	25
Ketone bodies	Extrahepatic tissues	15	6
	Total body	250	100

* Total liver O_2 consumption $\frac{1}{3}$ of that of the whole body.

Independent measurements have shown that 15 per cent of the CO_2 production of a fasting animal at rest comes from glucose. The precise figures must be quoted only with caution, and apply only to the state defined, but they are sufficiently striking. The ketone body oxidation, calculated for a blood ketone level of 4–5 mg/100 ml, is not negligible; in serious ketosis, the oxidation of these substances may supply most of the energy requirements of resting muscle.

It is impossible to discuss fat metabolism without considering carbohydrate metabolism.

1. If glucose is readily available, i.e. after a meal containing carbohydrate, fat synthesis will occur. Isotopically labelled carbon atoms from glucose are found to a considerable extent in fat, even if the organism is not laying down 'permanent' fat. The amount of glucose which can be oxidized depends on the activity of the tricarboxylic acid cycle, and thus is strictly limited by the energy needs of the tissues (and to some extent by the other substrates available). That glucose which cannot be oxidized must be converted to glycogen or catabolized to pyruvate. The only ways of utilizing the C atoms of pyruvate are to oxidize them by way of citrate, or use them for fatty acid synthesis. The alternative is unimportant in muscle, but very important in adipose tissue and liver. It is striking that very little fat synthesis takes place in the absence of insulin, even if there is hyperglycaemia (Chapter 18). Fat synthesis requires, besides the carbon atoms which actually form the acetyl residues, two molecules of $NADPH_2$ per acetyl. Some dehydrogenases of the common terminal pathway reduce NADP (isocitric, glutamic, malic enzyme), but probably the major source of $NADPH_2$ is the hexose shunt (p. 167). There is good, though not complete, correlation between the rate of hexose shunt oxidation in tissues and their rate of fatty acid synthesis. The final formation of triacylglycerols requires glycerol-3-phosphate, which comes (particularly in adipose tissue) much more from the catabolism of glucose than from the recycling

of glycerol. Since fatty acyl-CoAs inhibit their own synthesis, at the malonyl-CoA carboxylase level, continuous synthesis of fat depends very much on their incorporation into triglyceride, and thus on the oxidation of carbohydrate. This may be one reason why fat synthesis is suppressed when ketosis is evident (p. 210). The rate of formation of oleyl and other unsaturated fatty acyls, which are components of normal triglycerides, and the rate of hydrolysis of phosphatidic acid, may be other controlling factors in triglyceride synthesis.

2. If carbohydrate is ingested, fat oxidation is dramatically reduced. One investigator found that 45 per cent of a dose of albumin-bound palmitate was converted into CO_2 in an hour by fasted rats; after the animals had been fed glucose the percentage fell to 2. There seem to be several reasons for this. For example, the uptake of NEFA by tissues depends on the concentration in plasma. There is strong evidence that the rate of catabolism of fatty acids in liver, at least, is determined by the amount of available substrate, and that feedback control is rather inefficient. In carbohydrate lack of fatty acid content of the liver usually rises from its normal value of 5 per cent. In extreme cases it may reach as much as 30 per cent. In other tissues the imbalance may be less marked. Glucose, particularly if insulin is also present, inhibits NEFA secretion by adipose tissue, and there is a rapid fall in plasma NEFA after glucose administration. To some extent this may be due to an increased rate of triglyceride synthesis made possible by an increased rate of phosphoglycerol production, but it has been observed that ingestion of glucose causes a fall in NEFA uptake, and a reduction in fatty acid oxidation, which is greater than can be explained by this and which occurs in tissues in which fatty acid synthesis is negligible. No explanation of this is at present generally accepted.

Although increased lipolysis in adipose tissue is not always accompanied by acetoacetate production, severe ketosis is certainly often accompanied by high blood NEFA levels. Injection of fat—to avoid complications caused by gastric nausea—can cause ketosis in the absence of starvation or hormone deficiency. One of the most striking pieces of evidence is that ketosis is common in certain glycogen storage diseases, particularly type I, which is due to an absence of glucose-6-phosphatase. In this disease the liver can metabolize glucose perfectly well, and indeed raised concentrations of lactic and pyruvic acids in blood are frequent, showing that catabolism of glucose in liver is greater than necessary for the needs of that organ. The subjects cannot, however, secrete glucose from liver into blood, and suffer from fasting hypoglycaemia. This affects the adipose tissue, either directly, or more probably because of the secretion of adrenaline, a very potent lipolytic agent. Thus ketosis need not be associated with any reduction in the ability to break down glucose.

At the present time, therefore, the cause of ketosis appears to be due to

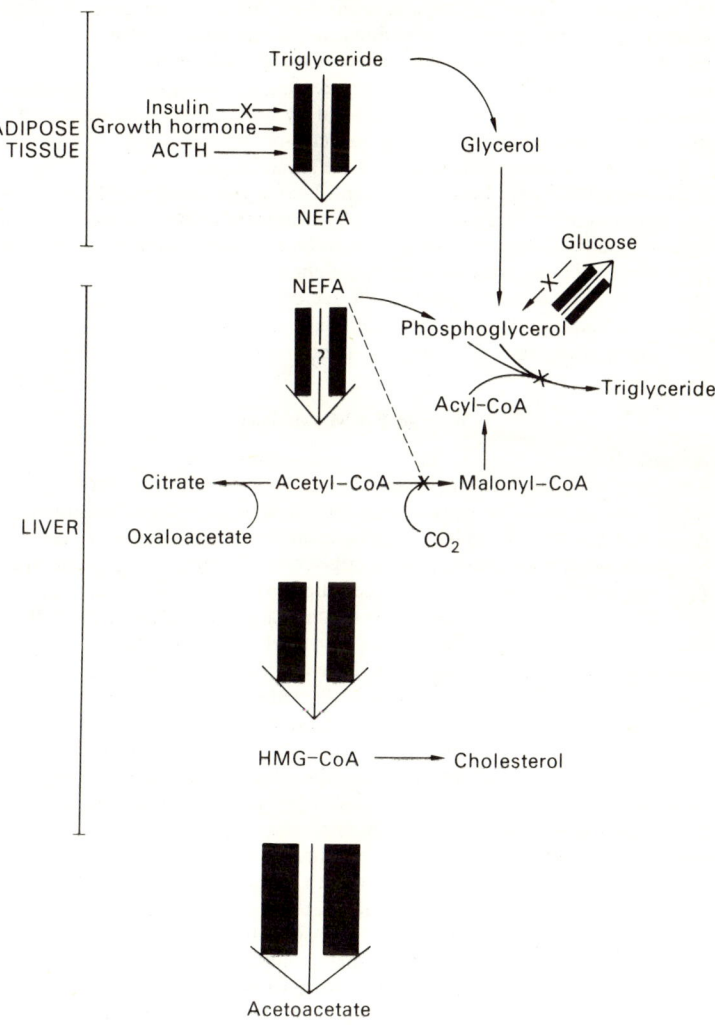

FIG. 10.4. Possible metabolic imbalances in ketosis. Thickened arrows indicate reactions whose rate is increased. X indicates reactions which are inhibited.

a relative or absolute overproduction of fatty acyl-CoA in liver, i.e. a rate out of balance with the hepatic utilization of acetyl residues through oxidation and fat synthesis combined, in the metabolic circumstances in which the organism finds itself. It seems likely that the small quantity of carbohydrate which is effective in suppressing ketosis acts by stimulating insulin secretion, especially if it is taken by mouth, and perhaps also by making possible some synthesis of glycerol-phosphate.

Fig. 10.4 gives a simplified view of the situation.

On the other hand, both NEFA and ketone bodies inhibit the utilization and oxidation of glucose. This effect appears to operate both at the level of glucose transport across the membrane (at least in muscle), and on pyruvate oxidation, and it is independent of those effects of fasting or diabetes which can be attributed to the decrease in active pyruvate dehydrogenase on fasting, or simply to insulin lack. Inhibition of pyruvate transport into mitochondria may be a contributory cause.

Sites of Fat Metabolism

Adipose Tissue

On a protein basis, adipose tissue is an active one. It can synthesize fatty acids from glucose by way of acetyl-CoA, and triglycerides from fatty acids and glycerol phosphate. The neutral fat content of the tissue also increases by the accumulation of triglycerides from both plasma lipoproteins and chylomicrons. Fat cell lipoprotein lipase appears to be required for this. The fat cells also contain a triglyceride lipase which is responsible for the production of the non-esterified fatty acids (and glycerol) which appear in plasma. These are the lipids which are oxidized by other tissues. This second lipase is affected by a number of factors.

Cyclic AMP (p. 83) activates the lipase in extracts of adipose tissue, and the lipolytic activity of fat cells themselves is very strongly correlated with their cyclic AMP content (for the mechanism of formation and destruction of this effector, see p. 374). The cyclic AMP content, and the rate of lipolysis, in adipose tissue in vitro rise very rapidly after the introduction of adrenaline and noradrenaline, ACTH, glucagon and TSH (Chapter 18). It appears that this is due to the activation of adenyl cyclase, and the roles of adrenaline and glucagon are very similar to those postulated for these hormones in the control of glycogenolysis. Growth hormone and adrenal cortical steroids, on the other hand, do not act immediately in vitro and the effect of growth hormone in particular is secondary to induction of RNA synthesis in the cells (and presumably of more lipase molecules).

Insulin, in the presence of glucose, strongly inhibits NEFA secretion from adipose tissue. It has been shown to lower the intracellular concentration of cyclic AMP when this has previously been raised by, e.g. adrenaline, but its main effect seems not to be associated with cyclic AMP. The actual rate of fatty acid loss from this tissue, as distinct from the rate of intracellular lipolysis, depends on the rate of resynthesis of triglycerides, and in this insulin can play an important part by accelerating the formation of dihydroxyacetone phosphate; and hence of glycerol phosphate from glucose. Adipose tissue contains both an active triglyceride lipase, which has a low activity against monoglycerides, and a powerful *monoglyceride lipase*, which liberates free glycerol as well as fatty acid. Since glycerol is poorly phosphorylated by adipose tissue, it is lost to the blood stream, and new triglyceride synthesis depends almost entirely on the provision of glycerol phosphate. Insulin also increases the rate of fatty acid synthesis *de novo*.

The in vivo effectiveness of these hormones appears to be much as predicted from the studies in vitro, except that growth hormone increases plasma NEFA levels more than would be expected, and the reduction in plasma NEFA induced by glucose feeding or insulin is very dramatic indeed. There is fragmentary evidence that other lipid-mobilizing factors, perhaps peptides of pituitary origin, exist.

Liver

This organ is the site of synthesis of most of the cholesterol and cholesterol esters in plasma, and of the degradation of endogenous and dietary cholesterol to bile acids. It also synthesizes much of the phospholipids and protein of the plasma lipoproteins. It absorbs some of the triglycerides from intestinal chylomicrons and rearranges them into the triglycerides carried by the lipoproteins. In view of the relation between hypercholesterolaemia and cardiovascular disease, there is much interest in the rate of secretion of the lipoproteins and in their sterol contents.

Apart from this, there is no reason to believe that the liver is responsible for all the fat metabolism in the body, or that fats from fat depots have to be metabolized in some way by liver before they can be oxidized by other tissues. Ketone body production seems to be the result of a badly controlled rate of β-oxidation in an organ which readily oxidizes fat.

Fatty livers occur in many conditions, such as poisoning by CCl_4, primary or induced choline deficiency, many kinds of malnutrition, particularly in infants, and ketosis. There appears to be no common factor in all these phenomena, other than the sensitivity of lipid metabolism in liver to disturbance.

Intestine

The mucosal cells of the small intestine, including the jejunum, are active sites of lipid metabolism. Apart from absorbing the products of digestion by pancreatic enzymes, they resynthesize triglycerides, possibly also synthesizing some fatty acids *de novo*. The intestine is the only tissue in the body which has an enzyme capable of acylating monoglycerides. Besides this, these cells are the site of perhaps 40 per cent of all the cholesterol and cholesterol ester synthesis in the body, and also synthesize the phospholipids and some of the polypeptides that are necessary to give stability to chylomicrons and very low density lipoproteins, which are secreted into the lymph.

Bile salts are reabsorbed particularly in the ileum.

11   Amino Acid Metabolism

The chemistry of amino acids has been dealt with in Chapter 4.

The Dynamic Equilibrium of Body Proteins

The concept of a dynamic equilibrium of protein introduced by Schoenheimer in 1939, is fundamental to an understanding of present views. The concept was a result of studies on the metabolism of amino acids labelled with ^{15}N. When glycine, for example, containing ^{15}N in the amino group was fed to animals, a large percentage of the isotope was excreted within the next 24 hours mostly in the amino groups of urea. The actual percentage depended on the level of protein intake, but a significant amount was excreted much more slowly, so slowly that it could hardly have gone through the same metabolic processes as the ^{15}N at first excreted. Study of the carcasses of animals fed labelled amino acid showed that a great deal of it had been incorporated into tissue proteins (Table 11.1), not only

Table 11.1
Distribution of Isotope Several Days after Feeding ^{15}N
Glycine to a Rat (on Diet Containing 15 per cent Protein)

Isotope recovered from	Percentage of isotope fed
Urine (as urea and NH_3)	30
Protein (as glycine)	20
Protein (as other amino acids)	40
Non-protein N in tissues (porphyrins, purines, etc.)	10
	100

into proteins such as those of the epithelium and of the digestive enzymes, which are known to be destroyed and re-formed throughout life, but also into the enzyme protein of non-dividing cells such as muscle, into connective tissue, and even into the collagenous matrix of bone. These experiments can be repeated with any other isotopically labelled amino acid; although the quantitative picture may be different, as different amino acids are needed in differing amounts for the synthesis of non-protein nitrogenous materials, the qualitative picture is the same. The only explanation for this must be that even in the adult animal, proteins are continually being broken down and re-formed; they are in *dynamic equilibrium* with amino acids, and the life of a protein molecule varies considerably among the various tissues. This is usually expressed as the *half-life*, i.e. the time in which half the protein will be replaced. The half-lives of a few enzymes in liver are only 1–2 hours; many are only a day or two; liver endoplasmic reticulum in the rat has a half-life of 2–3 days, and the mitochondria about 7 days. On the other hand, the half-life of the most abundant protein in the body, collagen, is about 6 months. The *turnover* of the major muscle proteins lies between these two extremes.

This dynamic equilibrium means that of all the amino acids being catabolized at any moment, only a proportion has come directly from the food proteins, the rest has come from the endogenous tissue protein. Conversely, only a proportion of the amino acids absorbed from the gut by the adult animal in nitrogen equilibrium will be directly catabolized; some will be incorporated into tissue protein. The catabolism of amino acids is irreversible beyond a certain point; in some instances the carbon skeleton of the amino acid cannot be resynthesized in animals, and in all cases the incorporation of the α-amino group into urea is an irreversible process. Unless new amino acid molecules are provided in the diet, therefore, the amount of tissue protein will always be decreasing. This argument is also important during growth, including repair of surgical injury. Although children, for instance, can use dietary amino acids quite efficiently, it is always necessary to provide more protein than would be calculated to be necessary from the rate of increase in weight (see Chapter 19).

The continual catabolism of amino acids, whether from dietary or endogenous protein, can be reduced to a minimum if the energy intake is generous. The so-called *protein-sparing* effect of carbohydrate, for instance, depends on the fact that gluconeogensis from amino acid carbon skeletons is unnecessary if sufficient carbohydrate is being supplied in the diet. (See also Protein–Calorie Malnutrition, Chapter 19.)

The Concept of a Metabolic Pool

If a single isotopically labelled substance is injected into an animal, and the amount of isotope appearing in other compounds is measured at various time intervals, it is possible, by making certain assumptions about kinetics, to calculate the total amount of substance—labelled and unlabelled—in the body at the time of injection. For instance, injection of an

[15]N-labelled amino acid, followed by measurement of rate of appearance of isotope in protein or urea, can give a figure for the total amount of this amino acid at zero time. This is called the *metabolic pool* of that amino acid.

It is important to realize that this is a mathematical abstraction which does not always correspond to an anatomical reality. The amino acid pool, for instance, consists of free amino acid in plasma, free amino acid in cells (in different concentration in different tissues), and activated amino acid on tRNA. In an extreme example, if acetate is injected into the blood stream, within a very few minutes there is no free acetate in the plasma, and the acetate pool will consist of molecules of acetyl-CoA in different cells, with no physical contact between them. It has been found that within cells not all the molecules of a single amino acid are in isotopic equilibrium; there is clear evidence of *compartmentation* even at this level. The term metabolic pool nevertheless has considerable usefulness in intermediary metabolism.

Amino Acid Catabolism: General

In the past the terms protein metabolism and amino acid metabolism meant essentially the same thing, since very little was known about the synthesis and breakdown of protein in vivo. We now know enough about protein synthesis for it to merit a chapter of its own. This chapter is concerned only with the metabolism of the amino acids. It may be divided into two sections: a consideration of pathways common to all amino acids, and of the ways in which the nitrogen is excreted, and then some discussion of the individual pathways, and some of the more important syntheses of non-protein nitrogenous compounds from amino acids.

Some eight of the 20 amino acids commonly found in proteins are *essential* (see Chapter 19); that is to say, their carbon skeletons cannot be synthesized in the animal body. A discussion of the way in which they are synthesized by plants or micro-organisms is therefore appropriate to a textbook of general biochemistry, and will not be given here.

The general catabolic pathways involving amino acids are indicated by the following diagram:

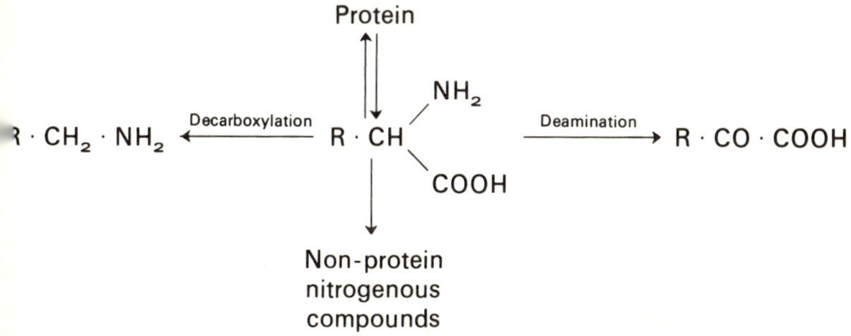

It is firmly established that the first stage in the catabolism of almost all amino acids is the formation of the corresponding oxo-acid

$$\underset{\overset{|}{\text{COOH}}}{\text{R}-\text{CH}-\text{NH}_2} + \tfrac{1}{2}\text{O}_2 \rightleftharpoons \underset{\overset{|}{\text{COOH}}}{\text{R}-\text{CO}} + \text{NH}_3 \qquad (1)$$

The last stage is the formation of urea, $\text{NH}_2 \cdot \text{CO} \cdot \text{NH}_2$, and of CO_2 and water.

Research has established the existence of an L-*amino-acid oxidase*, directly catalysing reaction (1) above, in animal tissues. The enzyme, however, is present in very low concentration even in tissues, such as liver or kidney, in which amino acid catabolism is vigorous. Most of the amino groups must thus be removed by another route, discussed in the next paragraph. Curiously, D-*amino-acid oxidase* is present in high concentration in kidney in particular. D-Amino acids occur infrequently in some peptides, but are not quantitatively important. It is probable that the D-amino acid oxidase is a detoxifying enzyme, since it is important, from the point of view of protein structure, to prevent the incorporation of un-natural isomers into peptide chains.

Transamination

If to a liver or kidney homogenate is added an amino acid and α-oxo-glutaric acid, the reaction mixture will soon contain, besides the original components, the oxo-acid corresponding to the amino acid and also glutamic acid; the enzyme carrying out this reaction is a *transaminase*:

$$\underset{\overset{|}{\underset{\text{COOH}}{\underset{\text{A}}{|}}}}{\text{CH} \cdot \text{NH}_2} + \underset{\overset{|}{\underset{\text{COOH}}{\underset{\text{B}}{|}}}}{\text{CO}} \rightleftharpoons \underset{\overset{|}{\underset{\text{COOH}}{\underset{\text{C}}{|}}}}{\text{CO}} + \underset{\overset{|}{\underset{\text{COOH}}{\underset{\text{D}}{|}}}}{\text{CH} \cdot \text{NH}_2} \qquad (2)$$

In the example mentioned, B was α-oxoglutaric acid. The reaction is reversible, so that no matter from which direction it is started, the reaction mixture will finally contain all four components.

This *transamination* system is now accepted as the most important way of deaminating amino acids. Confusion may arise unless certain points are understood:

1. The amino *donor* A may be almost any one of the normally occurring amino acids but the *acceptor* B can only be α-oxoglutaric acid, or, less generally, pyruvic acid. There are thus two groups of transaminases, each

group specific for one of these two oxo-acids. Confusion may occur if the reaction is thought of as running from *right to left* in equation (2); now the two groups of transaminases are specific for 2 amino donors (D)—glutamic acid or alanine—while the acceptor (C) may be any one of a number of oxo-acids.

2. Although this reaction produces an oxo-acid (C) from an amino acid (A), it is reversible. (A) will not continue to be deaminated unless the reaction products (C) and (D) are continuously removed. There must, therefore, be *another* mechanism for deaminating glutamic acid (or alanine), i.e. one which does not depend on a transamination.

It is significant that the amino group acceptors (B) for the two groups of transaminases are pyruvic acid, the end product of carbohydrate metabolism, and α-oxoglutaric acid, which is part of the tricarboxylic acid cycle. There will thus always be, so long as carbohydrate metabolism is functioning, ample supplies of the requisite acceptors.

Transamination is also important as a means of changing the proportions of amino acids available for synthesis. It is clear that the proportions of the constituent amino acids in the proteins of food are not necessarily related to the proportions, or absolute amounts, of each of the amino acids required by the body for synthetic purposes. If two transamination reactions are coupled together as shown below

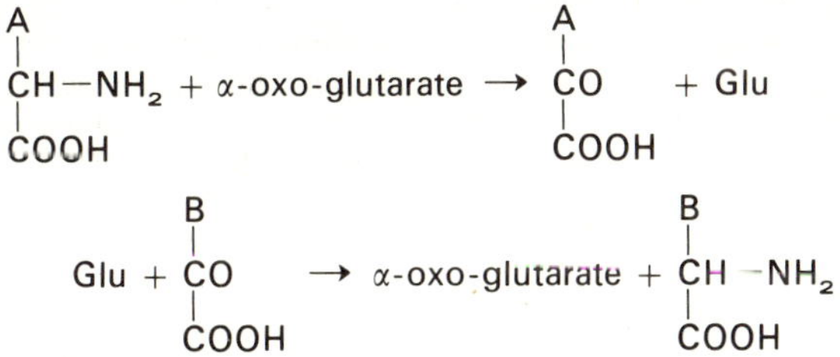

$$
\begin{array}{c}
\text{A} \\
| \\
\text{CH}-\text{NH}_2 \\
| \\
\text{COOH}
\end{array}
+ \alpha\text{-oxo-glutarate} \rightarrow
\begin{array}{c}
\text{A} \\
| \\
\text{CO} \\
| \\
\text{COOH}
\end{array}
+ \text{Glu}
$$

$$
\text{Glu} +
\begin{array}{c}
\text{B} \\
| \\
\text{CO} \\
| \\
\text{COOH}
\end{array}
\rightarrow \alpha\text{-oxo-glutarate} +
\begin{array}{c}
\text{B} \\
| \\
\text{CH}-\text{NH}_2 \\
| \\
\text{COOH}
\end{array}
$$

the overall effect has been to synthesize some of the amino acid with side chain B at the expense of the amino group of the amino acid with side chain A. This process must obviously depend on the availability of the oxo-acid $\text{B} \cdot \text{CO} \cdot \text{COOH}$, and cannot therefore be used to any significant extent for the formation of the essential amino acids. Glycine, too, although non-essential is not synthesized in this way. Nevertheless, the isotopic evidence suggests that amino transfer of this kind plays a part in the body's amino acid economy.

Transaminases require, as a prosthetic group, pyridoxal phosphate (vitamin B_6). The formula of this compound and the reaction mechanism is shown on p. 146.

The Products of Transamination: Glutamate and Aspartate

Glutamate dehydrogenase. This enzyme is not a flavoprotein like the other amino acid oxidases, but requires either NAD or NADP as coenzyme. It is present in high concentration in several tissues; no comparable dehydrogenase for any other amino acid is known. It is, like most NAD (or NADP)-linked dehydrogenases, reversible. The reaction which it catalyses is:

$$HOOC \cdot (CH_2)_2 \cdot CH(NH_2) \cdot COOH + NAD(P) + H_2O \rightleftharpoons$$

$$HOOC \cdot (CH_2)_2 \cdot CO \cdot COOH + NH_3 + NAD(P)H_2$$

The equilibrium lies fairly strongly in the direction of glutamate synthesis.

The enzyme is therefore of central importance in any scheme for the deamination of all amino acids (Fig. 11.1).

Fig. 11.2 shows that the amino group of aspartate is used directly in providing one of the $-NH_2$ groups of urea. Thus, so far as the synthesis of urea—the major end-product of amino acid catabolism—is concerned, glutamate and aspartate are of equal importance. It is significant that in isotopic tracer experiments, ^{15}N is always found in highest concentration in glutamate and aspartate. There is no amino acid-oxaloacetate group of transaminases, and thus the continued synthesis of aspartate depends on continuous transamination between glutamate and oxaloacetate. The enzyme catalysing this transfer—GOT—has the highest activity of all transaminases in most tissues. It is found both in mitochondria and in the cytoplasm; and may have another function also (see Chapter 12)

The Transport of Ammonia

Although the product of deamination of amino acids, by whatever mechanism, is shown as NH_3, this compound does not pass in the blood as such from the peripheral tissues to the liver but is probably transported as glutamine (see p. 235).

The concentration of alanine, but of no other amino acid, is higher in venous blood from skeletal and cardiac muscle than in arterial blood. It appears that an *alanine cycle* works to transport $-NH_2$ groups from muscle to liver where pyruvate and glutamate may be formed by the system shown in Fig. 11.1. If necessary the pyruvate may be converted to glucose and returned to the muscle. It is not known whether the alanine is formed by the operation of the pyruvate-specific transaminases mentioned earlier; the muscles of some species contain these enzymes, others do not. Nothing is known about man. In vitro, alanine production depends on the intracellular pyruvate concentration, and it seems likely that it is a product of the specific glutamate-pyruvate transaminase (GPT) found in most

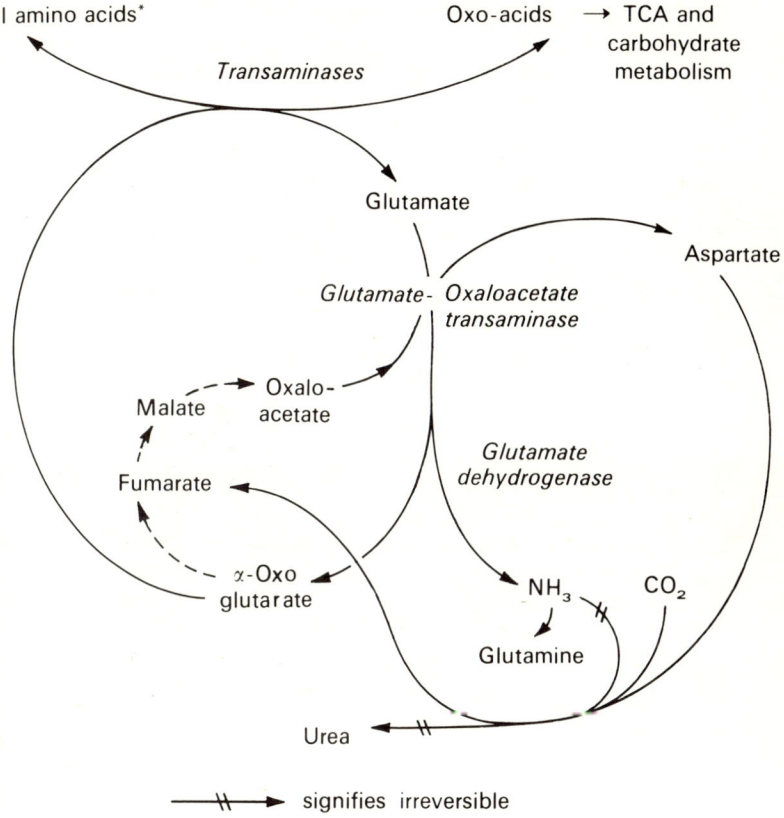

FIG. 11.1. Pathways by which amino groups of amino acids are converted to ammonia and urea.
(*glycine, lysine, threonine and tryptophan are probably exceptions)

tissues. Amino acid catabolism may be relatively rapid in muscle as several transaminases, especially those specific for the branched-chain amino acids, are found there.

Glutamine has other functions besides. It is readily taken up by most tissues, and may act as a source of amino groups, after hydrolysis by glutaminase (see p. 236). Amino acids may then be formed by reversal of the pathways shown in Fig. 11.1. Glutamine is also required for several specific transaminations (see Chapter 5).

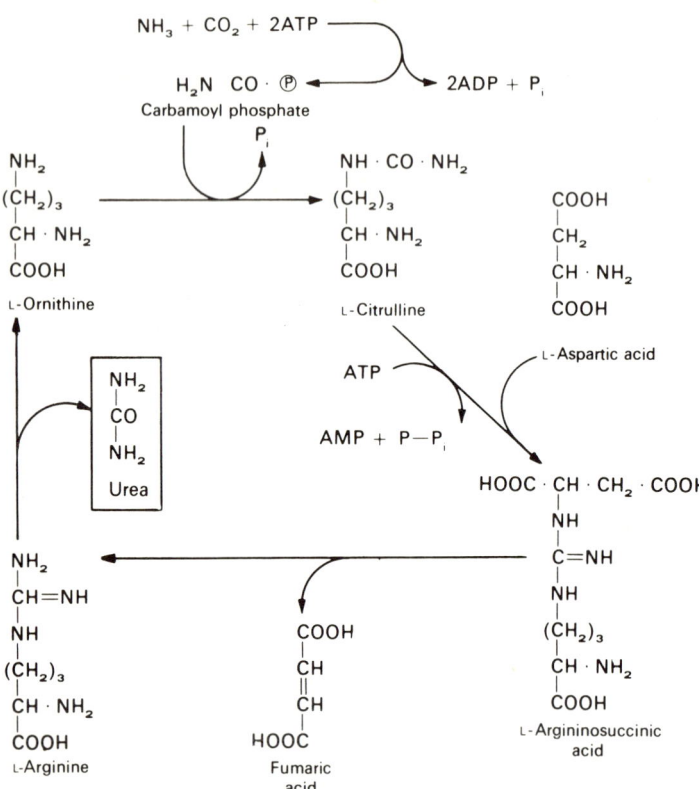

Fig. 11.2. The biosynthesis of urea.

Ammonium ion is toxic to brain, because when its concentration rises, the glutamate dehydrogenase reaction (p. 222) is forced backwards, and the concentration of α-oxoglutarate in the tissue drops, until the catalytic function of the TCA cycle is impaired. Oxygen consumption by the brain can be shown to fall as the blood ammonia concentration rises, and this is the dangerous consequence of impaired urea formation in failure of liver function (hepatic coma). Chronic ammonaemia leads to irreversible brain damage, especially in children. Curiously, the concentration of NH_4^+ in other organs is quite high (about 0·5 mM), much higher than in plasma, and they do not seem to suffer from TCA cycle inhibition. The intense ammonia production from AMP in anaerobic muscle has already been discussed (Chapter 5).

The Synthesis of Urea

The pathway of urea formation in mammals is shown in Fig. 11.2. It may be noted that the carbon atom and one of the nitrogen atoms of urea arise from CO_2 and ammonia while the second nitrogen atom comes via aspartic acid. The enzymes concerned in urea synthesis are all found in the liver, although they are not confined to this organ. Experimental hepatectomy has, however, shown that the liver is the site of production of blood urea. Urea synthesis is an energy-requiring process and 3 molecules of ATP are used in the formation of each molecule of urea.

By far the greatest proportion of nitrogen excreted in man is in urea. The actual amount of urea varies with the protein intake but is between 10 and 25 g/day. It is a very soluble substance, distributed throughout the body water, both intracellular and extracellular. It is completely harmless to tissues, even at higher than normal concentrations, and is so far as we can tell metabolically inert. *Uraemia* may be a symptom of renal dysfunction, or it may pose problems because of the retention of water which ensues.

Many plants and micro-organisms, however, contain an enzyme, *urease*, which will hydrolyse urea to $2NH_3$ and CO_2. The nitrogen thus becomes available for the growth of these organisms. Urea may be used as a source of protein nitrogen in ruminants, and a small amount may be tolerated by man, the flora in the alimentary tract hydrolysing it to CO_2 and NH_3, which is absorbed.

Decarboxylation

Removal of the $-COOH$ group as CO_2, leaving a primary amine, is not in man a major pathway of amino acid metabolism. Several of the resultant amines or derivatives of them, are, however, physiologically important. Examples are *histamine* from *histidine, adrenaline* from *tyrosine* by way of *tyramine, serotonin* or *5-hydroxytryptamine* (5-HT) from *5-hydroxytryptophan, ethanolamine* from *serine. Putrescine* and *cadaverine*, which give the characteristic smell and taste to tainted meat, are the amines corresponding to *ornithine* and *lysine* respectively. The most active of all these enzymes is probably brain *glutamate decarboxylase* (p. 236). The amino acid decarboxylases have pyridoxal phosphate as a prosthetic group. *Amine oxidases* (see Chapter 13) initiate the further oxidation of these compounds.

The Fate of the Oxo-acid Residues

As already described, an oxo-acid is the first product of amino acid catabolism in nearly all cases. Each oxo-acid or 'carbon skeleton' follows a unique metabolic pathway to a compound which can be completely oxidized by way of the tricarboxylic acid cycle. The detailed pathways for most of the amino acids will be given later in this chapter; in this section two general categories will be described.

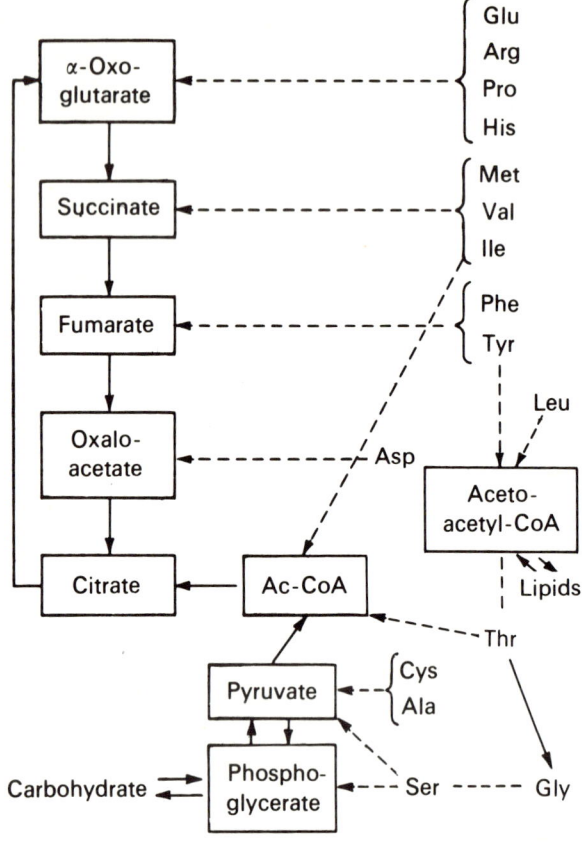

Fig. 11.3. The interrelationships of amino acids with the terminal common pathway.

It was observed many years ago that when single amino acids were fed to diabetic or phlorhizinized animals, some caused the excretion of extra glucose in the urine, others the excretion of extra ketone bodies. According to the results of this type of experiment, the amino acids then known were classified as *glucogenic* or *ketogenic*. It is now realized that the glucose excreted after feeding a glucogenic amino acid does not necessarily come directly from the carbon residue itself. The residue can be directly incorporated into the TCA cycle and so 'spare' glucose, i.e. leave more glucose in the blood to be (in diabetes) excreted. In starvation, on the other hand, the glucose found in the blood must have been formed by *gluconeogenesis*; almost all of it comes directly from the carbon skeletons of glucogenic amino acids.

The ketogenic amino acids give rise to acetoacetic acid which cannot *directly* be oxidized in the tricarboxylic acid cycle. It is, of course, oxidized completely, in normal conditions, unless carbohydrate is absent, just as in the oxidation of fats.

The usual classification is:

Glucogenic: Gly, Ala, Ser, Thr, CySH, Met, Val, Glu, Asp, Arg, His, Pro, Hypro.

Ketogenic: Leu.

Glucogenic and ketogenic: Ile, Phe, Tyr.

Unclassified: Lys, Trp.

The interrelationships between the pathways of amino acid metabolism and the terminal common pathway are shown in Fig. 11.3.

Metabolism of Individual Amino Acids

Glycine

1. *Conjugations:* the bile salt glycocholic acid, and also hippuric acid (benzoyl-glycine) and other detoxication products, are peptides formed between a $-COOH$ group and the $-NH_2$ group of glycine.

2. *Creatine*: the transfer of the guanidino group from arginine to glycine forms guanidino-acetic acid (glycocyamine)

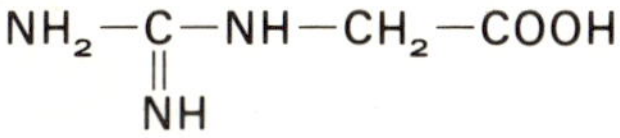

$$NH_2-\underset{\underset{NH}{\|}}{C}-NH-CH_2-COOH$$

Creatine is formed from this compound by methylation. Only one methyl donor exists, 'active' methionine (*S*-adenosyl methionine, see p. 230).

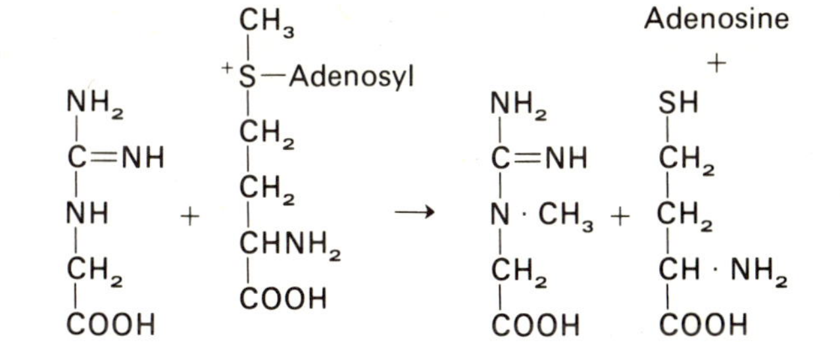

Glycocyamine   S-*Adenosyl methionine* → Creatine   Homocysteine

The reaction is not reversible.

Creatine is found in striated muscle in a concentration of about 1 g/100 g. About half of this creatine is combined with phosphate as creatine phosphate,

$$
\begin{array}{ccc}
CH_2 \cdot N(CH_3) \cdot C \cdot NH \cdot \textcircled{P} & & CH_2 \cdot N(CH_3) \cdot CNH \cdot PO_3{}^{2-} \\
| \qquad\qquad \| & or & | \qquad\qquad\quad \| \\
COOH \qquad\quad NH & & COO^- \qquad\qquad NH[H^+]
\end{array}
$$

at neutral pH.

which has been mentioned earlier (Chapter 9) as a phosphate energy store in this tissue. Smaller amounts of creatine and creatine phosphate are found in smooth muscle, and traces in other tissues.

Creatine phosphate slowly but spontaneously cyclizes to form *creatinine*

$$
\begin{array}{l}
CH_2 \cdot N(CH_3) \cdot C{=}NH \\
| \qquad\qquad\qquad\quad | \\
CO\text{————————}NH
\end{array}
$$

with simultaneous formation of inorganic phosphate. This internal amide is quite stable and diffuses from the muscle into the blood, from which it is excreted into the urine. The amounts found in urine per day depend on the muscle mass, and are reasonably constant from day to day for any one individual, so that it is common to express values for urinary constituents as 'x g/g creatinine' rather than 'x g/24 hours'.

An adult man excretes about 1–1·5 g creatinine per day, but only traces of creatine. Women may excrete more of the latter, since it is eliminated during the breakdown of the smooth muscle cells of the endometrium during menstruation. Creatine is also found in urine in other conditions of muscular wasting, such as muscular dystrophy and thyrotoxicosis. It is not uncommon to find creatine in the urine of children.

Creatine must therefore be continuously synthesized to replace the lost creatinine; it appears to be formed in liver and kidney and not in muscle itself.

3. *Porphyrin* synthesis: Glycine is incorporated into the precursor of protoporphyrin IX in such a way that its amino group provides all four N atoms (see p. 105).

4. *Purine* synthesis: The N–C–C skeleton of one molecule of glycine is incorporated into each molecule of the purines adenine and guanine (p. 95). There is no incorporation of glycine into the pyrimidines.

5. Glycine transamination seems to be important only in plants. In animals it is decarboxylated and transformed into a formimino unit (p. 246), and it is also oxidized by a flavoprotein glycine oxidase to *glyoxylic acid*, $OHC \cdot COOH$. This is further oxidized to CO_2 and formic acid, $H \cdot COOH$.

The latter is used for various synthetic purposes (see the one-carbon unit pool, p. 245). Formate formation appears to be blocked in *oxaluria*, when the glyoxylate is oxidized to oxalate.

6. Glycine can be formed from, and converted to, *serine*, see below.

Alanine

Alanine is glucogenic and non-essential. The alanine-α-oxoglutarate transaminase (GPT), and the alanine cycle, have already been mentioned on p. 222.

Serine

Serine is non-essential, being synthesized from 3-phosphoglycerate by oxidation to phosphohydroxypyruvate, followed by transamination to phosphoserine, and hydrolysis of the phosphate ester bond.

Transamination yields hydroxypyruvic acid, $CH_2OH \cdot CO \cdot COOH$, which can be converted to phosphoglyceric acid, and so to carbohydrate. Serine is also metabolized by a dehydratase which converts it to pyruvic acid:

$$CH_2 \cdot C(NH_2) \cdot COOH \xrightarrow{-H_2O} CH_2{=}C(NH_2) \cdot COOH$$
$$\underset{OH \quad H}{}$$

$$\xrightarrow{+H_2O} CH_3 \cdot CO \cdot COOH + NH_3$$

The most important fate of serine, however, is probably conversion to glycine, since the latter is required for so many syntheses. The reaction requires tetrahydrofolic acid (FH_4: formula on p. 146) as a coenzyme. The enzyme contains pyridoxal phosphate. The reaction:

$$CH_2OH \cdot CH(NH_2) \cdot COOH + FH_4 \rightleftharpoons$$

Serine	$FH_4 \cdot CH_2OH + CH_2(NH_2) \cdot COOH$
	Hydroxymethyl- Glycine
	FH_4

is an example of *one-carbon unit* metabolism, which will be discussed in more detail later.

The decarboxylation of combined serine in phosphatidyl-serine to form phosphatidyl-ethanolamine has already been discussed. The decarboxylation of free serine or of phosphoserine has not been reported.

Methionine

Methylation. The methylation of guanidinoacetic acid to form creatine has been described. The activation of methionine, i.e. the formation of

S-adenosyl methionine, a sulphonium ion which acts as a methylating agent, takes place by means of the following reaction:

$$\text{Met} + \text{ATP} \rightarrow CH_3 - \overset{+}{\underset{|}{S}} - CH_2 \cdot CH_2 \cdot CH(NH_2) \cdot COOH + P_i + P{-}P_i$$

$$\text{Ribose-Adenine}$$

Quantitatively the most important of the other methylations is the synthesis of *choline* from ethanolamine (in phosphatidyl-ethanolamine). This requires three successive methyl transfers from 3 molecules of S-adenosyl methionine. About 0·5–1 g of choline is usually present in the diet, and another 0·5–1 g is synthesized in the body each day during the synthesis of lecithin (Chapter 10).

Choline is degraded by oxidation to *betaine*

$$(CH_3)_3\overset{+}{N} \cdot CH_2 \cdot CH_2OH \overset{Ox.}{\rightarrow} (CH_3)_3\overset{+}{N} \cdot CH_2 \cdot COOH$$

One of the methyl groups may be transferred to homocysteine yielding methionine. The other two are probably oxidized, forming hydroxymethyl-FH_4. N-Methylglycine or *sarcosine* is found in muscle; complete demethylation of betaine gives glycine.

Several other methylations may be mentioned; among them are the formation of adrenaline from noradrenaline, several detoxication reactions (Chapter 13), and the formation of some 'unusual' methylated bases (but *not* thymine) in RNA (cf. p. 85). Transmethylation is also discussed in the later section of this chapter on one-carbon unit metabolism.

It seems probable that most of the dietary methionine is used in this way. The homocysteine which remains is converted to cysteine by the following reactions:

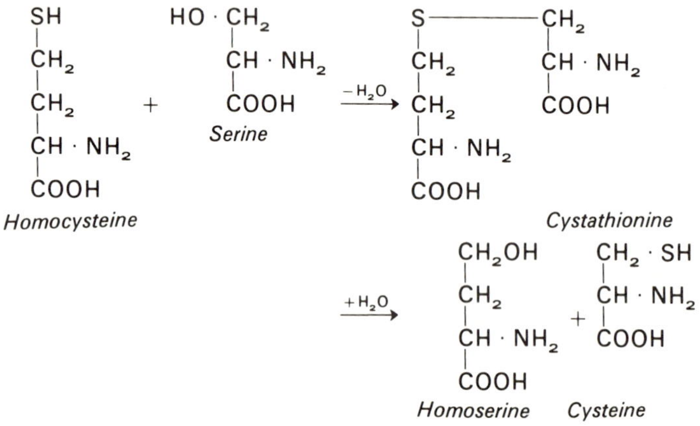

Homocysteine   Serine   Cystathionine

Homoserine   Cysteine

The reaction is not reversible, thus methionine is an essential amino acid, but cysteine is not if sufficient methionine is present in the diet.

Homoserine is converted to α-oxobutyric acid,

$$CH_3 \cdot CH_2 \cdot CO \cdot COOH$$

which is oxidatively decarboxylated to propionic acid (Chapter 10).

Cysteine

If enough methionine is present in the diet, cysteine is not an essential amino acid.

It is converted to pyruvic acid in liver by a 'desulphydrase' enzyme (cf. serine catabolism).

$$HS \cdot CH_2 \cdot CH(NH_2) \cdot COOH \xrightarrow{+H_2O}$$
$$CH_3 \cdot CO \cdot COOH + H_2S + NH_3$$

However, probably more important is the oxidation of the —SH of cysteine in situ, which can lead to the formation of *taurine*,

$$NH_2 \cdot CH_2 \cdot CH_2 \cdot SO_3H$$

This compound is conjugated with cholic acid to form the bile acid taurocholic acid (p. 46) and it is also found free in muscle, particularly heart. It does not seem to be further metabolized in the body but is probably degraded by bacteria in the gut.

Cysteine oxidation proceeds by the following pathways:

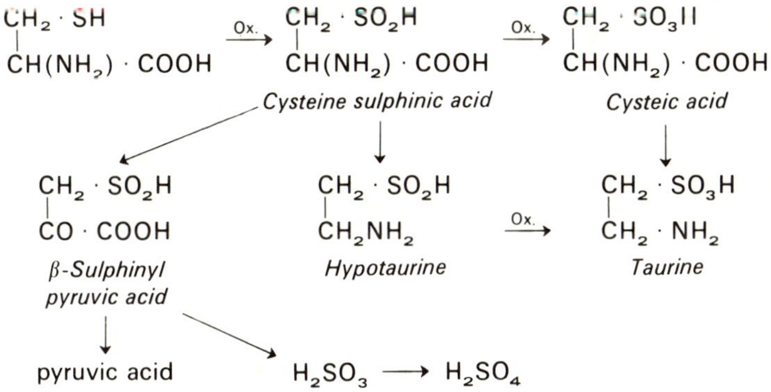

Sulphate metabolism. About 2–4 g of sulphates (as sulphuric acid) are excreted in the urine each day. In the equilibrium (adult) state this must be stoichiometrically equivalent to the S content of the dietary methionine *and* cysteine. About 10 per cent of this sulphate in normal urine is conjugated with various organic compounds (see Chapter 13); the rest is inorganic sulphate. Some of it may not have come directly from the cysteine of the preceding day's diet, since there are many sulphate esters of carbohydrate in the body. These include chondroitin sulphate (in cartilage and bone), mucoitin sulphate (in mucus), heparin, etc. (see Chapter 2). The turnover of chondroitin sulphate is probably very slow, so that there is not much demand for sulphate except in growth and wound repair.

A *sulphate transferring coenzyme*, 3'-phosphoadenosine-5'-phospho-sulphate (PAPS), is concerned in the synthesis both of mucopolysaccharides and urinary phenol sulphates.

$$\begin{cases} ATP + SO_4{}^{2-} \rightarrow \text{Adenosine} - \circled{P} - SO_3{}^- \text{ (APS)} + P - P_i \\ APS + ATP \rightarrow [3' - \circled{P}] \text{-Adenosine} - \circled{P} - SO_3{}^- + ADP \end{cases}$$

Cystine is excreted in a congenital abnormality of renal function. As it is very insoluble, it may form harmful kidney stones.

Valine

The three branched-chain amino acids valine, leucine, and isoleucine are all essential, and their catabolic pathways all begin by the same four reactions, (A) transamination followed by (B) oxidative decarboxylation, and then (C) a dehydrogenation followed by (D) addition of H_2O across the double bond, (cf. β-oxidation, cf. fatty acids, p. 188).

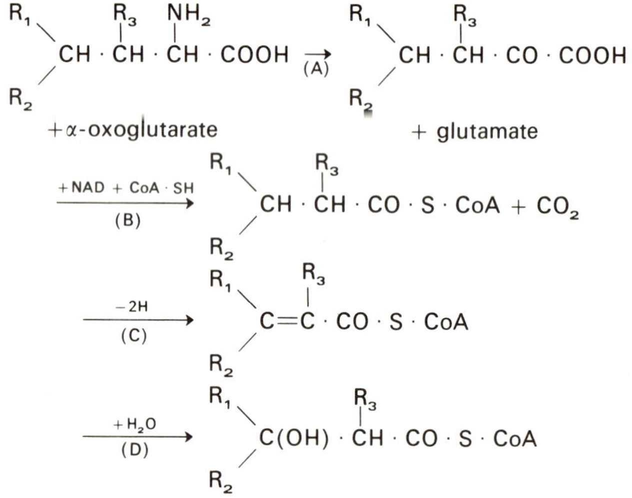

where, for valine, $R_1 = R_2 = H$, $R_3 = Me$; for leucine, $R_1 = R_2 = Me$, $R_3 = H$; for isoleucine, $R_1 = H$, $R_2 = R_3 = Me$.

There is a general 'branched-chain amino acid' transaminase which catalyses step (A), but separate specific transaminases are probably more important, because the congenital defect *hypervalinaemia*, in which valine accumulates in blood and urine, does not affect leucine and isoleucine metabolism.

There is a specific α-oxo-acid decarboxylase catalysing the oxidative decarboxylation (step (B)) of α-oxo-isovaleric acid, the product of valine transamination. The product of the subsequent steps (C) and (D) is β-hydroxyisobutyryl-CoA (I, below). CoA is split off, and the free acid (II) is then oxidized by two 2NAD-dependent steps to methylmalonyl-CoA (III). This compound is converted to succinate, as already described for the metabolism of propionate (p. 203). Valine is therefore glucogenic.

$$HOH_2C \cdot \overset{\overset{\displaystyle CH_3}{|}}{CH} \cdot CO \cdot S \cdot CoA \xrightarrow{\ -CoA \cdot SH\ } HOH_2C \cdot \overset{\overset{\displaystyle CH_3}{|}}{CH} \cdot COOH \xrightarrow{\ 2NAD\ }$$

$$\text{(I)} \qquad\qquad\qquad\qquad\qquad \text{(II)}$$

$$\xrightarrow{\ CoA \cdot SH\ } CoA \cdot S \cdot CO \cdot \overset{\overset{\displaystyle CH_3}{|}}{CH} \cdot COOH \rightarrow$$

$$\text{(III)} \qquad\qquad HOOC \cdot CH_2 \cdot CH_2 \cdot COOH$$
$$\textit{Succinate}$$

Isoleucine

This amino acid has two asymmetric carbon atoms (marked by asterisks in the formula), and only one of the four possible isomers will support growth in experimental animals.

$$\overset{\displaystyle CH_3}{\underset{\displaystyle CH_2}{\underset{\displaystyle \underset{\beta}{CH} \cdot \underset{\alpha}{CH}(NH_2) \cdot COOH}{\big\backslash}}}$$
$$\underset{\displaystyle CH_3}{\big/}$$

The molecule with the 'unnatural' configuration at the β-carbon atom is referred to as *allo*-isoleucine.

The oxo-acids produced by the transamination of isoleucine and leucine share a common oxidative decarboxylase. Both this and the oxidase for α-oxo-isovaleric acid (above), although not fully studied, require thiamine pyrophosphate, NAD, and probably lipoic acid, just like pyruvate and α-oxoglutarate oxidases (p. 144).

The product of this reaction (B) for isoleucine is α-methyl-β-hydroxy-butyryl-CoA (I below). This is further oxidized to α-methylacetoacetyl-CoA (II), which is cleaved by CoA to acetyl-CoA and propionyl-CoA; the latter is converted into succinate via methylmalonyl-CoA (p. 204). Because of this isoleucine is weakly glucogenic (3 carbon atoms out of 6), and is also slightly ketogenic, since in unfavourable circumstances the acetyl radicals can combine to form acetoacetate.

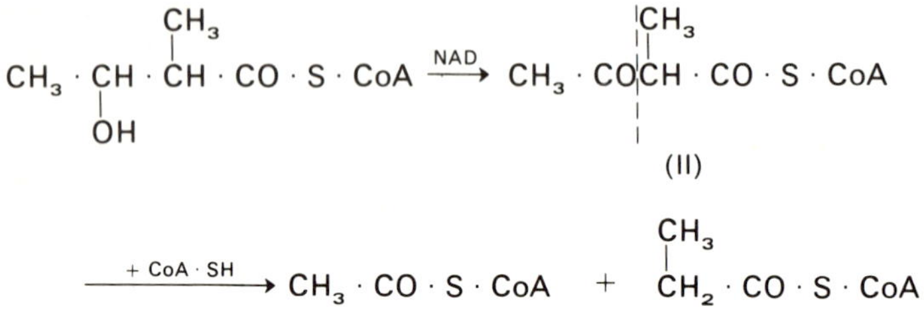

$$CH_3 \cdot \underset{\underset{OH}{|}}{CH} \cdot \underset{\underset{CH_3}{|}}{CH} \cdot CO \cdot S \cdot CoA \xrightarrow{NAD} CH_3 \cdot CO \cdot \underset{\underset{|}{|}}{\overset{CH_3}{\overset{|}{CH}}} \cdot CO \cdot S \cdot CoA$$

(II)

$$\xrightarrow{+ CoA \cdot SH} CH_3 \cdot CO \cdot S \cdot CoA \quad + \quad \underset{\underset{CH_2}{|}}{\overset{CH_3}{\overset{|}{}}} \cdot CO \cdot S \cdot CoA$$

Leucine

The reaction sequence is slightly different for the catabolism of this amino acid. The product of reaction (C), p. 232, is carboxylated on the methyl group R_1 by a biotinyl enzyme, and the unsaturated dicarboxylic acid (I) is then hydrated to β-hydroxy-β-methylglutaryl-CoA (HMG-CoA) (II). This compound is readily split to acetoacetate and acetyl-CoA (Chapter 10, p. 206) and leucine is therefore strongly ketogenic.

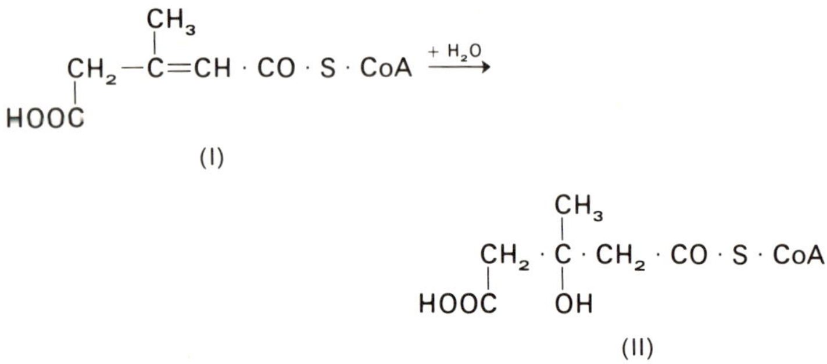

$$\underset{HOOC}{\overset{CH_3}{\underset{|}{CH_2{-}\overset{|}{C}{=}CH}}} \cdot CO \cdot S \cdot CoA \xrightarrow{+ H_2O}$$

(I)

$$\underset{HOOC}{\overset{CH_3}{\underset{|}{CH_2 \cdot \overset{|}{C} \cdot}}} \underset{OH}{\underset{|}{CH_2}} \cdot CO \cdot S \cdot CoA$$

(II)

Alternatively, HMG-CoA may be used in the synthesis of cholesterol (p. 198); it is in fact found that the carbon atoms of leucine are incorporated into cholesterol much more rapidly than those of acetate.

In an inborn error of metabolism called 'maple syrup urine disease' the common oxidase catalysing step (B) for leucine and isoleucine is absent. The corresponding α-oxo-acids accumulate, and as they inhibit the α-oxo-isovaleric acid oxidase (p. 233), this oxo-acid also accumulates and the defect comes to involve all three branched-chain amino acids. The α-oxo-acids also inhibit α-oxoglutaric acid oxidase, and thus impair the working of the TCA cycle. There is rapid degeneration of the CNS and the disease is usually fatal within a few months of birth. The three branched-chain amino acids and the oxo-acids accumulate in blood and urine, and the latter are partially reduced to the corresponding hydroxy-acids, which have a characteristic odour of maple syrup.

Threonine

The α-amino group of threonine is not transaminated.

Threonine aldolase can convert threonine into acetaldehyde and glycine:

$$CH_3 \cdot CHOH \cdot CH(NH_2) \cdot COOH + HO{-}H \rightarrow$$
$$CH_3 \cdot CH(OH)_2 + CH_2(NH_2) \cdot COOH$$

but this only accounts for a fraction of threonine catabolism. *Threonine dehydratase* oxidizes threonine to NH_3 and α-oxobutyrate (see p. 231). By either pathway threonine will be glucogenic.

Lysine

This is an essential amino acid and its α-amino group does not undergo transamination. The first step in its catabolism may be blocking off the ε-amino group. In any case, the α-amino group is lost first, and the ε-amino group is transferred to the α-position by means of a cyclic intermediate. The product of these reactions is α-aminoadipic acid which is the next higher homologue of glutamic acid. This is oxidatively decarboxylated to glutaryl-CoA

$$HOOC \cdot CH_2 \cdot CH_2 \cdot CH_2 \cdot CO \cdot S \cdot CoA$$

This is transformed by oxidation and decarboxylation, to acetoacetyl-CoA. Lysine ought therefore to be strongly ketogenic, but it is neither ketogenic nor glucogenic. Possibly the initial removal of the α-amino group is so slow that the final products can never accumulate in blood or urine.

Glutamic Acid and Glutamine

The formation of glutamic acid by transamination and its reversible conversion to α-oxoglutaric acid and ammonia have already been discussed.

Glutamine is synthesized in most tissues by the following reaction:

$$HOOC \cdot CH(NH_2) \cdot CH_2 \cdot CH_2 \cdot COOH + ATP + NH_3 \rightarrow$$
$$HOOC \cdot CH(NH_2) \cdot CH_2 \cdot CH_2 \cdot CONH_2 + ADP + P_i$$

Most of the free amino acid pool in the body, both in tissues and in plasma, consists of glutamine and glutamic acid.

A *glutaminase*, particularly concentrated in kidney, hydrolyses glutamine to glutamate and ammonium ion. It is part of the renal system for the control of acid–base balance (see Fig. 14.15). Besides this, the amide-N of glutamine is used in the synthesis of purines (see Fig. 5.7), and it is also used to transport NH_3.

Glutamate decarboxylase, which is very active in brain, converts glutamate to *γ-aminobutyric acid* (GABA), which is important in the function of the central nervous system. GABA is transaminated to succinic semi-aldehyde, which is oxidized to succinate.

Aspartic Acid and Asparagine

Aspartate is very readily metabolized by isolated tissues. It first undergoes transamination with α-oxoglutarate to oxaloacetate and glutamate. The latter is oxidized as described earlier (p. 222). Besides this, the nitrogen atom of aspartate is used specifically in the synthesis of purines (Fig. 5.9) and pyrimidines (Fig. 5.7), and of arginine from citrulline (Fig. 11.2). Thus half the nitrogen in all urea excreted has passed through the amino group of aspartate.

Asparagine, the amide of aspartic acid, occurs widely in plants but its concentration in plasma and tissues is very low. However, it is a normal constituent of proteins, with a separate codon (Chapter 15), and is found in much higher concentration in many proteins than tryptophan, methionine or cysteine, for example. More than one ATP-dependent asparagine synthetase appears to exist, but the total rate of synthesis may be very low.

Asparagine supplementation increases the growth rate of young rats which puts it in the semi-essential category of arginine and histidine. Nothing is known about human demands.

Some tumours appear to lack asparagine synthetases, and to depend entirely on the host. These tumours regress rapidly when the host is injected with bacterial *asparaginase*, but so far complete regression has not been achieved, partly because the rate of asparagine synthesis in the host responds slowly to the increased rate of loss.

Arginine and Ornithine

The synthesis of arginine from ornithine has been shown in Fig. 11.2. In adult man arginine is non-essential, but there is a demand for supplementation in the diet of the young. It appears, then, that ornithine can be made slowly by humans; other animals, e.g. the rat, are completely dependent on dietary arginine for their ornithine requirements.

Ornithine is known to be converted to glutamic acid by the following pathway

$$H_2N \cdot CH \cdot (CH_2)_2 \cdot CH_2 \cdot NH_2 + \alpha\text{-oxoglutarate}$$
$$\qquad | \qquad\qquad\qquad\qquad\qquad\qquad \Updownarrow$$
$$\qquad COOH$$

Ornithine

$$H_2N \cdot CH \cdot (CH_2)_2 \cdot CHO$$
$$\qquad |$$
$$\qquad COOH + glutamate$$

Glutamic acid $\quad$ NAD $\Updownarrow$
semi-aldehyde $\quad$ H_2O

$$H_2N \cdot CH \cdot (CH_2)_2 \cdot COOH$$
$$\qquad |$$
$$\qquad COOH$$

Glutamic acid

and it is presumed that it is formed, slowly, by reversal of the same pathway.

Histidine

This is another amino acid part of whose skeleton is converted to glutamic acid, but details of its metabolism are more complex than for ornithine and proline. It is not essential for man, but is so for many animals and for young children. It is synthesized from 5-phosphoribosyl-pyrophosphate and ATP (see Fig. 11.4).

Histidine is catabolized by opening the ring to produce formimino-glutamic acid. The formimino group $-HC=NH$, is transferred to FH_4, and converted into $FH_4 \cdot CHO$ ('active formaldehyde') and NH_3. See Fig. 11.5.

An important derivative of histidine is the local hormone *histamine*, $R \cdot CH_2 \cdot CH_2NH_2$, formed by decarboxylation of the amino acid. This is oxidized in tissues by *histaminase* (see p. 257) to $R \cdot CH_2 \cdot CHO$. It is further oxidized in liver to $R \cdot CH_2 \cdot COOH$, imidazolylacetic acid and excreted as the ribotide of the latter. In subjects suffering from folic acid deficiency, the methylated derivatives of histamine and imidazolylacetic acid are excreted in considerable quantities, suggesting that the pathway of histidine catabolism via formimino-glutamic acid is inhibited.

Proline and Hydroxyproline

These are both non-essential amino acids, required in quantity for the

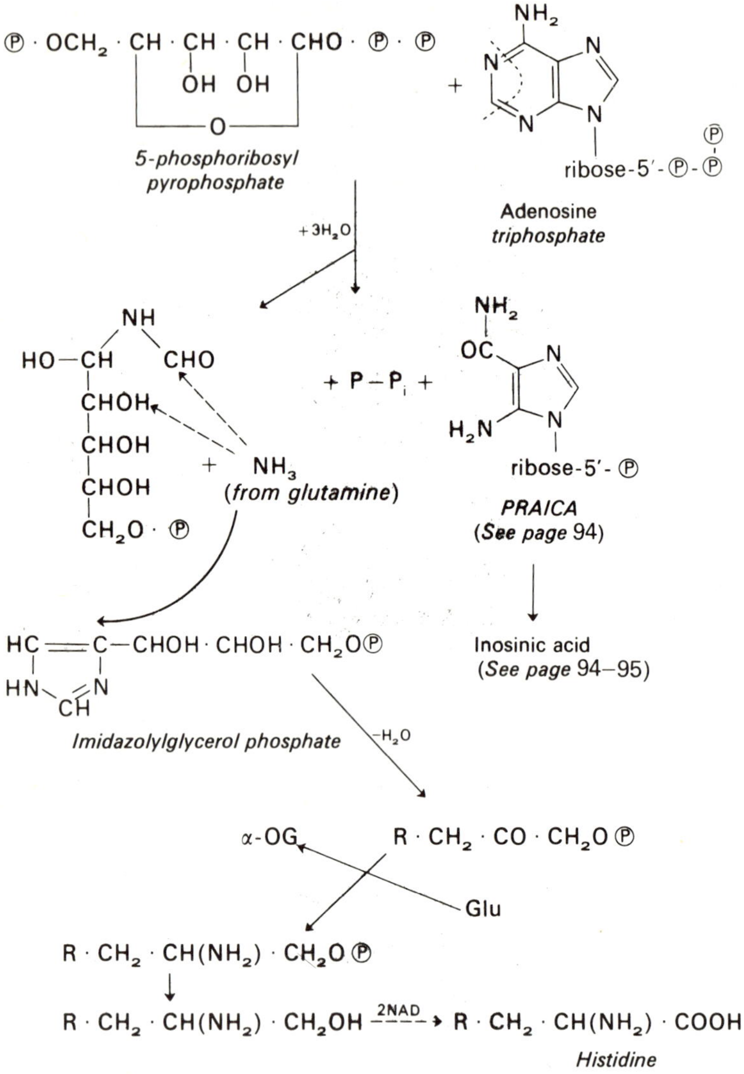

FIG. 11.4. Biosynthesis of histidine.

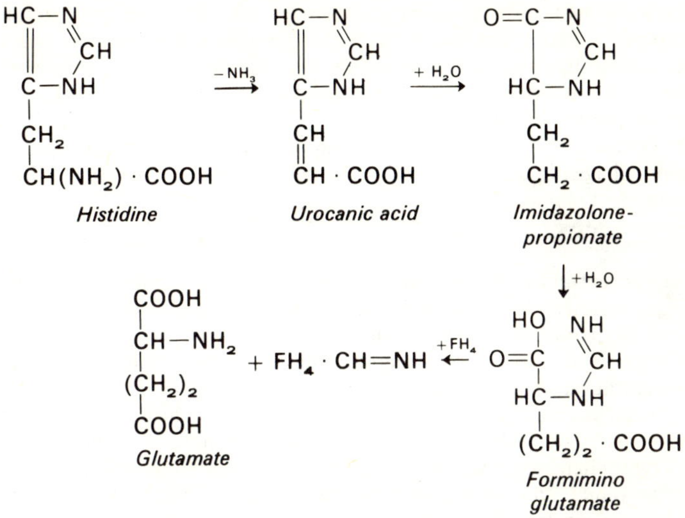

FIG. 11.5. Catabolism of histidine.

synthesis of collagen. *Proline* is convertible to glutamic acid semialdehyde:

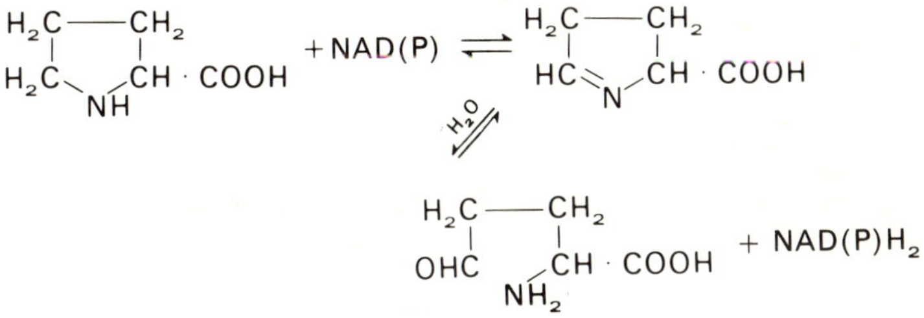

and the latter can be oxidized to glutamic acid as described above.

Hydroxyproline is found only in collagen; the intake from food is probably low. There is no RNA codon for hydroxyproline, and it is formed by oxidation of prolyl residues in pro-collagen. The enzyme is a mixed function oxidase in which α-oxoglutarate acts as oxidizable co-substrate (see Chapter 12).

A fraction of degraded collagen is excreted as small Pro-Hypro-peptides (mol. wt. $\leq 1\,000$). The rest of the hydroxyproline is generally thought to be oxidized in a similar way to proline, but one case of an inborn error of metabolism is known in which hydroxyproline is not oxidized, although proline metabolism is normal. There is severe mental retardation. From this case it can be estimated that the endogenous release of hydroxyproline from collagen in an adult is about 2 g per day; about one-sixth of this is excreted in peptide form.

Tryptophan

The concentration of tryptophan in tissues is very low, and in plasma it is almost undetectable; in liver the supply of tryptophan may be the limiting factor in the rate of protein synthesis.

Tryptophan does not undergo transamination unless the tissue concentration of keto acids is abnormally high, as in phenylketonuria or 'maple syrup urine' disease. The resulting keto acid is converted into *indolylacetic acid*, which is excreted in larger than normal amounts.

Tryptophan can be decarboxylated to form *tryptamine*, or first hydroxylated and then decarboxylated to form 5-hydroxytryptamine (5-HT, or *serotonin*), a local hormone. This substance, or its derivative 5-hydroxyindolyl acetic acid, is excreted in quite large amounts by subjects suffering from *carcinoid*, a tumour of the argentaffin cells of the intestinal mucosa. Estimation of these substances in urine is of great aid in the diagnosis of the disease.

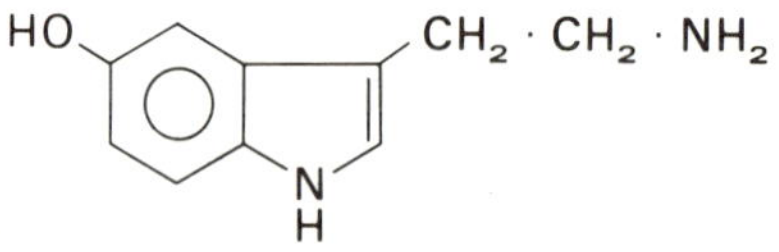

5-Hydroxytryptamine

A major pathway of tryptophan metabolism begins by opening the 5-membered ring to form formylkynurenine:

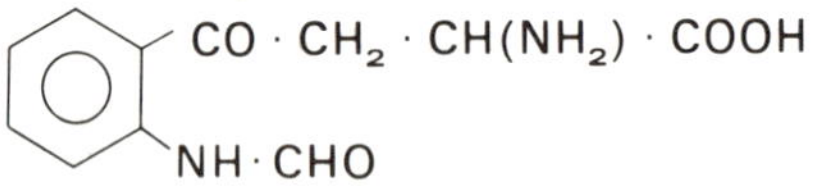

This involves the addition of 2 atoms of oxygen from a molecule of O_2 across the double bond in the pyrrole ring. The oxygenase responsible, *tryptophan pyrrolase* is one of the few mammalian enzymes that is inducible.

Formylkynurenine loses a formyl group to the one-carbon unit pool, leaving *kynurenine*. This is hydroxylated to form 3-hydroxykynurenine:

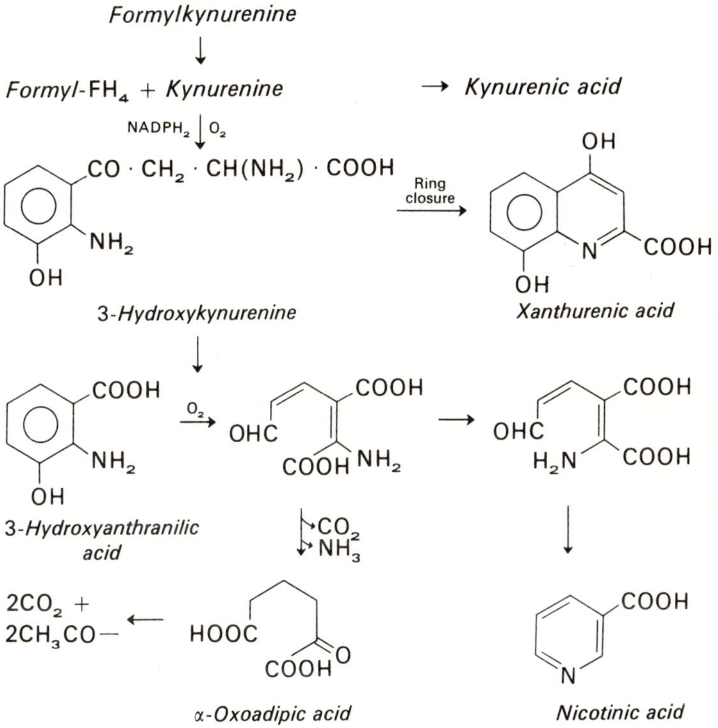

Formylkynurenine

↓

Formyl-FH$_4$ + *Kynurenine* → *Kynurenic acid*

NADPH$_2$ | O$_2$

CO · CH$_2$ · CH(NH$_2$) · COOH

NH$_2$

OH

3-*Hydroxykynurenine*

Ring closure →

OH

N COOH

OH

Xanthurenic acid

↓

COOH

NH$_2$

OH

3-*Hydroxyanthranilic acid*

O$_2$ →

OHC COOH

COOH NH$_2$

→

OHC COOH

H$_2$N COOH

↓

↓CO$_2$
↓NH$_3$

2CO$_2$ +
2CH$_3$CO —

← HOOC

COOH O

α-*Oxoadipic acid*

COOH

N

Nicotinic acid

The series of reactions from 3-hydroxyanthranilic acid to nicotinic acid then leads, through nicotinic acid ribotide, to nicotinamide ribotide, the immediate precursor of NAD. This indirectly provides enough nicotinamide, one of the B vitamins (see Chapter 20), in some species such as the rat, for the animal to be completely independent of dietary sources (other than tryptophan). Man does depend on his diet for the vitamin, but a fraction of his requirement can come from tryptophan. A deficiency of this amino acid can exacerbate a deficiency of nicotinamide; hence the peculiar prevalence of pellagra in maize-eating areas (the chief protein of maize is particularly deficient in tryptophan).

The conversion of 3-hydroxykynurenine into 3-hydroxyanthranilic acid requires another of the B vitamins, pyridoxine (B$_6$). Thus pyridoxine deficiency will accentuate the symptoms of pellagra (see Chapter 20).

In B_6 deficiency, 3-hydroxykynurenine and its precursors accumulate in the tissues or are excreted in greater quantity, and it is not uncommon to test for a pyridoxine deficiency by a tryptophan-load test. The excretion rate of an intermediate, usually xanthurenic acid, is measured after a standard dose of tryptophan.

Phenylalanine and Tyrosine

The main catabolic pathway of these essential amino acids is delineated in Fig. 11.6. The figure makes it clear why these two acids are both glucogenic *and* ketogenic. Tyrosine is also the starting point for the synthesis of several compounds of biological importance.

Peptide linked tyrosine is iodinated in the synthesis of *thyroxine* (Chapter 17).

The copper-containing enzyme *tyrosinase*, which occurs in melanocytes, and also in plants, catalyses the oxidation of tyrosine by oxygen to 3,4-dihydroxyphenylalanine ('dopa') and to the corresponding quinone.

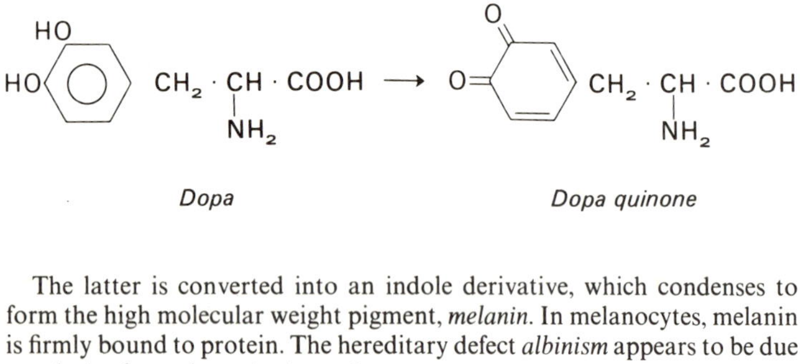

Dopa   Dopa quinone

The latter is converted into an indole derivative, which condenses to form the high molecular weight pigment, *melanin*. In melanocytes, melanin is firmly bound to protein. The hereditary defect *albinism* appears to be due to a lack of tyrosinase in the melanocytes.

Dopa is also the starting point for the synthesis, in the adrenal medulla, of *noradrenaline* and *adrenaline* (Chapter 17).

Phenylketonuria, alkaptonuria, and *tyrosinosis* are closely related inborn errors of metabolism in which the normal metabolism of aromatic amino acids is blocked at different points.

Phenylketonuria is characterized by mental deficiency and the excretion in the urine of phenylpyruvic acid in excessive amounts (0·5–1·0 g per day). In addition to phenylpyruvic acid there are abnormally large amounts of phenylalanine, phenyllactic acid, and phenylacetylglutamine in the urine. The body fluids contain a high concentration of phenylalanine, but not of the other substances listed above, which appear to be rapidly cleared from the blood by the kidneys. The basic defect is the inability of the body to

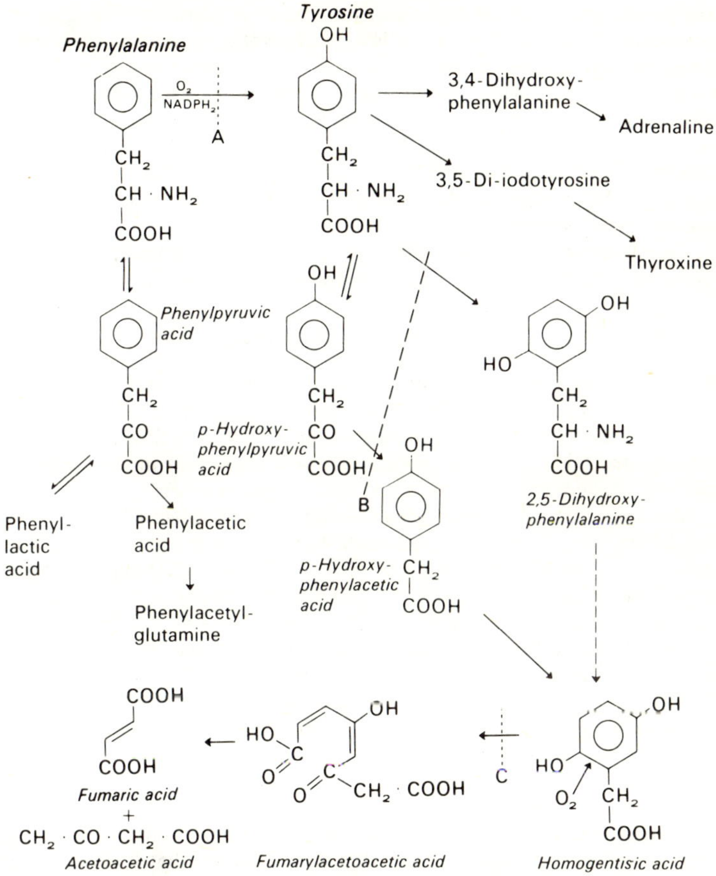

FIG. 11.6. Main metabolic pathways of aromatic amino acids. Blocks occur at: A in phenylketonuria, B in tyrosinosis, C in alkaptonuria.

hydroxylate phenylalanine (Fig. 11.6): it has been demonstrated directly that part of the liver enzyme-complex which normally catalyses this reaction is absent.

Alkaptonuria is characterized by the continuous daily excretion of several grams of homogentisic acid in the urine. The presence of this substance in urine causes it to darken on standing in contact with air, due to the oxidation of homogentisic acid to a pigment. The level of homogentisic acid in the blood is low, because of the low renal threshold. There are no serious effects on the sufferer. No other constituents are found in blood or urine in abnormal concentrations. Homogentisic acid is on the normal pathway of metabolism of tyrosine, the further catabolism of which is completely blocked. Such a block affects 80–90 per cent of the metabolism of phenylalanine and tyrosine, since the other pathways to dihydroxyphenylalanine, adrenaline, thyroxine, etc., are probably quantitatively limited.

Tyrosinosis is a very rare disease in which there is a continuous excretion in the urine of tyrosine and *p*-hydroxyphenylpyruvic acid.

One-carbon Unit Metabolism

By this is meant the transfer of radicals containing one carbon atom (in various states of oxidation) from one compound to another. This is not quantitatively important by comparison with the overall flow of materials through the body, but it is qualitatively very important in the synthesis of some vital compounds and in the detoxication of others.

Methanol, CH_3OH, formaldehyde, $H \cdot CHO$, and formic acid, $H \cdot COOH$, are one-carbon compounds, and might be expected to participate in this metabolism, but they do not occur in the body; indeed, the first two are toxic. The fixation of CO_2, as in the carboxylation of pyruvate (Chapter 9), or in the formation of malonyl-CoA (Chapter 10) might strictly be called a part of one-carbon unit metabolism but is not usually so regarded. The name is reserved for the transfer of methyl ($-CH_3$), hydroxymethyl ($-CH_2OH$), or formyl ($-CHO$) radicals by means of carriers.

There are two carriers: S-adenosyl methionine (pp. 227 and 230) and tetrahydrofolic acid (p. 146). Although there is a connexion between the two, they are essentially separate carriers concerned with the synthesis of different types of compound.

In the synthesis of compounds containing methyl groups, methionine itself is the most important donor. There is evidence that homocysteine can be remethylated. Isotopic evidence shows that hydroxymethyl -FH_4 can be reduced to 5-methyl-FH_4. It is not at all certain that the animal enzyme catalysing the transfer of the methyl group requires B_{12}, as the bacterial enzyme does. Thus a single carbon skeleton of methionine can be used for the transfer of more than one methyl group. The number of transfers is nevertheless not very large, because of the competing claims of cysteine synthesis (p. 230). Fig. 11.7 summarizes these transfers.

The most important substances synthesized by transmethylation from S-adenosyl methionine are creatine, choline, adrenaline, and the methylated bases in RNA, especially tRNA. Noradrenaline is rendered inactive by methylation of one of its phenolic $-OH$ groups, and there are other

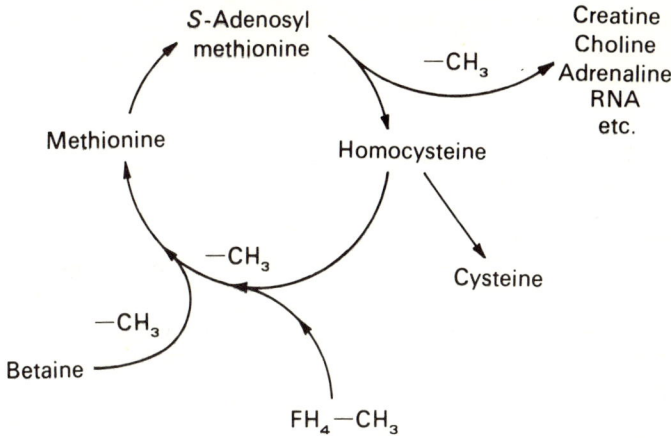

FIG. 11.7. Transmethylations involving methionine.

examples of such detoxifying methylations. In the plant world a large number of aromatic and aliphatic compounds contain methyl groups originating from the same donor.

Tetrahydrofolic Acid

The second aspect of one-carbon unit metabolism is more complicated. The coenzyme, formed by reducing the vitamin folic acid (Chapter 20) at C-5, -6, -7 and 8 on the pteridine ring, carries the one-carbon unit

Tetrahydrofolic acid (FH_4)

attached either to N-5 or N-10, or to both (in the latter case as a one-carbon bridge). It is helpful to consider the relationships between the formyl derivatives first. A reaction which is of little importance in higher animals leads to the formation of N^5-*formyl*-FH_4:

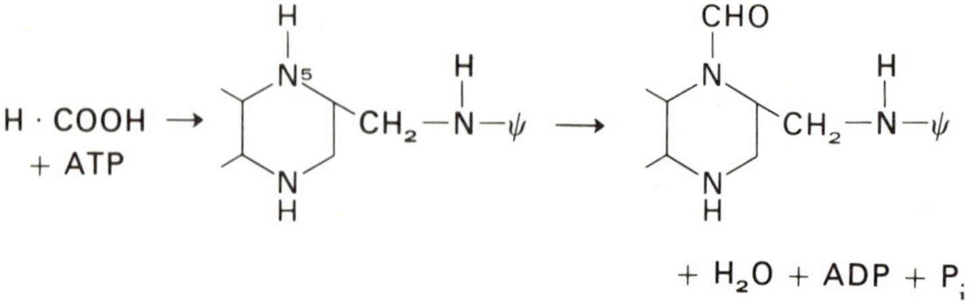

$$H \cdot COOH + ATP \rightarrow \cdots \rightarrow \cdots + H_2O + ADP + P_i$$

An ATP-dependent cyclase catalyses the formation of the N^5–N^{10} methenyl ring, which can be opened to form N^{10}-formyl-FH_4:

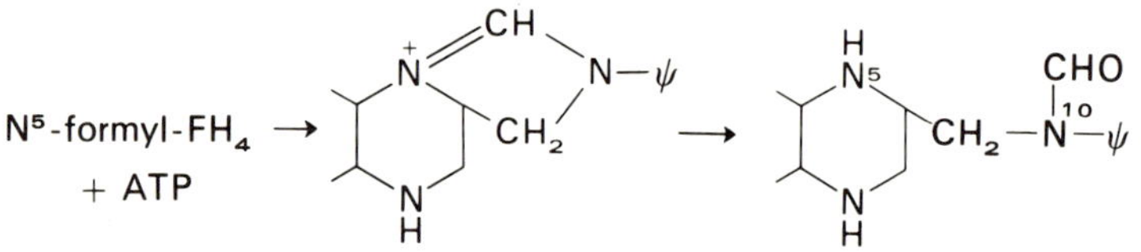

$$N^5\text{-formyl-}FH_4 + ATP \rightarrow \cdots \rightarrow \cdots$$

Although the free energy of hydrolysis of the N^{10}-derivative is considerably greater than that of the N^5-, there are examples of enzyme-catalysed transfer from all three forms of the coenzyme, and it is simpler to think of them collectively as FH_4—CHO. (In these formulae, ψ represents the rest of the folic acid molecule.)

Another form of the coenzyme at the formyl stage of oxidation is formimino-FH_4, the formimino group —CH=NH arising from intermediates in the catabolism of histidine and glycine.

In a similar way N^5-N^{10}-methylene-FH_4 is equivalent to N^{10}-*hydroxymethyl*-FH_4:

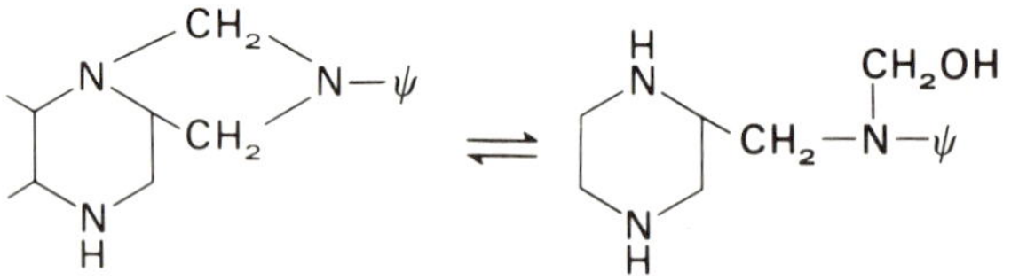

and although the latter is not the immediate hydroxymethyl donor, the abbreviation FH_4—CH_2OH expresses the fact that this form of the

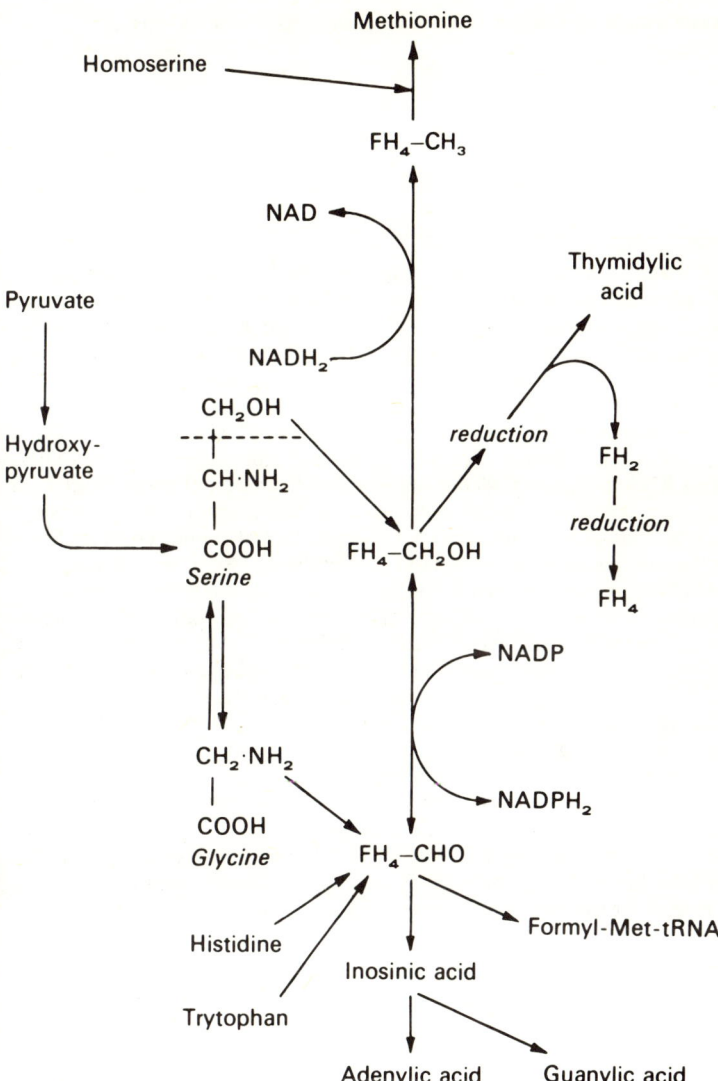

FIG. 11.8. Interrelationships between tetrahydrofolic coenzymes.

coenzyme is at a higher stage of oxidation than 5-methyl-FH_4,

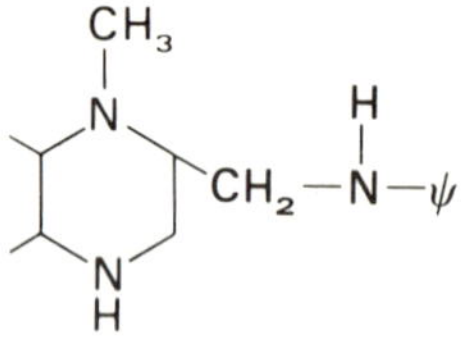

the final form of the coenzyme. It is probable that the $-CH_2OH$ group of serine is the major precursor not only of FH_4-CH_2OH, but also of formyl- and methyl-FH_4, by the interconversions which are shown in Fig. 11.8. This would also fit in with the fact that glycine, the other product of serine dismutation, is required for various syntheses (p. 227) in amounts probably larger than the supply in foodstuffs, particularly in uricotelic animals.

Inspection of Fig. 11.8 shows the importance of 1–C units in the synthesis of nucleotides. Since nucleotides are essential for synthesis of both DNA and RNA (Chapter 5), derangements of 1–C unit metabolism affect both protein synthesis and cell reduplication. Moreover, the formylation of methionine-tRNA is necessary for the initiation of peptide chain synthesis (Chapter 16), certainly in bacteria, and probably also in man. Antagonists of folic acid, such as aminopterin, which has an amino group instead of the hydroxyl at position 4 of folic acid, and methotrexate (amethopterin), which is the N^{10}-methyl derivative of aminopterin, markedly inhibit cell division. Both these compounds bind very tightly, though reversibly, to folic acid reductase thereby inhibiting this enzyme. Even high concentrations of folic acid do not reverse this inhibition. These drugs have been used as anti-tumour agents.

On the other hand, trimethoprim binds 10^4 times more strongly to the dihydrofolic acid reductase of bacteria than it does to the mammalian enzyme. Since FH_2 must be reduced to FH_4 before it can take part, or re-enter, the shuttle of 1–C units depicted in Fig. 11.8, drugs of this kind inhibit the growth of bacteria but not of animal cells, and can therefore be used for the control of bacterial infections in man.

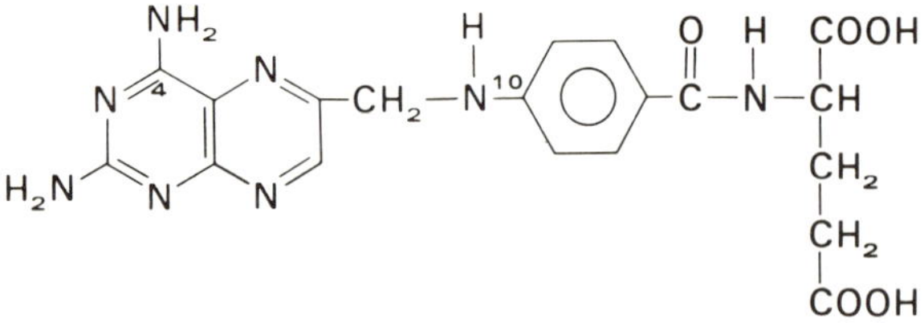

Aminopterin

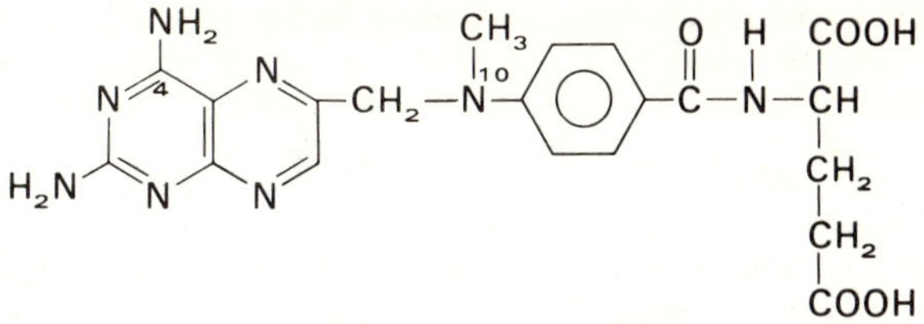

Methotrexate (*amethopterin*)

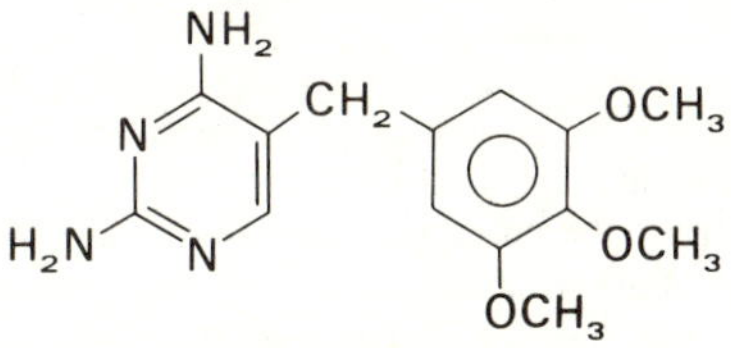

Trimethoprim

A third type of antagonism to folic acid is that produced by *sulphonamides* which competitively inhibit the incorporation of *p*-aminobenzoic acid into the folic acid molecule by bacteria normally able to synthesize the complete compound.

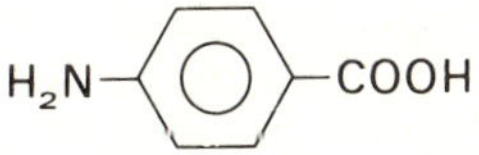

p-*Aminobenzoic acid*

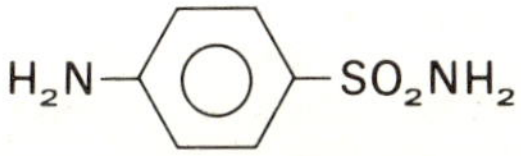

Sulphanilamide

The sulphonamide drugs are therefore also bacteriostatic agents.

General Considerations

There is a difference between carbohydrate and lipid metabolism, on the one hand, and amino acid metabolism, on the other, in that there is no storage polymer for amino acids. Reflection will show that the pattern of amino acids absorbed from the gut will vary according to the dietary protein, while every polypeptide is synthesized according to an inflexible sequence carried in the genome. Thus there is no possibility of synthesizing a protein of varying amino acid composition which would store varying proportions of amino acids. In this sense amino acid metabolism is immediate; if free amino acids are given by mouth, the major fraction is incorporated into protein, or catabolized, within 4 hours. There are several Na^+-dependent amino acid permeases in most tissues, and the free amino acid content of cells rises significantly after a protein meal, but the total storage of monomers is not large. One must remember, however, that the liberation of amino acids from protein in the gut is much slower than the release of glucose from polysaccharides, so that after a protein meal amino acids may be taken up into the bloodstream for many hours.

Although there is no specific storage protein, many enzymes are present in cells in amounts far beyond catalytic requirements. Analysis of genetic defects shows that cells may often lose 95 per cent of a particular enzyme, without serious harm to the metabolic pathway. As indicated in the earlier part of this chapter, many enzymes have short half-lives, and in consequence, liver may lose 30 per cent of its protein in a 2-day fast. This burst of catabolism is supplemented by a much longer lasting breakdown of more stable proteins.

This very large reservoir of protein explains the discrepancy between the immediate utilization of dietary amino acids, and the fact that protein deficiency diseases may take months to develop, and that adults may live in a negative nitrogen balance for years. It is also significant that there are so many inborn errors of amino acid metabolism. Some, indeed, are lethal within a few months of birth, but others are compatible with an indefinite life-span, suggesting that the pathways in which they occur are of secondary importance. There are no inborn errors of tricarboxylic acid cycle metabolism.

It would be unreasonable to expect the student to be familiar with the details of the metabolism of all 20 amino acids. Some, like alanine, glutamate and aspartate and their amides, glycine and serine, and methionine are worthy of intensive study. Others have been included for reference, because the laboratory findings in vitamin deficiencies or inborn errors of metabolism cannot be understood without a knowledge of the metabolic relationships.

12 The Common Terminal Pathway of Metabolism

In the preliminary stages of metabolism, carbohydrate, fat, and the carbon skeletons of amino acids are partially oxidized and produce, apart from some carbon dioxide and water, one of six substances: (a) *acetate* as acetyl-coenzyme A (acetyl-CoA) from two-thirds of the carbon of carbohydrate and glycerol, all the carbon of fatty acids, and a part of the carbon of amino acids; (b) *α-oxoglutarate* from glutamate, histidine, arginine, citrulline, ornithine, and proline; (c) *oxaloacetate* from aspartate; (d) *fumarate* from part of the benzene rings of phenylalanine and tyrosine; (e) *succinate* from methionine, valine, and isoleucine; and (f) *acetoacetate* from parts of leucine, phenylalanine and tyrosine.

These substances are metabolically closely related in that they all take part in a common metabolic pathway—the *tricarboxylic acid cycle* or *citric acid cycle* (see Fig. 12.1); acetoacetate enters the cycle through acetyl-CoA. In the cycle both carbon atoms of the acetyl residue are oxidized to carbon dioxide at the expense of the oxygens of two water molecules, the hydrogen atoms being transferred by dehydrogenases to the hydrogen transporting and oxidizing mechanism described later in this chapter. The cycle also has what has been called an *anaplerotic* function, that is, part of it can be used for the synthesis of metabolites needed for other functions in the cell. Although this role is more important in bacteria than in animals, the continuous provision in liver cells of oxaloacetate for the formation of the aspartate necessary for the urea cycle (p. 224) is an example of anaplerosis.

The major function of the cycle in animals is undoubtedly that it is the most important series of reactions in cellular economy in which energy from oxidative processes is released and trapped. One example of this importance is the brain damage that follows inhibition of α-oxoglutarate oxidation by ammonia (p. 224) or branched-chain oxo-acids (p. 234). Before discussing the reactions of the tricarboxylic acid cycle in detail, the theoretical background of biochemical energetics will be discussed.

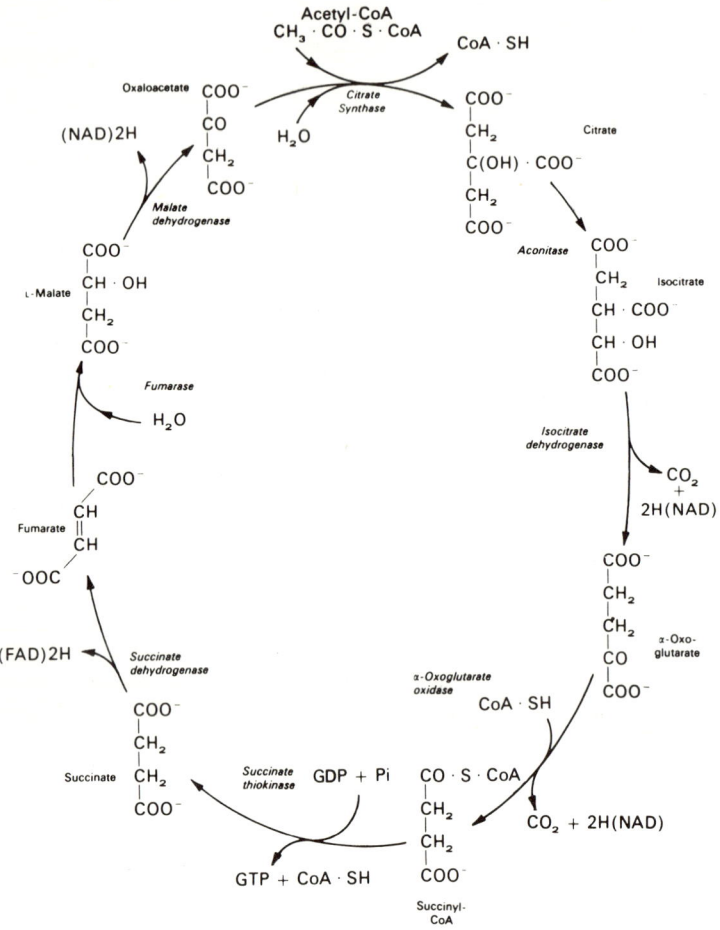

Fig. 12.1. The tricarboxylic acid cycle.

Biochemical Energetics

Living organisms require a constant supply of energy and materials to maintain their structure and to perform their numerous functions; these requirements are supplied by the food taken in. The processes of digestion and metabolism are directed towards supplying energy and compounds necessary for growth and replacement of the tissues. Energy is not only required for mechanical work, maintenance of body temperature, and osmotic work, but also to drive numerous synthetic reactions.

Free Energy

The free energy content of a substance, G, is a measure of the potential energy of that substance, but it is a quantity that cannot be measured directly. If, however, substance A reacts chemically and is converted into substance B:

$$A \rightleftharpoons B \tag{1}$$

the free energy content of B (G_B) probably differs from that of substance A (G_A) and a *change of free energy* (ΔG) has occurred as follows:

$$\Delta G = G_B - G_A$$

When ΔG is negative, then G_B is less than G_A, and when A reacts to produce B the free energy content of the system (A plus B) decreases. This means that an equal amount of energy has been given up by the system (as heat, work, or change in entropy). When ΔG is positive, G_B is greater than G_A, and when A reacts to give B the free energy content of the system increases, i.e. energy has been taken up. For such a reaction to proceed energy must be supplied to drive it.

When the free energy change in a reaction is negative (ΔG negative) the reaction is called *exergonic*, and when the free energy change is positive (ΔG positive) the reaction is called *endergonic*.

It should be emphasized that the sign and magnitude of ΔG give no information on the rate of a reaction. Thus for the complete oxidation of glucose according to the equation

$$C_6H_{12}O_6 + 6O_2 \rightarrow 6CO_2 + 6H_2O$$

$\Delta G = -2\,870$ kilojoules per mole of glucose; in spite of this large negative ΔG, glucose is quite stable in air (which contains 20 per cent oxygen), the rate of spontaneous reaction being negligibly small.

For reaction (1) the free energy change is given by:

$$\Delta G = \Delta G^0 + \mathbf{R}T \log_e \frac{[B]}{[A]} \tag{2}$$

where ΔG^0 is the *standard free energy change*, $\mathbf{R}$ is the gas constant, and T is the Kelvin temperature.

The standard free energy change may be evaluated as follows. At equilibrium reaction (1) goes from left to right at the same rate as from right to left, i.e. there is no net change of A into B or vice versa. The equilibrium constant, K, for the reaction is given by:

$$K = \frac{[B]}{[A]} \qquad (3)$$

At equilibrium there is no change in free energy, $\Delta G = 0$, therefore (2) becomes:

$$0 = \Delta G^0 + \mathbf{R}T \log_e \frac{[B]}{[A]}$$

which, using (3) and changing to logarithms to the base 10, becomes:

$$0 = \Delta G^0 + 2 \cdot 303 \mathbf{R}T \log K$$

or

$$\Delta G^0 = -2 \cdot 303 \mathbf{R}T \log K \qquad (4)$$

Equation (4) shows that if it is possible to determine the equilibrium constant for any reaction, the standard free energy change can also be calculated.

Table 12.1
Relationship Between the Equilibrium Constant (K) and the Standard Free Energy Change (ΔG^0) of a Reaction at 25°

K	$\log K$	$\Delta G^0 =$ $-2 \cdot 3 \mathbf{R}T \log K$ joule (J)
1000	3	-17121
100	2	-11414
10	1	-5707
1·0	0	0
0·1	-1	5707
0·01	-2	11414
0·001	-3	17121

The relationship between K and ΔG^0 for a number of values is shown in Table 12.1, from which it is apparent that when $K < 1$, ΔG^0 is positive, i.e. the reaction is endergonic, and when $K > 1$, ΔG^0 is negative, i.e. the reaction is exergonic. If, for reaction (1), $K = 100$, then since at equilibrium $[B]/[A] = 100$, the reaction favours the formation of B. Should the reaction be started with 101 parts of A, it will proceed to equilibrium at which there will be 1 part of A and 100 parts of B. That is to say, 100 parts of A (99 per cent) will have been converted to B. Similarly, if $K = 0·01$ and the reaction is started with 101 parts of A, only 1 part will be converted to B and at equilibrium $[B]/[A] = 1/100 = 0·01$; large-scale conversion of B to A in this instance is endergonic.

If the numbers of reactants and products are not equal, e.g., in the reaction $A \rightleftharpoons B + C$, their ratio is not a pure number, and the value of the logarithmic term in (2) depends very much on the absolute concentrations, i.e. whether they are molar, millimolar or micromolar. This, in turn, means that the actual value of ΔG can be very different from ΔG^0.

Coupled Reactions

Two or more reactions may be coupled in such a way that an exergonic reaction may be used to drive an endergonic one. Many instances of this sort of coupling are found in metabolism when the energy derived from an exergonic reaction (usually an oxidation) is used to drive an endergonic synthetic reaction. This sort of coupling requires both reactions to contain a common reactant. Thus if we have two reactions:

$$A \rightleftharpoons B, \qquad K = 0.01, \qquad \Delta G^0 = 11\,414 \text{ J/mol} \tag{5}$$

and

$$B \rightleftharpoons C, \qquad K = 1\,000, \qquad \Delta G^0 = -17\,121 \text{ J/mol} \tag{6}$$

reaction (5) is endergonic and [B]/[A] tends to 1/100; reaction (6) is exergonic and [C]/[B] tends to 1 000/1. Since both reactions can be coupled through the common reactant B, (5) and (6) can be added together to give:

$$A \rightleftharpoons C, \qquad K = \frac{[C]}{[A]} = 10, \qquad \Delta G^0 = -5\,707 \text{ J/mol} \tag{7}$$

If we start with 110.1 parts of A, reaction (5) would proceed until 1 per cent of A has been converted to B, but as soon as B begins to be formed 99.9 per cent of it will be converted to C by reaction (6), keeping reaction (5) going from left to right. This process will continue until all three components are in equilibrium when there will be 10 parts of A, 0.1 part of B, and 100 parts of C. Thus, although the reaction of $A \rightarrow B$ is endergonic and therefore unfavourable, the sequence $A \rightarrow B \rightarrow C$ proceeds to the right because the reaction $B \rightarrow C$ is exergonic and favourable. Reaction (5) has been driven to the right by the favourable free energy change of reaction (6).

Heat of Reaction

The change in the heat content of the reactants, ΔH, is related to the free energy change as follows:

$$\Delta G = \Delta H - T\Delta S$$

or

$$\Delta H = \Delta G + T\Delta S \tag{8}$$

where ΔS is the change in the *entropy* of the reactants. The entropy of a system is a measure of the degree of order of the system and a measure of the freedom of the components of the system to exist in all possible resonance forms, tautomers, etc.

When ΔH for a reaction is negative the heat content of the reactants diminishes as the reaction proceeds and heat is given out. Conversely, when ΔH is positive the heat content increases and heat is taken up. The magnitude and sign of ΔH for a given reaction may be measured in a calorimeter but, because of relation (8), the change in the heat content must not be equated with the free energy change.

Oxidation-reduction

It is appropriate at this point to give a short account of oxidation-reduction systems. *Oxidation* implies that the atom oxidized loses electrons and, conversely, *reduction* implies that the atom reduced gains electrons. Thus in the system:

$$Zn + Cu^{2+} \longrightarrow Zn^{2+} + Cu$$

the metallic zinc atom is the reducing agent which is oxidized to a zinc ion, while the cupric ion is the oxidizing agent which is reduced to metallic copper.

Another, more complex, example is the oxidation of hydroquinone by Fe^{3+} (ferric) ions in which the hydroquinone loses not only 2 electrons but also 2 protons (H^+), thus:

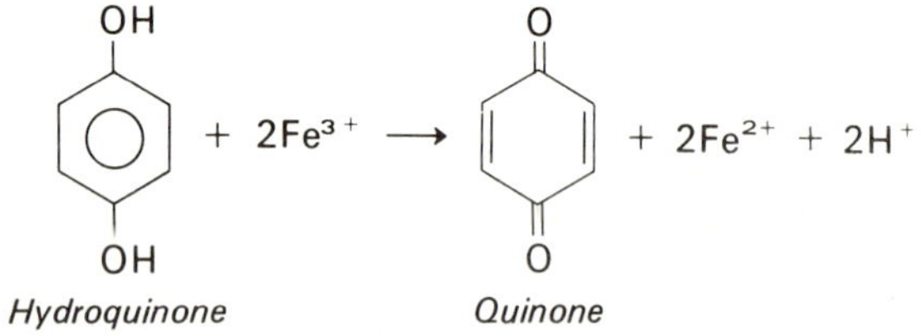

Hydroquinone   Quinone

In a third example, 2 atoms of hydrogen may be transferred, as in:

$$CH_2(NH_2) \cdot COOH + O_2 \longrightarrow CH(:NH) \cdot COOH + H_2O_2$$

Any reaction of this kind can be represented as if the transfer of $2H^+$ and 2 electrons had taken place.

A fourth type of oxidation is that in which the compound which is oxidized gains one or more oxygen atoms directly from molecular oxygen. This is uncommon in biological systems, but some important reactions of this type are known, and are considered later in this chapter.

Oxidation-reduction Potential

The tendency of any particular atom, ion, or molecule to lose an electron, thereby being oxidized, is measured by its oxidation-reduction potential,

E_0', which is the potential of a platinum electrode placed in a solution of equimolar concentrations of both the reduced and the oxidized forms of the substance and measured against the standard hydrogen electrode. The more strongly reducing the substance the more negative is its E_0'. Thus substance A with a more positive E_0' is able to oxidize substance B with a more negative E_0'. It is often, however, found in practice that some intermediate link or carrier is required to bring about the reaction.

The difference in oxidation-reduction potential, $\Delta E_0'$, between an oxidizing and a reducing agent is related to the free energy change, which occurs when oxidizing agent is reduced by the reducing agent, by the expression

$$\Delta G = -nF\Delta E_0' \qquad (1)$$

where n is the number of electrons transferred and F is Faraday's constant (96 230 J/v equivalent). For instance, the oxidation of $NADH_2$ by molecular oxygen (through the intermediate steps to be described) can be represented by the reaction

$$NADH_2 + \tfrac{1}{2}O_2 \rightarrow NAD + H_2O \qquad (2)$$

In this reaction $n = 2$ and $\Delta E_0'$ is given by the difference between E_0' for $NADH_2/NAD$ (-0.320 v) and for water/oxygen (0.816 v) both at pH 7.0; $\Delta E_0' = 0.816 - (-0.320) = 1.136$ v. Therefore for reaction (2) numerical values can be inserted into equation (1) giving:

$$\Delta G = (-2)(96\,230)(1.136) = -218.6\,kJ/mol$$

Since ΔG for this reaction is large and negative the oxidation of $NADH_2$ is strongly exergonic. This favourable free energy change does not, however, mean that $NADH_2$ is rapidly oxidized by molecular oxygen.

High-energy Compounds

A large number of the coupled reactions discussed on p. 255 involve ATP or another nucleotide triphosphate. That is, an endergonic reaction is forced to go by being kinetically coupled with the hydrolysis of ATP, either to ADP and P_i, or to AMP and pyrophosphate. Examples of such coupled reactions are the carboxylation of pyruvate (p. 165) and the formation of N^5,N^{10}-methenyl-tetrahydrofolic acid (p. 246). In reactions of this type it is remarkable that neither reactants nor products of the endergonic process contain phosphate. Whether coupled reactions that are theoretically possible actually occur depends on the presence of an enzyme to catalyse them, and in many instances it is known or suspected that a phosphorylated intermediate, not shown in the overall reaction, is formed. The extensive use of ATP rather than GTP, UTP and CTP is striking.

In the metabolic pathways so far outlined, ATP has always been formed by transfer of a phosphoryl group from a suitable donor; in the processes described later in this chapter it is shown that ATP is synthesized in processes coupled to the transport of reducing equivalents from oxidizable substrates to oxygen, coupled reactions in which the synthesis of ATP is the endergonic process. Thus ATP can take part in two-fold coupled reactions, effectively transferring chemical energy between oxidative and synthetic reactions. This phenomenon is so widespread, and so important to the economy of the cell, that it is convenient to speak of the nucleotide triphosphates as 'high-energy compounds', or as 'the common energy currency of the cell'. The former term has been extended to the acyl thioesters of coenzyme A and acyl carrier protein, and to a few phosphate-containing compounds (e.g., creatine phosphate) for which the equilibrium constant approaches 1 for phosphoryl exchange with ADP.

In many cases the large free energy of hydrolysis may be explained on structural grounds (pp. 143 and 148), but several instances may now be quoted of apparently simple compounds which could be classed as 'high-energy' on equilibrium grounds. Acetylcarnitine, aminoacyl-tRNA and N^{10}-formyl-FH_4 all fall into this category.

It should also be remembered that endergonic reactions need not always be driven by ATP. The synthesis of long-chain fatty acids, for example, is largely driven by the energy inherent in the redox potential of the $NADPH_2/NADP$ couple.

Reactions of the Tricarboxylic Acid Cycle

Citrate Synthase (Condensing Enzyme)

In this reaction the acetyl residue of the energy-rich thioester, acetyl-CoA, is condensed with oxaloacetate to form citrate, free $CoA \cdot SH$ being liberated. This is an exergonic reaction, ΔG being about $-33\,600\,J/mol$, and the equilibrium lies very much in the direction of citrate synthesis (see p. 254). The coupling of the hydrolysis of acetyl-CoA with citrate synthesis is very necessary because the condensation of free acetate with oxaloacetate to form citrate (as catalysed, for example, by a bacterial enzyme) is not nearly so favourable to citrate synthesis. The exergonic nature of this and other reactions in the cycle makes it effectively uni-directional.

Aconitase

It was formerly believed that *cis*-aconitate

$$HOOC \cdot CH{=}C \cdot CH_2 \cdot COOH$$
$$\underset{\textstyle COOH}{|}$$

was an intermediate between citrate and isocitrate. It is known that aconitase, in the presence of Fe^{2+}, catalyses the isomerization between all three acids and that citrate can be isomerized to isocitric acid without pass-ing through the stage of *cis*-aconitate. As has been discussed on p. 122, the citrate formed by the condensing enzyme does not behave as a symmetrical molecule and in the isocitrate formed by the aconitase reaction the hydroxyl group is on the methylene carbon atom which was derived from oxaloacetate.

Isocitrate Dehydrogenase

This enzyme catalyses the oxidative decarboxylation of isocitrate to α-oxoglutarate, CO_2 being liberated and 2H being transferred to a pyridine nucleotide, which may be either NADP in the cytoplasmic enzyme of liver or NAD in the mitochondria of most tissues. The former reaction is reversible, the latter almost irreversible (because of a very high K_m for CO_2).

Although this reaction can be considered to occur in two stages, free oxalosuccinate

$$HOOC \cdot CO \cdot \underset{\underset{\displaystyle COOH}{|}}{CH} \cdot CH_2 \cdot COOH$$

is not an intermediate. For the NAD-linked mitochondrial enzyme, there is no evidence that enzyme-bound oxalosuccinate is an intermediate.

α-Oxoglutarate Oxidase

This enzyme complex catalyses the oxidative decarboxylation of α-oxo-glutarate with the formation of succinyl-CoA. The mechanism is analogous to the oxidative decarboxylation of pyruvate to give acetyl-CoA (see p. 165). NAD, thiamine pyrophosphate (TPP), lipoic acid, and CoA-SH are the necessary coenzymes (see p. 144). In this reaction an energy-rich thioester is formed and the reaction is not readily reversible.

Succinate Thiokinase

In this coupled reaction the energy-rich thioester succinyl-CoA, formed in the oxidative decarboxylation of α-oxoglutarate, is hydrolysed, and at the same time the energy-rich compound guanosine triphosphate (GTP) is formed by the phosphorylation of GDP:

$$\text{succinyl-CA} + \text{GDP} + \text{P}_i \rightleftharpoons \text{succinate} + \text{GTP} + \text{CoA} \cdot \text{SH}$$

The reaction is readily reversible, K being approximately 4.

The GTP can be used directly in a number of energy-requiring reactions which are specific for this nucleotide. Alternatively, the terminal phos-phate may be transferred to an adenine nucleotide. It is probable that in mitochondria the relevant reaction is:

$$GTP + AMP \rightleftharpoons GDP + ADP$$

since the enzyme *nucleoside diphosphate kinase*, which catalyses the general reaction

$$NDP + ATP \rightleftharpoons NTP + ADP$$

appears to be purely cytoplasmic.

Succinate Dehydrogenase

The next step in the cycle is the dehydrogenation of succinate to form the unsaturated compound fumarate. The oxidizing agent is FAD which is, in this enzyme, firmly bound to the dehydrogenase. The succinate dehydrogenase enzyme is very strongly inhibited by the next lower homologue to succinate, i.e. malonate, and also by oxaloacetate.

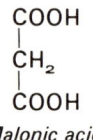

Malonic acid

Fumarase

Water is next added across the double bond of fumarate to form the hydroxy compound L-malate. The free energy change in this step is small; K is between 4 and 5.

Malate Dehydrogenase

In this oxidative step two hydrogen atoms are removed from L-malate to form oxaloacetate; NAD is the hydrogen acceptor. At pH 7·0 the equilibrium is very much in favour of malate; K is about 10^{-5}. The reaction is, however, driven towards oxaloacetate because this compound reacts with acetyl-CoA in the exergonic citrate synthesis.

Oxidative Steps in Metabolism

During the preliminary stages of metabolism, before the tricarboxylic acid cycle, there are a number of oxidative steps where 2 hydrogen atoms are removed from a substrate under the influence of a dehydrogenase. In the preliminary metabolism of carbohydrate there are two such steps:

$$\text{Triosephosphate} \rightarrow \text{Phosphoglycerate} + 2H$$
$$or\ \text{Lactate} \rightarrow \text{Pyruvate} + 2H$$

and
$$\text{Pyruvate} \rightarrow \text{Acetyl-S-CoA} + CO_2 + 2H$$

There are also in fat metabolism:

$$\text{Acyl-S-CoA} \rightarrow \alpha,\beta\text{-unsaturated acyl-S-CoA} + 2H$$

and $\quad \beta\text{-Hydroxyacyl-S-CoA} \rightarrow \beta\text{-Oxo-acyl-S-CoA} + 2H$

The glycerol of neutral fat requires, in addition to phosphorylation, an oxidative step:

$$\text{Glycerol 3-phosphate} \rightarrow \text{Triosephosphate} + 2H$$

Most amino acids require at least one oxidative step (usually mediated by glutamic acid):

$$\alpha\text{-Amino acid} + H_2O \rightarrow \alpha\text{-Oxo-acid} + NH_3 + 2H$$

though some require several.

During the oxidation of the two-carbon fragment in the tricarboxylic acid cycle there are four points at which 2 hydrogen atoms are removed, these are:

$$\text{Isocitrate} \rightarrow \alpha\text{-Oxoglutarate} + 2H$$

$$\alpha\text{-Oxoglutarate} \rightarrow \text{Succinate} + 2H$$

$$\text{Succinate} \rightarrow \text{Fumarate} + 2H$$

and $\quad\quad\quad\quad\quad \text{Malate} \rightarrow \text{Oxaloacetate} + 2H$

The Transport and Oxidation of Hydrogen

Nicotinamide-adenine Dinucleotides

In a great number of the dehydrogenations listed above the 2 hydrogen atoms are first transferred to a nicotinamide-adenine dinucleotide, either NAD or NADP (structures on p. 140) by a dehydrogenase specific both for the substrate and coenzyme. The reaction may be represented thus:

$$\underset{\text{(or NADP)}}{AH_2 + NAD} \xrightleftharpoons{\text{Dehydrogenase}} \underset{\text{(or NADPH}_2)}{A + NADH_2}$$

Table 12.2 shows some dehydrogenases known to be linked to NAD and NADP respectively.

In the reduction of nicotinamide-adenine dinucleotides by a dehydrogenase there are transferred to the coenzyme not 2 hydrogen atoms (each consisting of a proton and an electron) but 1 hydrogen atom and 1 electron (ε); the other proton, which is an oxidized hydrogen atom, is taken up by the buffers of the cell.

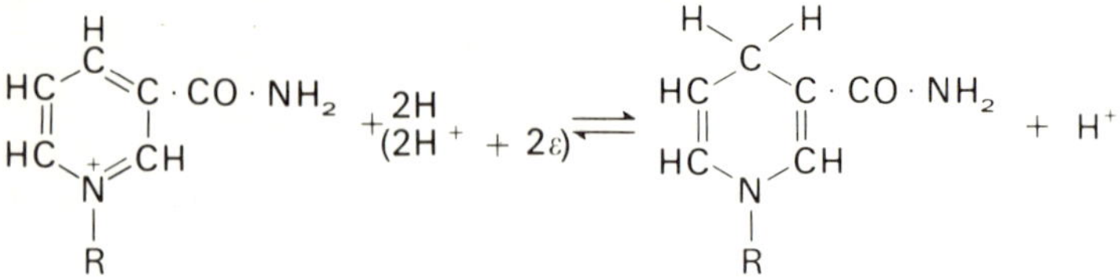

For this reaction E_0' is -0.320 v for NAD and -0.324 v for NADP (at pH 7.0).

Table 12.2
Dehydrogenases

NAD-linked	NADP-linked
Isocitrate	Isocitrate
Triosephosphate	Glucose 6-phosphate
Malate	6-Phosphogluconate
Glutamate	Glutamate
β-Hydroxybutyrate	Malic enzyme
β-Hydroxyacyl-CoA	
Lactate	
Pyruvate (to acetyl-CoA)	
α-Oxoglutarate	
Ethanol	

Dehydrogenases not Linked to Nicotinamide-adenine Dinucleotides

Two important dehydrogenation reactions of intermediary metabolism are not linked to either NAD or NADP.

Succinate dehydrogenase, which catalyses the removal of 2 hydrogens from succinate to yield fumarate, is a flavoprotein. The E_0' of this couple is less negative than that of the others in the citric acid cycle, and it is thermodynamically unlikely that it would be linked to a nicotinamide-adenine dinucleotide.

Acyl-CoA dehydrogenase, which catalyses the α,β-unsaturation of acyl-CoA (the first step in the breakdown of fatty acids) is also a flavoprotein.

The Flavoproteins

There are a number of these proteins, mostly yellow, which are concerned with oxidation-reduction. The flavin is usually very tightly bound as a prosthetic group. There are two such groups, *flavin mononucleotide* (FMN, 6,7-dimethyl-isoalloxazine ribityl-5′-phosphate), and *flavin adenine dinucleotide* (FAD), whose formula is given on p. 141.

The reduction of the flavoproteins occurs by the addition of 2 hydrogen atoms to the *iso*-alloxazine ring as follows:

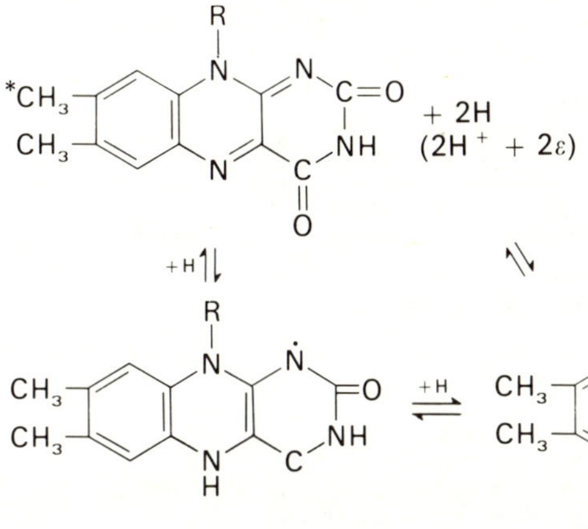

In some instances there is a semi-quinone intermediate (bottom left, with unpaired electron) between the fully reduced and the fully oxidized forms. These semi-quinones are stabilized by the apo-enzyme, and it is also probable that Fe, or other metal ion with variable valency, is often an enzyme-bound participant in the 2H transfer.

A major interest in flavoproteins is that they form an essential part of the respiratory chain described below, connecting reduced NAD with the cytochromes. When reduced nicotinamide-adenine dinucleotides are the substrates, the reaction may be represented by:

$$NAD(P)H + H^+ + FP \rightleftharpoons NAD(P)^+ + FPH_2$$

the proton coming from the cell buffers.

Other flavoproteins are 'aerobic dehydrogenases', that is to say they can transfer 2H directly from a substrate to molecular oxygen, which is reduced to hydrogen peroxide, H_2O_2 (cf. glucose oxidase, p. 175). H_2O_2 is toxic, and it is presumed that these enzymes usually work in conjunction with *catalase*, a haem-containing enzyme widely distributed in tissues, which catalyses the reaction

$$2 H_2O_2 \rightarrow 2 H_2O + O_2$$

However, in congenital acatalasia lesions are confined to the oral and nasal mucosa, and may be quite absent in some individuals.

A list of some flavoprotein enzymes is given in Table 12.3.

Table 12.3
Flavoprotein Enzymes

Substrate	Enzyme	Other groups
(a) FMN-containing enzymes		
$NADPH_2$	'old yellow enzyme' ($NADPH_2$ oxidase of red cells)	
$NADPH_2$	cytochrome c reductase	
$NADH_2$	$NADH_2$ dehydrogenase	Fe, labile S
(b) FAD-containing enzymes		
lipoic acid (enzyme bound)	lipoyl dehydrogenase	
succinate	*succinate dehydrogenase	Fe, labile S
L-amino acids	L-amino acid oxidase	
glycine	glycine oxidase	
glucose	glucose oxidase	
xanthine	xanthine oxidase	Mo, Fe
aldehydes	aldehyde oxidase	Mo, haem
acyl-CoA	acyl-CoA dehydrogenase	

* Flavin covalently bound to the enzyme protein through the methyl marked with an asterisk on p. 263.
E_0' for the yellow enzyme is -0.122 V at pH 7.

The Cytochromes

There exists in all aerobic cells a group of respiratory pigments called cytochromes consisting of a protein with various iron-containing haem-like prosthetic groups, see p. 114. It is believed that there are at least 4 cytochromes concerned in the 'mainstream' transfer of electrons from

Table 12.4
The Cytochromes

		E_0' (pH 7.0)
* Cytochrome b	Firmly bound, not autoxidizable when bound	$+0.05$ v
Cytochrome b_5	Not autoxidizable, microsomal	$+0.02$ v
Cytochrome P_{450}	Autoxidizable, forms a compound with CO. Microsomal	$+0.24$ v
* Cytochrome c_1	Not autoxidizable, bound	$+0.22$ v
* Cytochrome c	Not autoxidizable, soluble	$+0.25$ v
* Cytochrome a	Not autoxidizable, firmly bound	$+0.29$ v
* Cytochrome a_3	Firmly bound. Autoxidizable, forms compounds with CO, HCN, and H_2S. Contains Cu	

* Component of mitochondrial respiratory chain.

oxidizable substrate to oxygen. Many others are known. The properties of some cytochromes are shown in Table 12.4.

The concentration of cytochrome c, and also of the cytochromes a, in tissues closely parallels the rate of oxygen uptake by these tissues in vitro. Because of this, and because 85 per cent of the oxygen consumption of many tissues is stopped by HCN or CO, it is thought that the greatest part of the oxidation processes in most cells involve cytochromes.

The cytochromes are easily reduced and oxidized, not by the addition and removal of hydrogen atoms but by the addition and removal of electrons to the iron atom, making it ferrous (Fe^{2+}) in the reduced state and ferric (Fe^{3+}) in the oxidized state. The change may be represented thus:

$$\text{(reduced) } Fe^{2+} \text{ cyt} \rightleftharpoons \text{(oxidized) } Fe^{3+} \text{cyt} + \varepsilon$$

This is in marked contrast with the mode of action of haemoglobin as an oxygen carrier where the iron atom always remains in the ferrous state.

Cytochrome oxidase will bring about the reduction of molecular oxygen by a reaction which may be represented:

$$4\varepsilon + O_2 + 4H^+ \rightleftharpoons 2H_2O$$

The absorption spectrum of cytochrome oxidase apparently shows the presence of two haem groups with slightly different properties. One (a_3) reacts with CO, HCN and H_2S to form compounds which do not permit the passage of electrons to oxygen. The other haem (a), as judged by the spectra, does not so react. It is quite likely, however, that the two haem groups are identical, but react differently because of different environments in the protein. In fact, the cytochrome oxidase complex contains several ($a + a_3$) haem groups and several atoms of Cu. This complexity is not unexpected, since it is the oxygen *molecule* which accepts 4 electrons from cytochrome c, at least two at a time. This latter requirement is very important, because O_2 molecules possessing 1 or 3 extra electrons have a very low redox potential, and are powerful oxygen and electron donors. It has recently been found that all aerobic cells possess a copper-containing enzyme (previously called haemocuprein) which catalyses the reaction:

$$2O_2^{1-} + 2H^+ \rightarrow O_2 + H_2O_2$$

It is presumably the absence of this *superoxide dismutase* that makes oxygen toxic to obligate anaerobes, and the formation of O_2^- may explain part of the toxicity of hyperbaric oxygen to animals. Note that from this point of view, hydrogen peroxide is the product of reduced toxicity.

From the values of the E_0' and for other reasons it is believed that the cytochromes are linked together and with the flavoproteins as follows:

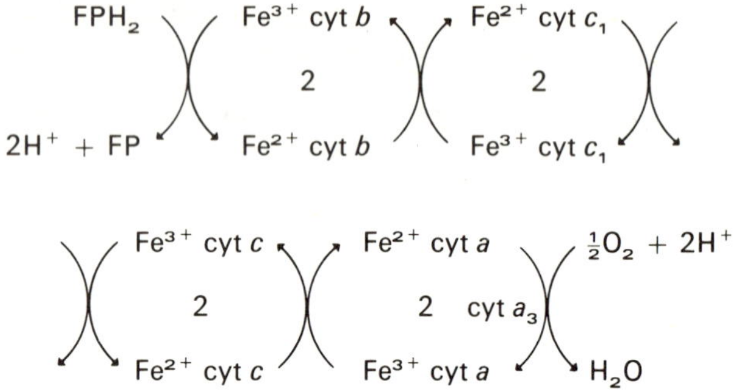

Note that as this scheme is written, and as it in fact occurs, each molecule of cytochrome is transferring a *single electron*. The associated protons are released to, and taken up from, the cell buffers.

Some cytochromes are not associated with respiration in mitochondria. Their possible functions are discussed later in this chapter (p. 268).

Non-haem Iron

Several flavoproteins listed in Table 12.3 contain Fe which is not held in a porphyrin ring. A small protein containing 2 atoms of non-haem Fe per molecule has been isolated from mitochondrial respiratory chain assemblies. Several polypeptides containing up to 8 Fe atoms per molecule have been isolated from bacteria and plant chloroplasts and some animal mitochondria. On adding acid to each of these proteins (including the Fe-containing flavoproteins) H_2S is liberated, which makes it seem likely that the Fe is bound in conjunction either with S^{2-}, or particularly labile cysteine residues, or both. All proteins of this type are now known as *ferredoxins*. Their role is much more clearly established in photosynthesis than in the respiratory chain of animal cells. The ferredoxins of photosynthetic reactions have standard redox potentials more negative than the pyridine nucleotides, which would not be appropriate to their postulated roles in electron transport. There is, however, good evidence that some animal ferredoxins are connected with mitochondrial oxidations not in the respiratory chain (p. 270).

Lipid Hydrogen Acceptors

The oxidation of intermediates of the tricarboxylic acid cycle, and of many other substrates, takes place to a very large extent in the subcellular particles known as mitochondria, and the transport of hydrogen atoms or electrons to oxygen by the carriers described in the preceding pages is organized in *respiratory assemblies* actually located in the lipoprotein inner membrane of the mitochondria. Since lipid makes up about 50 per cent by weight of mitochondria, it is reasonable to ask whether any lipids take part in hydrogen transport. A number of fat-soluble substances can be reversibly oxidized and reduced, particularly quinones (see p. 256). Among naturally occurring substances of this type are the vitamins K and vitamin E, but the latter only occurs in traces in mitochondria, the former not at all. A family of quinones with long polyisoprenoid side-chains is widely distributed in animal mitochondria, in plants, and in bacteria, in amounts 5–10 times greater than the concentration of cytochrome *c*. These compounds have been called the *ubiquinones*, or coenzymes Q (formula on p. 142). The most common has a side-chain with 10 isoprenoid units, $-C_{50}H_{81}$.

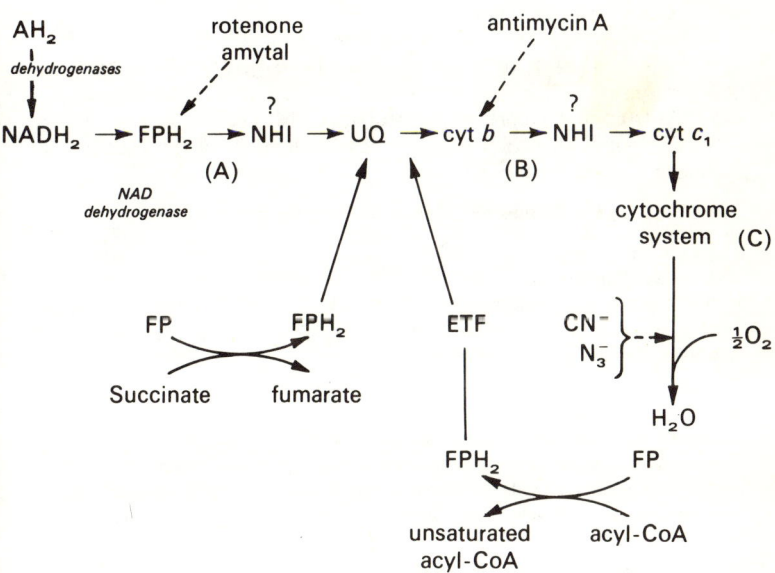

Fig. 12.2. The complete oxidative pathway.

Ubiquinone is easily reduced by flavoproteins and oxidized by cytochromes, and it is reasonably well-established as a link between these two groups of respiratory pigments.

There is so far no evidence for a semi-quinone intermediate.

The Complete Oxidative Pathway

We are now able to see the complete mechanism for the transport of hydrogen atoms or electrons from the substrates until the reduction of molecular oxygen which is carried to the cells by the haemoglobin of the blood. The K_m of the system for oxygen is extremely low, about $0.5 \mu M$. The whole pathway is blocked by cyanide, azide, carbon monoxide, or hydrogen sulphide at the oxidase step. Only cytochrome a_3 reacts with these reagents.

A scheme of the oxidative pathway is shown in Fig. 12.2.

Notes

1. The physiological hydrogen acceptor for the purified $NADH_2$ dehydrogenase studied in vitro has never been established.

2. NHI is non-haem iron proteins; it is not clear if these are ferredoxins.

3. Part of the mitochondrial cytochrome *b* does not respond to the passage of ε down this pathway.

4. ETF is the electron transferring factor, probably flavoprotein.

5. It is thought that $NADPH_2$ may be largely oxidized by other mechanisms.

6. (A), (B), and (C) represent the probable sites of ADP phosphorylation.

7. The points of action of well-known inhibitors of the transport of reducing equivalents are indicated by dotted arrows.

Oxidations not Linked to the Respiratory Chain

It has already been mentioned that several flavoprotein dehydrogenases reduce oxygen directly. *Xanthine oxidase* oxidizes not only xanthine and hypoxanthine, but aldehydes and $NADH_2$, with production of some superoxide ion. It is found in liver and in milk. The various amino acid oxidases, however, produce only H_2O_2.

A few examples are known of reactions catalysed by *oxygenases*, in which *both* atoms of a molecule of O_2 are directly incorporated into the substrate. Examples are the oxidation of tryptophan (p. 240) by *tryptophan pyrrolase*, of 3-hydroxyanthranilic acid (p. 241), and of homogentisic acid (p. 243). These reactions all involve the opening of an aromatic ring by adding the oxygen across a localized double bond. This seems to be a characteristic feature.

A third class of oxidations is that in which *one* atom of a molecule of oxygen is incorporated into a molecule of substrate by mono-oxygenases or *mixed function oxidases*. The latter name is given because the simultaneous dehydrogenation of a co-substrate is obligatory in the reaction:

$$S-H + XH_2 + O_2 \rightarrow S-OH + X + H_2O$$

the essential function of which is to disrupt the rather strong bonds between the two oxygen atoms in O_2. By now, several co-substrates XH_2 are known, and the following list cannot be exhaustive.

An interesting co-substrate is α-oxoglutarate, which is oxidatively decarboxylated to succinate:

$$S-H + \alpha\text{-OG} + O_2 \rightarrow S-OH + \text{succinate} + CO_2 + H_2O$$

The most important example of this class in aminals is *proline hydroxylase*, which catalyses the conversion of the first proline in the sequence —Pro—Pro—Gly— to hydroxyproline, and so converts pro-collagen to collagen (p. 74). An ingenious example is the enzyme which converts p-hydroxyphenylpyruvic acid to homogentisic acid, since the α-oxo acid reductant is on the same molecule. Enzymes of this class require Fe^{2+} ions and ascorbate, although the latter is not the co-substrate. It is only known to be the co-substrate in the hydroxylation of dopamine to adrenaline.

In the hydroxylations of phenylalanine → tyrosine, tyrosine → DOPA and tryptophan → 5-OH-tryptophan, the co-substrate is reduced pteridine, i.e. the heterocyclic moiety of tetrahydrofolic acid up to, but not including, N^{10} (p. 146).

In the conversion of kynurenine to 3-OH-kynurenine (p. 241), the co-substrate is $NADPH_2$. Several hydroxylations with $NADH_2$ as co-substrate are now known. A rather unspecific aromatic hydroxylase (Chapter 13) uses reduced flavoprotein, while the 11β-hydroxylation of steroids in the adrenal uses a ferredoxin. Both these co-substrates are presumably themselves reduced by electron donors of lower redox potential. Cytochrome P_{450} takes part in the mechanism of hydroxylation with these latter two co-substrates.

In a large number of other reactions, such as squalene epoxidase (p. 200), prostaglandin synthase (p. 203), and many steroid hormone hydroxylations, both on the ring system and in the side-chain, the co-substrate is not yet known.

Among other reactions of this kind are the hydroxylations of cholesterol to bile acids (p. 203), and of cholesterol, progesterone and various other steroids in the formation of the adrenal cortical and sex hormones. Other enzymes catalyse the detoxication of barbiturates and other drugs (p. 276).

Many of these hydroxylations are carried out by enzymes associated with the smooth endoplasmic reticulum, particularly in liver. (On homogenizing the tissue, the disrupted reticulum spontaneously forms vesicles called *microsomes*.) Microsomes contain both an $NADH_2$ and an $NADPH_2$ dehydrogenase. Both arc flavoproteins. The latter can reduce cytochrome c directly, the former through another haemoprotein called cytochrome b_5. However, the cytoplasm does not contain cytochrome e, and microsomes do not contain cytochrome oxidase, so the oxidation of reduced pyridine nucleotides in the cytoplasm does not terminate simply in water as it does in mitochondria. One or two hydroxylation reactions in mitochondria have been shown to involve (a) a $NADPH_2$ dehydrogenase; (b) a non-haem iron protein; (c) an unusual haemoprotein called P_{450} which is autoxidizable, i.e. which is the proximate oxygen-activating component. Some microsomal hydroxylations also require P_{450}, but there is at present no evidence for a microsomal ferredoxin in the pathway of electron flow between co-substrate, substrate and oxygen.

The plant enzyme *peroxidase* catalyses the oxidation of a wide range of substrates, including many aromatic dyes, by H_2O_2.

Oxidative Phosphorylation

Much evidence makes it clear that in animal cells the function of the electron transport chain, and of the oxidative processes coupled to it, is the resynthesis of ATP from ADP and P_i. The energy-transferring functions of nucleoside triphosphates have been considered earlier in this chapter, and numerical values for the free energy of hydrolysis of ATP and ADP are given on p. 148. It now remains to consider the mechanism of synthesis and its implications.

During the oxidation of a mole of reduced NAD by $\frac{1}{2}O_2$, a considerable amount of energy is liberated (217·5 kJ). There are, however, several stages in the oxidation process (Fig. 12.2), and the decrease in free energy associated with a single step may be quite small. As pointed out on p. 225, the hydrolysis of ATP has one reactant and two products, so that the standard free energy change does not necessarily apply. In fact, it may be calculated that the reversal of the hydrolysis, at the usual millimolar concentration of the reactants in vivo, requires 54–58 kJ, instead of the standard 31 kJ. (It is worth noting that this figure would be much reduced if the concentration of H_2O, implicit in the reaction, were lower than 55 M, which it is in dilute aqueous solutions. One of the reasons for the location of oxidative phosphorylation in a lipid-rich membrane may be to reduce the water concentration, and so favour the synthesis of ATP.) From these energy considerations, it appears that there are rather few reactions in the respiration chain which could effectively be coupled with phosphorylation. The experimental evidence shows clearly that the number of molecules of ATP synthesized per *atom* of oxygen reduced ($\equiv 2H \equiv 2\varepsilon$ transferred along the chain), is 3 for NAD-linked substrates, but only 2 for succinate and other FP-linked oxidations. As a result of such considerations, the three sites of phosphorylation from $NADH_2$ to O_2 have been assigned to the approximate sites shown in Fig. 12.2. Other kinds of experimental evidence confirm this.

About the mechanism of oxidative phosphorylation there is much less agreement. It is established that no phosphoryl-intermediates of the respiratory chain carriers are formed (indeed, with 1-electron carriers in the cytochrome system, but $1 \sim P$ formed per 2ε transferred, it would be quite diffi-cult to imagine such intermediates). At the other end of the process, there has been isolated from mitochondrial membranes a large insoluble protein called F_1, which catalyses the final condensation of ADP with P_i. Molecules of this protein can be seen in electron micrographs as knobs projecting from the membrane into the matrix; they were identified because they also catalyse the reverse reaction, acting as a mitochondrial *ATP-ase*. No agree-ment about the intervening stages has been reached, largely because it has proved impossible to isolate the components from the membrane for study. This may mean simply that the catalysts require a hydrophobic, or water-poor, environment, or it may mean that the membrane itself is involved in the coupling process. The *chemical coupling* theory postulates that a labile high-energy intermediate (probably a modified enzyme) is the transmitter between oxidation and phosphorylation, or, in a sophisticated version of the theory, that a concerted change in the conformation of membrane proteins leads to phosphoryl incorporation. There is certainly at least one well-defined intermediate because certain mitochondrial processes, notably accumulation of metal ions, and transhydrogenation between $NADH_2$ and NADP, can use the energy trapped in this interme-diate and so reduce or circumvent ATP synthesis. The *chemi-osmotic* theory postulates that the primary process is extrusion of the H^+ ions formed during the transport of reducing equivalents to the outside of the membrane, while the electrons stay inside. The secondary exergonic reaction is then, in effect, the recombination of H^+ and OH^- ions to form water. This theory runs into some difficulty in the cytochrome part of the chain where H^+ ions are not involved. There is, however, no clear-cut experimental evidence at present which would decide between the two hypotheses.

There are three types of drugs affecting oxidative phosphorylation; *inhibitors of respiration*, such as those depicted in Fig. 12.2, which block the transfer of reducing equivalents down the respiratory chain; *uncouplers* like 2,4-dinitrophenol, which prevent the synthesis of ATP, but may even stimulate respiration; some of them activate mitochondrial ATP-ases. *Oligomycin-like* inhibitors (also triethyltin) block the coupling process, so that both respiration and phosphorylation are inhibited.

General Considerations

Apart from the reactions in which oxygen is directly incorporated into organic compounds, the pathways considered in this chapter are almost all concerned with energy production and trapping in the chemically useful form of ATP. Anaplerotic use of the TCA cycle is of little importance in animals other than for gluconeogenesis.

Many facts show that oxidative phosphorylation is of major importance in the energy metabolism of most tissues, for example, the sensitivity of the brain to anoxia, the rapidly toxic effects of cyanide, or of atractyloside, which prevents the egress of ATP from mitochondria. There are, of course, cell types which obtain their energy from glycolysis, notably erythrocytes, and the cells of the cornea and the retina, but elsewhere fatty acid oxidation, even to ketone bodies, must always be kept going by respiratory chain activity.

It may be calculated that the conversion of one mole of glucose to two moles of pyruvate gives rise to 2 moles of ATP by substrate-level phosphorylation. At the same time 2 moles of $NADH_2$ are formed which can yield 6 moles of ATP by oxidative phosphorylation. Conversion of the pyruvate to acetyl-CoA yields another 2 moles of $NADH_2$ and thus 6 moles of ATP. The oxidation of each acetyl group in the TCA cycle gives rise to 12 moles of ATP, from $3 \times NADH_2$, $1 \times FPH_2$, and one substrate level phosphorylation. Thus the total ATP yield from the oxidation of 1 mole of glucose is $(2 \times 12) + (2 \times 6) + (2 \times 6) + 2 = 38$ moles of ATP, of which only 1/19 is provided by the Embden–Meyerhof pathway. The results of a similar calculation for fatty acid oxidation have been given in Chapter 10, and show for palmitic acid oxidation a total of 131 moles of ATP formed (less 2 for acyl-CoA formation), of which 96 are synthesized in the oxidation of the acetyl-residues on the TCA cycle. For amino acids the returns generally are lower and more variable, since not all the reactions are coupled to the respiratory chain.

This considerable synthesis of ATP may be directly coupled to synthetic reactions. For example, it may be calculated that 40 per cent of all ATP synthesized in growing *E. coli* is used for protein synthesis, particularly for the highly endergonic synthesis of short-lived messenger RNA. In animal cells, however, much of the ATP is coupled to more complex energy-requiring processes, such as muscular contraction, the sodium-potassium pump and tubular reabsorption processes in the kidney.

Most of the processes described in this chapter take place within or on the membrane of the mitochondria and biochemists become accustomed to taking into account the extremely restrictive and selective nature of mitochondrial permeability. The outer membrane is moderately permeable, but the inner membrane, although permeable to O_2, H_2O and CO_2, is impermeable to cations, and to most anions other than monocarboxylates (except pyruvate). Inorganic phosphate, di- and tri-carboxylate ions and amino acids are transported by complex permeases on which one ion has to be exchanged for another of the same charge. ADP has to be exchanged for ADP, and this permease is absolutely specific for the adenine nucleotides. The impermeability of the membrane to CoA and acyl-CoA has already been mentioned (Chapter 10).

One of the more remarkable aspects of this character of the mitochondrial membrane is that it is not permeable to $NADH_2$, so that the redox potential of the cytoplasm ([NAD]/[$NADH_2$] about 1 000:1) is quite different from the mitochondria ($[NAD]/[NADH_2]$ about $7:1$). Moreover, although E' for NAD and NADP are almost identical, the ratio $[NADP]/[NADPH_2]$, both mitochondrial and cytoplasmic, is much lower (about $2:1$). Complex shuttles of redox pairs have been proposed to explain how reducing equivalents are transported in and out of mitochondria, but it is difficult to prove their existence unequivocally, because the concentrations of metabolites can only be estimated indirectly. This *compartmentation* within the cell can have unexpected results, such as, for example, the massive glucose production by liver in vitro caused by ethanol. This is due to the decrease of the $[NAD]/[NADH_2]$ ratio in the cytoplasmic compartment, which favours gluconeogenesis, as a result of the NAD-linked oxidation of ethanol in the cytoplasm at a rate faster than the mitochondrial shuttle system can cope with.

13   Detoxication and Excretion

This chapter is mainly concerned with the chemistry of substances found in the urine and faeces; the mechanics of excretion, i.e. renal function, are outside the scope of this book.

Detoxication

This is a complex subject because there is an infinite number of compounds 'foreign' to the body which may be altered chemically and excreted more or less efficiently. The fact of chemical change implies enzyme action, but it cannot be supposed that specific enzymes exist to detoxify each of the compounds which may find its way into the body. It is more likely that 'foreign' chemicals are substrates for enzymes of relatively low specificity which are present in tissues for the metabolism of 'naturally occurring' compounds. This view is supported by the fact that the products of the reactions are sometimes more toxic than the original compounds. It must be emphasized that detoxication is a relatively inefficient process, in that the maximum amounts of foreign chemicals which can be dealt with are frequently measured in milligrammes, and are rarely more than 1 or 2 grammes per day.

The term is also extended to cover chemical attack on drugs which may not be toxic, and also on naturally occurring compounds, such as plant essential oils, hormones, and some amino acid derivatives. The common logic here is that the detoxified products are excreted and not completely oxidized in the body. Such a definition will not, however, include the detoxication of ethanol, and one or two other alcohols. There are thus no general rules for detoxication mechanisms, and there are many species differences. Only some important examples can be dealt with here.

Oxidation

For the detoxication of *oxygen* free radicals, see Chapter 12.

Oxidation may be divided into two categories, on the basis of the enzymes involved.

Dehydrogenases transfer reducing equivalents to an orthodox hydrogen acceptor. Thus alcohols (including the toxic ethanol) are oxidized to the corresponding aldehyde by *alcohol dehydrogenase*, which does not attack methanol. An aldehyde dehydrogenase will complete the conversion to the corresponding acid.

Cyanide in small quantities (e.g., arising from the hydrolysis of amygdalin in almond essence) is oxidized to thiocyanate by an enzyme *thiosulphate: cyanide sulphurtransferase*. Thiosulphate, present in small quantities in plasma, is the oxidizing agent.

$$HCN + Na_2S_2O_3 \rightarrow NaCNS + NaHSO_3$$

Amine oxidases fall into two classes: *the monoamine oxidases* (MAO) and *diamine oxidases*. Monoamine oxidases oxidize amines formed from many of the amino acids, particularly in the gut, and more especially tyramine and serotonin (5-hydroxytryptamine) formed in the central nervous system. MAO is a flavoprotein. *Diamine oxidase* contains pyridoxal phosphate. It oxidizes histamine (see p. 225), putrescine (from ornithine) and cadaverine (from lysine). In all cases the hydrogen acceptor is oxygen

$$RCH_2 \cdot NH_2 + H_2O + O_2 \rightarrow RCHO + H_2O_2 + NH_3$$

Oxidases are responsible for the direct addition of oxygen (from O_2) on to an organic molecule (see Chapter 12). In this context the microsomal *mixed function* oxidases are important, and have been discussed on p. 268. Apart from their normal functions (e.g., hydroxylation of steroids and aromatic amino acids), the enzymes attack foreign chemicals in a variety of ways.

Hydroxylation of aromatic rings. This is thought to occur by addition of an atom of oxygen across a double bond (epoxidation), e.g.

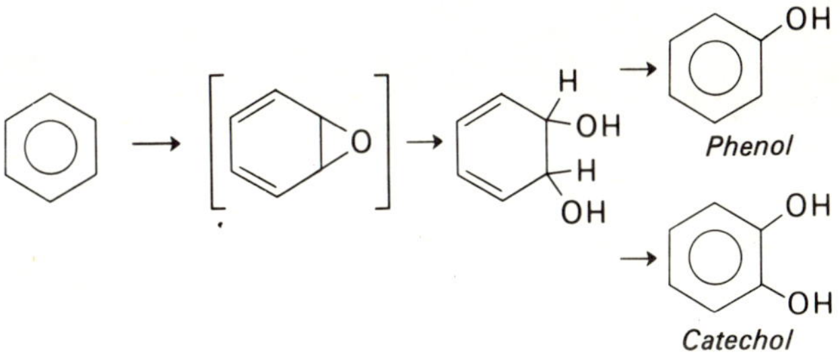

Phenol

Catechol

The insecticide *aldrin* is metabolized into the more toxic (to animals) and persistent *dieldrin* by epoxidation.

Many carcinogenic hydrocarbons are detoxified by epoxidation or direct hydroxylation, e.g., benzopyrene to (1:3) or (1:6) dihydroxybenzo-

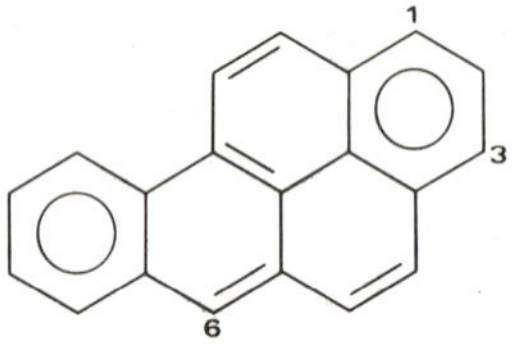

Benzopyrene

pyrene. *Phenobarbitone* (I) is converted into the inactive compound II.

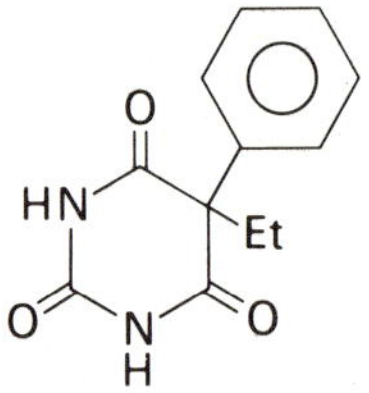

I 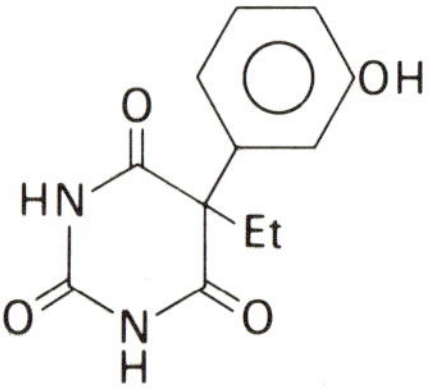   II

Acyclic hydrocarbons may be attacked at the terminal methyl group (ω-oxidation).

N-hydroxylation. The primary product may well be a so-called *N-oxide* ($R_3N \rightarrow O$). If the N is in a ring and is also methylated, this may be the first stage towards demethylation. Amine oxides will rearrange to give hydroxylamines, and this is an instance where the term detoxication is an unfortunate one, because arylhydroxylamines, in particular, are often potent carcinogens. For example, it is the products of *β-naphthylamine* (below) which produce bladder cancer.

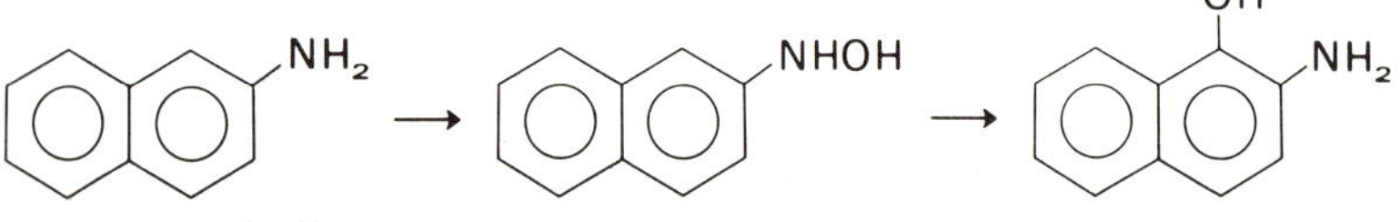

2-Naphthylamine   2-Naphthyl-   2-Amino-
                         hydroxylamine   1-naphthol

De-alkylation (specifically demethylation) of N-containing derivatives has already been mentioned. De-alkylation of O- and S-ethers is also common. *Tetra-ethyl* lead, $Pb(Et)_4$ is oxidized to the toxic free radical tri-ethyl lead $Pb(Et)_3^{\cdot}$ by a microsomal system.

Desulphuration. The sulphur-containing insecticides *malathion* and *parathion* are rapidly desulphurated by insects to products which are potent inhibitors of cholinesterase

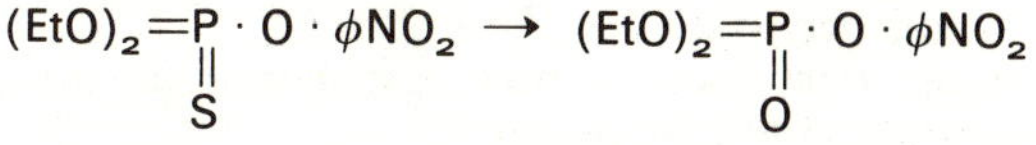

$$(EtO)_2{=}\underset{\underset{S}{\|}}{P} \cdot O \cdot \phi NO_2 \quad \rightarrow \quad (EtO)_2{=}\underset{\underset{O}{\|}}{P} \cdot O \cdot \phi NO_2$$

Parathion   Paraoxon

$$(MeO)_2{=}\underset{\underset{S}{\|}}{P} \cdot S \cdot \underset{\underset{CH_2 \cdot COOEt}{|}}{CH} \cdot COOEt \quad \rightarrow \quad (MeO)_2{=}\underset{\underset{O}{\|}}{P} \cdot S \cdot \underset{\underset{\cdot CH_2 \cdot COOEt}{|}}{CH} \cdot COOEt$$

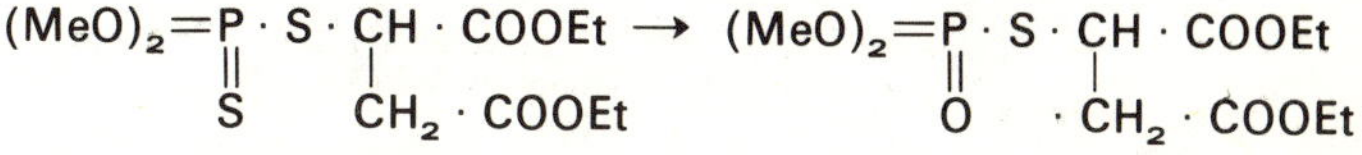

Malathion   Malaoxon

However, while mammals rapidly desulphurate parathion, which is therefore toxic, they only slowly attack malathion, and this is relatively safe.

The mixed function oxidases are being intensively investigated because they are inducible, quite often by their foreign substrates. Moreover, it has been found that some inducers will increase the activity of some, but not all, of the other oxidases. The induction may be immediate or delayed, and may, after chronic drug treatment, last for months after treatment has been stopped. Whether liver endoplasmic reticulum oxidases are increased by phenobarbitone appears to be genetically determined.

Infants have very little mixed function oxidase activity, and are therefore more susceptible to the action of many foreign compounds. Full activity is reached about 8 weeks after birth.

Reduction

Chloral, $CCl_3 \cdot CHO$, is reduced to the corresponding alcohol, $CCl_3 \cdot CH_2OH$, which is said to be the active hypnotic agent. *Aromatic nitro compounds* are often partially reduced to amines, e.g., picric acid to picramic acid:

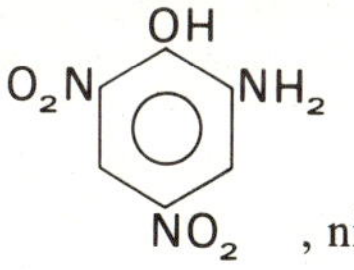

, nitrobenzene to *p*-aminophenol

Dehalogenation

Small amounts of *mono-* and *di-iodotyrosine* are formed in the thyroid during the synthesis of thyroxine. Normally, these are reductively de-iodinated by a *deiodinase*, with re-formation of I^-. The congenital absence of this enzyme leads to continuous loss of body iodine and the development of chronic goitre. The toxic *carbon tetrachloride* CCl_4 is similarly reductively dechlorinated to chloroform, which is metabolically stable, while the anaesthetic *halothane* $CF_3 \cdot CHBrCl$ is converted to $CF_3 \cdot CH_3$ (the $C-F$ bond is very stable).

The most important (but very slow) first stage in the detoxication of *DDT* by mammals is reductive dechlorination

$$\begin{array}{c} Cl\phi \\ \diagdown \\ \diagup \\ Cl\phi \end{array} CH \cdot CCl_3 \rightarrow \begin{array}{c} Cl\phi \\ \diagdown \\ \diagup \\ Cl\phi \end{array} CH \cdot CHCl_2$$

DDD

On the other hand, flies remove HCl directly to a non-toxic product; resistant strains can do this much more rapidly

$$\begin{array}{c} Cl\phi \\ \diagdown \\ \diagup \\ Cl\phi \end{array} CH \cdot CCl_3 \rightarrow \begin{array}{c} Cl\phi \\ \diagdown \\ \diagup \\ Cl\phi \end{array} C{=}CCl_2 + HCl$$

DDE

$[\phi = \text{phenyl (here } C_6H_4)]$

Conjugation

Many of the compounds which have been discussed, and a great number of other compounds, are further acted on before they are excreted. The reaction which takes place is *conjugation*, i.e. either *esterification* or *peptide bond* formation. It is obviously necessary that the compounds which are conjugated should have either a hydroxyl group (aliphatic or aromatic) or a carboxylic acid group, or in rare instances an amino group. The compounds used for this detoxication reaction are glucuronic acid, sulphuric acid, acetic acid, glycine, and glutamine. They all come from endogenous sources, and the amounts available are limited by the rates of metabolism. This is particularly true for H_2SO_4, which can come only from oxidation of sulphur-containing amino acids. The production of glycine and glutamine can be increased to a limit set by the daily nitrogen intake. Glucuronic acid production can be increased almost indefinitely, as it comes from glucose. Glucuronides therefore replace sulphates and glycine conjugates if there is much material to be excreted.

Almost all the conjugation processes occur in the liver; the rate at which various easily estimated substances such as bromosulphthalein are removed from the circulation as they are conjugated, is used as a liver function test. Many of the conjugates appear to be excreted in bile and partially reabsorbed into the blood from the intestine. This is true, for example, of some derivatives of sex hormones and thyroxine. The parallel with bile pigments may be noted.

The usual effect of conjugation is to increase the solubility of the detoxified compound; a polar group is almost always left free.

Glucuronic acid. The compounds usually formed are glucuronides, i.e. conjugated through the aldehydic —OH on C_1, leaving the —COOH group free, e.g. phenylglucuronide

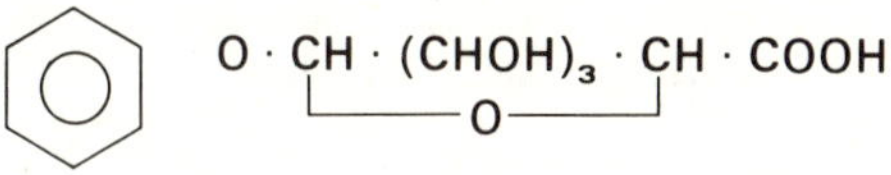

$$\text{O} \cdot \underset{|\underline{\hspace{1.2cm}} \text{O} \underline{\hspace{1.2cm}}|}{\text{CH}} \cdot (\text{CHOH})_3 \cdot \text{CH} \cdot \text{COOH}$$

Other compounds so conjugated are benzoic acid, butyl alcohol, bilirubin, and various steroids. The donor is UDP-glucuronate (see p. 149).

Glucuronates, esterified through the —COOH group, occur to a small extent.

Sulphuric acid. The most commonly occurring sulphate esters are those of phenolic compounds, phenol itself, cresol, and indoxyl:

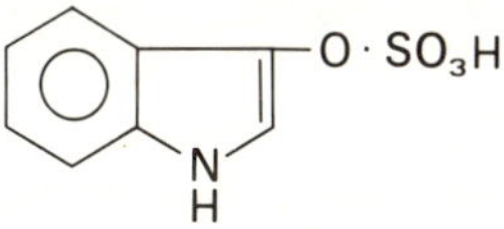

$$\text{O} \cdot \text{SO}_3\text{H}$$

The donor is 3′-phosphoadenosine-5′-phosphosulphate (PAPS). See p. 232.

Acetic acid. Acetylation of aromatic amino groups takes place to some extent. About half of a dose of sulphanilamide may be acetylated before it leaves the body:

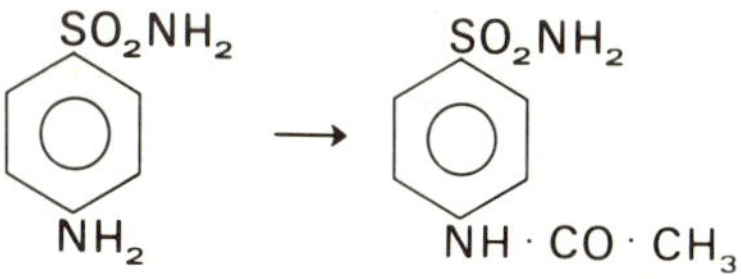

$$\text{SO}_2\text{NH}_2 \qquad \qquad \text{SO}_2\text{NH}_2$$
$$\rightarrow$$
$$\text{NH}_2 \qquad \qquad \text{NH} \cdot \text{CO} \cdot \text{CH}_3$$

This may be an important cause of reduced bacteriostatic efficiency of the drug in the blood-stream. The acetyl donor is acetyl-CoA.

Glycine. The best-known glycine conjugate is hippuric acid, the benzoic acid derivative:

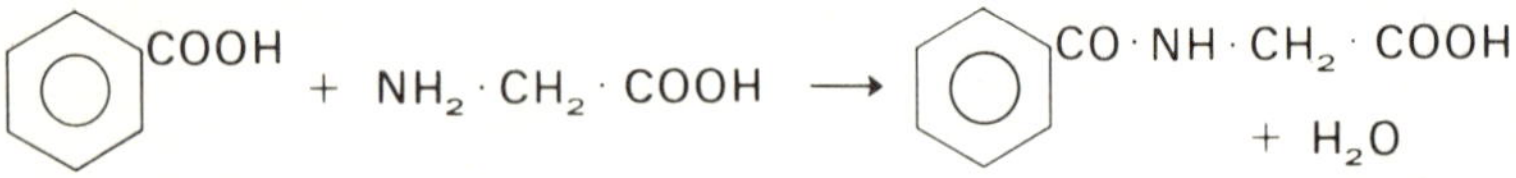

$$\phi\text{-COOH} + NH_2 \cdot CH_2 \cdot COOH \longrightarrow \phi\text{-CO} \cdot NH \cdot CH_2 \cdot COOH + H_2O$$

Salicylic acid, and other aromatic acids, may also be excreted in part as glycine conjugates.

Glutamine. Phenylacetylglutamine is found in considerable quantities in the urine of phenylketonurics, as a detoxification product of phenyl-acetic acid, itself formed from phenylpyruvic acid:

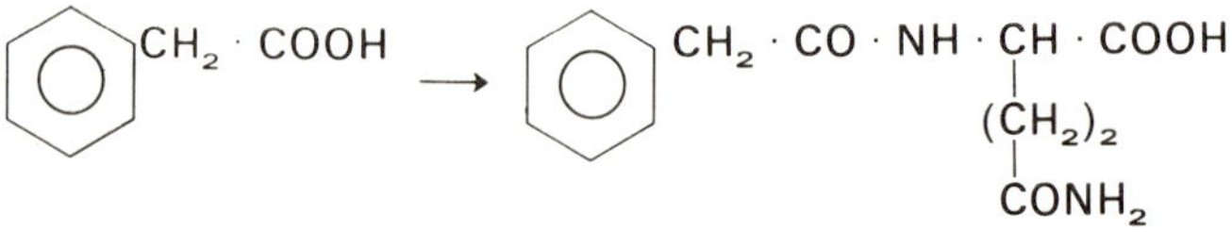

$$\phi\text{-CH}_2 \cdot COOH \longrightarrow \phi\text{-CH}_2 \cdot CO \cdot NH \cdot CH \cdot COOH$$
$$(CH_2)_2$$
$$CONH_2$$

In conjugation both with glycine and glutamine, the *acid* is activated by esterification with coenzyme A, thus

$$\phi \cdot COOH + ATP + CoA \longrightarrow \phi \cdot CO \cdot S \cdot CoA + AMP + P\!-\!P_i$$
$$\phi \cdot CO \cdot S \cdot CoA + H_2N \cdot R \longrightarrow \phi \cdot CO \cdot NH \cdot R + CoA \cdot SH$$

Methylation

Several foreign compounds, as well as the natural catecholamine hormones (see p. 400) can be methylated. The donor is S-adenosyl methionine.

Hydrolysis

Many drugs, e.g., aspirin, procaine, and atropine, are esters; other drugs and natural products, e.g., the cardiac glycosides, are glycosides. It is usual for these compounds to be hydrolysed; one or both of the reaction products may then be further acted on. The muscle relaxant *succinyl-dicholine (succamethonium)* is detoxified by hydrolysis by serum *pseudo-cholinesterase*. For genetic reasons there is an almost continuous range of activity of this enzyme against succamethonium, from fast hydrolysis to very slow.

Urine

The constituents of urine are discussed in three sections: the *normal* constituents are those whose complete absence from urine would be decidedly abnormal. The *occasional* constituents include substances like glucose, which may be present in small quantities in urine from healthy

persons, even though the excretion of large quantities would have pathological significance. This group may also include many common drugs, and
flavouring or preservative substances in food; for example, the present
production of aspirin by one British manufacturer is many millions of
tablets per day, so that acetylsalicylic acid may almost be said to be a
normal constituent of urine. *Abnormal* constituents are those which are
always absent from normal urine; they may occur in harmless inborn
errors of metabolism as well as in disease.

The amounts of the inorganic constituents of urine are expressed
throughout in *milliequivalents* (expressed as mEq), rather than in grammes.
This practice has the advantage that a balance can be made between the
total anions and cations of the urine.

The volume of a normal daily output is 600–1 600 ml, although this
must depend on the fluid intake. The specific gravity varies between 1·001–
1·030. The extremes of pH which have been observed are 4·8 and 8·6, but
normal urine is usually pH 5·3–7·0. It is held between these limits largely
by the buffering of HPO_4^{2-} and $H_2PO_4^-$; the total amount of phosphate
excreted does not vary very much, only the ratio between the two ionic
species. Bicarbonate ions are also often present in alkaline urine.

Colour

Light yellow	Diuretics, heat stroke, diabetes insipidus, or mellitus
Orange-yellow	'Urobilin'; some drugs.
Orange-brown	Senna, rhubarb.
Brownish-red	Porphyrinuria; methaemoglobin.
Red	Blood; some drugs; beetroot.
Purple	Phenolphthalein (in alkaline urine).
Dark brown	Haemoglobinuria; porphyrinuria.
Black-brown	Haemoglobinuria; phenol poisoning; alkaptonuria.
Greenish	Biliverdin; methylene blue, some phenolic compounds.
Blue	Methylene blue; indigotin.

Normal Constituents

Inorganic	Normal values mEq/24 hours	Comments
Sodium	170–260	Governed partly by intake and partly by activity of adrenal cortex; thus excretion is diminished in fever.
Potassium	65–90	As for Na, except that adrenal cortex stimulation increases K output.
Calcium	0·5–15	Affected by parathyroid activity and vita-min D.
Magnesium	14–24	
Ammonia	30–60	Increases somewhat in acidosis. Must be estimated in fresh urine.
Chloride	170–250	Depends partly on intake (chiefly as salt). Reduced if extracellular fluid volume increases.

Normal Constituents—(continued)

Inorganic	Normal values mEq/24 hours	Comments
Phosphate (mEq P)	10–50 (this may include some organic phosphate esters).	The proportion of HPO_4^{2-} to $H_2PO_4^-$ will depend on the pH of the urine.
Sulphate	35–80	Often lowered after poisoning (see organic sulphates).
Bicarbonate		Present in alkaline urine.

Organic	Normal values g/24 hours	Comments
Urea	20–35	Reduced in some renal diseases. Otherwise depends on nitrogen in food. About 85 per cent of total N excretion in urine.
Uric acid	0·1–1·5	
Creatine	0–0·06	Maximum normal value 50 mg in men, up to 100 mg in women. Greatly increased in thyrotoxicosis and muscular dystrophy.
Creatinine	1·1–3·2	Dependent on muscle mass, and fairly constant from day to day in individual subjects (see, however, creatine).
Organic sulphates	Equivalent to 0·18–0·36 g H_2SO_4	Usually esters of phenolic compounds.
Organic acids	Up to 1·0	Includes lactic and citric acids, and oxalic acid coming from foodstuffs.
Hippuric acid	0·7	Benzoic acid is frequently added as a preservative to foods.
Phenols	0·1–0·3	Increased by putrefaction.
Indican	0·004–0·02	Increased in intestinal obstructions.
Amino acids and peptides	1·1	Those excreted in greatest concentration are histidine, cystine, aspartic and glutamic acids, but the two latter compounds are almost completely excreted conjugated, i.e. as small peptides or phenolic esters. Up to 50 mg hydroxyprolyl-prolyl-peptides, coming from collagen breakdown. May be increased in protein–calorie malnutrition. There are two types of cystinuria in which cystine is excreted in abnormal quantities, in one type with arginine, lysine, and ornithine. Abnormalities of amino acid excretion are present in many inborn errors and renal lesions.

Organic	Normal values g/24 hours	Comments
Proteins	0·02–0·1 including enzymes, etc.	Amylase, 8 000–30 000 units, is increased in acute pancreatitis.
Vitamins		Significant traces of all water-soluble vitamins.
Hormones		Significant traces of sex and adrenal cortical hormones or their metabolites, either free or conjugated.
Coproporphyrins	0·1 mg	Increased in liver diseases.
Urobilinogen	0·2–3 mg	
Urobilin	10–130 mg	Increased in liver diseases.

Occasional Constituents

	Normal values g/24 hours	Comments
Reducing substances	0·5–1·5	By Fehling's, Benedict's, or Nylander's tests. It is very rare for urine to be completely free of reducing substances. Besides mono- and di-saccharides, uric acid, creatinine, glucuronides, salicylic acid, and homogentisic acid will reduce Fehling's solution. The other two reagents are less affected. *Protein must always be removed.*
Lactose		In late pregnancy and lactation.
Pentoses		Heavy fruit-eaters; inborn pentosuria.
Fructose		Inborn fructosuria.
Galactose		Inborn galactosuria.
Glucose	0–0·5	After heavy carbohydrate meal; renal glycosuria or diabetes cause increase. Glucose and galactose may be identified by means of their respective oxidases.
Acetone bodies	20–50 mg	Includes acetone, acetoacetic acid and β-hydroxybutyric acid. The latter does not respond to any simple test, but is often 50 per cent of the total. Increased in starvation, carbohydrate deprivation, diabetes and acidotic vomiting in children.

Pathological Constituents

	Normal values g/24 hours	Comments
Glucose	Up to 600 g Usually about 100 g	In severe diabetes the urine contains 3–5 per cent glucose with considerable polyuria. In mild diabetes, fasting glycosuria may be absent.

Pathological Constituents—(continued)

	Normal values g/24 hours	Comments
Acetone bodies	Rarely more than 50 g	Gerhardt's test positive. (Salicylic acid and other compounds also give a colour with $FeCl_3$, but one which is not destroyed by boiling.)
Fat		Extractable with ether from alkalinified urine. The urine is usually milky. Disturbances of the lymphatic system; renal lesions.
Peptides		Peptides of various sizes, apparently unfinished protein chains, are excreted in protein malnutrition.
Protein		Abnormal proteins, such as Bence–Jones protein, are sometimes produced by tumours.
Plasma proteins		Usually albumins, globulins may also be present. Identified by electrophoresis. Benign proteinuria, postural or after severe exercise, otherwise usually nephritis.
		Semi-quantitative estimations make use of the change in colour of a buffered indicator when it is bound to the protein.
Haemoglobin		Rare. Hb in the blood outside the corpuscles is usually rapidly oxidized. Incompatible transfusions, blackwater fever.
Methaemoglobin		More usual excretion form of Hb.
Methaemalbumin		The haem group from haem pigments in the plasma is fairly readily detached and adsorbed by albumin.
Myoglobin		Severe muscular destruction.
Blood		Renal lesions, poisoning, parasitic infections.
Bilirubin		Only in jaundice.
Phenylpyruvic acid	up to 2	Green colour with $FeCl_3$. Phenylketonuria.
Homogentisic acid	up to 2	Alkaline urine turns black on standing. Alkaptonuria.
Porphyrins	0·3–150 mg	Toxic (liver diseases, haemolytic infections, barbiturate and other drugs), acute or congenital porphyria.

Faeces

The average daily weight of faeces excreted by an adult on a mixed diet is about 160–250 g; 75 per cent of this is water. In children the range is much greater, from 4–120 g. On a vegetable diet the weight of faeces may rise to over 350 g, while on a meat diet it may fall to 60 g. Even in prolonged fasting there is still an excretion of up to 20 g; this is largely made up of bacteria and desquamated epithelium from the intestine. In pathological conditions the weight may rise to over 1 kg per day. Usually this includes a greatly increased volume of water; it must be remembered that the total daily production of digestive secretions, including saliva and bile, is estimated to be 7–8 litres. Normally all this fluid, excepting for about 150 ml, is reabsorbed in the large intestine. Even a slight inefficiency in the reabsorptive process will produce copious excreta. Pathologically increased faeces may also contain a large quantity of fat (pancreatic diseases, sprue, or coeliac disease), or protein (pancreatic diseases—scarcity of digestive enzymes).

The composition of normal faeces is approximately:

	Per cent	Per cent
Water		75
Solid matter:		
Food residues		
Cellulose	2	
Nitrogenous (largely collagen and indigestible plant proteins)	2	
Lipids (plant sterols, and about equal quantities of neutral fats and free fatty acids)	6	
Bacteria	8	
Salts, mucus, stercobilin	7	25
		100

The large percentage of fat is noteworthy; it comes from bacteria and desquamated mucosal cells.

The brown colour of normal faeces is due to stercobilin and related pigments, decomposition products of haemoglobin (see Chapter 6). Changes in colour may be associated with variations in haemoglobin catabolism, or may be due to drugs or foodstuffs.

Colour

Very pale	Obstructive jaundice.
Grey	Undigested fats, calcium soaps.
Yellow	Unchanged bilirubin (breast-fed infants), rhubarb, senna, santonin.
Green	Chlorophyll, unchanged biliverdin, calomel.
Grey-black	Plant juices, iron (as sulphide).
Pitch black	Haematin (black puddings or haemorrhage into the digestive tract).
Blue	Methylene blue.

Inorganic Constituents

	Daily output	Percentage of total excretion
Sodium	0·12 g (5 mEq)	5
Potassium	0·47 g (12 mEq)	20
Calcium	0·64 g (16 mEq)	90
Magnesium	0·20 g (6 mEq)	70
Iron	up to 12 mg	ca. 100
Other heavy metals		ca. 100
Chloride	0·09 g (2·5 mEq)	2
Phosphorus	0·51 g	40
Sulphur	0·13 g	17

A good deal of the potassium, phosphorus and sulphur are present intracellularly in the bacteria of the excretion (the latter two, of course, in organic combination).

Chemistry of
         Blood

The blood consists of the liquid plasma in which are suspended the so-called formed elements, erythrocytes, leucocytes, and platelets. The chief functions of the blood are to provide for the nutrition of and excretion from the tissues by the carriage of oxygen, carbon dioxide, metabolites, products of digestion, hormones, etc. It also has a very considerable acid–base buffering action and provides an environment of very constant composition for the cells of the body.

When whole blood, containing an anti-coagulant, is centrifuged, the formed elements sediment into a volume equal to about 45 per cent of the total volume of blood; the supernatant is known as *plasma*. When, on the other hand, blood is allowed to clot and the clot to retract, the fluid left is known as *serum*. Essentially serum is plasma without the protein fibrinogen and other clotting factors.

Coagulation

The detailed mechanism of blood coagulation is very complex, and not yet completely understood. The major processes involved are the following:

(*a*) Shed blood releases from itself Factor XII (the Hageman factor), perhaps from platelets, which initiates the clotting process. Blood collected without platelet damage (in siliconed vessels) from which the platelets are subsequently removed by centrifugation, will not clot for long periods.

(*b*) There then follows a series of reactions in which enzymes are successively set free from protein pro-enzymes in the plasma, as shown in Fig. 14.1. This process is usually called a *cascade* (see also Chapter 17), because each step amplifies the effect of the previous one until, after a lag of 5–7 minutes, the activation of prothrombin becomes explosively fast.

The structure of thrombin (Factor IIa) has been worked out in detail. It is a proteolytic enzyme similar to trypsin, of the type known as 'serine-esterases' (see p. 135). Convincing evidence is accumulating that Factors Xa and IXa (and also VIIa) are serine-esterases of rather similar structure,

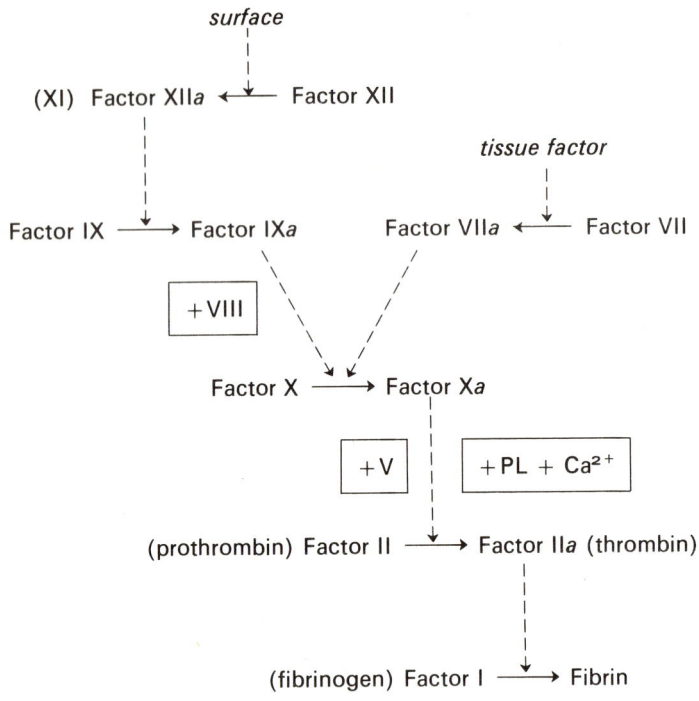

Fig. 14.1. Major stages in the initiation of blood coagulation. In accordance with current practice, the active forms of the enzymes have been indicated with the suffix *a*. Older names have been given in brackets where appropriate.

perhaps accompanied by some polysaccharide and lipid. The specificity of the reactions they catalyse may be accounted for by the fact that each requires a non-enzymic plasma protein co-factor, Factor VIII for IX*a*, and Factor V for X*a* (compare the effect of lactalbumin in determining the specificity of lactose synthetase, p. 366). In addition Factor X*a* requires Ca^{2+} ions and either phosphatidyl-serine or -ethanolamine, for full activity. This whole lipoprotein complex is probably identical with the entity usually known as thromboplastin, or Factor III.

The explosive activation of thrombin is not indefinite in extent. That would be very dangerous if it happened in the vascular bed. It is probably confined by the limited availability of Factors V and VIII, and these factors are inhibited by products of the clotting reaction (negative feedback).

Possibly many clot initiation events in vivo are attenuated, rather than amplified, by the cascade process.

(c) A 'tissue factor' present in injured tissues will convert another plasma pro-enzyme, Factor VII, to a serine esterase type of protease. This provides an alternative way of activating Factor X, as shown in Fig. 14.1. If tissue extracts are added to blood, clotting is almost instantaneous.

(d) Thrombin attacks the plasma protein fibrinogen, releasing a small peptide. The major product polymerizes into the insoluble fibrin. The network of fibrin fibrils enmeshes blood cells and platelets, forming a clot. Subsequently, inter-chain peptide bonds form between glutamyl and lysyl residues producing a *hard clot*.

(e) The platelet membranes contain actomyosin, which contracts, releasing ADP. The contraction causes the clot to shrink, expressing serum.

(f) After a few days, clots are lysed by plasmin (fibrinolysin), formed from a plasma precursor plasminogen. The bacterial proteolytic enzyme streptokinase is also used to activate plasminogen. Fibrinolysin is a protease, preferentially hydrolysing arginine-lysine bonds.

Coagulation may be prevented by the addition to blood of heparin, or of citrate, oxalate, or fluoride which form complexes with calcium ions.

Specific gravity of blood. This may be determined by a buoyancy method in copper sulphate solutions. The range of values found are:

(a) For whole blood, 1·055–1·062 (men), 1·050–1·056 (women).

(b) For plasma, 1·025–1·029.

(c) For serum, 1·024–1·028.

The specific gravity of whole blood rises and falls with the total number of cell elements, the specific gravity of erythrocytes being 1·084–1·117; the changes in specific gravity of plasma and serum reflect the plasma protein level.

pH. Blood is an excellent buffer and this property is discussed fully later in this chapter. The pH of blood is in the region of 7·3–7·5, being towards the lower value in the veins and the higher in the arteries.

Plasma Proteins

The total protein concentration in plasma is about 7 g/100 ml; it is distributed among the various fractions as shown in Table 14.1.

Fractionation of the proteins may be effected by precipitation at different salt concentrations, or by alcohol precipitation at low temperatures. A rapid semi-quantitative method of determining the relative amounts of the various proteins present in a plasma is by paper electrophoresis (Fig. 14.2). Zone electrophoresis in starch or polyacrylamide gel gives more and sharper bands.

Albumins. These have the lowest molecular weight of the plasma proteins, 69 000, and are formed mainly in the liver. Plasma albumin has a half-life

Table 14.1

Plasma proteins	Normal values g/100 ml
Total	6·3–7·8
Albumins	3·2–5·1
α_1-Globulins	0·06–0·39
α_2-Globulins	0·28–0·74
β-Globulins	0·69–1·25
Immunoglobulins (or γ-globulins)	0·8–2·0
IgA	0·15–0·35
IgG	0·8–1·8
IgM	0·08–0·18
IgD	approx. 0·003
Fibrinogen	0·2–0·4
Mucoprotein	approx. 0·135
Haptoglobins	0·03–0·19

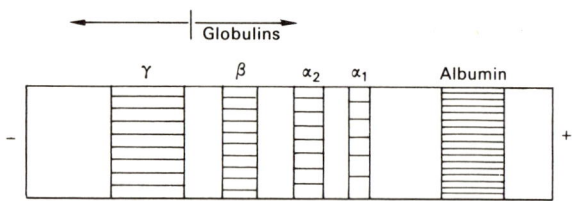

FIG. 14.2. Paper electrophoresis of normal serum. Barbiturate buffer pH = 8·6, $I = 0·05$. The arrows show the directions in which the protein bands move from the starting line.

of 17–20 days. It is responsible for the major part of the colloid osmotic pressure of the plasma. The albumin content of plasma is lowered in a great variety of abnormal conditions. Concentrations below 2 g/100 ml are almost always associated with oedema.

Globulins. These have molecular weights between 90 000 and 1 000 000, are separable into a large number of sub-fractions, and their functions are equally diverse. Large amounts of the plasma lipids are carried on the α- and β-lipoglobulins (lipoproteins, see Chapter 10), while antibodies occur in the γ-globulins (see below). The level of plasma globulins is increased in a great number of diseases and the relative changes of the various fractions are often of importance in diagnosis. The globulins are formed in the reticuloendothelial system, in the macrophages, and in the lymphocytes.

The Immune Response

Antigens are high-molecular-weight substances recognized as 'foreign' when introduced into the body. Such 'foreign' substances are usually protein or carbohydrate and, although they are made up of the same amino acids and saccharides as the body's own proteins and carbohydrates, presumably at least part of their molecules present configurations which are unfamiliar to the animal. Antigens enter the body as invading bacteria, viruses, or other parasites, or they may be introduced into the body through the skin (immunization or injury) or through the gut (food allergies). *Auto-immune* reactions arise when antibody production is stimulated by one of the body's own macromolecules, which is for some reason treated as a 'not-self' substance.

Antibodies. These are γ-globulins which are characteristically formed when antigens are introduced into the body; this is known as an immune response. They can be detected by their ability to combine with antigen. This combination can be demonstrated in a number of ways of which the simplest is by the formation of a precipitate when a solution of a soluble antigen is mixed, in a test tube, with some serum from an animal immunized with that antigen. A more dramatic demonstration of the presence of antibodies in the blood occurs if the antigen is a lethal toxin or a pathogenic micro-organism. When such an antigen is injected into animals, those which are immune (with specific antibody in their blood) live and those that are non-immune (no antibody in the blood) die. The most remarkable aspect of this phenomenon is that the antibody formed is specific for the antigen which caused its formation, and will combine only with that particular antigen or with substances of closely related structures. As far as is known, there is no limit to the number of different specific antibodies one animal can make in response to different antigens.

All antibodies are found in a group of closely related plasma proteins known as immunoglobulins (formerly γ-globulins) which can be divided into different classes known as IgG (immunoglobulin gamma), IgM (macroimmunoglobulin), IgA, IgD, and IgE. These can be distinguished chemically from one another by size, carbohydrate content, and amino acid analysis. There is, however, no correlation between class and specificity; antibodies of any specificity can be found in any class.

All these immunoglobulins react with antigens, and they all have the same basic four-peptide chain structure—two light chains and two heavy chains—shown in Fig. 14.3. The heavy chains are a little larger in IgA and IgM, and the polysaccharide attached to them is longer, 10–12 per cent of the total mol. wt. rather than the 3 per cent in IgG. Thus the molecular weight of the tetra-peptide unit varies from 150–180 000. IgA is made up of two units that readily dissociate, while IgM contains five units held together by $-S-S-$ bonds.

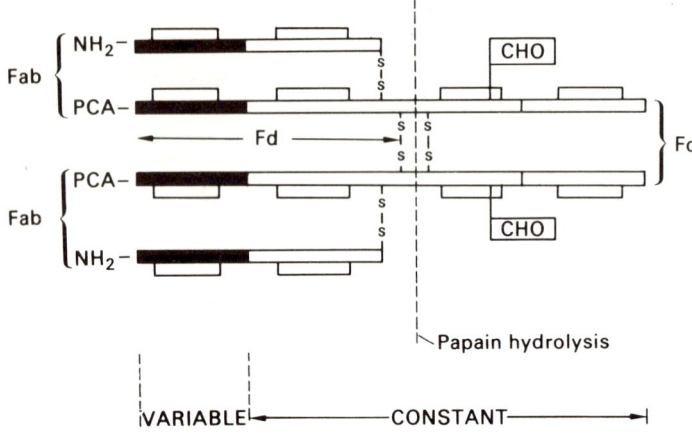

Fig. 14.3. Structure of IgG immunoglobulin. The two light chains are identical, as are the two heavy chains. PCA is the N-terminal residue of the heavy chains. The rectangular loops on each segment of each chain indicate intra-chain disulphide bonds. Papain hydrolysis produces two identical Fab fragments and one Fc fragment. Each Fab fragment contains one antigen-binding site. The variable portion of each peptide chain is indicated by heavy shading.

IgG is the principal immunoglobulin in human plasma; IgM increases sharply in *macroglobulinaemia*. IgA is present in relatively high concentration in saliva, milk and intestinal secretions, and is then associated with another polypeptide called the *secretor piece*, and with a separate *J protein*.

Proteolytic enzymes such as papain and pepsin bring about a limited hydrolysis of IgG. Papain splits it into three fragments, the —S—S— bond on the left of the dotted line in Fig. 14.3 being easily broken. Two are apparently identical, and are known as Fab (fragment antigen-binding); the third, Fc (fragment crystalline), is quite different (Fig. 14.3). Each Fab fraction contains one antigen-binding site; it can combine with antigen but will not form a precipitate. Amino acid sequence analysis has shown that the C-terminal half of all light chains so far examined has a constant composition, and the C-terminal three-quarters of all heavy chains is also constant. This is indicated in Fig. 14.3. This part of the heavy chain seems to contain three not dissimilar sequences, each with an intra-chain —S—S— bond. It does not combine with antigen, but is biologically active, being responsible for complement fixation, transport across the placental membrane, and binding to skin, etc.

The N-terminal segment, about 105–115 residues long, of each peptide chain has a variable amino acid composition. There are no similarities between N-terminal segments either of light chains or of heavy chains in different antibodies of whatever class (apart from the identity between the pairs in a single molecule). It appears that the very wide range of conformations which is made possible by this variability in primary structure, is responsible for the very great specificity of the antigen-antibody binding.

It is at present thought that plasma cells, although they all possess the genetic information to form immunoglobulins, exist in many cell types (clones), each of which is able to make only one specific antibody. As there must be at least 10^{12} different sequences possible in the variable part of each chain, it is an open question whether there are 10^{12} clones, or whether there are fewer cell types, with particularly labile DNA which is able to undergo extensive point mutation and recombination. It is at least clearer that the stimulus to extensive antibody formation is the absorption of the antigen to a particular clone, which then rapidly multiplies.

Pepsin hydrolyses IgG in such a way that the two Fab fragments are left held together by a disulphide bond. This structure is known as $(Fab')_2$ which is able to combine with two antigenic sites, and a precipitate will then form. This suggests that a three-dimensional lattice of antigen and antibody molecules, aided by the tendency to crystallize of Fc, builds up until the complex is no longer soluble. No covalent bonds are formed, because antigen and antibody can be separately recovered by suitable physical treatment. Fig. 14.4 shows precipitation of intact antibody in diagrammatic form.

Immunoglobulins are produced by plasma cells of all stages of maturity. An interesting aberration of the biosynthesis of antibody occurs in the disease *multiple myeloma* or plasmacytoma, a neoplastic condition in which there is uncontrolled production of immunoglobulin molecules of apparently a single variety, which appear in large quantities in the blood and contain only one type of light chain and one type of heavy chain. In about half of all myeloma patients an abnormal protein, known as Bence–Jones protein, appears in the urine. Bence–Jones proteins are precipitated from urine by heating to 50–60°: on further heating above 80° they redissolve but reprecipitate on cooling. It has been shown that Bence–Jones protein is identical with the light chains of the myeloma immunoglobulin in the plasma of the same patient but no two Bence–Jones proteins are identical, because of the different sequences in the variable segments.

When antibodies combine with cells and produce lysis, an additional factor known as *complement* is necessary. Complement is heat-labile and contains a number of enzymes which are concerned in the rupturing of cell walls.

Antibody production does not begin until well after birth; indeed recognition of substances as antigens can be permanently suppressed if

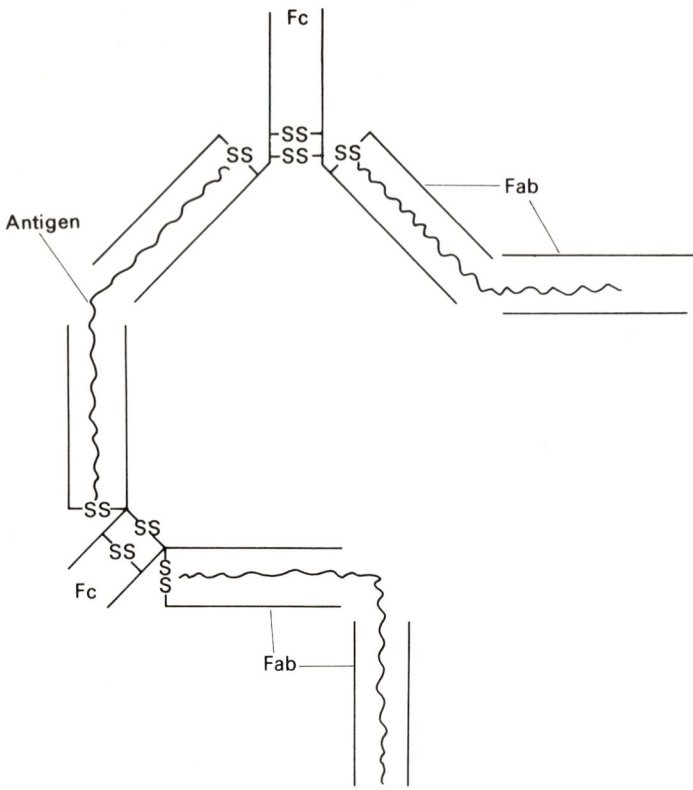

Fig. 14.4. Structure of a possible antigen-antibody complex. Here parts of four immunoglobulin molecules are shown reacting with three divalent antigen molecules. The peptide chains of the Ig molecules are shown as solid lines. For an explanation of Fab and Fc see text and Fig. 14.3. The binding has been made asymmetrical to suggest how a three-dimensional insoluble precipitate may be formed. The sizes of the antigens and antibodies are not to scale.

they are injected just after birth. *Passive immunity* for up to 12 months (in man) is provided by maternal antibodies transmitted into the infant's blood. In many species these antibodies are secreted into the milk (colostrum), but in man they are chiefly transmitted through the placental membrane.

Fibrinogen. This is essentially a globulin of very high molecular weight, 400 000, which is rather easily precipitated. It is the precursor of fibrin of the blood clot and it is almost exclusively formed in the liver. Increased fibrinogen concentration in the plasma is associated with a rapid erythrocyte sedimentation rate.

Haptoglobins, of which four are known, are normally present singly or in different combinations. These proteins firmly bind small amounts of haemoglobin which may be set free intravascularly. Up to 135 mg/100 ml haemoglobin can be bound in this way. This haemoglobin is then metabolized in the reticuloendothelial system.

Haemoglobin content of erythrocytes. The mean corpuscular haemoglobin concentration is 31·3–39·6 g/100 ml erythrocytes, which corresponds to 12–17 g/100 ml whole blood.

Other organic constituents. The principal non-protein organic constituents of whole blood and plasma or serum are shown in Table 14.2.

Hormones and vitamins. These include adrenaline, noradrenaline, androgens, oestrogens, pituitary hormones, insulin, adrenocortical hormones, thyroxine, and all the vitamins.

Coenzymes. These include coenzyme A, thiamine pyrophosphate, flavin-adenine dinucleotide, and the nicotinamide-adenine dinucleotides; these are mainly in the cells.

Enzymes. A very great number of enzymes are found in plasma. Some, such as pseudocholinesterase, lipoprotein lipase, and the blood-clotting enzymes, are specific to plasma, others are non-plasma specific and are formed in the tissues. The second group is by far the largest, and in recent years there has been a great development of interest in its members because changes in their levels occur in various diseases. It must suffice here to mention only a few which have aroused particular interest. Aldolase, lactate dehydrogenase, malate dehydrogenase, and glutamate-oxaloacetate transaminase are increased in myocardial infarction and liver disease. Isocitrate dehydrogenase, phosphoglucomutase, glucose-6-phosphatase, phosphohexoseisomerase, ornithine transcarbamylase, and glutamate-pyruvate transaminase are increased in liver disease. Lipase, amylase, and leucine aminopeptidase are increased in diseases of the pancreas. An increased level of acid phosphatase is associated with carcinoma of the prostate, while increased alkaline phosphatase is a sign of bone or liver diseases. Increased levels of creatine phosphokinase are found in myocardial infarction and muscular dystrophy. In addition, the presence of some particular isoenzymes (see p. 139) may give information on the site of the disease.

There are many enzymes in the red cells concerned with their metabolic processes. Care must therefore be taken to prevent haemolysis in blood intended for plasma enzyme estimations.

Table 14.2
Non-protein Organic Constituents of Blood
Values in mg/100 ml

	Whole blood	Plasma or serum
Non-protein nitrogen		
Total-N	28–39	22–29
Urea-N	8·9–15·2	9·6–17·6
Amino acid-N	4·6–6·8	4·3–7·7
Creatine-N	1·0–1·6	0·3–0·4
Creatinine-N	0·4–0·6	0·4–0·5
Uric acid-N	0·3–1·3	0·7–1·3
Nucleotide-N	4·4–7·4	0·2–0·4
Ammonia-N	0·01–0·02	0·01–0·02
Bilirubin (total)	—	0·26–1·4
Bilirubin (direct)	—	0·1–0·5
Carbohydrates		
Glucose (capillary, as true glucose)	50–90	max. 180
Glycogen	1·2–16·2	0
Hexosamine (total)	—	83·4
Pentoses (total) after hydrolysis	—	1·81–3·29
Hexuronic acids	4·1–9·3	0·4–1·4
Neuraminic acid	—	60
Lipids		
Total (ether soluble)	397–722	400–700
Fats (neutral)	85–237	0–450
Fatty acids (total)	290–420	200–450
Non-esterified fatty acids	—	10–17
Cholesterol (total)	129–228	150–250
Free cholesterol	80–110	30–60
Bile acids	2·5–6·0	0·2–3·0
Bile salts	—	5–12
Phospholipids (total)	—	150–250
Products of intermediary metabolism		
Acetone bodies (total)	0·5	0·3–0·9
α-Oxo-acids (total)	0–3·1	0·6–2·1
α-Oxoglutaric acid	0·05–0·27	0·8
Pyruvic acid	0·41–1·11	0·5–1·0
Lactic acid	4·7–15·1	6·1–16·9
Citric acid	1·3–2·3	1·6–3·2
Malic acid	0·24–0·75	0·1–0·9

Blood Electrolytes

The average concentrations of various anions and cations in the cells and the plasma are shown in Table 14.3.

Table 14.3
Principal Blood Electrolytes
1 litre whole blood = 550 ml plasma + 450 ml cells
(The figures in brackets indicate the normal range)

	Plasma constituents		Cell constituents		Total
	mEq/l plasma	*Equivalent to: mEq/l whole blood*	*mEq/l cells*	*Equivalent to: mEq/l whole blood*	*mEq/l whole blood*
Cations					
Na$^+$	145 (135–155)	79·8	21 (16–25)	9·5	89·3
K$^+$	4·2 (3·1–5·5)	2·3	95 (92–100)	42·8	45·1
Ca^{2+}	2·5 (2·1–2·7)	1·4	—	—	1·4
Mg^{2+}	1·6 (1·3–1·8)	0·9	2·5	1·1	2·0
Total	153·3	84·3	118·5	53·3	137·6
Anions					
Cl$^-$	103 (100–107)	56·7	58·9	26·5	83·2
HCO$_3$$^-$	27·0	14·9	16·7	7·5	22·4
Phosphate	2	1·1	1·5	0·7	1·8
Sulphate	1	0·6	—	—	0·6
Plasma proteins	15·6	8·6	—	—	8·6
X$^-$	4·7	2·4	4·7	2·1	4·5
Haemoglobin	—	—	36·7	16·5	16·5
Total	153·3	84·3	118·5	53·3	137·6

Discrepancies in the totals are due to rounding off.

Before any discussion of the changes in the concentrations of ions in the blood or in any other system can be followed, two fundamental points must be clearly understood. The first of these is the *requirement of electroneutrality*. This implies that the total number of anionic charges must equal the total number of cationic charges; i.e. the solution is electrically neutral. Thus, should the total number of cationic charges remain constant, then the total number of anionic charges must also be constant and equal to the number of cationic charges. The distribution of the charges among the various species of anions and cations may, however, be varied. The second important point is the *independence of ions of strong electrolytes*; this means that in solution we have, say, sodium ions (Na$^+$), chloride ions (Cl$^-$) and bicarbonate ions (HCO$_3$$^-$), *not* sodium chloride and sodium bicarbonate as such. Thus a part of the total number of cationic Na$^+$ charges is balanced by Cl$^-$ anions and (in this instance) the rest by HCO$_3$$^-$ anions.

Cations. From Table 14.3 it will be seen that in the blood the sodium is mainly extracellular while the potassium is intracellular. The maintenance of this distribution has been shown to be an active process, dependent on energy-yielding metabolic processes in the erythrocytes. The control of the blood level of cations is by the kidney, which is influenced in this function by various hormones and the requirements of acid–base balance which are discussed below.

Anions. The blood anions may conveniently be divided into two groups as follows:

(*a*) Anions which are not able to combine with hydrogen ions at the pH of the blood. These are the anions of the strong mineral acids, Cl^-, SO_4^{2-}, and also of such strong organic acids as lactic and pyruvic, which exist in blood as lactate and pyruvate anions.

(*b*) Anions the concentration of which varies with pH. These are the blood buffers and are the anions of acids the pK s of which are near the pH of the blood; they include the plasma proteins, haemoglobin, oxyhaemoglobin, HCO_3^- and, of minor importance, phosphate.

Plasma proteins as anions. Since the isoionic points of the plasma proteins are below the blood pH, the molecules carry a residual net negative charge, hence they may be considered as anions. They are also weak acids and therefore act as buffers. From titration curve data we have the relation:

mEq anions/l of plasma represented by plasma proteins

$$= 0.104 \times \text{g plasma proteins} \times (\text{pH} - 5.08)$$

From this relation we see that, at pH 7·4, 6·45 g/100 ml plasma proteins is equivalent to 15·6 mEq anions/l.

Haemoglobin and oxyhaemoglobin as anions. Again, since the isoelectric point of these proteins is below blood pH, they are anions, and they are also buffers. Haemoglobin has, in addition, the property that on oxygenation at blood pH there is an increase in its net negative charge. Alternatively, we may consider that, near the blood pH, haemoglobin is a weaker acid than oxyhaemoglobin. This change in the acid strength on deoxygenation is of prime importance in the buffering of the acid metabolite CO_2.

At pH 7·4, 1 mmol of oxyhaemoglobin (16·1 g, or approximately the amount present in 100 ml blood) is ionized so as to be equivalent to 2·4 mEq anions. The same amount of haemoglobin, at the same pH, is equivalent to only 1·7 mEq anions. Thus for every mmol of oxyhaemoglobin deoxygenated there is a loss of 0·7 mEq anions and hence 0·7 mEq cations are set free and have to be accommodated, because of the electroneutrality requirement, by other anions. Such anions are provided by the CO_2 entering the blood, and being hydrated to HCO_3^- ions and H^+ ions at the same time as the oxygen leaves the oxyhaemoglobin for the tissues. The H^+ ions are simultaneously taken up, to a large extent, by the now deoxygenated haemoglobin. The details of this system are discussed below.

Carbon Dioxide Transport and the Carbonic Acid Buffering System

This system is of special and central importance in living organisms because it removes one of the chief products of oxidation of the food which is also an acid anhydride; about 25 000 mEq H^+ ions are formed every day as the result of CO_2 production. At the same time the system is able to buffer effectively against acids and to a less extent against alkalis. Numerical values for the constants defined below are given in Table 14.4 for 38° in a solution like plasma.

Table 14.4
Values of Constants Relating to the Carbonic Acid Buffering System Appropriate to Blood Composition and Temperature

$[H_2CO_3]/[CO_2] = K_h'$	≈ 0.00296	$pK_h' \approx 2.53$
$[H^+][HCO_3^-]/[H_2CO_3] = K_{H_2CO_3}$	$= 1.66 \times 10^{-4}$	$pK_{H_2CO_3} = 3.78$
$[H^+][HCO_3^-]/[CO_2 + H_2CO_3] = K_1'$	$= 7.94 \times 10^{-7}$	$pK_1' = 6.1$
$[H^+][CO_3^{2-}]/[HCO_3^-] = K_2'$	$= 1.58 \times 10^{-10}$	$pK_2' = 9.8$
$[H^+][OH^-] = K_w$	$= 2.51 \times 10^{-14}$	$pK_w = 13.6$
$[CO_2 + H_2CO_3]/P_{CO_2} = q*$	$= 3.16 \times 10^{-2}$	$-\log q = 1.5$

* For P_{CO_2} in mm Hg and $[CO_2 + H_2CO_3]$ in mmol/l.

Hydration of Dissolved CO_2

Carbon dioxide is formed in tissue cells. As such it is dissolved in water where some of it is hydrated to carbonic acid according to the following reaction:

$$CO_2 + H_2O \rightleftharpoons H_2CO_3 \tag{1}$$

This reaction is slow but in the red cells there is an enzyme, *carbonic anhydrase*, which greatly speeds up the attainment of equilibrium. Because of this enzyme and the ability of CO_2 and H_2CO_3 to penetrate the red cell membranes, in cells and plasma dissolved CO_2 is in equilibrium with its hydrated form H_2CO_3. The concentration equilibrium constant for the hydration equation (1), K_h', is defined by:

$$K_h' = \frac{[H_2CO_3]}{[CO_2]} \tag{2}$$

Since it is difficult to determine H_2CO_3 separately from dissolved CO_2, it is usual in blood analysis to determine the sum of the concentrations of dissolved CO_2 and H_2CO_3.

Solubility of CO_2

Carbon dioxide obeys Henry's law and its concentration in solution is proportional to its partial pressure, P_{CO_2}, in equilibrium with the solution.

Remembering that CO_2 is partially hydrated, we have:

$$[CO_2] + [H_2CO_3] = qP_{CO_2} \tag{3}$$

where q is the Henry's law coefficient. An important consequence of equation (3) is that the concentration of *both* CO_2 *and* H_2CO_3 will vary, in plasma, with the partial pressure of CO_2 in equilibrium with the plasma (in the alveolar air).

Ionization of H_2CO_3

The 'true' first ionization of carbonic acid is:

$$H_2CO_3 \rightleftharpoons H^+ + HCO_3^-$$

for which the acid dissociation constant, $K'_{H_2CO_3}$, is given by:

$$K'_{H_2CO_3} = \frac{[H^+][HCO_3^-]}{[H_2CO_3]} \tag{4}$$

The 'Overall' First Ionization Constant

Because, in blood, H_2CO_3 is in equilibrium with dissolved CO_2, we have the overall equilibrium:

$$CO_2 + H_2O \rightleftharpoons H_2CO_3 \rightleftharpoons H^+ + HCO_3^- \tag{5}$$

for which the equilibrium constant, K_1', is given by:

$$K_1' = \frac{[H^+][HCO_3^-]}{[CO_2] + [H_2CO_3]} = \frac{[H^+][HCO_3^-]}{qP_{CO_2}} \tag{6}$$

Combining (2), (4), and (6) we obtain:

$$K_1' = K'_{H_2CO_3}\left(\frac{K_h'}{K_h' + 1}\right) \tag{7}$$

From Table 14.4 we see that carbonic acid is a fairly strong acid with $pK'_{H_2CO_3} = 3.78$. However, because of the dehydration of the acid and the volatility of its anhydride, for the 'overall' first dissociation in plasma $pK_1' = 6.1$.

The Second Ionization Constant

The bicarbonate ion, HCO_3^-, can lose another proton

$$HCO_3^- \rightleftharpoons H^+ + CO_3^{2-} \tag{8}$$

to form the carbonate ion, CO_3^{2-}. The acid dissociation constant for equilibrium (8), K_2', is given by

$$K_2' = \frac{[H^+][CO_3{}^{2-}]}{[HCO_3{}^-]} \tag{9}$$

Since $pK_2' = 9.8$ this dissociation is unimportant at physiological pH s.

Buffering by the Carbonic Acid System

From equations (3) and (6) we have:

$$[HCO_3{}^-] = \frac{K_1'}{[H^+]}qP_{CO_2} \tag{10}$$

and from (3), (9), and (10) we have:

$$[CO_3{}^{2-}] = \frac{K_1'K_2'}{[H^+]^2}qP_{CO_2} \tag{11}$$

Assuming respiratory control is adequate to keep the alveolar air composition constant with $P_{CO_2} = 40$ mm Hg, we can, from (10) and (11), and the values in Table 14.4, calculate the concentration of $HCO_3{}^-$ and $CO_3{}^{2-}$ ions which would occur in plasma at different pH s. These concentrations are shown in Figs. 14.5 and 14.6. The curve in Fig. 14.6 for $[HCO_3{}^-]$ is the same as in Fig. 14.5 but on a larger scale and for a more restricted pH region. From it one can see that for $P_{CO_2} = 40$ mm Hg, $[HCO_3{}^-] = 25.2$ mEq/litre at pH 7.4, this value rises to 31.8 mEq/litre at pH 7.5 and falls to 20.0 mEq/litre at pH 7.3. The shape of the curve is not the same as that of the usual titration curve (see Chapter 1) because it is plotted on a different basis; here the acid concentration is held fixed (P_{CO_2}) and that of the conjugate base $(HCO_3{}^-)$ varies with pH. At pH 7.4 and $P_{CO_2} = 40$ mm Hg, the carbonate ion concentration, $[CO_3{}^{2-}] = 0.105$ mmol/l or 0.21 mEq/litre, less than 1 per cent of the bicarbonate ion concentration. The curve in Fig. 14.5 shows that $[CO_3{}^{2-}]$ only begins to become significant above pH 8.

Near the pH of blood (7.4) we have therefore only to consider the 'overall' first ionization of carbonic acid and we have the most important relationship:

$$pH = 6.1 + \log\frac{[HCO_3{}^-]}{qP_{CO_2}} \tag{12}$$

usually known as the *Henderson–Hasselbalch equation*.

Carbon Dioxide Dissociation Curves

Since an important function of blood is to transfer CO_2 from the tissues for excretion through the lungs, it is necessary that small differences in P_{CO_2}, between the tissues and the alveolar air, shall be sufficient to cause

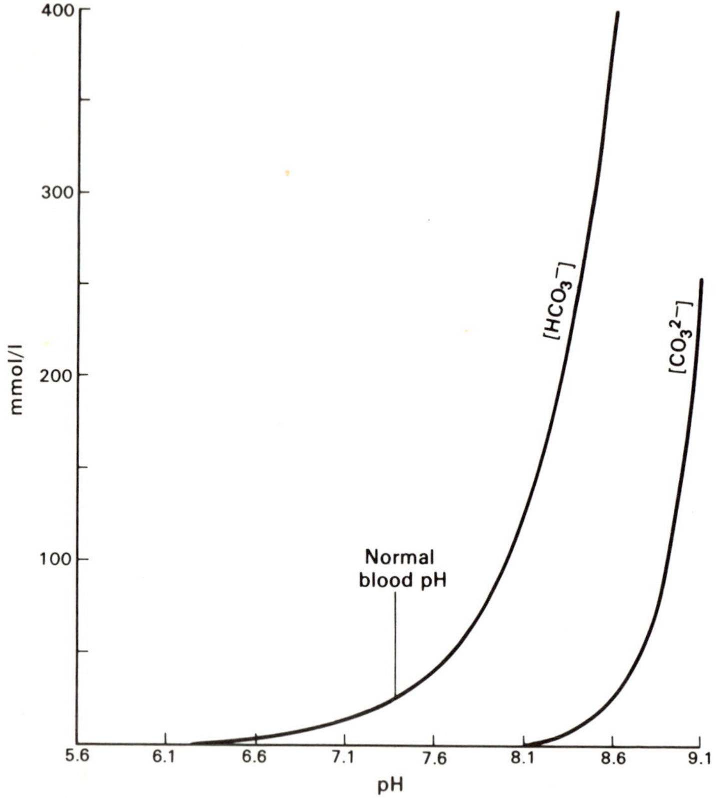

FIG. 14.5. Concentration of bicarbonate and carbonate ions in a solution in equilibrium with gas containing a fixed partial pressure of carbon dioxide (P_{CO_2} 40 mm Hg), as a function of pH. Small scale, over a wide pH range.

a relatively large change in the total carbon dioxide content of the blood during its passage through the lungs.

The total carbon dioxide concentration, extractable from solution in the presence of strong acid, may be represented by T where

$$T = [CO_2] + [H_2CO_3] + [HCO_3^-] + [CO_3^{2-}]$$
$$= qP_{CO_2} + [HCO_3^-] + [CO_3^{2-}] \tag{13}$$

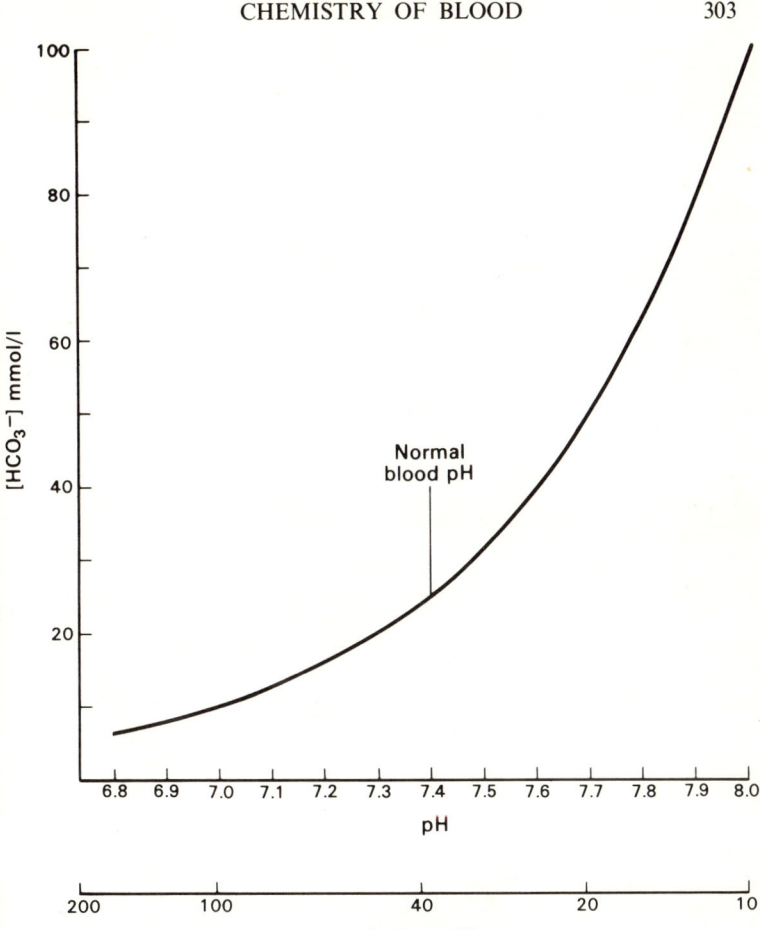

Fig. 14.6. The concentration of bicarbonate ions, in the same conditions as for Fig. 14.5, plotted on a large scale near the pH of blood.

It is possible, using the equations given above, to determine T in terms of P_{CO_2}, the pH, and known constants. It is also possible to define P_{CO_2} in terms of the amount of base present in solution and the pH. We can therefore construct a curve relating T to P_{CO_2}. Such curves are shown in Fig. 14.7.

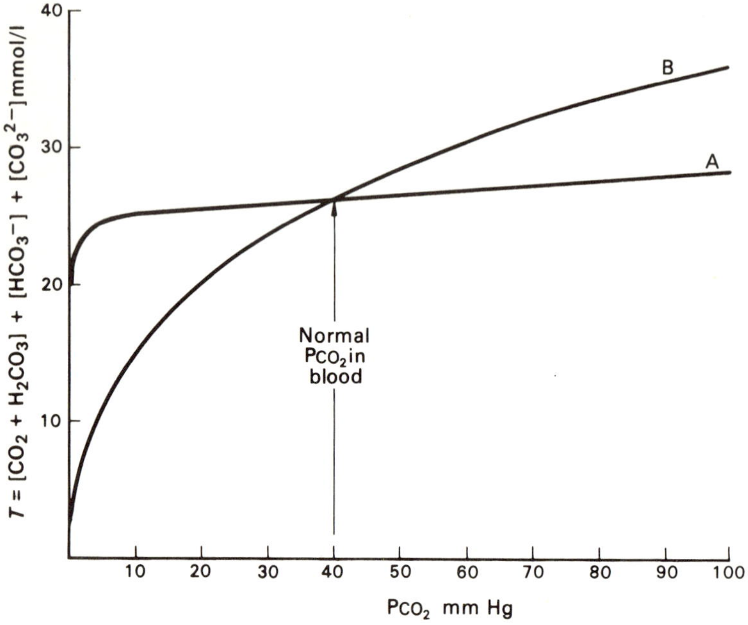

FIG. 14.7. Carbon dioxide dissociation curves. Curve A: for a solution containing 25 mEq $NaHCO_3$ and neutral salts only. Curve B: for a solution containing 25 mEq $NaHCO_3$ + 25 mEq 'imidazolium' chloride + 25 mEq 'imidazole'; the pK of the 'imidazolium' being 7·4.

When there are only $NaHCO_3$ and neutral salts in the solution we get a curve A of Fig. 14.7. This is very poorly adapted in the physiological range of P_{CO_2} to the unloading of CO_2 with small changes in P_{CO_2}, since T remains very nearly constant over a wide range of P_{CO_2}.

To understand this we must consider the equilibrium:

$$HCO_3^- + H_2O \rightleftharpoons H_2CO_3 + OH^- \qquad (14)$$

When P_{CO_2} is reduced H_2CO_3 is dehydrated and the resultant CO_2 is removed from the solution. This causes equilibrium (14) to move to the right and, since $[OH^-]$ increases, the pH rises. The hydroxide ion formed in (14) can react with a bicarbonate ion as follows:

$$OH^- + HCO_3^- \rightleftharpoons CO_3^{2-} + H_2O \qquad (15)$$

This means that as the pH rises so some bicarbonate is converted into carbonate (see Fig. 14.5). Although both equilibria (14) and (15) are reversible, only if P_{CO_2} is lowered to zero will all the bicarbonate and carbonate ultimately be removed from solution leaving an equivalent amount of hydroxide ion in their place. The pH is then high.

In blood, however, the carbon dioxide dissociation curve is very much like curve B of Fig. 14.7. This is because in blood there are buffer substances, principally haemoglobin, with pK s near 7·4. Haemoglobin contributes a quantity of imidazolyl and amino groups, with pK s near 7·4, equivalent to 40 millimoles 'imidazole' per litre of blood. The plasma proteins contribute another 10 millimoles 'imidazole' per litre, making a total of about 50 millimoles 'imidazole' per litre of blood. If we assume that a solution (such as blood), at pH 7·4, contains 25 millimoles $NaHCO_3$, 25 millimoles 'imidazolium' (HIm^+), and 25 millimoles 'imidazole' (Im) per litre, and that p$K_{HIm^+} = 7·4$, we can calculate the shape of the curve relating P_{CO_2} to T (the total carbon dioxide of all forms) in the solution. This curve is curve B of Fig. 14.7.

This system behaves very differently from the simple bicarbonate solution. When P_{CO_2} is reduced carbonate ions are again formed as in reactions (14) and (15), but they are reconverted to bicarbonate by reacting with 'imidazolium' ion thus:

$$CO_3^{2-} + HIm^+ \rightleftharpoons HCO_3^- + Im \tag{16}$$

The bicarbonate again reacts as in (14) and (15), the carbon dioxide going off to the gas phase and the carbonate again reacting as in (16). If there is sufficient 'imidazolium' to react with all the carbonate produced, reactions (14), (15), and (16) can proceed until all the bicarbonate (and carbonate) has been removed from solution. It is apparent from curve B of Fig. 14.7 that such a mixed buffer system (as in blood) is well adapted to take up or give up CO_2 near physiological values of P_{CO_2}.

The Cell–Plasma System: the Chloride–Bicarbonate Shift

The principal ions of the cells and the plasma are shown in Fig. 14.8.

Plasma		Cells
$[HCO_3^-]_p$		$[HCO_3^-]_c$
$[Cl^-]_p$		$[Cl^-]_c$
$[OH^-]_p$		$[OH^-]_c$
$[H^+]_p$		$[H^+]_c$
$[Protein^-]_p$		$[Hb^- + HbO_2^-]_c$
$[Na^+]_p$		$[K^+]_c$

cell membrane

Fig. 14.8.

Because of the semi-permeable nature of the cell membrane only the first three ions (HCO_3^-, Cl^-, and OH^-) are able to pass freely across it. It is effectively impermeable to the others, the relative concentrations of which cannot therefore alter greatly.

Since the concentration of anions due to the haemoglobin and oxyhaemoglobin inside the cells is greater than that due to the plasma proteins outside, it follows that:

$$[Cl^-]_p > [Cl^-]_c, \qquad [HCO_3^-]_p > [HCO_3^-]_c$$

and

$$[OH^-]_p > [OH^-]_c$$

Also from the Gibbs–Donnan equilibrium we have:

$$\frac{[HCO_3^-]_c}{[HCO_3^-]_p} = \frac{[Cl^-]_c}{[Cl^-]_p} = \frac{[OH^-]_c}{[OH^-]_p} = \frac{[H^+]_p}{[H^+]_c} \tag{17}$$

the last equality being a consequence of the fact that:

$$[H^+]_p[OH^-]_p = K_W = [H^+]_c[OH^-]_c$$

where K_W is the ion product of water.

Given these relationships we can now consider what happens to the concentrations of the various ions in the blood when in the tissues and in the lungs. These changes are shown schematically in Fig. 14.9.

In the capillaries CO_2 diffuses from the tissues into the plasma and into the cells where its hydration to H_2CO_3 is catalysed by carbonic anhydrase. This has the effect of reducing $[CO_2]_c$ and encouraging the diffusion of more CO_2 from the plasma into the cells. The H_2CO_3 formed partially dissociates into H^+ and HCO_3^-. The H^+ thus formed tends to lower the pH of the cells but this effect is minimized by the simultaneous deoxygenation of oxyhaemoglobin to haemoglobin. As has been said above, haemoglobin is a weaker acid than oxyhaemoglobin; hence on deoxygenation H^+ ions are taken up by the haemoglobin. Electroneutrality is maintained because, although the number of mEq anions due to the cell protein (haemoglobin + oxyhaemoglobin) is reduced, some HCO_3^- has been formed and balances the positive charges of the intracellular K^+ and H^+ ions.

Now we have the situation that $[HCO_3^-]_c$ has been increased and the equalities in (17) no longer hold. In order to restore the Gibbs–Donnan equilibrium, HCO_3^- diffuses *out* of the cells into the plasma, and in order to maintain electroneutrality Cl^- and OH^- diffuse *into* the cells at the same time. This latter movement of ions also helps to restore the Gibbs–Donnan equilibrium.

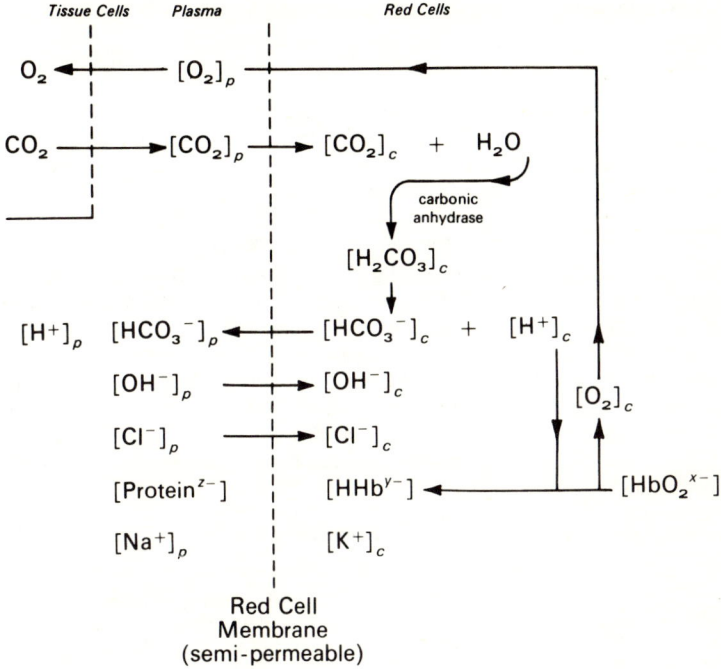

FIG 14 9 The arrows indicate the changes occurring in capillary blood during the liberation of oxygen to the tissues and acquisition of carbon dioxide from the tissue. In the lungs the changes are reversed. x, y, and z are numbers, and $x > y$.

In the lungs, when CO_2 is being given up to the alveolar air and the haemoglobin in the cells is being reoxygenated, the reverse sequence of events occurs. Because of these shifts, the osmotic pressure inside venous cells is higher than in arterial ones and the cells in venous blood increase in size by imbibing water.

In spite of the powerful buffering mechanism described above, the pH of venous blood is slightly lower (more acid) than that of arterial blood. The change, however, is much less than would occur if the blood were composed solely of plasma.

Carbamido Compounds

CO_2 combines reversibly at physiological pH with the free N-terminal amino groups of proteins, particularly haemoglobin and to a lesser extent oxyhaemoglobin, to form carbamido compounds, thus:

$$R-NH_2 + CO_2 \rightleftharpoons R-NH-COO^- + H^+$$

This is a very rapid reaction and possibly an important one in the transport of CO_2 in the blood. It does not, however, contribute much to the buffering of the carbonic acid since about one H^+ ion is produced, at blood pH, per molecule of CO_2 thus combined.

Acid–Base Status of the Blood

Blood is a buffer solution in equilibrium with a gas phase which contains the volatile acid anhydride CO_2. In normal conditions the hydrogen ion concentration of blood is near 40 nanomoles per litre (pH $\approx$ 7·4). The acidity ($[H^+]$ or pH) can be altered by two essentially independent means: either by changing the P_{CO_2} of the gas phase or by addition (or removal) of non-volatile acid or base. The hydrogen ion concentration of blood depends, therefore, on a combination of a respiratory component (the $CO_2-HCO_3^-$ system) and a non-respiratory (non-volatile buffers) component.

Control by the Carbon Dioxide–Bicarbonate System

When acid metabolites accumulate in the blood there is a fall in pH, and as a consequence the ratio $[HCO_3^-]/[CO_2 + H_2CO_3] = [HCO_3^-]/qP_{CO_2}$, also falls (Equation (12)). More precisely, the acid metabolites, by definition, are a source of protons which are taken up by the strong Bronsted base, bicarbonate ion.

$$HA \rightleftharpoons H^+ + A^-$$
$$H^+ + A^- + HCO_3^- \rightleftharpoons A^- + H_2CO_3 \tag{18}$$

These equilibria are very far to the right. H_2CO_3 dissociates almost completely into H_2O and CO_2 (p. 299), and an accumulation of acid metabolites therefore has the effect of increasing P_{CO_2} (Equation (3)) which increases the amount of CO_2 blown off in the lungs. The ratio

$$[HCO_3^-]/qP_{CO_2}$$

returns to normal, as does the pH. The subject's extracellular fluids (including the blood) are not normal, however, since although the *ratio* $[HCO_3^-]/qP_{CO_2}$ is normal, i.e. equation (12) is satisfied for a pH of 7·4, the *absolute concentration* of HCO_3^- in the extracellular fluids is below normal. The importance of this can be seen if the effect of a further accumulation of acid is considered; the reaction above would again occur, moving from left to right, and a normal pH would become re-established only at the cost of a further fall in $[HCO_3^-]$. Obviously, this process could not continue indefinitely; at some point $[HCO_3^-]$ would fall to zero, and control of pH by this buffer system would no longer be possible. This reasoning, backed by experimental results, led to the idea of the bicarbonate ion in the extracellular fluids as a reserve of base against intrusions of H^+ into the blood; it was formerly called the 'Alkali Reserve'.

Accumulation of alkaline metabolites can hardly occur in the body, with the exception of NH_4^+, and this is so toxic for other reasons (p. 224) that its effects on acid–base balance are secondary. However, there are other ways of raising the blood pH, and in these circumstances the opposite changes take place. When a base, such as an hydroxide ion, enters the blood the following reactions occur:

$$OH^- + H_2CO_3 \rightleftharpoons H_2O + HCO_3^- \qquad (19)$$

and

$$CO_2 + H_2O \rightleftharpoons H_2CO_3$$

The P_{CO_2} falls at first and less CO_2 is blown off. Since there is now a higher than normal $[HCO_3^-]$, the system can only become restabilized at $[H^+] = 40$ nmol/l (pH = 7·4) by an accumulation of CO_2 and H_2CO_3, i.e. the P_{CO_2} becomes restabilized at a value higher than normal.

Non-volatile Buffers

There are other buffers in blood and extracellular fluid which increase the resistance of the body to changes in $[H^+]$ beyond that to be expected from the CO_2—HCO_3^- system alone. Of these, haemoglobin is the most important. Each molecule contains 36 histidyl residues, many of which have pK s close to 7·4 (p. 12); only 4 of them are involved in the change of acidity of Hb on oxygenation, and the buffering power of the others is quite independent of the state of oxygenation of the corpuscles. The plasma proteins also contain histidyl residues, although they are not quantitatively so important as those of haemoglobin.

Relation between pH and log P_{CO_2}. Equation (12), which describes the CO_2—HCO_3^- buffer system, can be written in the form:

$$pH = 6·1 - \log q + \log [HCO_3^-] - \log P_{CO_2} \qquad (20)$$

If the only buffer in a solution with a pK near the pH of the blood is the CO_2—HCO_3^- system, addition or removal of small amounts of CO_2 will not significantly affect $[HCO_3^-]$. This is because HCO_3^- is the only anion present (ignoring the very small OH^- concentration) and, because of the electroneutrality condition, its concentration cannot change without changing the cation concentration (changes in $[H^+]$ being insignificantly small). Therefore, since q is a constant, it follows from equation (20) that a graph of log P_{CO_2} vs. pH will be a straight line of slope -1. The actual position of the line will depend on the magnitude of log $[HCO_3^-]$. This is shown in Fig. 14.10.

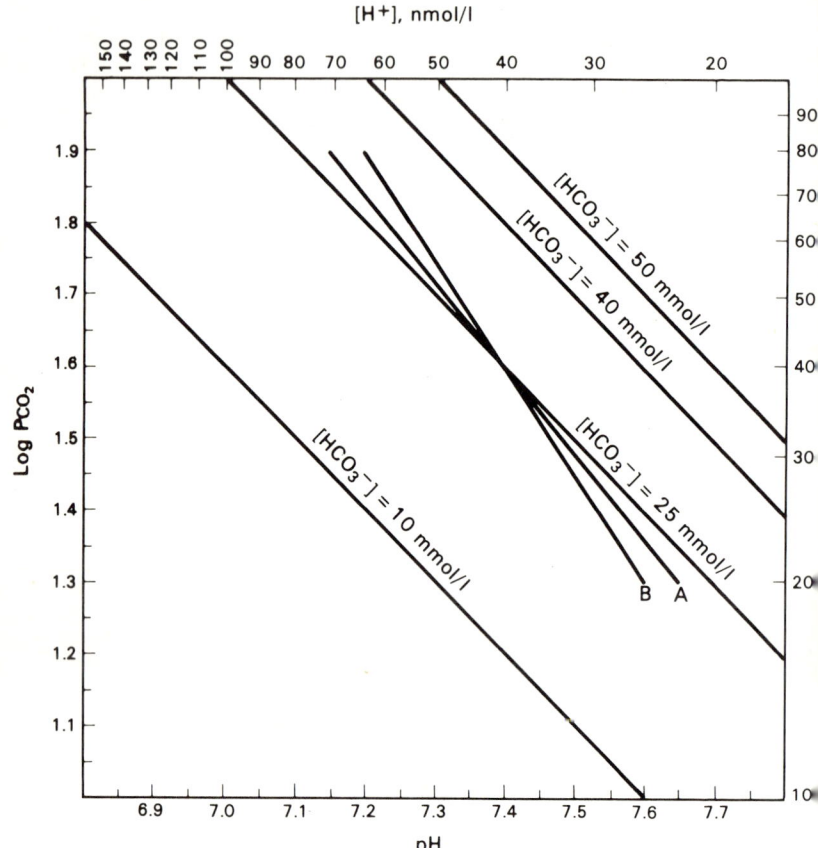

Fig. 14.10. Relationship between pH and log P_{CO_2}. The parallel lines are for solutions containing bicarbonate and CO_2 only; they have a slope of -1. Lines A and B are for solutions containing CO_2, bicarbonate (25 mmol/l), and a non-volatile buffer of p$K = 7.4$ at contractions of 40 mmol/l and 60 mmol/l respectively; the slopes of these lines are more negative than -1.

Blood, however, contains non-volatile buffer systems as well as the $CO_2-HCO_3^-$ buffer (see p. 298). If we represent the protein buffers as HPr for the acid form and Pr^- for the base, then, since H_2CO_3 is a stronger acid than HPr, on addition of CO_2 the following reactions will occur:

$$CO_2 + H_2O \rightleftharpoons H_2CO_3$$

and

$$H_2CO_3 + Pr^- \rightleftharpoons HPr + HCO_3^-$$

Thus, in the presence of a non-volatile buffer system, the addition of CO_2 has not only increased $[H_2CO_3]$ but also $[HCO_3^-]$. The graph of $\log P_{CO_2}$ vs. pH is still a straight line, over a limited range of P_{CO_2}, but its slope is more negative than -1, the slope of the graph for the $CO_2-HCO_3^-$ system alone. This effect is also illustrated in Fig. 14.10.

It follows from equation (20) that the three parameters pH, $\log P_{CO_2}$, and $\log[HCO_3^-]$ are related and that the determination of any two of them will allow us to calculate the third.

Characterization of the Acid–Base Status of Blood

The following are the main variables (parameters) used in defining the acid–base status of the blood:

The *pH* can be determined by glass electrode measurements, either at the P_{CO_2} existing in the artery, the vein, or the capillaries from which the blood is drawn, or the specimen may be equilibrated with gas of known P_{CO_2} before its pH is measured.

The *standard bicarbonate* is the plasma bicarbonate-ion concentration, $[HCO_3^-]$, of the whole blood that has been equilibrated to a P_{CO_2} of 40 mm Hg at 37°. It is expressed in mmol/l. It may refer to blood as drawn or to the blood oxygenated in vitro; clearly it is independent of P_{CO_2}, but it is dependent to some extent on the haemoglobin concentration.

The *base excess* is the base concentration in whole blood as measured by titration with strong acid to pH 7·40 at a P_{CO_2} of 40 mm Hg at 37°. If the pH of the blood is below 7·4 the titration can be made with strong base, the base excess is then negative and is sometimes known as *base deficit*. Base excess (or base deficit) is expressed in mEq/l.

The *buffer base* is the sum of concentrations of the buffer bases of whole blood including bicarbonate, plasma proteins, haemoglobin, etc. It is expressed in mEq/l.

Normal values of some of these parameters are shown in Table 14.5.

Table 14.5
Normal Values for Some Parameters of the Acid–Base
Status of Blood

Hydrogen-ion concentration:	
pH	$7 \cdot 35 - 7 \cdot 45$
$[H^+]$, nmol/l	35–45
P_{CO_2}, mm Hg	34–46
Standard bicarbonate, mmol/l	22–26
Base excess, mEq/l	0 ± 2

Determination of the Acid–Base Status of Blood

Since the graph relating $\log P_{CO_2}$ to pH is a straight line, the position of which depends on the bicarbonate concentration and the slope of which depends on the concentration of non-volatile buffers, it is possible by measuring the pH of blood as it is taken and also the pH s of the same blood after equilibration at two known but different P_{CO_2} values to obtain a complete picture of the acid–base status. The two latter measurements allow the straight line relating $\log P_{CO_2}$ to pH for that particular specimen of blood to be established and from the pH of the blood as drawn it is then possible to obtain its P_{CO_2}. This is illustrated in Fig. 14.11.

In recent years micromethods, which require only 100–200 μl of blood, have been developed for the equilibration of blood with gases of different P_{CO_2} values and for pH measurements. These methods, which make repeated determinations on capillary blood a relatively simple procedure, have greatly improved the possibilities of control during treatment of patients with acid–base disturbances. A nomogram based on the $\log P_{CO_2}$ vs. pH graph of Fig. 14.11, from which the P_{CO_2}, the standard bicarbonate, the base excess, and the buffer base of a specimen of blood may be determined, is shown in Fig. 14.12. Although it is not possible in a book of this size to describe in detail how this nomogram is used, its value in clinical practice is such that it was thought well to illustrate it. The two parabolas and the standard bicarbonate abscissa are of course fixed, since they represent the necessary pH and P_{CO_2} values of blood which has a given value of buffer base (or of base excess or of standard bicarbonate). For example, the standard bicarbonate line must be parallel to the X-axis and go through $P_{CO_2} = 40$ mm Hg, because *by definition* this is the partial pressure of CO_2 at which the standard bicarbonate is measured. The line ACB, on the other hand, is variable in slope and position, since it refers only to the observations made on a particular sample of blood, as explained in the legend to Fig. 14.11.

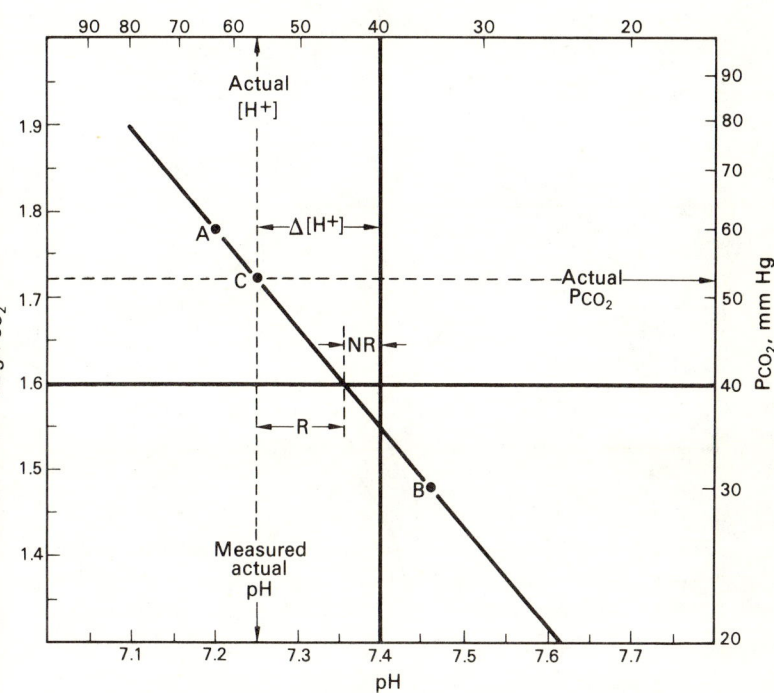

FIG. 14.11. Diagram showing how the actual P_{CO_2} of a blood specimen may be determined from pH measurements. The pH of the blood as obtained was found to be 7·25. After equilibration at 60 and 30 mm Hg CO_2, the pHs were those shown by points A and B respectively. The actual P_{CO_2} of the blood is the P_{CO_2} coordinate of the point C which lies on the line joining A and B. The excess hydrogen ion concentration over normal is given by $\Delta[H^+]$, of which the amount R is contributed by CO_2 (respiratory acid) and the amount NR by non-volatile (non-resporatory) acids. R and NR are found from the intersection of AB with the horizontal line running through $P_{CO_2} = 40$ mm Hg.

Acidosis and *Alkalosis* are terms which are used in the physiological sense to describe abnormal processes or conditions which would cause a deviation of pH if there were no secondary changes in response to the primary aetiological factor. These terms may be modified by such general adjectives as 'respiratory' or 'metabolic' or by more specific adjectives such as 'renal' or 'diabetic'. This usage does not necessarily imply that a change of pH has occurred.

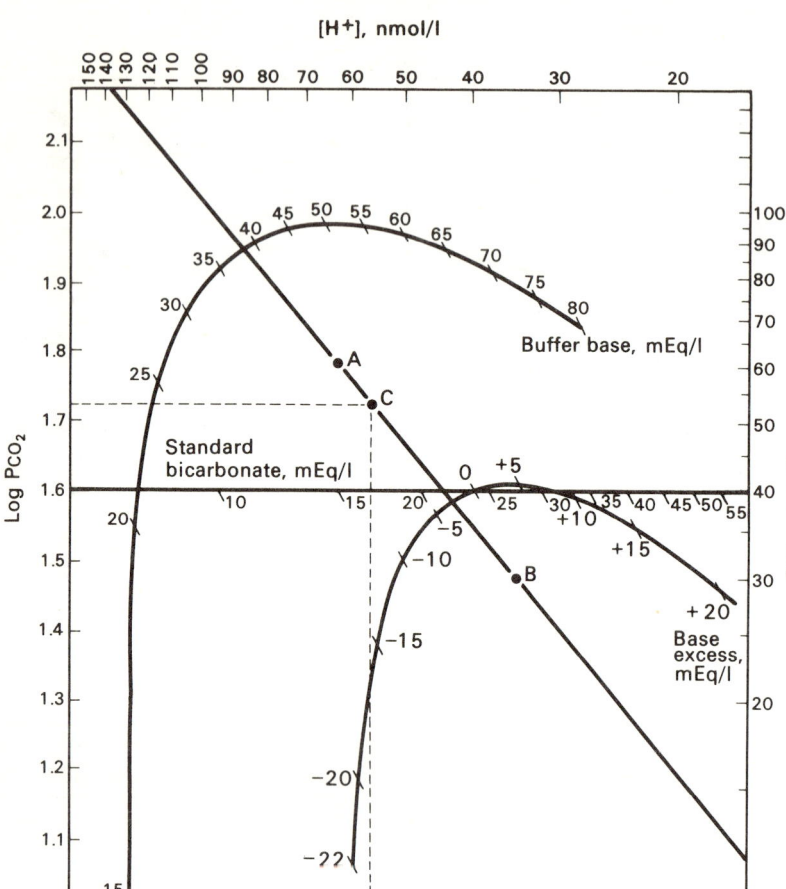

FIG. 14.12. pH, log P_{CO_2} nomogram for blood. Points A and B represent the measured pH of a blood sample at P_{CO_2} of 60 and 30 mm Hg respectively. The point of intersection of the line joining A and B with the base excess curve shows the base excess to be -3 mEq/l, and the intersection with the buffer base curve shows the buffer base to be 38·5 mEq/l. The standard bicarbonate (at $P_{CO_2} = 40$ mm Hg) can be seen to be 21·5 mEq/l. Point C represents the measured pH in the arterial blood and shows the log P_{CO_2} in the arterial blood to be 1·726 ($P_{CO_2} = 53·2$ mm Hg). The data are the same as in Fig. 14.11.

In severe disturbance of acid–base status there is some change in the pH as well as in the values of $[HCO_3^-]$ and $[CO_2 + H_2CO_3]$. It is not necessary for HCO_3^- to disappear completely from the extracellular fluids for acidosis to be serious; a pH of 6·9 (when the $[HCO_3^-]$, given a normal P_{CO_2}, is still 8 mEq/litre (see Fig. 14.5)) is barely compatible with life. Similarly, at pH 7·8 the carbon dioxide–bicarbonate system is still buffering, but insoluble $CaCO_3$ and $CaHPO_4$ begin to form until the plasma ionized Ca^{2+} level falls and *tetany* (muscular spasms) develops. This is fatal if prolonged.

Disturbances in respiratory function which prevent excretion of CO_2 whether by failure of the respiratory centre, high atmospheric P_{CO_2}, or pulmonary congestion, cause an increase in P_{CO_2} and blood $[CO_2 + H_2CO_3]$ and a consequent fall in pH. This is *respiratory acidosis*. Partial compensation is achieved by an increase in the blood $[HCO_3^-]$ and fall in $[Cl^-]$, but the pH tends to remain below normal. Respiratory acidosis is therefore characterized by high blood $[CO_2 + H_2CO_3]$ *and* $[HCO_3^-]$.

Hyperpnoea gives rise to *respiratory alkalosis* from lowering of P_{CO_2}, removal of CO_2 and H_2CO_3 from the blood, and a consequent rise in the $[HCO_3^-]/[CO_2 + H_2CO_3]$ ratio and in the pH.

Acid–Base System and Renal Control

Metabolism continually produces acids other than CO_2. By far the most important of these is H_2SO_4 from the catabolism of the sulphur-containing amino acids (see Chapter 11). It is a normal function of the kidney to excrete H^+ (together with the SO_4^{2-} and other unwanted anions), and urine is therefore normally more acid than blood (pH 5·5–6·0). It is generally thought that the reaction

$$H_2CO_3 \rightleftharpoons H^+ + HCO_3^-$$

takes place in the tubular cells, and that the H^+ is passed through the tubular membrane by an unknown mechanism, while the HCO_3^- is transferred to the plasma. There is a high concentration of carbonic anhydrase in the kidney, localized in the tubular cells, and inhibitors of this enzyme prevent acidification of the urine.

Before the role of the kidney in the regulation of acid–base balance can be appreciated, three general principles must be firmly understood.

1. *Electroneutrality must be preserved.* In any solution the number of positive charges must equal the number of negative charges. With reference to the kidney, this means that ions of one sign passing through a membrane must be balanced *either* by an exactly equivalent amount of ions of the opposite sign moving in the same direction, *or* by an exactly equivalent

amount of ions of the same sign moving in the opposite direction (see Fig. 14.13). Blood and urinary electrolytes are expressed in milliequivalents to enable this necessary calculation to be performed.

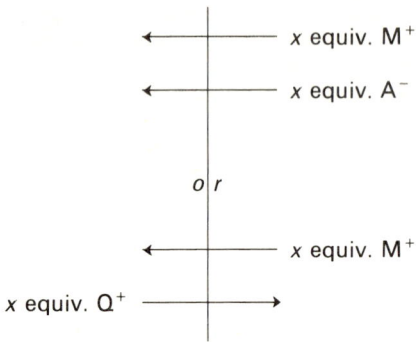

FIG. 14.13. Electroneutrality.

2. *The osmotic pressure of the body fluids is preserved.* All the evidence shows that the kidney maintains the crystalloid osmotic pressure of the extracellular fluids by regulating electrolyte excretion (controlled by ADH) even if this conflicts with acid–base balance. The normal extracellular fluid *volume* of a normal subject is determined by the total amount of body electrolytes outside the cells; in practice, because of its relative concentration, and because of principle (1), this means Na^+. The excretion of large quantities of electrolytes, unrelated to intake, will be accompanied by the loss of an equivalent volume of water, quite apart from the inability of the kidney to excrete a very concentrated urine. This will lead to dehydration (reduced total extracellular fluid volume). On the other hand, any regulation of acid–base status which requires the reabsorption of larger than normal quantities of electrolytes from the glomerular filtrate will lead to oedema (increased total extracellular fluid volume). Both these states can lead to heart failure.

Further, if the extracellular fluids become hyperosmotic for any reason, the kidney will fail to rectify any acid–base disturbances by acidifying or alkalinifying the urine for the simple reason that no urine will be secreted.

3. *In buffered systems, pH is determined by the ratio* [*base*]/[*acid*], not by the absolute amounts of either.

Acidosis

From the description of respiratory control above it is clear that a return to normality from acidosis can only occur if the plasma $[HCO_3^-]$ returns to normal. Since the concentrations of $Na^+ + K^+$ in the extracellular fluids are fixed, according to principle (2), $[HCO_3{}^-]$ can only rise to normal levels if the acid anion A^- is removed—usually by renal excretion. By principle (1) some cation must accompany it. The kidney assists by excreting larger than normal quantities of H^+, which is exchanged for Na^+ and K^+ (although it must be remembered that the reabsorption of most of the Na^+ from the glomerular filtrate is balanced by the—probably passive— reabsorption of Cl^-). The $HCO_3{}^-$ formed at the same time as the H^+ in the tubule cells is transferred to the plasma, and the *apparent* effect is the reabsorption of $NaHCO_3$ and the excretion of HA.

In respiratory acidosis there is no abnormal anion A^- to be excreted. The kidney secretes extra H^+ and reabsorbs less Cl^-; the plasma $[HCO_3{}^-]$ rises above normal levels, sometimes to 30 mEq/l, and continues high until respiratory function returns to normal. This is difficult to understand if the existence of a renal threshold for $HCO_3{}^-$ is believed in. Fig. 14.5 shows that even at the normal pH of urine the concentration in it of $HCO_3{}^-$ must be negligible.

The secretion of H^+ by the kidney is limited by its inability to excrete urine more acid than pH 4·6. Some organic acids will be partly undissociated at this pH (e.g., β-hydroxybutyric acid, pK 4·39), so that the anions of these acids in the glomerular filtrate can pick up H^+. Lactic acid (pK 3·72) and acetoacetic acid (pK 3·58) are mainly dissociated at this pH, and of course H_2SO_4 and HCl are completely so.

Phosphate also plays a part in keeping $[H^+]$ low in acid urine. At pH 7·4 the ratio of $[HPO_4{}^{2-}]/[H_2PO_4{}^-]$ is 4:1, whereas at pH 4·6 only $H_2PO_4{}^-$ exists. Each millimole of $HPO_4{}^{2-}$ converted into $H_2PO_4{}^-$ removes 1 mmol H^+ from the tubular fluid, or to put it in another way, spares the excretion of 1 mmol Na^+. Unfortunately the total phosphate excretion per day is only about 50 mmol, and it does not change very much unless there is severe chronic acidosis, when cellular and skeletal phosphate is mobilized.

In severe acidosis the amount of A^- to be excreted can rise to as much as 500 mmol/day, and a final defence mechanism to minimize the loss of Na^+ and H^+ (which would, by principle (2), lead to dehydration) is the exchange of $NH_4{}^+$. Ammonium ions can be formed from glutamine in the tubule cells. This hydrolysis is catalysed by the enzyme *glutaminase*. At the cellular pH, of near 7·4, the reaction is

The ammonium ion is in equilibrium with NH_3 and hydrogen ion according to the reaction

$$NH_4^+ \rightleftharpoons NH_3 + H^+ \tag{21}$$

pK_a for NH_4^+ is 8·89 at 37°, so that at the pH of the cells the equilibrium for reaction (21) will be well to the left.

There is some evidence that NH_3 can pass freely through the cell walls into the tubule, while NH_4^+ does so only with difficulty. As usually formulated, therefore, the transfer proceeds as follows:

$$\begin{array}{ccccc} & & \textit{Cell} & & \\ & \textit{Cell} & \textit{membrane} & \textit{Tubule} & \\ NH_4^+ & \longrightarrow NH_3 & \longrightarrow & NH_3 \longrightarrow & NH_4^+ \\ & + H^+ & & & + H^+ \end{array}$$

This does not contribute at all to the relief of acidosis, since for every NH_3 molecule which is formed so that it can enter the tubule, one hydrogen ion is produced in the cell, and must be buffered there. The continual displacement of reaction (21) from equilibrium continues because the urine is very much more acid than the cell sap, and the concentration of free NH_3 correspondingly lower. The cell membrane thus behaves as if it were a non-return valve. In this formulation the main function of the NH_3 when it has been secreted into the tubule, is to lower the hydrogen ion concentration in the urine. The proton which is used up when reaction (21) goes from right to left in the lumen presumably comes mainly from undissociated acids there (e.g., β-hydroxybutyric, $H_2PO_4^-$). The change in the ratio base/acid would raise the pH.

There is much experimental evidence, however, that the secretion of ammonia is mainly a device to reduce the loss of cations, particularly Na^+. Fig. 14.14 makes this point very clearly. The daily excretion of NH_4^+ ions rose during the fasting period from about 20 mEq/day to 90, while the excretion of metal cations actually *fell*, from about 95 mEq/day to 30. No doubt the actual fall was due to the loss of the normal cation intake in food during the fast (the inorganic anion output also fell markedly), but this only emphasizes the importance of the NH_4^+—inorganic cation defence mechanism. For example, rats fed on diets very low in inorganic salts may excrete a urine in which as much as 95 per cent of the cations are NH_4^+ ions.

Nothing is known of how this exchange mechanism works, nor is it known how the ammonia secretion is activated. Very large quantities of glutaminase can be extracted even from non-acidotic kidneys, and the tissue always contains 6–10 mM glutamine as do most other tissues in the body. The maximum excretion of NH_4^+ is about 200 mEq/day, which may still not be enough to prevent the loss of some Na^+ and K^+.

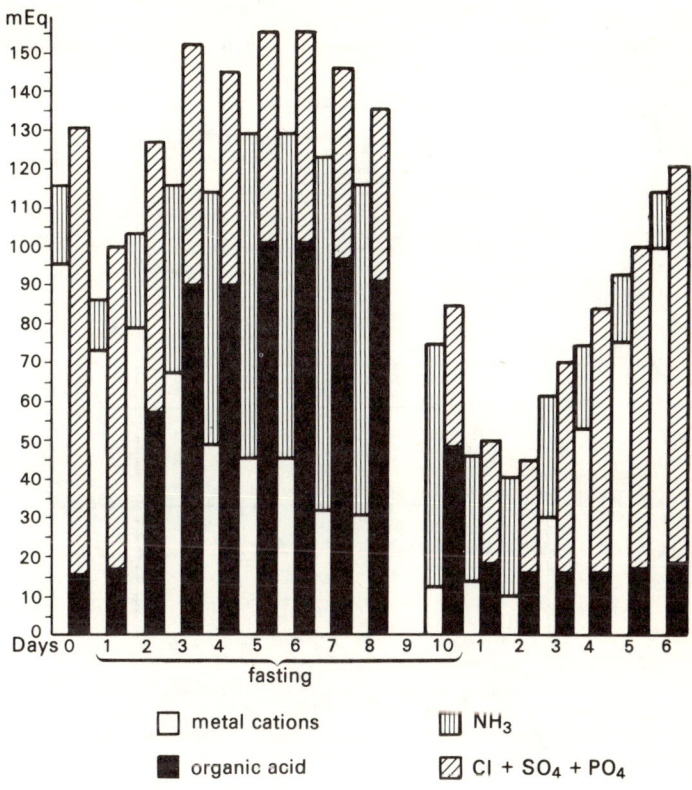

Fig. 14.14. Excretion of electrolytes in urine during a 10-day fast.

The role of the kidney in acidifying urine is summarized in Fig. 14.15.
In acute acidosis the excretion of NH_4^+ ions takes about four days to reach a maximum, and until it does so cations (Na^+, Ca^{2+}, etc.) from body fluids have to be excreted to 'cover' the extra acid anions being excreted. Extracellular body water contains $145 \, mEq/l \, Na^+$, $115 \, mEq/l \, Cl^-$, and $30 \, mEq/l \, HCO_3^-$. Thus its excretion provides $30 \, mEq/l$ cations on the assumption that the HCO_3^- can be excreted through the lungs. Similarly, some intracellular water can be excreted, its phosphate being used to buffer, its protein being metabolized, and its K^+ being used as base. Thus acute

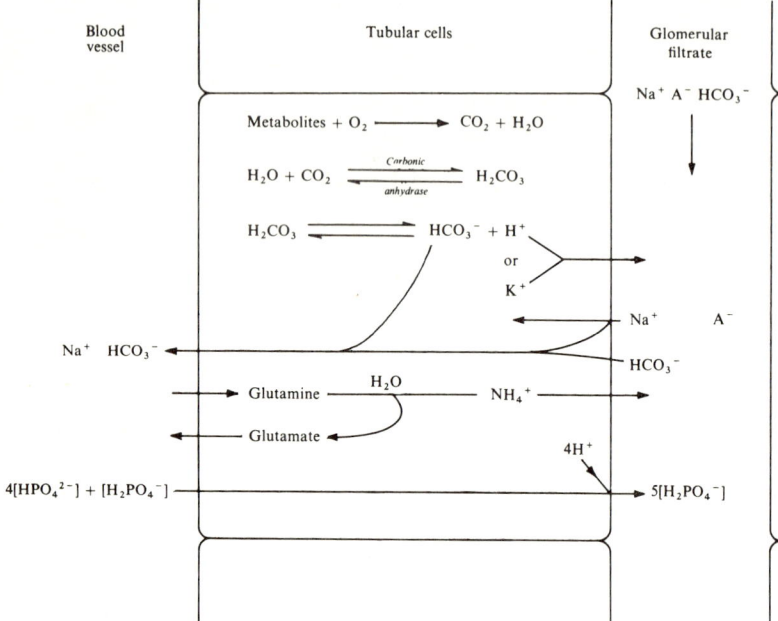

FIG. 14.15. Acidification of urine and ammonium ion excretion by the kidney. The exchange of H^+, NH_4^+ (and K^+) for Na^+, the conversion of HPO_4^{2-} to $H_2PO_4^-$, and the reabsorption of HCO_3^- are shown.

acidosis is accompanied by dehydration while in chronic acidosis Na^+ and Ca^{2+} may be mobilized from the skeleton and excreted in the urine.

If the kidneys are damaged their ability to excrete an acid urine and to form NH_4^+ ions may be impaired. Then the extracellular fluid and the skeleton are called on more extensively.

Alkalosis

The renal defences against alkalosis are not so efficient as they are against acidosis. The limit is set by the kidney's inability to excrete a urine of pH more alkaline than 7·9. In alkalosis, the secretion of H^+ falls and HCO_3^- appears in the urine. Although the CO_2—HCO_3^- buffer system is still quite efficient at pH 7·4–7·9, it can only work if the concentration of HCO_3^-, which means, in effect, of Na^+ as well as HCO_3^-, is raised quite considerably (Fig. 14.5). The fall in H^+ secretion also means that less Na^+

can be reabsorbed. Moreover, tubular NH_4^+ production falls to zero with alkaline urine, and thus alkalosis is always accompanied by the loss of Na^+ and K^+ and rapid dehydration.

If the alkalosis has been caused by loss of fluid from the alimentary tract, the dehydration from both sources can be rapidly fatal.

Causes of Acidosis

(a) *Metabolic*. Accumulation of organic acids such as lactic (temporary), acetoacetic, and β-hydroxybutyric. All produced by tissues. Lowered plasma $[HCO_3^-]$ and P_{CO_2}, accompanied by over-breathing. Acid urine, increased excretion of cations, and NH_4^+ in chronic acidosis, together with the anion of the acid responsible.

(b) *Respiratory*. Accumulation of $CO_2 + H_2CO_3$. Raised plasma P_{CO_2} and $[HCO_3^-]$. Acid urine, increased excretion of Cl^-.

(c) *Alimentary*. Ingestion of dilute acids, or of NH_4^+ salts of mineral acids. (In the latter case, the formation of urea

$$2NH_4^+ + CO_2 \rightarrow CO \cdot (NH_2)_2 + H_2O + 2H^+$$

makes this equivalent to the ingestion of strong acid.) Blood electrolyte picture as in (a).

(d) *Diarrhoea* (*cholera, dysentery, enteritis, etc.*). Loss of a fluid containing a greater ratio of HCO_3^- to Cl^- than is found in plasma. Lowered plasma $[HCO_3^-]$ and P_{CO_2}. Over-breathing caused by lowered blood pH. Renal control may be rendered difficult by dehydration, with consequent tendency to anuria.

(e) *Addison's disease* (*secondary importance*). Loss of Na^+ in urine. Exchange of H^+ for Na^+ in kidney tubules made more difficult. Urine pH may be normal or alkaline. Plasma picture complicated by reduction in total electrolyte concentration.

Causes of Alkalosis

(a) *Respiratory*. Hyperpnoea. Initially, normal $[HCO_3^-]$ with lowered P_{CO_2}. Renal control by reducing H^+ secretion and excretion of HCO_3^- in an alkaline urine.

(b) *Alimentary*. Ingestion either of $NaHCO_3$ directly, or the Na (or K) salt of a metabolizable organic acid. The anion of such an acid is completely oxidized to $HCO_3^- + H_2O$, and the process is equivalent to the absorption of $NaHCO_3$. Initially, raised plasma HCO_3^- and retention of CO_2. Renal control by excreting an alkaline urine containing HCO_3^-. Common in a mild form when much fruit and vegetables are eaten.

(c) *Vomiting*. Loss of gastric juice or of a fluid containing a greater ratio of Cl^- to HCO_3^- than is found in plasma. Raised plasma $[HCO_3^-]$ and lowered $[Cl^-]$, with under-breathing. Control by the excretion of an alkaline urine, but this may be difficult because of dehydration.

(d) *Hypokalaemia*. In the distal tubules, K^+ and H^+ compete with each other for exchange with Na^+. If plasma K^+ is low more H^+ is exchanged and urine becomes acid, while plasma becomes alkaline and $[HCO_3^-]$ rises. Cells may also become acid as intracellular K^+ is exchanged for Na^+ and H^+. Only relieved by the administration of K salts.

15 Nucleic Acid Synthesis: Transcription

The Organization of the Chromosome

The genetic material of all cells whether eukaryote or prokaryote, and of viruses is contained in one or more chromosomes. These have a complex composition—human chromosomes contain 15 per cent DNA, 10 per cent RNA and 75 per cent protein—but the genetically important component is almost always double-stranded DNA. Exceptions are the chromosomes of some very simple viruses (bacteriophages) which contain single-stranded DNA, and some plant and animal viruses (see Fig. 15.2) which contain RNA, which is always single-stranded. The *genes*, which correspond to functional units of RNA, as described below, are linearly arranged along one strand of the DNA. The chromosomes of viruses, bacteria, mitochondria and chloroplasts are circular, as is the nucleolar DNA of oocytes, that is, each strand is joined in an endless chain, but the linear arrangement of genes still holds good. In simple cells each gene (segment of DNA) is found just once on its chromosome and is separated from the next by only a short length of non-informative nucleotides, or perhaps by an operon or promoter site (see Chapter 17). In the chromosomes of complex organisms, however, there seem to be long stretches of non-informative DNA, and it has also been shown that some segments of DNA exist in hundreds, or even thousands, of copies along one chromosome. This is notably true of the nucleolar (extrachromosomal) DNA of oocytes, which codes hundreds of replicate copies of ribosomal RNA. In general, the fine structure of the chromosome has hitherto been established more by genetic techniques than by biochemical or biophysical methods. Newer methods, e.g., (staining with) mepacrine hydrochloride localization by quinacrine, or in situ by hybridization, are beginning to show the location of chromosome regions where multiple copies of genes, for example those coding for ribosomal RNA or immunoglobulin (cf. p. 293) synthesis are located.

The fundamental process which must occur before cell division can take place is therefore the replication of chromosomal DNA; here we shall consider only the replication of double-stranded DNA. By a variety of ingenious experiments it has been shown that the replication is *semi-conservative*, that is, that each strand of the double helix serves as a template for the synthesis of a new strand more or less simultaneously, and the original strands persist intact in the daughter cells for several generations (Fig. 15.1). Before considering the conceptual difficulties to which this gives rise, we will discuss the enzymes involved in DNA synthesis.

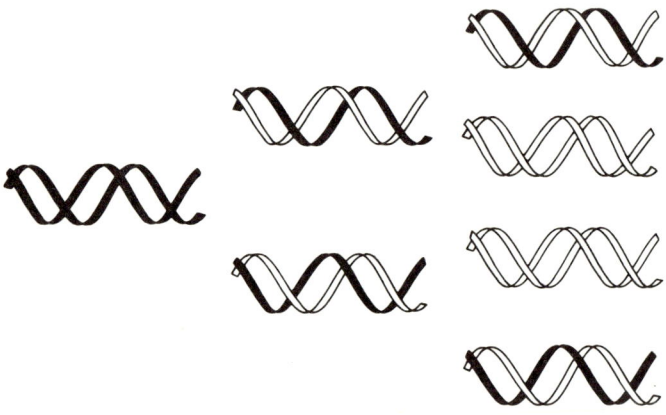

FIG. 15.1. Schematic representation of semi-conservative DNA replication.

Enzymic Synthesis of Deoxyribo-polynucleotides

Two enzymes are known which can form phosphodiester bonds between deoxyribonucleotides. The first to be discovered was *DNA polymerase*; several enzymes of this class have now been characterized. They catalyse the addition of deoxyribonucleotides to the free $-$OH at the 3′ end of a chain:

$$pppN_1pN_2pN_3-OH + dNTP \rightarrow$$

$$
\begin{array}{c}
\qquad\qquad\qquad\qquad \overset{O}{\underset{O^-}{pppN_1pN_2pN_3-O-\overset{\uparrow}{\underset{|}{P}}-O-N-OH}} + P-P_i
\end{array}
$$

The enzymes do not readily initiate a chain. Suspicion is growing that the primer in vivo is a short length of *RNA*, so that in this sense replication and transcription are closely related. DNA polymerases possess type *a* exonuclease activity, the reason for which is unknown.

The enzymes have two outstanding properties. First, they require a mixture of deoxynucleotide triphosphates to function, and secondly they will not function without a segment of DNA, which acts as a template. In principle this must be a single strand, but polymerase I binds strongly to the ends of double-stranded DNA, or to a nick in a circular chromosome.

Many kinds of analysis have shown that the newly synthesized stretch of DNA has its base arranged in a *complementary* sequence (p. 88) to the template DNA. 'Nearest-neighbour' analysis—which utilizes the fact that the (labelled) phosphate binding two nucleotides comes from the incoming nucleotide, but the bond can be split by type *b* nuclease to leave the phosphate with the acceptor nucleotide, the 'nearest neighbour'—has been a particularly powerful tool in this work. It is also found that the newly synthesized strand has opposite polarity (see Chapter 5) to the template strand, and when about 12 nucleotides have been added, the strands will spontaneously intertwine to form a complementary double-helical strand.

No complete sequences for a significant stretch of double-stranded DNA have yet been published, but Fig. 15.2 shows the complete sequences of the complementary $(+)$ and $(-)$ strands synthesized by an RNA virus, together with some idea of the secondary structure formed by intra-chain helices. The chromosomes of animal cells are some 10^7 times longer than this.

A second enzyme, *DNA ligase*, will form a phosphodiester bond between the 3' and 5' ends of two large segments of DNA. It will not sequentially lengthen a single segment. The reaction is unusual in that an adenylyl group is used as an intermediate, thus

$$\text{Enzyme} + \text{ATP} \longrightarrow \text{E-AMP} + \text{P--P}_i \qquad (1)$$

$$\text{E-AMP} + \text{DNA} \ldots \text{p (5' end)} \longrightarrow \text{E} + \text{DNA} \ldots \text{p-p-adenosine}$$

$$\text{DNA} \ldots \ldots \text{p-p-adenosine} + \text{HO} \ldots \text{DNA (3' end)} \longrightarrow$$

$$\text{DNA} \ldots \text{p} \ldots \text{DNA} + \text{AMP} + \text{Enzyme}$$

In bacteria NAD is the adenylyl donor, and NMN is produced in reaction (1) in place of P--P_i. The reaction does not occur unless the two segments of DNA to be joined are both attached to the same intact complementary strand of DNA. This is apparently necessary to bring the two free ends in juxtaposition. Thus at first sight, DNA ligase is only useful in repairing breaks in one strand of double-helical DNA, but it may have another equally important function, as discussed below.

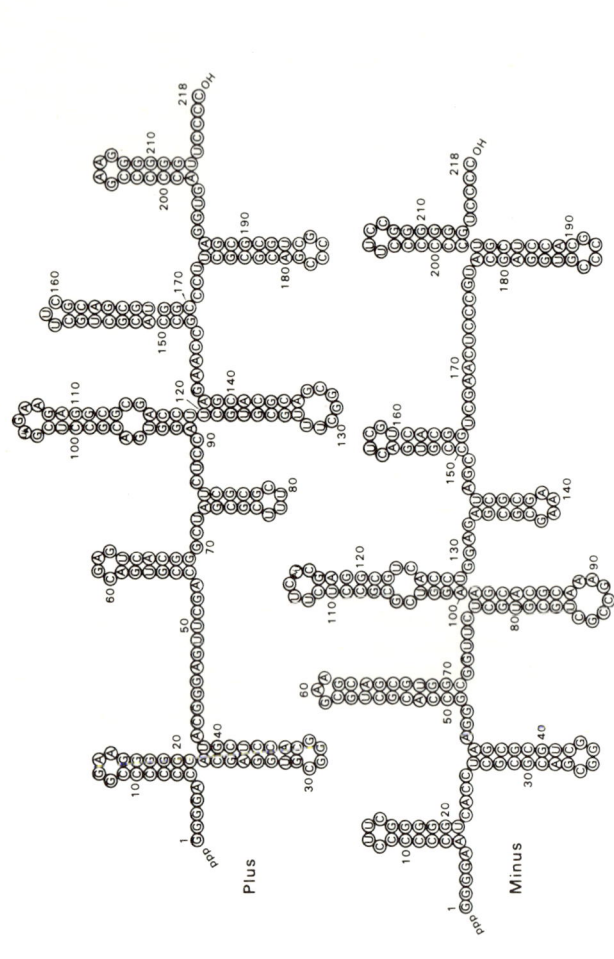

FIG. 15.2. The complete sequence of the complementary strands of MDV-1 RNA. The chromosome of this virus is a single strand (the (+) strand) of RNA. When it infects a cell the (−) strand is synthesized as a template for replication of the chromosome.

Notice that the complementarity is anti-parallel, i.e. nucleotides 218 → 1 on the (−) strand are complementary to 1 → 218 on the (+) strand.

The Semi-conservative Replication of DNA

In the mechanism of DNA polymerase outlined above, synthesis only occurs from the 5′ end of the new polymer towards the 3′ end, since nucleotides are always added to the free 3′ hydroxyl of an already bonded nucleotide. An intensive search has failed to reveal, in any organism, a polymerase which works in the opposite sense. It follows that the *template* must always be copied from the 3′ end, since the old and the new strand run antiparallel. The experimental evidence shows definitely, however, that *both* strands of double-stranded DNA are copied at the same time, without the double helix being completely unwound, although one strand—that beginning in a 5′ phosphate—cannot possibly act in any simple way as a template (Fig. 15.3a). This dilemma has been solved by the discovery that newly synthesized DNA is formed in short segments in the 5′ → 3′ direction, and that these are later joined by DNA ligase. Since the latter requires that the strands to be joined are aligned on an intact DNA strand, complementarity is preserved. The mechanism is outlined in Fig. 15.3(b–d). The

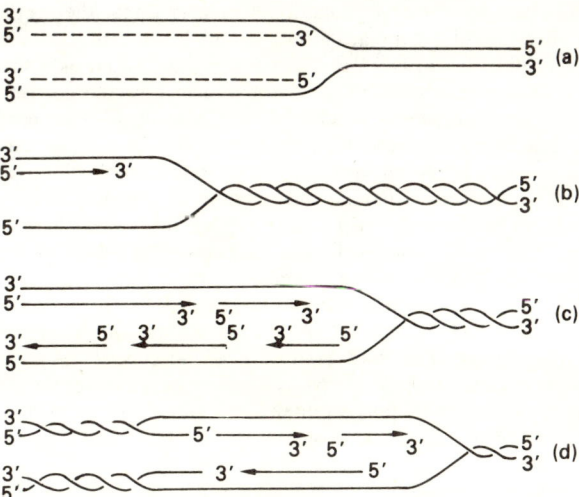

FIG. 15.3. Discontinuous synthesis of two new DNA chains. (a) The opposite polarity of the two new chains at the growing point, forced by the anti-parallel complementarity of the double helix. (b) The helix partially unwinds and synthesis begins catalysed by DNA polymerase. (c) Short lengths of new DNA accumulate. (d) The new chains are joined by DNA ligase and spontaneously start to form new helices.

The primer needed by DNA polymerase has been omitted for the sake of clarity. Note how the replication is semi-conservative.

rate of DNA replication in *E. coli* is rather fast, about 90 000 diester bonds or 2·7 μm per minute. The rate of chromosome replication at a single site in mammals, including histone synthesis and incorporation, is much slower, about 0·5 μm per minute. Coincidentally with this, there are many initiation points in eukaryotic chromosomes; autoradiographic evidence suggests that they are perhaps 50 μm apart, and that the rate of chromosome replication depends on the number of active initiation sites. In bacterial chromosomes, on the other hand, there is usually only one initiation point. The DNA in a human chromosome is about 40 mm long, compared with 1·3 mm for *Escherichia coli*.

The Genetic Code

It is not easy to define a gene unequivocally. For the present purpose a gene may be thought of as a section of DNA that will act as a template for the synthesis of a complete, independent polymer with one and only one function, while at the same time recognizing that this definition will not cover the regulatory genes described in Chapter 17. These may never be transcribed. This limitation apart, the genetic DNA directs the synthesis of RNA, which may itself be functional or may direct the synthesis of protein. Because of the polarity inherent in RNA synthesis, only one strand of double-helical DNA can be the template. So far as is known, it is always the same strand, but it is not known how the correct one is chosen. In higher organisms, much of the chromosomal DNA appears not to carry genetic information, not even repressed code. The various kinds of single-stranded RNAs are complementary to the template, thus adenine dictates the insertion of uracil, and guanine of cytosine. The complementarity can be shown by hybridization experiments (see Chapter 5).

Of the 3 kinds of RNA whose function is known, the nucleotide sequence of one, *ribosomal RNA* (rRNA), presumably dictates only its secondary and tertiary structure. This is also true to a large extent for the various species of *transfer RNA* (tRNA), but each species has two specific recognition sites, as described below. The third kind of RNA, *messenger RNA* (mRNA) contains in its linear sequence of bases a code which dictates the sequence (primary structure) of all polypeptide chains manufactured in ribosomes. The way in which mRNA and protein synthesis are linked is described in Chapter 16.

This sequence of bases on mRNA is known as the *genetic code*. It is a triplet code, each set of 3 bases specifying one amino acid. There are no punctuation marks indicating the space between one amino acid and the next—the code is *commaless*. Because a triplet chosen from 4 different bases can be arranged in $4^3 = 64$ different ways, the code is *degenerate*, i.e. some amino acids are coded for by more than one triplet *codon* (6 for serine and leucine). The code is nevertheless unambiguous, and as far as is known it is universal—the same for all protein-containing organisms.

The genetic code is given in Table 15.1. The 3 codons UAA, UGA and UAG specify termination of a peptide chain. Initiation takes place by a different mechanism (see pp. 340–42).

Table 15.1
The Genetic Code for the Codons on Messenger RNA

First base	Second base				Third base
	U	C	A	G	
U	PHE	SER	TYR	CYS	U
	PHE	SER	TYR	CYS	C
	LEU	SER	'ochre'	'umber'	A
	LEU	SER	'amber'	TRP	G
C	LEU	PRO	HIS	ARG	U
	LEU	PRO	HIS	ARG	C
	LEU	PRO	GLN	ARG	A
	LEU	PRO	GLN	ARG	G
A	ILE	THR	ASN	SER	U
	ILE	THR	ASN	SER	C
	ILE	THR	LYS	ARG	A
	MET	THR	LYS	ARG	G
G	VAL	ALA	ASP	GLY	U
	VAL	ALA	ASP	GLY	C
	VAL	ALA	GLU	GLY	A
	VAL	ALA	GLU	GLY	G

A: adenine; C: cytosine; G: guanine; U: uracil. The abbreviations for the amino acids are given on pp. 52–55. The terms 'amber', 'ochre' and 'umber' referred originally to certain mutant bacterial strains, but are in fact signals for terminating polypeptide chains.

Synthesis of Poly-ribonucleotides

The synthesis of RNA, whatever its function, from a DNA template is known as *transcription*.

One enzyme normally catalyses the synthesis of RNA, a nuclear enzyme called *DNA-directed RNA polymerase*. Like DNA polymerase, it uses nucleotide triphosphates as precursors, so that the reaction may be formally written

$$\ldots pY{-}OH(3') + XTP \rightarrow \ldots pYpX{-}OH(3') + P{-}P_i$$

where X and Y here stand for any of the 4 bases that occur in RNA. This, and all other enzymes mentioned in this chapter, is Mg^{2+}-dependent. The enzyme needs a template of DNA, and it catalyses the synthesis of RNA only from the 5' end of the RNA chain. In these respects it resembles DNA polymerase. The rate of formation of new diester bonds is rather slow, only 9 000 per minute at 37°. There must thus be more than one starting point along the DNA chain, but this is to be expected, since RNA need only be transcribed in rather small, functionally related units (see Chapter 17), the synthesis always proceeds from the 5' end of the RNA chain, only one of the two strands of the duplex chromosomal DNA is transcribed, and there must therefore be some way in which multiple starting points on the correct chain can be recognized.

DNA-directed RNA polymerase is in fact a rather large enzyme (mol. wt. approx. 500 000), with a complex structure, composed of several sub-units (α, β', β and ω). Moreover, it has been shown that a fairly small peptide, called a sigma factor (σ) must be bound to the enzyme before it can start transcribing. After transcription has begun, the σ factor is released and can bind to other enzyme molecules. The supply of σ peptides thus in some measure regulates the rate of transcription of chromosomal DNA. It is not known what are the start signals on the DNA which are recognized by the polymerase with σ factor bound to it. The duplex DNA strand does not have to be completely unwound for transcription to be accomplished. The segments of newly formed RNA do not have to be linked by an RNA ligase; on the contrary, they may have to be cut by an endonuclease (see especially p. 333). A protein termination factor (p), not regarded as part of the polymerase, is required for transcription to cease.

Viruses whose genetic information is contained in a single RNA strand must replicate first a complementary copy, and then a copy of this which duplicates the original (see Fig. 15.2). These syntheses are carried out by an *RNA*-directed RNA polymerase. The invaded cells do not have such an enzyme, and the first section of the viral RNA carries the information necessary to synthesize the enzyme protein.

It is a central dogma of molecular genetics that heritable information only goes from DNA to RNA, and never vice versa, but there is now good evidence that some cancer-inducing RNA viruses can direct the synthesis of DNA by a *reverse transcriptase*. The new DNA may be used to direct the immediate synthesis of viral RNA, or it may be incorporated into the host cell's chromosomes.

Function and Structure of Ribonucleic Acids

Although RNA is never found as a duplex (complementary double-stranded) chain, the possibilities of secondary structure and of a precise three-dimensional form for RNA molecules, no doubt related to their function, have been discussed in Chapter 5.

Ribosomal RNA

So far as is known, the sequence of bases in rRNA is only important in so far as it determines the secondary structure (as yet unknown), and the possibility of interaction with ribosomal proteins.

rRNA appears to be chiefly synthesized in the nucleolus (with extra-chromosomal DNA as template). There are at least 300 copies of the code in each haploid human genome, so that many copies of the template can be transcribed simultaneously. The RNA for the larger ribosomal sub-unit is synthesized as a single strand with sedimentation velocity 45S. It is later cut by an endonuclease into the two strands of 28 and 18S which are found in the larger and smaller sub-units, respectively. The small (5S) rRNA molecule, for which there are some 500 sites probably not in the nucleolus, is synthesized separately. There is some evidence that ribosomes are not 'exported' from the nucleus unless they are attached to a strand of messenger RNA, but the significance of this is not at all clear.

Transfer RNA

All tRNA molecules are relatively small, about 75–85 nucleotides (5S). The secondary structure is conventionally expressed by a 'clover-leaf' pattern, although the true structure is probably cylindrical and L-shaped, with the anti-codon site at the end of one arm (see Fig. 5.2).

The astonishing number of 'unusual' bases in tRNA has already been mentioned in Chapter 5. These must be made by chemical attack on the normal bases after the polynucleotide has been synthesized, for the corresponding DNA does not have any bases to which the variant bases could be specifically complementary. The function of the unusual bases is unknown, but it may be to prevent too extensive development of intra-chain helix formation, and thus to preserve a unique secondary/tertiary structure.

The anti-codon. This is a sequence of 3 bases on the middle lobe of the clover-leaf (see Fig. 5.1), which is complementary to the triplet code on messenger RNA. Since there are 64 possible triplets (Table 15.1), there ought to be 64 tRNAs, each with a specific anti-codon. It seems likely, however, that significantly fewer than this exist, since it has been observed that one tRNA may bind to more than one of the codons which specify a single amino acid. This is possible because the third base in the anti-codon (counting from the 3' end) is often inosinic acid (formula on p. 95) instead of the expected adenylic acid (hypoxanthine instead of adenine). Hypoxanthine can form sterically satisfactory hydrogen bonds with 3 other bases, namely uracil, cytosine and adenine (Fig. 15.4), unlike adenine which can only pair (in RNA) with uracil. Thus an anti-codon which terminates with I is not so specific as one ending with A, G, U, or C, an observation usually known as the 'wobble effect'. For example, serine has 6 codons (Table 15.1), which may be paired with only 3 anti-codons as shown in Table 15.2.

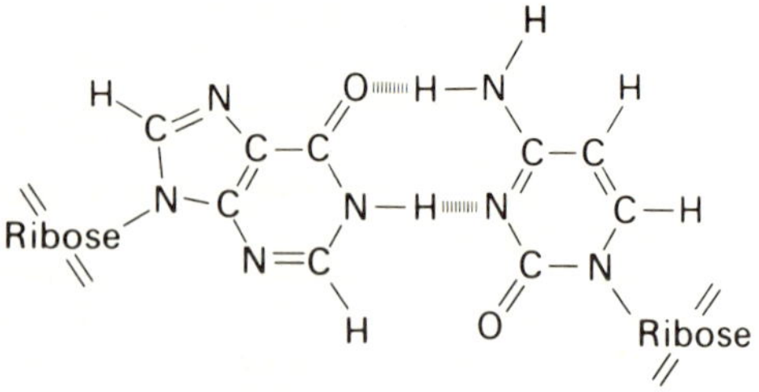

Hypoxanthine————Cytosine

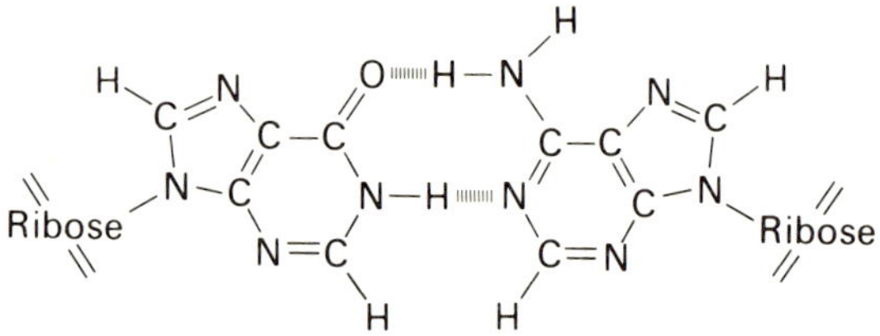

Hypoxanthine————Adenine

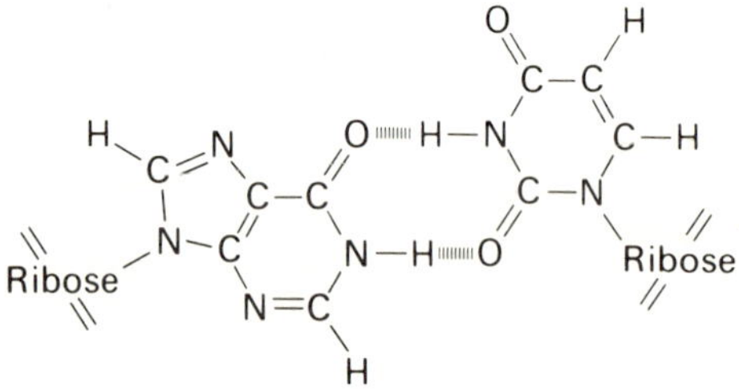

Hypoxanthine————Uracil

FIG. 15.4. Base-pairing in the 'wobble' effect.

Table 15.2
'*Wobble*' *Pairing with Inosinic Acid in the Anti-codon*

Codon (5′ → 3′)	UCU	UCC	UCA	UCG	AGU	AGC
Anti-codon (3′ → 5′) {	AGI	AGI	AGI			
				AGC		
					UCI	UCI

Within a given organism, it appears that some of the possible codons do not appear in the messenger RNA, so that although two or three tRNA species may theoretically be necessary to cope with all possible codons for one amino acid, within one organism only one will actually occur in any quantity. Apart from the possible significance of suppressor tRNAs, the reasons for this limitation are not yet clear.

All tRNAs have the sequence Adenylic acid—Cytidylic acid—Cytidylic acid— at the 3' end. The amino-acyl residue is attached to the 3'–OH of the ribose in the terminal adenylic acid group. This sequence of three nucleotides is enzymically added to each tRNA from an A-C-C-ppp donor after transcription is complete.

As discussed in detail in Chapter 16, the ribosomal protein-synthesizing machinery recognizes a nucleic acid molecule, not an amino acid. The attachment of the correct amino acid to each tRNA is accomplished by enzymes each specific for a given amino acid and a tRNA molecule. These enzymes must therefore unambiguously recognize the appropriate tRNA by means of a sequence or conformation which is separate from the anti-codon. The enzyme recognition sites have not so far been established.

Messenger RNA

When mRNA is being translated by ribosomes, it is in an extended thread-like conformation, but there is evidence, from the order in which enzymes coded on multi-cistronic stretches of bacterial RNA sometimes appear, that segments of the RNA code for one protein are inaccessible to ribosomes unless a ribosome is attached to a preceding section of the mRNA. Such evidence argues for secondary structure which can be altered by ribosome binding.

In bacteria, messenger RNA is usually short-lived, as is evident from the rapidity with which the bacterial protein-synthesizing system is turned over to the synthesis of viral proteins after infection by viruses (phages). Animal mRNA, however, seems to have a wide range of stabilities, from hours to perhaps weeks, with the more stable species perhaps predominating.

Another difference between prokaryotic and eukaryotic mRNA is that the former appears to be usable as it is formed, almost without modification. It has been known for a long time, however, that the stimulation of protein synthesis in nucleated cells is preceded by the synthesis of a great deal of rapidly turning-over RNA which never leaves the nucleus (hetero-geneous nuclear RNA, or Hn-RNA). Evidence is accumulating that the mRNA which is functional in cytoplasmic protein synthesis is a fraction of this Hn-RNA, presumably excised from it by endonucleases. The rest is degraded.

Yet another, at present inexplicable, difference is that almost all the mRNA has attached to it at the 3' end, after transcription, about 200 adenylic acid residues. One exception is the mRNA which codes for histone synthesis. The addition is probably sequential and may be made first to the Hn–RNA, but it is not yet certain that functional mRNA is always at the 3′ end of the latter. Poly-A addition does not seem to be related to the 'export' of mRNA from the nucleus, as RNA which is synthesized in the cytoplasm, e.g., poliomyelitis virus RNA, also has poly-A attached.

Fig. 15.5 shows the overall relation between nucleic acids and protein synthesis.

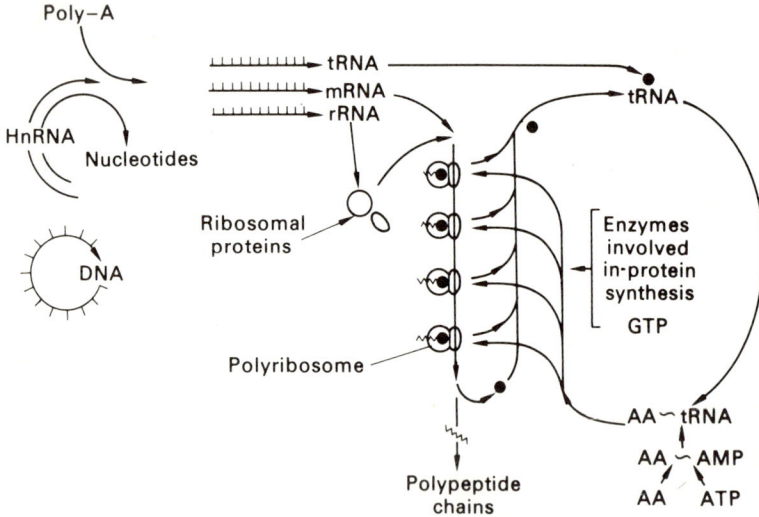

FIG. 15.5. Nucleic acid relationships.

Mutations and Mutagens

Four types of single-point mutations may be defined;

1. *Transitional mutations.* In these a purine in one chain is replaced by the other purine, or a pyrimidine is replaced by another pyrimidine. For example, nitrous acid, HNO_2, will deaminate adenine to hypoxanthine, which pairs with C more readily than with T. Thus if the original chain is

$$-X-A-X-$$

the modified chain is

$$-X-I-X-$$

The complementary chain will be

$$-Y-C-Y-$$

and a new chain synthesized upon the complementary chain as substrate will be

$$-X-G-X-$$

Other mutagens causing transitions are *2-aminopurine*, which can be read either as A or as G, and *5-bromouracil*, which is a thymine replacement that pairs both with G and with A.

2. *Transverse mutations.* In these a purine is replaced by a pyrimidine, or vice versa. No special agent is known, but analysis of polypeptide sequences shows that they are rather common.

3. *Insertion mutations* are caused by the insertion of an extra nucleotide, originally into one chain only of a double helix. *Acridine*

will intercalate between the two bases in a double-stranded DNA, and cause the insertion of an extra base in the complementary chain. When the original chain is replicated, it too will gain an extra base, probably by a repair mechanism (see below).

4. *Deletion mutations* arise from the loss of a base in one strand, which may cause the loss of the complementary base in the other strand on replication.

Mutations of types 1 and 2 will give rise to so-called *mis-sense* mutations, in which one amino acid is replaced by another. Very often these are *silent* mutations, i.e. they have little biological effect; indeed, if the original and the replacement amino acids have neutral side-chains, the replacement may be very difficult to detect. However, if the mutation gives rise to one of the chain termination codons, the peptide will be prematurely released from the ribosome. This is then called a *nonsense* mutation.

Mutations of types 3 and 4 are called *frameshift* mutations. Since the code is commaless (p. 328), all the amino acids beyond the mutation point will be incorrect. If one of these aberrant codons happens to be a termination codon, the effect will apparently be a nonsense mutation. More frequently, the peptide will remain attached to the ribosome, because the terminating codon can never be read. Because of this, and because the primary structure of the protein is severely defective, frameshift mutations are usually lethal.

It has been experimentally observed that mis-sense mutations do not lead to premature termination as often as expected, and this has led to the definition of *suppressor genes*. These appear to determine the synthesis of tRNA species that normally are not found in the particular organism, but which would have the effect of translating the accidental termination codon as a harmless amino acid. The subject is too complex to be dealt with here. However, it is of great interest that one or two tRNA species that will suppress insertion mutations have been characterized. They have four bases instead of three in the anti-codon, e.g., $-C-C-C-C-$. This shows definitively that it is the structure of the tRNA molecule (and presumably the conformation of the ribosome that accepts it) which determines that the genetic code is triplet. There is in this sense, no 'reading frame' in the ribosome.

Repair and Recombination

Single strand defects in DNA may arise from many causes, e.g., irradiation by ultra-violet light, which produces mainly dimers between two neighbouring pyrimidine residues. Nitrogen mustards, e.g.,

$$(Cl \cdot CH_2CH_2)_2N \cdot CH_3$$

mainly form covalent links between guanine residues. X-rays break one strand of the duplex, leaving a gap.

All such defects are very frequently repaired by a standard mechanism:

1. If the damaged strand is not broken, it is cut by a damage-specific type b endonuclease (p. 158).

2. A 3′ phosphatase removes the phosphate on the free 3′—OH at the break.

3. A type b exonuclease (which can be DNA polymerase, see p 325) removes the damaged nucleotides, and a few more, on the 5′ side of the break.

4. DNA polymerase inserts the missing nucleotides, using the intact strand as a template.

5. DNA ligase joins the break.

In normal human cells, these repair mechanisms provide considerable protection against mutagens and carcinogens. In a rare inborn error of metabolism, *xeroderma pigmentosum*, enzymes required for one of the stages 1–3 above are missing, and there is excessive sensitivity of the skin to sunlight, and increased incidence of skin cancer. Other such congenital defects may exist.

DNA damage involving both strands, e.g., cross-linking by nitrogen mustards or by *mitomycin C*, is more difficult to repair, unless there is a homologous duplex strand which can act as a template for both strands. Some mechanism is, however, needed to explain recombination, i.e. the exchange of genetic material between one chromosome and another that is very frequent in bacteria, but not common in animal cells. Various mechanisms have been proposed.

Drugs Inhibiting Nucleic Acid Synthesis

Mitomycin C, and the nitrogen and sulphur mustards, which are mutagenic agents, have already been mentioned.

Actinomycin D inhibits DNA-directed RNA synthesis by binding to the DNA helix. It is a widely used experimental drug, but has toxic side effects in vivo.

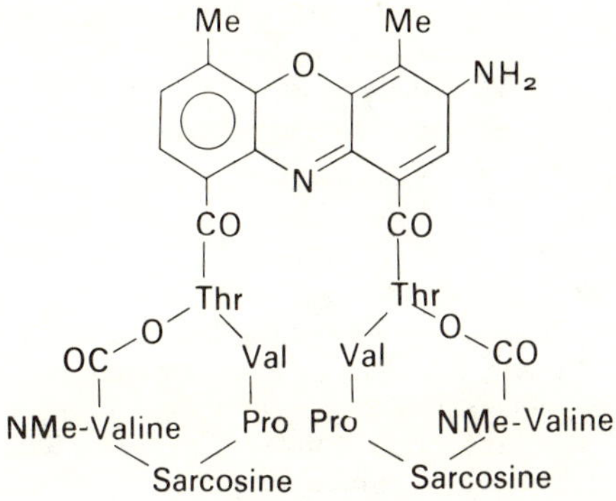

Actinomycin D

Rifamycin inhibits RNA polymerase I by binding specifically to the β- sub-unit, while the toadstool poison *α-amanitin* inhibits polymerase II.

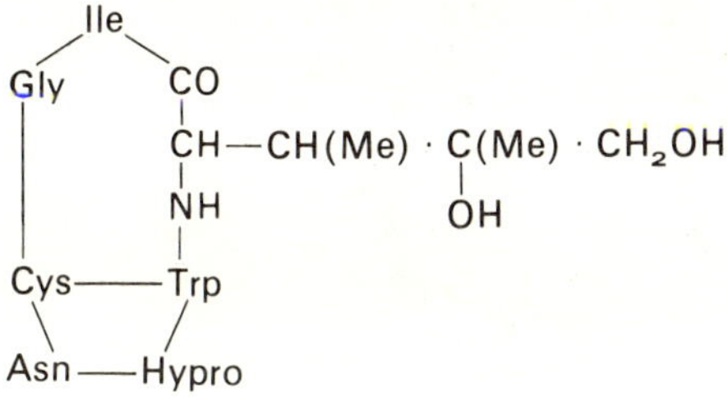

α-Amanitin

16 Protein Synthesis: Translation

The formation of a polypeptide chain requires two essentially different mechanisms to be brought together—the linking together of α-amino acids by elimination of water, and a device to ensure that each amino acid is linked in correct sequence to ensure the unique primary structure for each protein. To convert the polypeptide chain into the biologically active molecule may subsequently require formation or cleavage of special bonds, assemblage of sub-units and other processes.

The information about the sequence of amino acids in a polypeptide is normally carried to the protein-synthesizing machinery by a special ribonucleic acid called messenger RNA or mRNA, but it is now clear that many proteins are also involved in the assemblage of the proper amino acids in the correct order. Protein synthesis in all types of cells is carried out in organelles called *ribosomes*. These are mostly found in the cytoplasm, attached to the rough endoplasmic reticulum, but there are also ribosomes in nuclei, and in mitochondria. Ribosomes vary in size; those of *Escherichia coli*, which have been most studied, are 70S (Chapter 5), which is equivalent to $2 \cdot 8 \times 10^6$ daltons. Mammalian ones are usually larger, about 80S or $4 \cdot 2 \times 10^6$ daltons. All ribosomes can be dissociated into two sub-units, a larger one, about 50S (*E. coli*), and a smaller one, about 30S (the corresponding sizes for animal ribosomes are 60S and 40S). The two sub-units can be dissociated in vitro by lowering the Mg^{2+} concentration in the suspending medium; they are known to dissociate at a certain stage in the cycle of transcribing mRNA, as discussed below. Formation of peptide bonds is associated with the larger unit, whereas binding of mRNA, and of the successive aminoacyl-tRNA s, is associated with the smaller unit. The latter contains (in bacterial ribosomes) one polynucleotide of about 16S, and about 20 proteins; the former contains one large (23S) and probably two small (5S) polynucleotides and up to 34 proteins. There is good evidence that some of these proteins are found in all ribosomes from a given source, while others are unevenly distributed so that there is on average only one molecule of them in every third ribosome. Ribosomes are therefore heterogeneous, although this may in part only reflect the cytoplasmic protein factors (see below), which happened to be on each ribosome when it was harvested.

The formation of a polypeptide chain can be divided into four stages: activation of the amino acids; chain initiation; elongation of the chain (transpeptidation); and chain termination.

Amino Acid Activation

The formation of a peptide bond is energetically not particularly difficult. The free energy of formation is only about 21 kJ of which the suppression of ionization of the $-NH_3^+$ and $-COO^-$ groups forms the major part. Peptide detoxication products (Chapter 13) are made by activating the $-COO^-$ group with Coenzyme A. However, the ribosomes recognize only polynucleotides, and attaching an amino-acyl residue to a suitable polynucleotide is more complicated. The carboxyl group of the amino acid is activated by forming a mixed anhydride with adenosine mono-phosphate (AMP); the amino-acyl residue is transferred from this inter-mediate to an $-OH$ residue (probably the 3′) of the ribose of the terminal adenyl residue at the 3′ end of a specific tRNA (for structure of transfer RNA s, see Chapter 5). The evidence is that the intermediate acyl-adenylate remains bound to the enzyme, called an *aminoacyl-tRNA synthetase*, which therefore recognizes both the amino acid and one of the tRNA s carrying the appropriate anti-codon for that amino acid.

This reaction may be summarized:

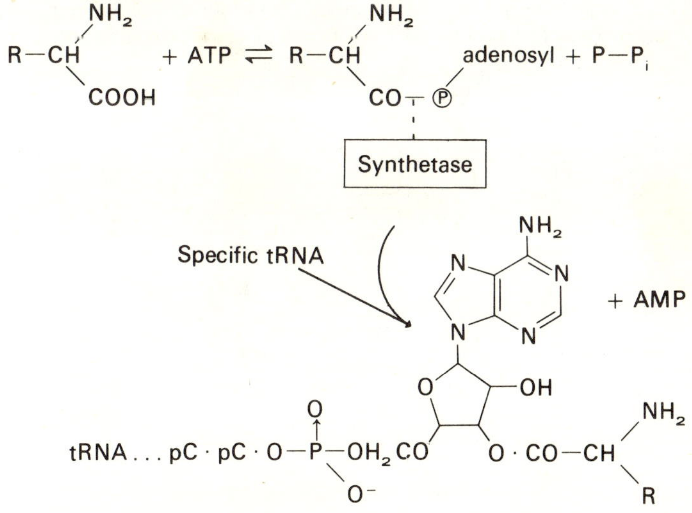

The almost absolute specificity of these enzymes is very important, because there is no subsequent check that the correct amino acid is attached to the tRNA. Only one instance of 'mischarging' has ever been discovered, although some of the enzymes will charge their tRNA s with certain synthetic analogues, e.g., p-fluorophenylalanine for phenylalanine. There could be more than one synthetase for each amino acid, since certain amino acids can have several tRNA s because of the degeneracy of the genetic code (p. 328), but not all cells seem to possess all possible tRNA s, nor, presumably, all synthetases.

Peptide Bond Formation

Fig. 16.1 shows the essential process in elongating a peptide chain, which is an attack by the free —NH_2 group of the newly bound amino acyl-tRNA, on the carboxyl ester group which links the growing peptide to another tRNA. As a result of this attack, the whole of this peptide is transferred to the amino group, and a new peptide bond is formed. This reaction in itself has no coupled energy requirement. The site from which the peptide chain moves is called the *donor* site; the site containing the free —NH_2 group is the *acceptor* site.

Fig. 16.1 also shows the function of the tRNA s in aligning the amino acids on the messenger RNA. A frameshift mutation (e.g., starting at CAA) would be possible as the diagram stands, but is prevented by proper alignment in the initiation process.

Chain Initiation

This is important not only to ensure that translation does not commence in the middle of a length of mRNA, but also to see that the first anti-codon

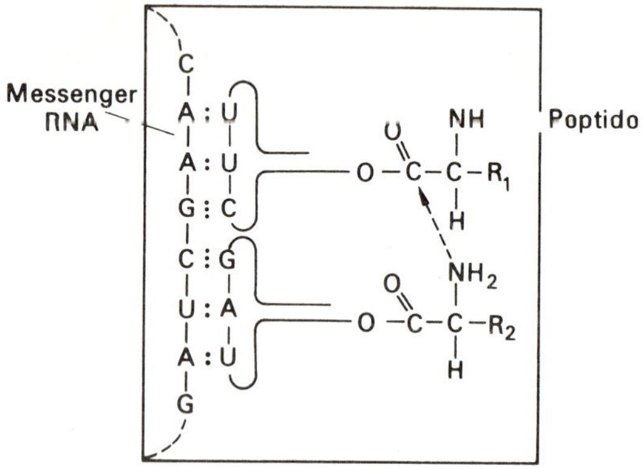

Fig. 16.1. Peptide bond formation in the ribosome.

is correctly aligned with the first triplet codon of the message. Misalignment at the start by even a single base will cause a continuous reading of false triplets of nucleotides, a so-called frameshift mutation, since the anti-codon of the tRNA is the 'reading frame'. There is therefore a special series of events leading up to chain initiation, perhaps most easily understood by referring to the process as it occurs in *E. coli*.

In this organism, and in bacteria in general, the NH_2-terminal amino acid of the nascent peptide chain is always N-formyl methionine. This is attached to a special tRNA as follows:

$$Met + tRNA^{f-Met} + ATP \rightarrow Met\text{-}tRNA^{f-Met} + AMP + P—P;$$

$$Met\text{-}tRNA^{f-Met} + FH_4 \cdot CHO \rightarrow CHO \cdot Met\text{-}tRNA^{f-Met} + FH_4$$

($FH_4 \cdot CHO$ is formyl-tetrahydrofolic acid, see Chapter 11).

In bacteria the 50S and 30S sub-units of the ribosome always dissociate after translating a cistron. The 30S unit picks up a new length of mRNA containing the 'initiator' codon AUG (or GUG). 'Frameshift' codons could clearly contain such a sequence by accident, and it appears that a non-translated string of nucleotides before the first codon may assist in registering the mRNA correctly. The f-Met-tRNA then binds to the donor site (Fig. 16.2) in a process requiring GTP, and assisted by several protein factors, both ribosomal and cytoplasmic. These initiation factors are usually in short supply and therefore limit the rate of protein synthesis. When these events are completed, the 50S sub-unit binds on, and the synthesis of the first peptide bond takes place. It happens that if the segment of messenger does not contain the AUG codon, f-Met-tRNA is not required, thus synthetic polynucleotides can be translated. The synthesis of poly-Phe as a result of translating a length of poly-U was of immense importance in the research which led to the determination of the genetic code.

In eukaryotic cells events follow the same general pattern, although in mammalian cells there may be a much greater proportion of free (undissociated 80S) ribosomes than in bacteria. The methionine on the initiation tRNA is not formylated; nevertheless this tRNA is always a particular one of the two $tRNA^{Met}$ species, and the other species is concerned only with the incorporation of internal Met. The occurrence of PCA at the N-terminal end of several peptides (cf. p. 56) and of the heavy chains of immunoglobulins shows that $tRNA^{Met}$ is not a universal initiator in animal cells as $tRNA^{f-Met}$ is in bacteria.

Each functional protein has a characteristic N-terminal amino acid, and it follows that the 'initiator' amino acid (or sometimes a short peptide, cf. pro-collagen, p. 347) is normally hydrolysed off the newly formed peptide after its release from the ribosome.

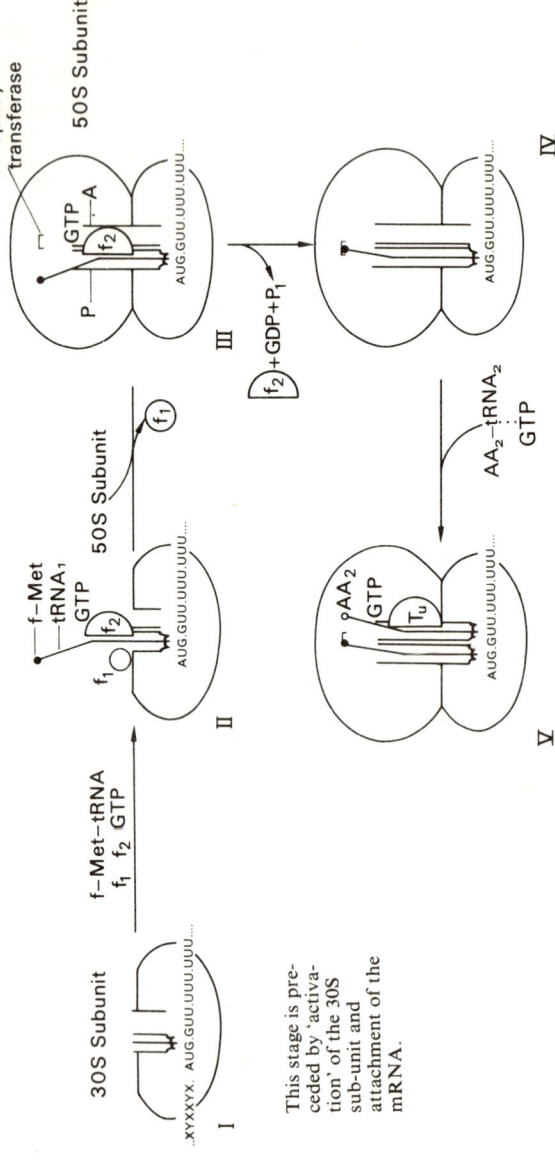

Fig. 16.2. Initiation of protein synthesis in bacteria. The mRNA has been given some 'pre-initiation signal' nucleotides in Step I, which are subsequently omitted for clarity. Step V is analogous to Step I in Fig. 16.3. P and A in Step III refer to donor site and acceptor site.

Chain Elongation

Although the 'initiation' aminoacyl-tRNA binds to the smaller sub-unit (which has previously bound mRNA) as described above, peptide bond formation (transpeptidation) takes place in the larger sub-unit, and if ribosomes are dialysed against low Mg^{2+} concentrations, they dissociate into the two sub-units with all the peptide chains attached to the larger one. It has been possible to show, particularly with the aid of puromycin (see p. 353), that the larger (50S or 60S) sub-unit contains two distinct sites at which peptide chains are bound. The *aminoacyl-binding, acceptor* or *A* site will also bind puromycin, and peptides can be prematurely terminated by transfer to this antibiotic from the *peptidyl-binding, donor* or *P* site. Peptides which are attached to the A site (as in Fig. 16.3 II) cannot be linked to puromycin until they are transferred back to the P site, see below. The basic stages in the process are shown in Fig. 16.3, and the part played by the cytoplasmic protein factors T_u, T_s and G, and by GTP are indicated in Fig. 16.4.

The cytoplasmic protein factor T_u may act as a kind of filter, since it will not bind tRNA s not carrying a 'normal' aminoacyl group, nor tRNA^{f-Met}. The aminoacyl-tRNA is only bound if GTP is also bound to the T_u; although the energy of the terminal phosphate of the GTP is not formally required for transpeptidation, correct alignment on the ribosome and subsequent dissociation of the protein only occur with T_u-GDP. Protein T_s displaces GDP from T_u; it may be substituted by pyruvate kinase + PEP (which presumably rephosphorylates the GDP).

In bacteria, cytoplasmic protein G is responsible for translocating the newly-formed peptidyl-tRNA from the A site to the P site on the 50S sub-unit, and also for moving the mRNA along one codon on the 30S sub-unit. A similar factor called T2 has been isolated from eukaryotic cells (see p. 354). It seems clear that energy from the hydrolysis of one GTP is needed for this translocation.

The acceptor amino acid is in vivo attached to a tRNA, but aminoacyl-adenosine will act as an acceptor (although not as a donor).

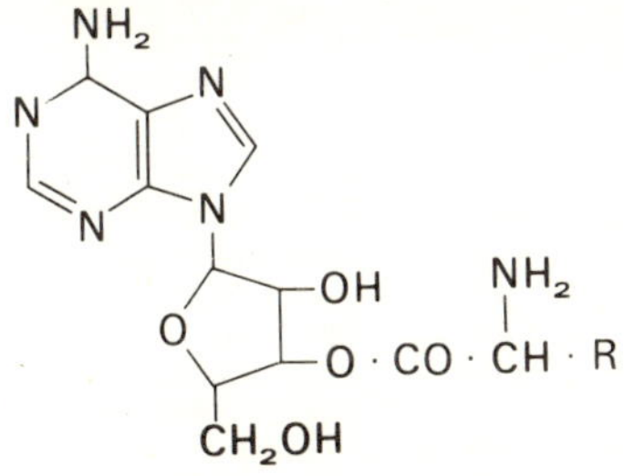

Aminoacyl-adenosine

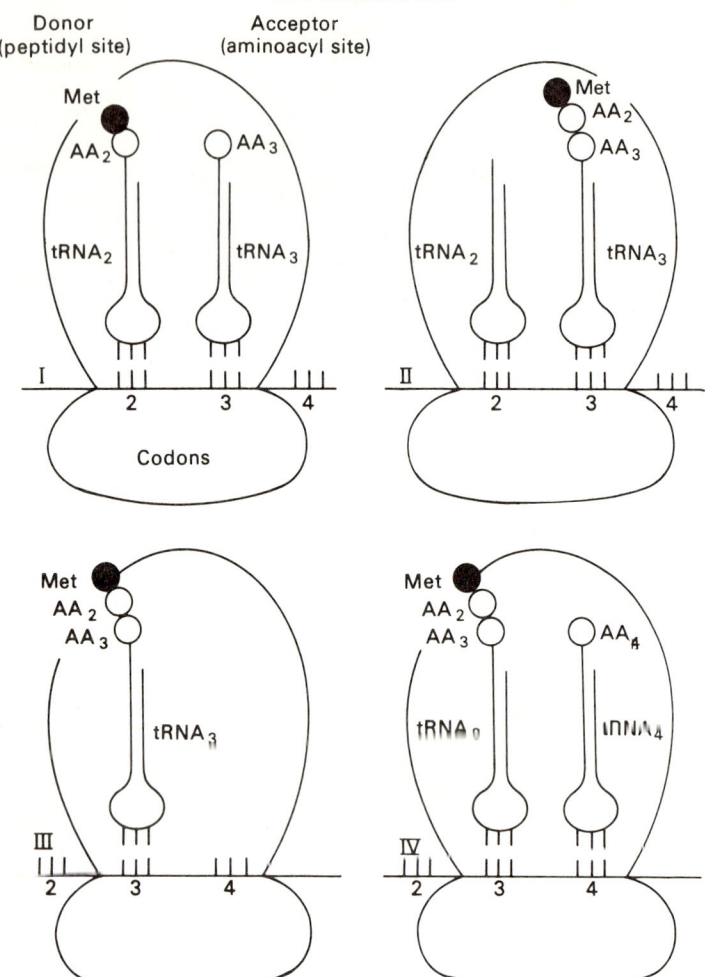

FIG. 16.3. The growth of a polypeptide chain on a ribosome. (I) A peptidyl and an aminoacyl-tRNA on the ribosome. (II) The peptidyl residue on tRNA$_2$ has been transferred onto the amino residue on tRNA$_3$; one peptide bond has been formed. (III) The peptidyl-tRNA$_3$ has been moved to the donor site by peptidyl translocase. At the same time tRNA$_2$ is lost, and the mRNA moves along one codon. (IV) The tRNA$_4$ molecule is in position ready to receive the peptide chain attached to tRNA$_3$.

The gaps shown between the codons on the mRNA are for clarity only.

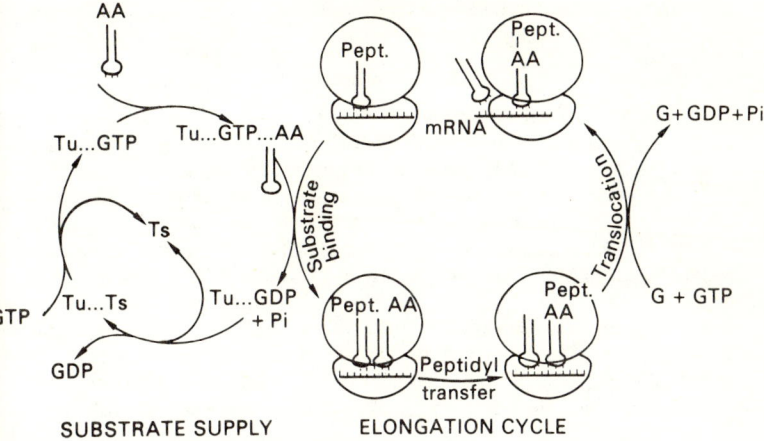

FIG. 16.4. Factors involved in peptide elongation in bacterial cells. Protein synthesis in animal cells is similar, but factor G is replaced by factor T2.

Similarly, f-Met-Adenylyl-Cytidylyl-Cytidylate will act as an artificial fMet donor, but not as an acceptor.

Polyribosomes—Longevity of mRNA

Gentle treatment of cell homogenates leads to the isolation of linear clusters of ribosomes, attached by a common thread of mRNA. (For schematic diagram see Fig. 15.2). Similar *polyribosomes* can be identified in electron micrographs, and it appears that it is the rule for more than one peptide chain to be synthesized at one time on one strand of mRNA; in some way the synthesis is more efficient. Bacterial messenger may contain the code for more than one gene on one strand (polycistronic mRNA), but the ribosomes detach when each protein is completed, not when the complete strand has run through the polysome cluster.

It may be calculated, from the experimental evidence, that an mRNA strand about 150 nm long is attached on the average to 5 or 6 ribosomes each 22 nm in diameter, leaving 3 nm between each. The mRNA within the ribosomes appears to be protected from ribonuclease attack. In bacteria, the average messenger only codes for 10–20 peptide chains, which suggests that it becomes susceptible to degradation (random nuclease action from the 5′ end) if ribosomes do not continually attach themselves. As the time taken to synthesize the average protein is about 2 minutes the expectation of life of a strand of bacterial mRNA is not much longer than this. Messenger from animal cells has perhaps 10–100 times as high a yield of peptide chains per strand, and there is some suggestion that degradation is determined by clock time, and not by the number of ribosomes which have traversed the strand.

Chain Termination

The existence of three termination codons has been deduced, namely UAA, UAG and UGA. These are recognized not by anti-codons on a tRNA, but by a soluble R (releasing) protein, which works in conjunction with an S protein. The completed peptidyl-tRNA has to be on the donor (P) site, and in effect it is transferred to water, rather than to the usual $-NH_2$. In keeping with this, ribosomes will also transfer a peptidyl residue to ethanol, i.e. they will catalyse the synthesis of a peptidyl-ethyl ester.

After the chain has been terminated, the two sub-units of the ribosome dissociate (Fig. 16.5) even if the mRNA contains another stretch of translatable material, and the free tRNA and the mRNA dissociate from it at about the same time, but it is not known how this is accomplished.

Met–tRNA   tRNA

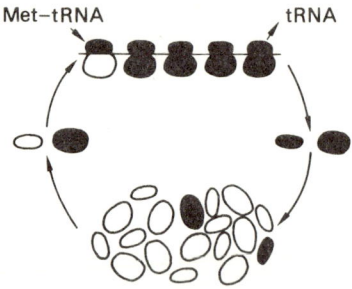

FIG. 16.5. Association and dissociation of ribosomal sub-units. (The pool of free sub-units is less evident in animal cells.)

Protein Synthesis in Mitochondria

The mitochondria of eukaryotic cells possess a double strand of circular DNA, which contains about 30 genes. Most of these code for the RNA of mitochondrial ribosomes, which are 70S and not 80S, and for mitochondrial tRNA. Unless a considerable quantity of mRNA comes from the cell nucleus, it is therefore unlikely that the apparatus of protein synthesis in mitochondria makes many proteins. Recent research suggests that, amongst other things, one of the sub-units in cytochrome aa_3, and one in cytochrome b, are synthesized in mitochrondria and the rest of the sub-units in cytoplasmic ribosomes. Perhaps these peptides are involved in binding the complete proteins to the membrane.

The circular DNA, the smaller size of mitochrondrial ribosomes, and the fact that they, like bacterial ribosomes, are inhibited by chloramphenicol, whereas 80S ribosomes are not, lend support to the view that mitochondria have evolved from parasitic bacteria.

Secondary Structure of Newly Synthesized Proteins

It is generally considered that, with few exceptions, the primary sequence of amino acids in a peptide chain determines a unique secondary and tertiary structure. The complete peptide chain should therefore take up its biologically 'native' conformation as soon as it leaves the ribosome, without the need for 'folding' templates. Even intra- or inter-chain disulphide bonds may form spontaneously if the native conformation is adopted. However, the polypeptide released from the ribosome is almost never identical with the biologically active protein. A formyl-methionyl, methionyl, or pyroglutamyl (PCA) residue is usually removed from the N-terminal end, and oligopeptides may also be excised. This may make a great difference to conformation, as may be seen in the conversion of fibrinogen to the linear fibrin (p. 289). The conversion of pro-insulin (after spontaneous formation of $-S-S-$ bonds) to insulin, has been intensively studied (Fig. 16.6).

An extreme example of post-synthesis manipulation is the synthesis of collagen fibrils (Fig. 16.7), where the processes include hydroxylation of proline residues; triple helix formation, for which an extra 20 per cent length of peptide at the N-terminal end is necessary; attachment of oligosaccharide chains; oxidation of the terminal $-NH_2$ of certain lysine residues to $-CHO$; scission of the N-terminal peptide; and the formation of covalent inter-chain lysaldehyde-lysine bonds and of inter-helix desmosine links

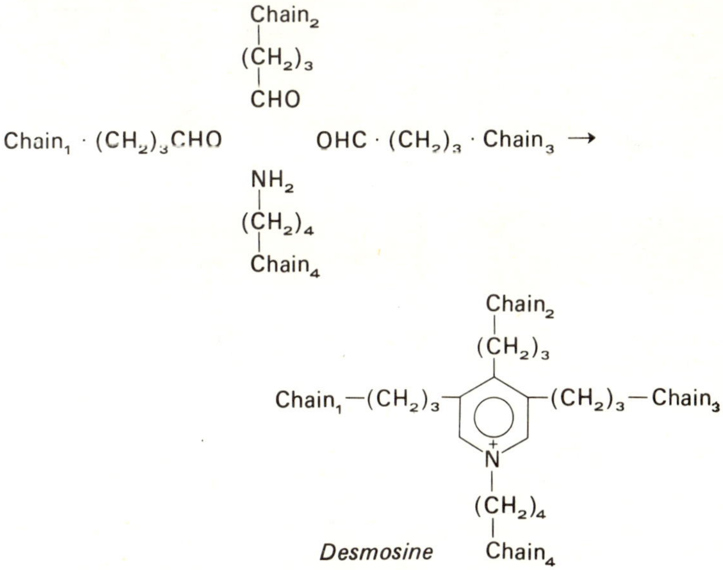

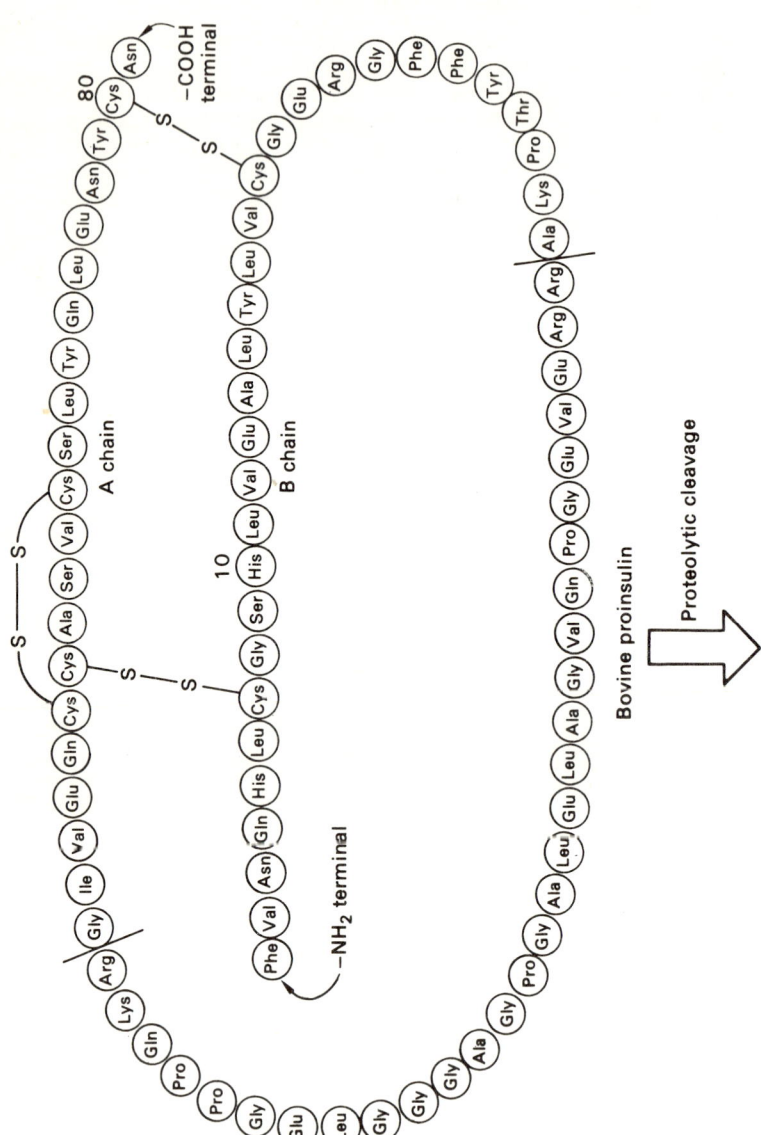

Bovine proinsulin

Proteolytic cleavage

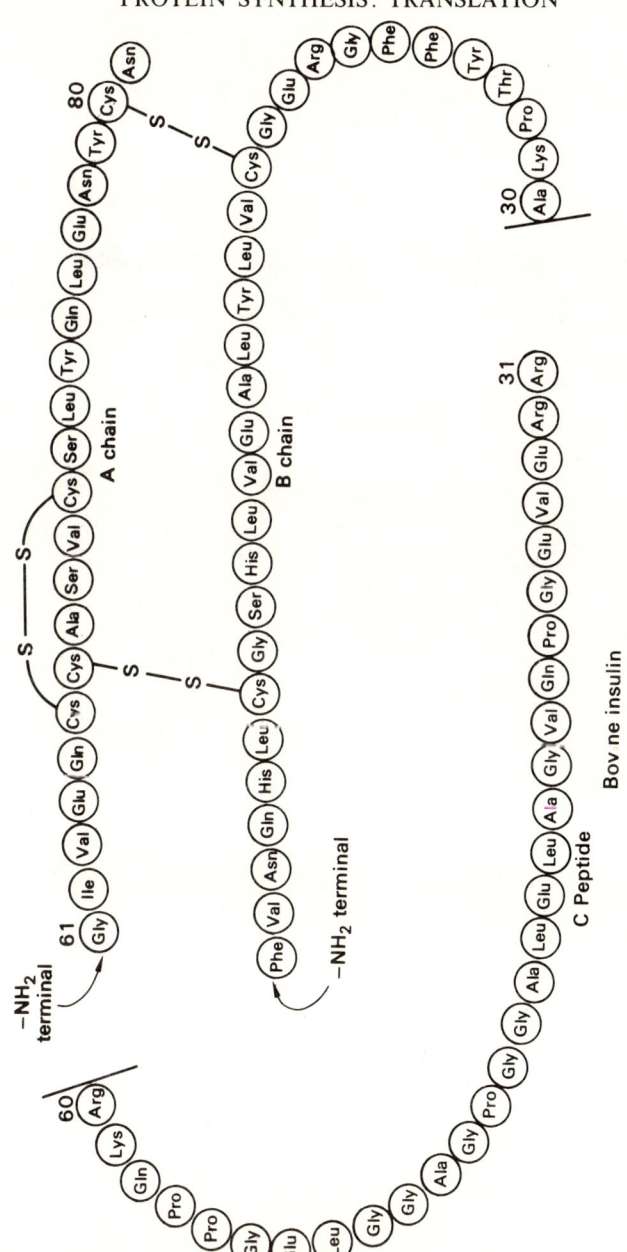

Fig. 16 6. Conversion of pro-insulin to insulin.

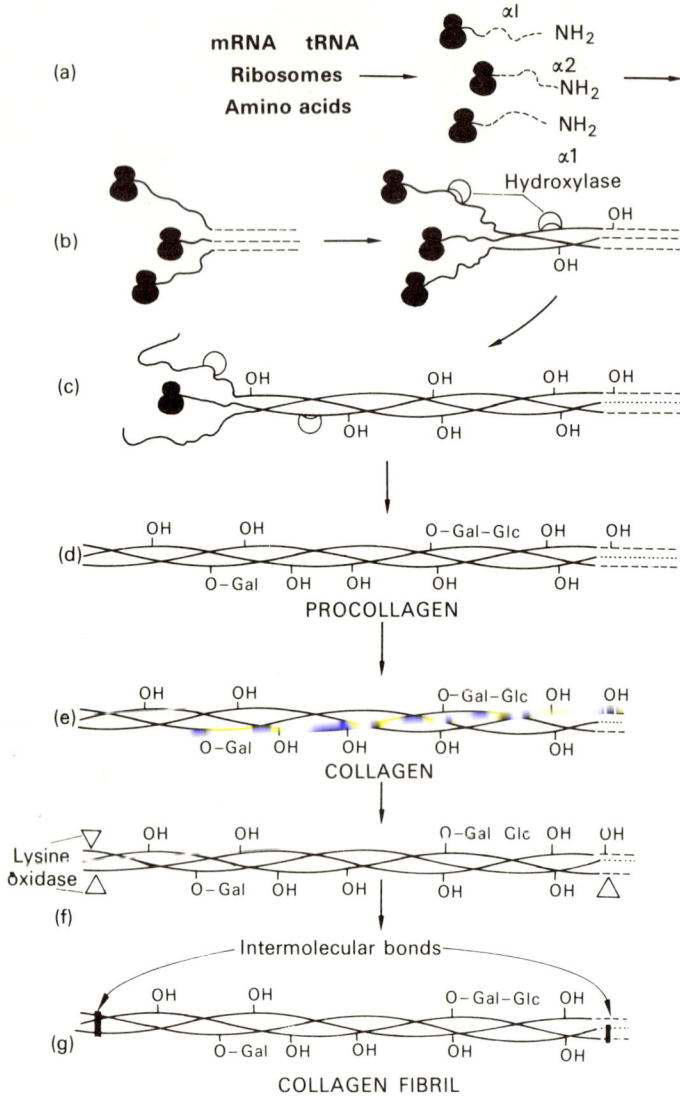

FIG. 16.7. Stages in the synthesis of collagen fibrils. The solid symbols represent ribosomes, the open symbols proline hydroxylase and lysine ε-amino oxidase. Steps (a)–(d) take about 6 minutes in vivo.

Quaternary structure. Assembly of sub-units may be spontaneous, but there are clearly differences. The sub-units of lactic dehydrogenase will hybridize, but those of haemoglobin will re-assemble into the original protein even in the presence of sub-units of Hb from a closely related species of animal. Further, in animals, proteins which are exported from the cell (e.g., in secretory tissues) are often deposited in small granules, Golgi bodies, and the like, which seem to be surrounded by a phospholipid membrane. Such proteins are often synthesized by ribosomes attached to the highly lipoid endoplasmic reticulum. It is not known whether the proteins in these 'packages' have their normal conformation.

Inhibitors of Protein Synthesis

Action on aminoacyl-tRNA synthetase. Unlike most other amino acid analogues *5-methyltryptophan* competitively inhibits the formation of

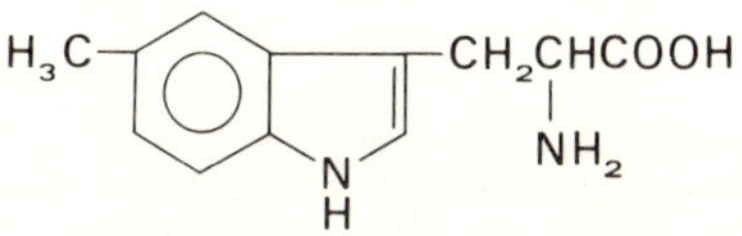

5-Methyltryptophan

tryptophany-tRNA, thereby preventing the synthesis of all tryptophan-containing polypeptides.

Action on ribosomes. Members of the *tetracycline* family, antibiotics synthesized by various *Streptomyces* and their derivatives, inhibit the

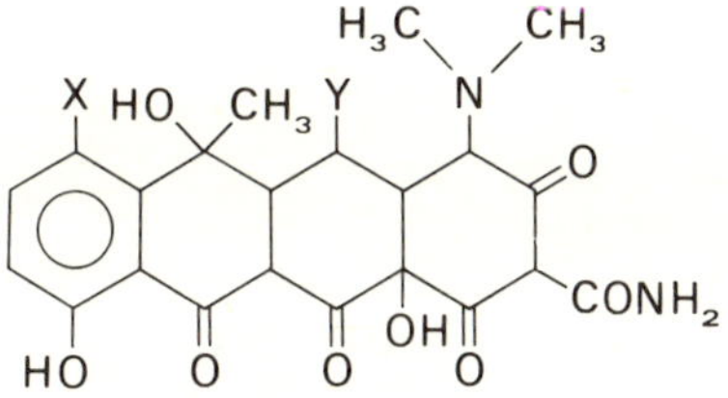

Tetracyclines

tetracycline X = Y = H
chlorotetracycline (aureomycin) X = Cl, Y = H
oxytetracycline (terramycin) X = H, Y = OH

binding of aminoacyl-tRNA to the acceptor site of the ribosome (see Figs 16.1 and 16.3). The drug *edeine*, on the other hand, blocks the donor site.

Chloramphenicol (chloromycetin), an antibiotic, originally obtained from *Streptomyces* but now produced synthetically, is bound by the 50S

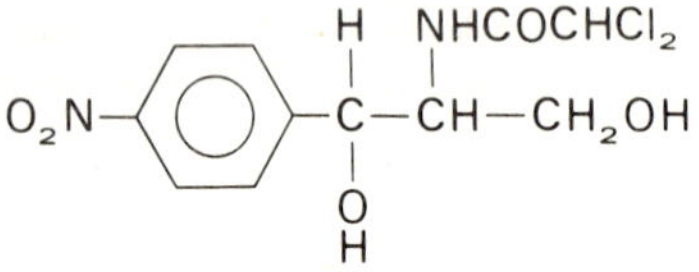

Chloramphenicol

ribosomal sub-units of sensitive 70S ribosomes of bacteria at very low concentrations. It appears to interfere with peptide bond formation on the 70S ribosome mRNA complex, perhaps by causing breakdown of one of the initiation factors. It also inhibits protein synthesis by mitochondrial ribosomes, so it can be toxic to animal cells in high concentrations.

Cycloheximide (actidione), a powerful fungicide antibiotic from *Streptomyces*, and related glutarimides act in a similar way to chloramphenicol but bind to 80S (animal, yeast) ribosomes and not to 70S (bacterial ribosomes).

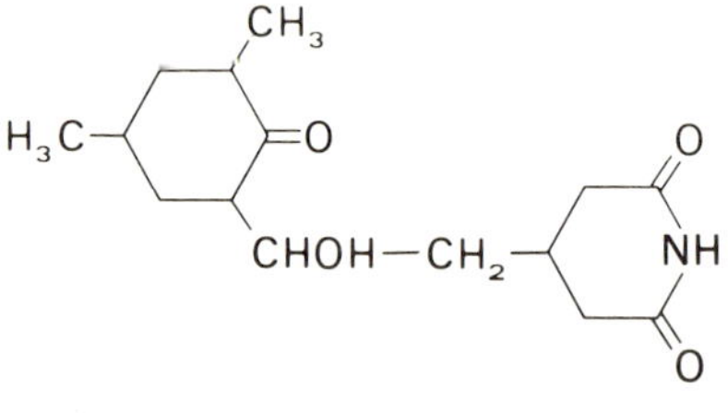

Cycloheximide

Streptomycin and related aminoglycoside antibiotics such as *kanamycin* bind to the 30S ribosomal sub-units from sensitive bacteria and lead to faulty translation of the mRNA. They also induce release of f-Met-tRNA from intact 70S ribosomes.

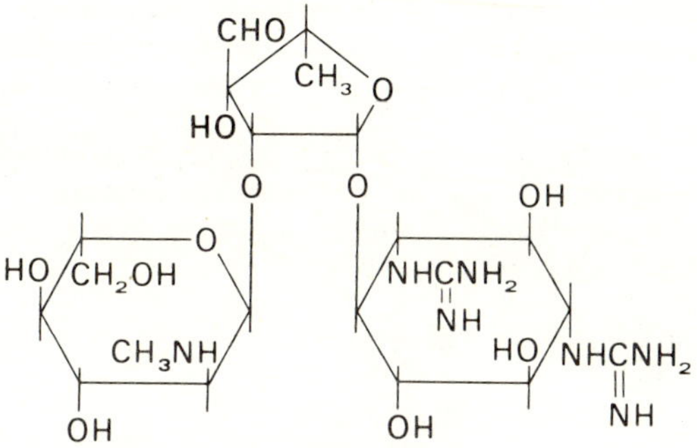

Streptomycin A

Puromycin has a structure which is similar to the aminoacyl end of an aminoacyl-tRNA molecule. It replaces an incoming aminoacyl-tRNA on the ribosome during polypeptide synthesis. A peptide bond is formed by

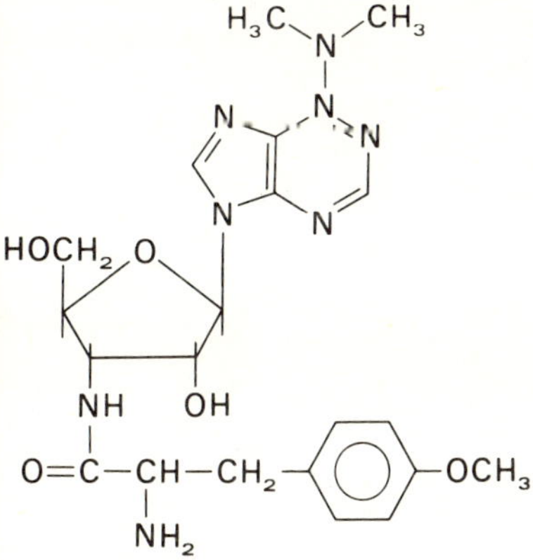

Puromycin

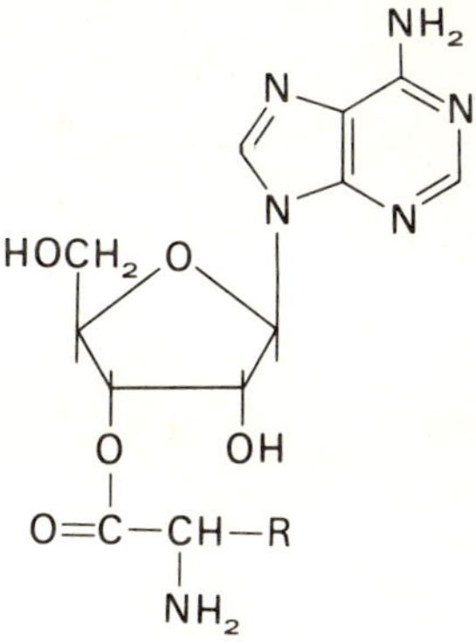

Terminal adenosine of an aminoacyl-tRNA molecule

the transfer of the polypeptide chain already formed on the ribosome to the free —NH$_2$ group of puromycin. The polypeptidyl-puromycin thus formed does not remain bound to the ribosome but is released, so that puromycin becomes the *C*-terminus of the incomplete polypeptide chain.

This antibiotic has become an important tool in research, both in elucidating the fine structure of the peptide-synthesizing site (Fig. 16.3), and also in identifying biological effects dependent on protein synthesis, but not on prior synthesis of new mRNA (i.e. not sensitive to actinomycin D, p. 337).

Diphtheria toxin acts by inactivating the T2 translocation factor of eukaryotic cells (see Fig. 16.4), but not the similar protein of prokaryotes. It does this by catalysing the transfer of adenosine-diphosphate-ribose from NAD (formula p. 140) to a site on the protein.

T2 + NAD $\rightarrow$ T2-adenosine-diphosphate-ribose + nicotinamide

17 Control Mechanisms

The Control of Protein Synthesis

In principle this can be exerted through control either of transcription or of translation of nucleic acids. Earlier work concentrated on control of protein synthesis by variations in the rate of formation of messenger RNA, but there is now evidence that the concentration of one or another species of tRNA, or of the ribosomal nucleic acids, may sometimes be rate-limiting. The latter seems especially likely in animal cells, where the life-time of ribosomes is on average probably not greater than that of mRNA. The rapid synthesis of protein in fertilized eggs appears to depend on a high concentration of ribosomes whose RNA was pre-formed, and stored in the nucleolus until fertilization.

The control of transcription, particularly in eukaryotes, is in itself not a simple matter. Initiation factors probably govern the start of transcription, and a sigma factor is essential for the binding of DNA to RNA polymerase (p. 330), while the concentration of the polymerase itself, and in particular the relative amounts of the β and β' sub-units, are under feedback control, and this may be rate-limiting. In addition, it appears that in nucleated cells a short stretch of poly-adenylic acid (see p. 334) is attached to the 3'-end of most newly formed mRNA, and this in some way slows transport of the mRNA out of the nucleus, and may also protect it in the cytoplasm. Histone mRNA does not contain poly-A, and is both very rapidly translated and (by eukaryote standards) short-lived. Moreover, some newly formed RNA is cut into segments by endonucleases in the nucleus or nucleolus. Thus in eukaryotic cells there are many mechanisms by which the rate of transcription of individual segments of DNA, or the conversion of the products to fully effective mRNA, may be varied. The nucleus of a single animal cell may also contain several thousand copies of certain stretches of DNA. So far as is known at present, these are either non-translated or code for ribosomal RNA, so that multiple DNA copies relate to the overall velocity of protein synthesis rather than to specific polypeptides.

In bacterial cells, on the other hand, the synthesis of enzymes often proceeds either at a maximal rate (given that amino acid supply is not limiting) or not at all, and bacterial enzymes are divided into *constitutive* (those whose synthesis is always occurring) and *inducible* (those whose presence can only be detected in certain conditions). Many lines of evidence have shown that when inducible enzymes cannot be detected in bacterial cells, it is because their synthesis has been indirectly prevented by blocking the synthesis of the corresponding mRNA (which is generally short-lived in these organisms). The blockage is effected by a *repressor* molecule, which binds to the transcribable DNA chain at the point corresponding to the beginning of an mRNA segment. The few repressor molecules which have been purified to date are relatively small proteins. Certain small molecules, usually substrates or related to substrates, can bind specifically to a repressor molecule and (presumably) alter its conformation so that it no longer binds to the DNA. Messenger RNA synthesis can then begin at this point, followed by translation and synthesis of the enzyme protein. These small molecules are known as *de-repressors*; alternatively, since they induce enzyme formation, they may be called *inducers*. This kind of control is a negative one; there is evidence that substances called *promoters* may bind to *promoter sites* in the neighbourhood of the repressor sites. Not much is known in general about the chemical nature of promoters, but the full mRNA-directing activity of the *lac* operon (see below) is inhibited if a 'catabolite-gene activator' protein is not available. To be effective, the protein must have bound to itself cyclic-AMP. This cAMP-protein appears to be a rather general promoter, and nothing comparable may exist in mammalian cells.

These ideas are rendered more complex by the observation that a single de-repressor may frequently induce the synthesis of more than one enzyme. The classical case is that of lactose and other β-galactosides, which induce the coordinated formation of three enzymes concerned with lactose metabolism, namely galactoside permease, galactosidase and thio-galactoside acetylase (see Fig 17.1). This is explicable if the genes for the three enzymes are juxtaposed, and if the mRNA for all three is transcribed at one time. Then a single repressor/promoter site will serve to control the synthesis of all these proteins. Genetic mapping shows that the genes are indeed contiguous, and the hypothesis becomes plausible. Such a set of genes operating as a unit is known as an *operon*; the individual genes are sometimes called *cistrons*, and the RNA as *polycistronic mRNA*. These ideas are expressed diagrammatically in Fig. 17.1. Many examples of such control are known in bacteria and fungi.

There is much less need for enzyme induction of this all-or-none kind in animal cells, living in the constant environment provided by the *milieu intérieur*. It is difficult to find evidence for operons, except for the ribosomal RNAs, which are synthesized in a single strand that is later cut. In some

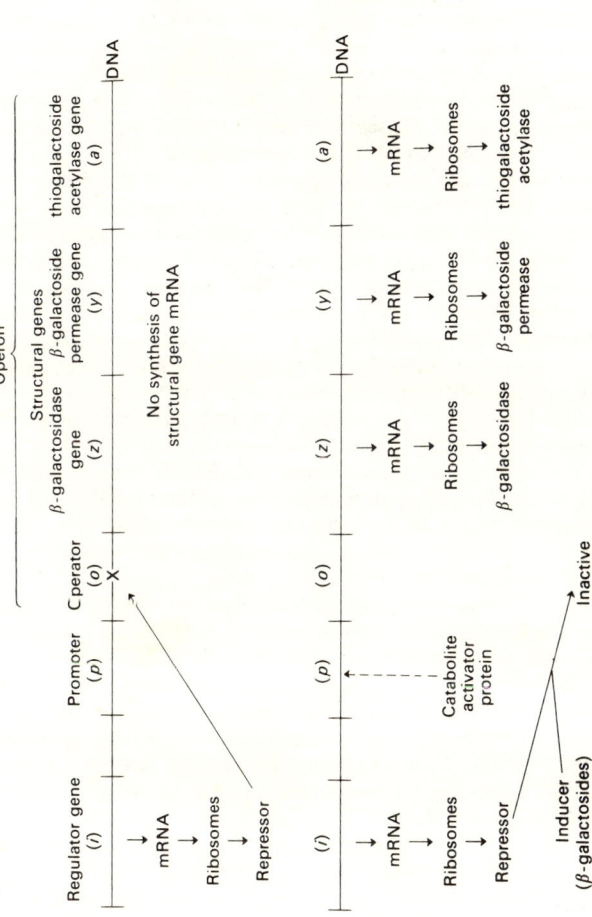

FIG. 17.1. Schematic representation of the control of synthesis of lactose enzymes in *E. coli.* (*Top*) no inducer present, repressor combines with operator and prevents enzyme synthesis. (*Bottom*) inducer present, repressor inactivated, enzyme synthesis proceeds. The catabolite activator protein permits full expression of the operon if it is bound to the promoter region.

instances genes for sub-units of single proteins which might be expected to occur together, e.g., for the α and β sub-units of haemoglobin, occur on separate chromosomes. In *thalassaemia* synthesis of the β-chains is severely depressed, while synthesis of the α-chains continues at a normal rate. Although the genetic evidence is difficult to interpret, it does not point to failure of control by an operon.

There is by now good evidence for repressor/depressor control of several animal proteins. The replacement of HbF by HbA soon after birth (which involves suppression of γ-chain synthesis and initiation of β-chain synthesis), the rapid increases in concentrations of several enzymes (especially in liver) at this time, and the very rapid synthesis of a few proteins in response to hormones (particularly oestrogens) are consistent with such a mechanism. In nocturnal-feeding animals (rats) the synthesis of HMG-CoA-reductase, a key enzyme in cholesterol biosynthesis, appears to be limited to a few hours around midnight. For many enzymes, however, the changes in rate of synthesis of the enzyme protein appear to be relative rather than absolute, for example acetyl-CoA carboxylase, or the microsomal hydroxylase (p. 269) whose activity in liver is increased by many substrates, e.g., phenobarbitone. It is interesting that changes in the rate of synthesis and degradation of the former enzyme, and probably of several others, are deleted in minimal deviation hepatomas.

It is now generally agreed that in animal cells, at all events, control of *translation* is just as important as control of transcription in determining the rate of protein synthesis. Experimentally, the strongest evidence is provided by showing that variations in protein synthesis continue in the presence of *actinomycin D* or *cycloheximide*, drugs which inhibit DNA-directed RNA synthesis, but this evidence must be interpreted with care, because actinomycin D, in particular, has toxic side-effects.

Mechanisms which would provide for control of translation have been elucidated. In Chapter 16 it was shown that several cytoplasmic protein factors are required for the initiation, elongation and termination of peptide chains. The ribosomal sub-units dissociate from the messenger after each termination codon has been read, and the stretches of non-translated RNA before each initiation codon may affect the ease with which the sub-units are re-bound to the mRNA. In addition, it is known that some aminoacyl-tRNA species bind less readily than others to the active ribosomes.

There is also the possibility of changes in the activity of the aminoacyl-tRNA synthetases, and the certainty of variations in the availability of amino acids. In protein malnutrition, unfinished peptide chains are apparently lost from ribosomes, and are certainly excreted (p. 413). Experimentally, much work has shown that the availability of tryptophan, in particular, markedly affects the efficiency of protein synthesis in liver, and this agrees well with other evidence showing that protein synthesis is greatly reduced if tryptophan is fed separately from the other amino acids. The effect appears not to be due solely to the concentration of tryptophanyl-tRNA, but also to a direct effect of the amino acid on ribosomal efficiency. It is possible that in other tissues other amino acids may exert the same kind of control, and thus control of translation depends to some extent on the availability of amino acid substrates. It may also depend on the availability of mRNA, since in the cytoplasm of animal cells much RNA may be temporarily stored in small protein-containing granules (sometimes called 'informosomes').

Degradation of Enzymes

In rapidly-growing bacteria, enzymes whose synthesis is repressed are simply diluted out of the culture as a whole, rather than destroyed. In animal cells, the random inactivation and degradation of enzyme protein is much more important. A simple mathematical analysis shows that if the rate of synthesis of an enzyme is K_1, while the first-order rate constant for degradation is K_2, the concentration of the enzyme in any steady state is K_1/K_2, but the time taken to attain the steady state *depends only* on K_2. Thus even when complete repression reduces K_1 to zero, the enzyme will only slowly disappear if K_2 is small. It is presumably this that makes it difficult to provide unequivocal evidence for repression and derepression in animal cells. The 'half-life' of hepatic enzymes varies over a very wide range from about 80 minutes to over a week, and there must be correspondingly wide variations in the rates of irreversible inactivation. Sometimes the substrate of an enzyme protects it against degradation, as tryptophan protects tryptophan pyrrolase (p. 240), thus causing an apparent induction

A single example of the importance of this effect must suffice. In rats on a normal or fat-free (high-carbohydrate) diet, K_2 for liver fatty acid synthetase is about 0·25 $(t_{1/2} = 2\cdot8$ days$)$, while in liver from fasted rats it is about 1·0. On the other hand, the rate of synthesis of the proteins of the enzyme complex is six times greater in liver from rats fed a fat-free diet than in liver from fasted rats. The ratio of synthetase activities between fasting and a fat-free diet ought thus to be:

$$[E]_{FF}/[E]_S = \frac{K_2'}{K_1'} \cdot \frac{K_1}{K_2} = \frac{1}{1} \times \frac{6}{0.25} = 24$$

A 20-fold difference in activity was found experimentally.

The constant K_2 is related to the *system time constant*, well known in physiological and engineering theory of control mechanisms. Similar considerations to those developed above apply to concentrations of chemicals that may act as allosteric effectors, hormones or intracellular 'messengers' (Chapter 18). That is to say, unless the rate of removal or destruction of the chemical is large in comparison with its concentration at a given site, the rate at which its concentration changes may be very slow, even if the rate of synthesis alters greatly, and the usefulness of the substance as an effector of the process it is supposed to control may be very small.

The Control of Enzyme Activity

Mechanisms which affect enzyme activity, without altering the concentration of enzyme protein, are:
1. Hyperbolic substrate binding.
2. Binding of activators and inhibitors.
3. Allosteric binding of (a) substrates; (b) other effectors.
4. Covalent changes to the enzyme.

Hyperbolic Substrate Binding

All enzymes can be saturated with their substrates, and an increase in substrate concentration does not then perceptibly increase the rate of reaction. This is often the case for the nucleotide substrates ATP and NAD, but the concentrations of most intracellular reactants are such that the enzymes which act on them are working in the first-order region of the Michaelis–Menten curve (Chapter 7). In laboratory conditions, if the substrate concentration of a single-substrate enzyme is low by comparison with the Michaelis constant, the velocity of reaction is proportional both to amount of enzyme and to substrate concentration, i.e.

$$v = [E_t] \cdot k_3 \cdot [S]/(K_m + [S])$$

reduces to

$$v = [E_t] \cdot [S] \cdot k_3/K_m$$

but with a two-substrate enzyme in which the concentrations of both substrates are much less than their K_ms, doubling the concentration of one substrate alone will increase v by a factor much less than 2, although doubling $[E_t]$ will still double the rate of reaction. However, within cells the concentration of enzymes is often equal to or greater than that of the substrates, and it is then difficult to predict the results of changes in concentrations of reactants.

Most enzymes are in principle reversible, so that the 'products' bind to the enzyme, as they are substrates for the reverse reaction. Any molecule of product which binds to an enzyme reduces the amount of enzyme available to catalyse the forward reaction. Such *product inhibition* can be important in determining metabolic rates, e.g., the coenzyme products ADP and $NADH_2$ are present intracellularly in low and non-saturating concentrations, so that small changes in the amounts present may cause large changes in their effectiveness as inhibitors.

Binding of Activators and Inhibitors

Many enzymes are affected by specific activators, particularly metal ions. Ca^{2+} is an activator for enzymes of muscle contraction, and release and sequestration of ionized calcium by the sarcoplasmic reticulum triggers the contraction-relaxation cycle in muscle. Many more enzymes, especially those for which a phosphate compound is a substrate, are activated by Mg^{2+}, which works in two ways. It is strongly bound to ATP, and the true substrate of many phosphorylating enzymes is not ATP^{4-}, but $MgATP^{2-}$. In addition the metal activates directly a few enzymes such as pyruvate kinase. Much intracellular Mg is bound to nucleotides, and the concentration of the free ion is low enough for changes in it to be of importance, as for example in the control of hexokinase activity.

Allosteric Binding

The activity of a small but important group of enzymes is affected by binding of a molecule (called a *ligand*) to a site distinct from the catalytic site whose activity is under consideration. Such enzymes are called *allosteric* enzymes. Several possibilities may be distinguished.

Homotropic effectors. In the simplest case, the ligand is a second (or subsequent) molecule of substrate. Many enzymes have several active sites per molecule, typically one per sub-unit (polypeptide chain), see Chapter 7. Usually events at one active site either do not affect, or affect in a very limited way, events at another. In the small class of allosteric enzymes, when a ligand (in this case a substrate) is bound at one site the affinity at other sites changes markedly, because changes in the conformation of the sub-unit that has bound the ligand induce changes in the other polypeptides. Changes in affinity of substrate show up as changes in K_m (*K-type effects*). Changes in maximum velocity—*V-type effects*—are much rarer (but see adenyl cyclase, p. 374).

Very often binding of substrate at one site promotes binding at others. This is known as *positive cooperativity*, and the effect is to give the velocity–substrate curve a sigmoid shape (Fig. 17.2a). The binding of O_2 to haemoglobin also shows positive cooperativity (see Fig. 6.7, p. 112), while binding to myoglobin, which has only one sub-unit, does not. Note that the classical Michaelis–Menten equation does not apply to allosteric enzymes, and K_m no longer has any meaning; it is usually replaced by S_{50}—the substrate concentration at which 50 per cent of maximal activity is elicited.

Not infrequently binding of substrate to one catalytic site is found to hinder binding at another (usually not by a direct steric effect); this is known as *negative homotropic cooperativity*. In the biologically most interesting cases the velocity–substrate curve reaches a maximum and then declines (Fig. 17.2b).

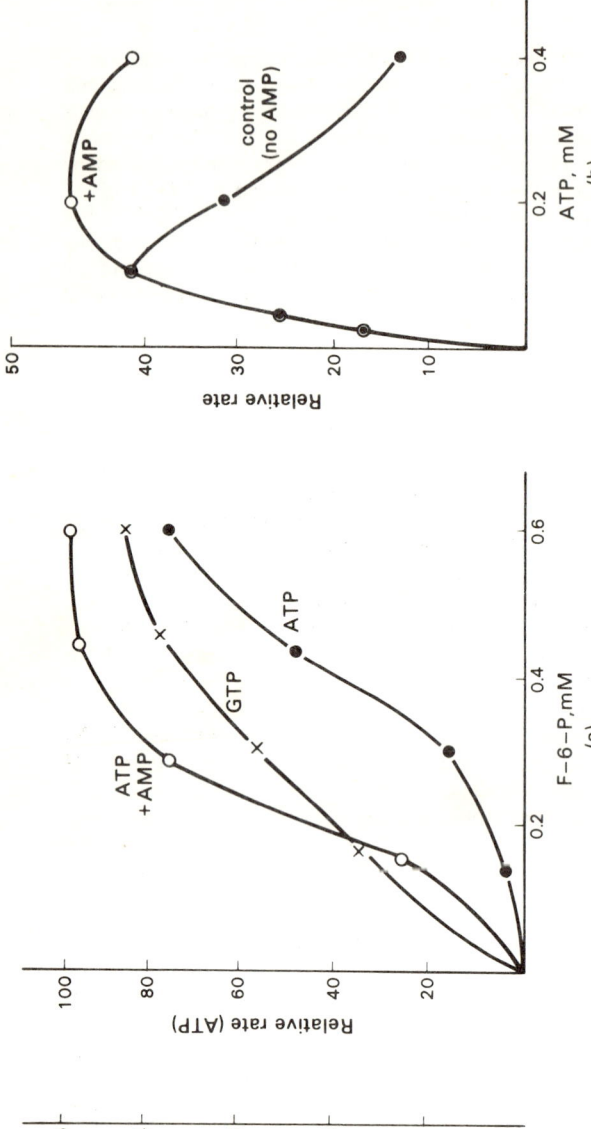

FIG. 17.2. Allosteric effects on phosphofructokinase. (a) shows the velocity/activity curve at fixed concentrations of nucleotides when the F-6-P concentration varies. Note that the activity curve is abnormal (sigmoid) with respect to F-6-P, although it is ATP which is the allosteric effector. The curve drawn through the open circles shows that the simultaneous presence of AMP makes the curve much less sigmoid. Although GTP is a much less effective phosphoryl donor than ATP, note that the curve is almost normal, implying that GTP is not found at the second site; (b) shows the very marked inhibition of the enzyme at physiological concentrations of ATP, and the extensive relief of the inhibition by AMP.

The substrate concentrations were (a) ATP or GTP, 0.2 mM; (b) F-6-P, 0.3 mM. These concentrations are not necessarily similar to those which might be found in a cell.

Table 17.1
Some Allosteric Enzyme Systems

Enzyme	Pathway	Activators	Inhibitors
Haemoglobin			Diphospho-glycerate
Aspartate transcarbamoylase (*E. coli*)	Pyrimidine synthesis	Aspartate, ATP	CTP
L-Threonine deaminase (*E. coli* and yeast)	Isoleucine synthesis	Valine	Isoleucine
Deoxycytidylate aminohydrolase	Cytidylate → thymidylate	Deoxy-CTP	Deoxy-TTP
Glutamate dehydrogenase	Glutamate ⇌ α-oxo-glutarate	ADP, leucine, methionine	ATP, GTP, NADH₂, oestrogens
Phosphorylase *b*	Glycogen breakdown	AMP, G-1-P	ATP
Glycogen synthase	Glycogen synthesis	G-6-P	
Phosphofructokinase	Embden–Meyerhof	AMP, cAMP ADP, F-6-P	ATP, creatine phosphate, citrate
Fructose diphosphatase	Gluconeogenesis		AMP
Pyruvate kinase	Embden–Meyerhof	FDP	
Pyruvate carboxylase	Oxaloacetate synthesis	Acetyl-CoA	
NAD-isocitric dehydrogenase	Citric acid cycle	ADP	
Acetyl-CoA carboxylase	Fatty acid synthesis	Citrate	

Heterotropic effectors. In the case of phosphofructokinase, Fig. 17.2*a* shows that F-6-P sites do not interact with one another when GTP is the co-substrate, but they do when ATP is present. The explanation is that ATP binds both it its substrate site, and also to a second non-catalytic or *regulatory site*. Since it is altering the behaviour of the enzyme with respect to the substrate F-6-P, and not itself, it is *heterotropic*. Most heterotropic ligands binding to regulatory sites are unrelated to substrates or products, but they can often be identified as terminal products of the metabolic pathway in which the enzyme lies, thus introducing the possibility of feedback control of that pathway.

Negative feedback effectors must increase the sigmoidicity of the velocity–substrate curve, which in effect inhibits the enzyme at low substrate concentrations. It is, however, not uncommon for heterotropic effectors to transform a sigmoid curve into a hyperbolic one (presumably by hindering the cooperative change in conformation produced by the substrate). In this case they are activators; an example is the activation of pyruvate kinase by FDP (Table 17.1).

Two theories have been advanced to explain the observations described above, they apply to enzymes (or proteins such as haemoglobin) having more or less identical sub-units. Both imply reversible conformational changes in each sub-unit from a less active (high S_{50}) to a more active form. The *concerted transition* theory requires that binding of an effector to one sub-unit initiates a change of conformation in all sub-units; no intermediate states are possible. The *induced fit* hypothesis predicts that each sub-unit may change conformation separately. Experimental evidence for both hypotheses exists; for example, yeast triose phosphate dehydrogenase may be described by concerted transition, while the analogous enzyme in muscle shows induced fit.

Some regulated enzymes have sub-units which are strikingly different from one another; in particular, one unit may have no catalytic site at all, although it has a binding site for a heterotropic effector. The classic example of an enzyme having a *regulatory sub-unit* and a *catalytic sub-unit* is *aspartate transcarbamoylase* (p. 93). When various terminal products of pyrimidine nucleotide synthesis are bound to the regulatory sub-unit, the activity of the catalytic sub-unit is inhibited. *Protein kinase* is regulated in a different way; when cyclic AMP binds to the regulatory sub-unit, the latter dissociates from the catalytic unit, which is then enabled to function. If the regulatory unit loses its cAMP, it (presumably) changes conformation and can re-associate with the catalytic unit.

Changes in conformation in allosteric enzymes are not necessarily instantaneous; characteristic time lags ranging from a few seconds to almost an hour have been observed.

Covalent Changes in the Enzyme Molecule

A phosphorylation/dephosphorylation cycle has been found to regulate the activity of several important enzymes of metabolism (see the control of glycogen metabolism, below). In every case the hydroxyl of one or more seryl residues in the molecule is phosphorylated, the donor being ATP, and there is a subsequent change in the conformation of the molecule. The seryl phosphates may be hydrolysed, with release of P_i, by a phosphatase.

Not much is known about the phosphatases, but the *protein kinases* have been much studied in recent years. Some of them are highly specific, like dephospho phosphorylase phosphokinase, and pyruvate dehydrogenase kinase, which attack only the protein of a specific enzyme. Other proteins, however, are phosphorylated by a general cellular protein kinase, or a small family of such enzymes. One in particular has excited interest because it is activated by cAMP, indeed the protein kinase appears to be part of the mechanism by which cAMP acts as a 'second messenger' within the cell. As already briefly mentioned above, the cAMP-sensitive protein kinase is an allosteric enzyme, the postulated mechanism is as follows:

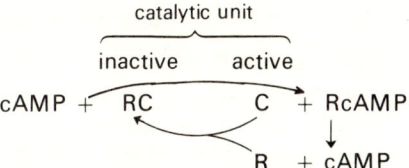

RcAMP is otherwise known as 'the cAMP-binding protein'.

The enzyme does not have a high turnover number, and only one or two of the available serines in each protein are phosphorylated. It is not at all specific; casein, and also histones, are substrates. The latter finding has led to speculation that protein kinase is linked in some way to gene expression.

A peculiarity of enzymes regulated in this way is that sometimes phosphorylation activates and dephosphorylation inhibits, and sometimes vice versa. A summary of present knowledge with regard to several well-known enzymes is given in Table 17.2.

Table 17.2

Enzyme	Phosphorylating enzyme	Effect of phosphorylation on activity	Sensitivity to cAMP
Phosphorylase b	phosphorylase b kinase	+	−
Phosphorylase kinase	protein kinase	+	+
Glycogen synthase	protein kinase	−	+
Pyruvate dehydrogenase	specific kinase	−	−
Fructose diphosphatase		−	
Adipose tissue lipase	protein kinase	+	+

The phosphoryl group is not the only radical that can be covalently bound to an enzyme to affect its activity (cf. the inactivation of the T2 protein, or peptidyl-tRNA translocase, by diphtheria toxin, p. 354).

Other Protein–Protein Interactions

Enzymes are known which have two quite disparate catalytic activities. An enzyme widely distributed in animal cells catalyses the transfer of a galactosyl residue to N-acetylglucosamine; the product is used in the formation of mucopolysaccharides. When to this 'B' protein is bound an 'A' protein, the enzyme becomes specific for glucose as an acceptor, i.e. it becomes a lactose synthetase. The 'A' protein is α-lactalbumin, found in milk, and synthesized only in milk-producing mammary tissue.

Control of Metabolism by Translocation

Cell membranes are not freely permeable to many substances, with liver cells being more permeable and nerve cells less permeable than the average. Perhaps the majority of nutrients have to be transported into the cell by 'translocases' or 'permeases', often with an associated expenditure of energy. Enzymes which initiate or regulate metabolic pathways are often associated with cell or other membranes—an example is the scavenger mechanism for purines (p. 100) which appears to be strongly associated with the cell membrane.

Within the cell, particular metabolites may be restricted to the cytoplasm, or to one or other of the organelles in the cell. Alternatively, the distribution of the enzymes may be such that a complex mechanism for transporting metabolites has to be envisaged to account for the metabolic activity known to occur. Examples relating only to mitochondria and cytoplasm are the absence of lactic dehydrogenase and the fatty acid synthetase complex from the mitochondria, and the impermeability of the mitochondrial membrane to CoA and to $NADH_2$. The ratio $NADH_2/NAD$ in the cytoplasm is far lower than it is in mitochondria, and this may have a marked effect on cytoplasmic gluconeogenesis (see p. 169), without necessarily affecting the rate of mitochondrial transport of reducing equivalents to oxygen.

General Principles and Examples

Most enzymes, even rate-limiting ones, are present in cells in concentrations far higher than necessary to support the flux of reactants through their metabolic pathways. This can be shown, in general, by recalling that 30 per cent of the protein of the liver (overwhelmingly enzymic) may be lost in a few days of fasting, and in particular by considering inborn errors affecting urea synthesis (see Chapter 11). Subjects are known with deficiencies of ornithine transcarbamoylase or argininosuccinic acid synthetase (normally the rate-limiting enzyme), or argininosuccinase. In the latter defect only 5 per cent of the normal activity of the enzyme is present. However, in each inborn error, urea is apparently formed at the normal rate although the precursor for each defective enzyme may accumulate in plasma and urine, especially after a heavy protein meal. Thus the phrase 'rate-limiting' must be used with care; it may only mean 'rate-limiting at the low concentrations of intermediates usually found in cells'. If these intermediates cannot readily escape from the cells the compensatory mechanism discussed on p. 360 may come into play.

Four general mechanisms for controlling the rates of enzymic pathways without altering the concentration of enzyme protein may be mentioned:

supply and demand, negative feedback, positive feedforward, and cascade (amplification) effects.

Supply and demand control is best exemplified by considering the control of intracellular energy production. It has already been pointed out in Chapter 12 that when mitochondria from vertebrate tissues are studied in vitro, the rate of O_2 consumption depends on the rate at which ADP and P_i (inorganic phosphate) are made available, i.e. oxidation and phosphorylation are tightly coupled. Only if these two substances are present in optimum concentration will the rate of O_2 uptake become dependent solely on the amount of substrate available, and of course on the activity of the various dehydrogenases and electron-transferring enzymes. If excess substrate is added to fresh mitochondria, with excess P_i but a fixed amount of ADP, it is possible to titrate the phosphate acceptor by means of the increase in oxygen uptake (Fig. 17.2). A similar experiment can be done with excess ADP, and a fixed amount of P_i. Alternatively, if an enzyme which uses ATP, together with its substrate, is added to the system just described, the O_2 uptake can be made continuous.

In cells so far examined essentially all the oxygen-utilizing enzymes, i.e. those of the respiratory chain, are found in the mitochondria, so that it is reasonable to expect that the rate of O_2 uptake by cells and tissues is controlled in much the same way. We know that ATP is required by many endergonic processes. If this hypothesis is correct ATP would act as the

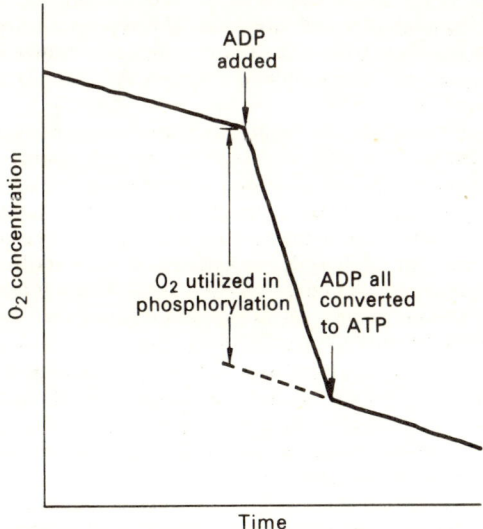

Fig. 17.3. 'Titration' of ADP using mitochondria.

immediate energy donor for the cell as a whole, while the ADP, arriving at the mitochondrial electron transport chain, would represent the integrated demand for oxidation to supply that energy, and its rate of arrival would control the actual rate of oxidation. Something like this must indeed occur; it is striking that in contrast to micro-organisms, the oxygen consumption of man is, apart from the energy required for digestion, independent of the level of nutrients in the bloodstream.

It is not, however, known whether the coupling of oxidation to phosphorylation is more or less efficient in vivo than it is in freshly prepared mitochondria in vitro. There may be occasions, as for instance in the protection of the body from cold, when an increased rate of direct heat production is required, with less regard to the efficiency of energy trapping. An alternative way in which this could be achieved is by the activation of enzymes hydrolysing ATP as it is formed. Such *ATP-ases* are widely distributed in cells, with no other known function. Shivering as a response to cold is another version of this process, using the actomyosin of skeletal muscles as an ATP-ase. There is, nevertheless, good evidence that the efficiency of oxidative phosphorylation may be under direct control in vivo. The varying efficiency of energy utilization in animals with various states of thyroid function (p. 387) provides one example of this.

Negative feedback is a very common type of control, in which a proximate or distant product of a reaction inhibits the enzyme which catalyses the reaction. Thus as the concentration of the product falls, the enzyme is released from the inhibition, and vice versa. This type of control is stable, unlike positive feedback, where the product stimulates its own formation, leading to an inherently uncontrollable situation. An example of negative feedback is shown in Fig. 17.4.

Positive feedforward, on the other hand, leads to departures from a steady-state which is quickly regained. In the best-known example, fructose diphosphate activates pyruvate kinase, the terminal enzyme of the glycolytic pathway. This facilitates the passage of a 'slug' of hexose carbon down the glycolytic pathway both directly and by providing an acceptor (pyruvate) for the $NADH_2$ produced in a reaction (the oxidation of triose phosphate) which could otherwise become rate-limiting in the pathway. The process cannot get out of control, however, because the concentration of FDP must ultimately fall, and pyruvate kinase return to its more inhibited state. This type of control helps to overcome the problem of slow changes in the concentrations of intermediates mentioned on p. 360.

In both negative feedback and positive feedforward, there may be a delay before the concentrations of effectors build up, which may lead to oscillations round a steady-state.

Cascade processes are sequences of catalysed reactions so arranged that the product of one reaction is the catalyst or activator for the next. Since each molecule of catalyst may easily produce a hundred molecules of the next activator, it is easy to see that an amplification of 10^6 may result from only a few stages. The *blood-clotting* process (Chapter 14) is an excellent example of cascade amplification. The control of glycogen breakdown and synthesis provides another example, which also shows features of control by covalent modification of enzymes and by effector ligands (discussed earlier in this chapter). For enzymes involved in glycogen metabolism see Chapter 9, p. 170.

It has been known for many years that glycogen phosphorylase exists in two forms, one effective in the absence of activators, the other (phosphorylase *b*) requiring 5′-AMP for activity. The latter is converted to phosphorylase *a* when seryl residues in it are phosphorylated by a *specific* phosphorylase *b* kinase and, as a result of this covalent change, the molecules aggregate to form dimers or tetramers. The changes are reversed by a specific phosphatase. Thus we have in muscle

$$2 \text{ phosphorylase } b \xrightarrow[\text{kinase}]{2\text{ATP}} \text{ phosphorylase } a + 2\text{ADP}$$

and in liver

$$4 \text{ phosphorylase } b \xrightarrow{4\text{ATP}} \text{ phosphorylase } a + 4\text{ADP}$$

and in the reverse direction

$$\text{phosphorylase } a \rightarrow 2(4)\text{phosphorylase } b + 2(4)\text{P}_i$$

The cascade effect arises because phosphorylase *b* kinase is itself activated and inactivated by addition and removal of a covalent phosphate group. The agent in this case is the generalized cyclic AMP-sensitive protein kinase (p. 365). When one takes into account the amplification involved in the activation of adenyl cyclase by adrenaline or glucagon, a five-stage amplification may be achieved (Fig. 17.4).

There are various partial shunts in this cascade. The activation of glycolysis on stimulation of muscle in vitro is faster than can be accounted for by activation of the cAMP-sensitive protein kinase, and it is suspected that the increase in concentration of intracellular Ca^{2+} as a result of stimulation is sufficient to activate phosphorylase *b* kinase without phosphorylation of its seryl residue(s). A more comprehensive shunt would be provided by 5′-AMP activation of phosphorylase *b*, but ATP competes with AMP for this allosteric effector site, and one would predict that in vivo a favourable AMP/ATP ratio would only arise in anaerobic conditions. However, inherited deficiency of phosphorylase *b* kinase, both in humans and in mice, does not prevent slow mobilization of glycogen.

Glycogen synthase (UDPG-glycosyl transferase) also exists in two forms. One, known as the 'dependent' (D) form, requires glucose-6-phosphate for activity, but the intracellular concentration of G-6-P is

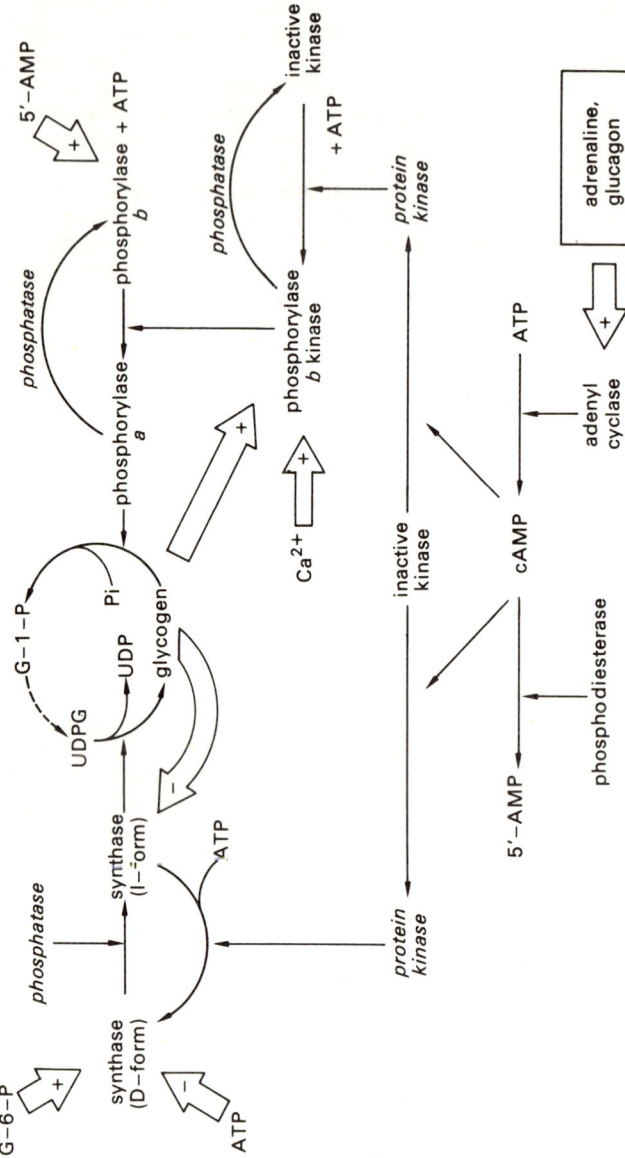

Fig. 17.4. Stages in the control of glycogen synthesis and degradation. Stages in the amplification (cascade) process are shown by the vertical arrows, starting from the bottom of the page. Feedback and heterotropic effector controls are indicated by the broad arrows.

normally too low for the D-form to be effective. The 'independent' (I) form of the enzyme does not require G-6-P; this form is converted to the (inactive) D-form by phosphorylation and the synthase is reactivated by a phosphatase. Thus a covalent activation/inactivation process is involved in the opposite sense to the phosphorylase *a* cascade. The enzyme converting the I-form to the D-form is the cAMP protein kinase, so that the amplification is less great than for phosphorylase; nevertheless, it is clear that in a given tissue at any point in time, the concentrations of catalysts and effectors must either favour glycogen synthesis at the expense of degradation, or vice versa, but not both at the same time.

There are other controlling factors in glycogen metabolism. Glycogen itself activates phosphorylase *b* kinase, and inhibits glycogen synthetase phosphatase, particularly in muscle. These may be regarded as feedforward and feedback controls respectively. It is noteworthy that at the present time much less is known about the control of phosphatase activity than about protein kinases. Finally, insulin increases markedly the proportion of glycogen synthase I-form in muscle, but the mechanism is not known.

The Adenylate Charge

The frequent involvement of AMP and ATP as allosteric effectors in the glycolytic pathway, and the tendency of MgATP-requiring enzymes to be inhibited by their product ADP, have led to suggestions that the general flux of metabolites may be regulated by the ratio of the 'energy-rich' adenine nucleotides to the total, i.e. by the ratio

$$\frac{2[ATP] + [ADP]}{[ATP] + [ADP] + [AMP]}$$

This is attractive because the concentrations of these nucleotides are interrelated by the action of *adenylate kinase*, which catalyses the reaction

$$ATP + AMP \rightleftharpoons 2ADP$$

It does not seem likely, however, that this may be a general regulator, since earlier beliefs that ATP is of first importance in controlling the rate of flux of the citric acid cycle have not been substantiated. Moreover adenylate kinase is absent from mitochondria, while the permease which transports adenine nucleotides into mitochondria is absolutely specific for ADP inwards and ATP outwards. Thus [AMP] in mitochondria is usually extremely low, and the effect of the ratio [ATP]/[ADP] on mitochondrial metabolism can best be explained by supply and demand control of electron transport (p. 367).

18   Hormones

In a book of this size it is impossible to deal adequately with all aspects of action of all mammalian hormones, some of which have very complex and long-lasting actions (for example, the thyroid and sex hormones). We shall limit ourselves to the main hormones, (see Table 18.1), their chemistry, where this is known, their metabolic functions, and some general principles.

Table 18.1
Classification of the Principal Hormones of the Body

Origin	General Hormones	Trophic Hormones
Hypothalamus		Releasing and inhibiting factors for pituitary hormones
Pituitary:		
anterior	Growth	Gonadotrophins (LH and FSH) Thyrotrophin Adrenocorticotrophin (ACTH). Prolactin
posterior	Vasopressin	Oxytocin
Thyroid	Thyroxine	Calcitonin
Parathyroid		Parathormone
Pancreas	Insulin	Glucagon
Adrenals:		
cortex	Steroid hormones	
medulla	Adrenaline	
Gonads	Oestrogens Testosterone	Progesterone
Placenta	Oestrogens Progesterone	Gonadotrophin, lactogen

Multiplicity of Actions

The majority of hormones found in man are found also in all other vertebrates, but their functions are not necessarily identical. For example, the salt and water balance problems of freshwater and marine fish are different from each other and from those of terrestrial animals. In the cold-blooded amphibia the thyroid hormones regulate the pace of metamorphosis, but have no effect on metabolic rate, which is an important aspect of thyroid function in higher animals. The effects of prolactin are necessarily different in non-mammals and in mammals. Thus during the course of evolution the hormones, whose chemical compositions are remarkably constant throughout the vertebrates, have gained new molecular sites of interaction, and may have lost others (see also p. 385). It would be dangerous to suppose that each hormone acts only on a single specialized cell type, or that when applied to a single type of cell, it has only one function. To suggest as an alternative that each hormone has only one receptor site within each cell, and that a multiplicity of observed effects is due to the lack of specificity of secondary processes, is to export the problem rather than to solve it. It is sometimes expedient to concentrate on one aspect of the action of a hormone for the purposes of research or application.

The 'Second Messenger'

It is evident from Table 18.2, which shows representative values for the concentrations of hormones in plasma, that most, if not all, hormone effects must pass through an amplification or 'cascade' process of the kind described in the previous chapter. In one instance—glycogen breakdown—the cascade is known with some precision (p. 369); in other instances—lipolysis in adipose tissue (p. 214), glycogen synthesis (p. 369), adrenal steroid formation (p. 381)—the general outline is now clear. A common feature of these and some other hormone effects is the intervention, at one stage, of cyclic adenosine monophosphate. So far as is known at present, in animal cells cyclic AMP does not bring about the unmasking of promoter sites on DNA (p. 357), so that the chain of amplification has the general form shown in Fig. 18.1. Protein kinase is rather unspecific, but in adipose tissue and muscle, at least, the stimulation of transcription and translation implied by phosphorylation of ribosomal protein and histones is unimportant.

In animals adenyl cyclase is bound to the cell membrane, from which it may with difficulty be separated, but the extracted enzyme is not sensitive to hormone stimulation. It appears that there are receptors on the cell surface which are each specific for a single hormone, and the spectrum of response of a given tissue to hormones depends on the presence or absence of the appropriate receptors. Thus liver phosphorylase responds to glucagon, while muscle phosphorylase responds only to adrenalin. It is interesting that if this hypothesis is true adipose cell membranes must contain all the receptors, since almost all the cAMP-linked hormones stimulate lipolysis.

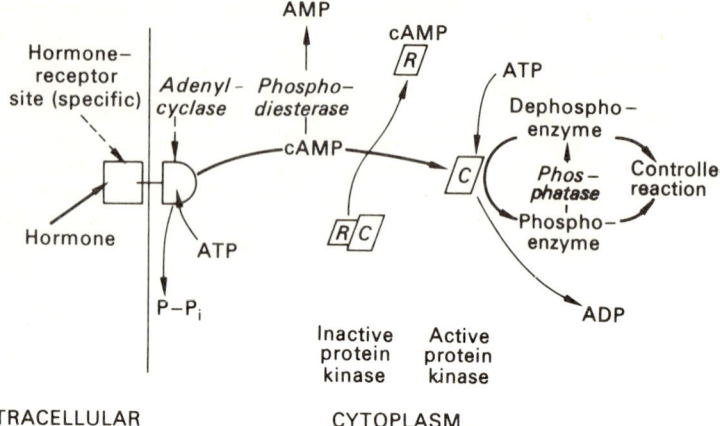

FIG. 18.1. Generalized amplification system in animal cells in which cyclic AMP takes part as 'second messenger'.

Between the receptor and the cyclase is some kind of coupling factor; in some cells it is phosphatidyl-serine, in others Ca^{2+}. The effect of hormone binding to the receptor is to increase V, not to lower the K_m for the substrate, ATP. Adenyl cyclase itself requires Mg^{2+} for activity, but is nevertheless strongly activated by fluoride ion. It is inhibited by nicotinic acid (which is strongly anti-lipolytic) and, in some cells at least, by prostaglandins.

In accordance with the principles outlined in Chapter 16, it is necessary for the intracellular cAMP to be rapidly destroyed if it is sensitively to amplify hormone binding to the external receptor. This is done by a *phosphodiesterase*, usually a cytoplasmic enzyme. It is Mg^{2+}-requiring and is inhibited by puromycin, by methylxanthines (Fig. 18.2), and by

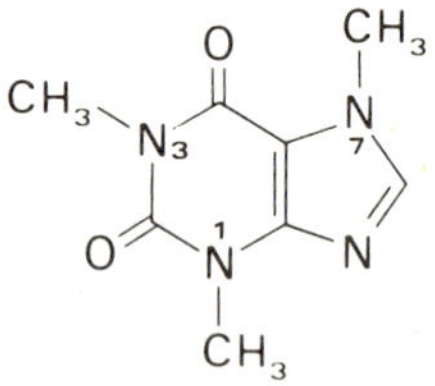

FIG. 18.2. Caffeine (1,3,7-trimethylxanthine). Theophyllin (1,3-dimethylxanthine) and 3-methylxanthine are also hinibitors of cyclic AMP hydrolysis.

Table 18.2
The Concentration of Some Hormones in Blood

The values quoted cannot be used to characterize a 'normal' individual, as the hormone concentration may fluctuate considerably for no apparent reason (particularly growth hormone). Moreover the values depend to some extent on the method of assay. They are quoted here only to give an idea of the order of magnitude.

All figures relate to adult fasting males.

Hormone	Concentration per 100 ml blood	Method of assay	Approx. no. of molecules per ml blood
ACTH	5 ng (0·7 milli-unit)	Biological	70×10^8
Growth hormone	55 ng	Radio-immunological	16×10^9
TSH	250 ng (13 milli-units)	Release of ^{131}I from thyroid tissue in vitro	60×10^9
Pituitary gonadotrophins (FSH + LH)	Not detectable (<5 HMG units)	Biological	
Vasopressin	11 ng (4 milli-units)	Biological	60×10^9
Insulin	85 ng (2 milli-units)	Radio-immunological	55×10^9
Adrenaline	7 ng ⎱	Fluorimetric	23×10^{10}
Noradrenaline	35 ng ⎰		13×10^{11}
Thyroxine	5 μg	Chemical	39×10^{12}
Aldosterone	2 ng ⎱		36×10^9
Cortisol	9 μg ⎬	Isotope dilution	16×10^{13}
Corticosterone	1 μg ⎰		20×10^{12}
*Oestradiol	5 ng	Isotope dilution	11×10^{10}
Testosterone	800 ng	Isotope dilution	17×10^{12}
†Androsterone	25 μg ⎱	Chemical	53×10^{13}
†Dehydroepiandrosterone (sulphate)	48 μg ⎰		10×10^{14}

* Also true of non-pregnant woman, but about 4 × higher just before menstruation.
† Not very much lower in woman, i.e. mainly of adrenal origin.

dibutyryl-cAMP (below); both the latter intensify adenyl cyclase-dependent hormone effects.

Cyclic GMP, like cyclic AMP, is found in urine and may also be a 'second messenger'.

Not all hormone effects are mediated through cAMP; if the effects can be mimicked by cAMP alone, it is reasonably certain that a cAMP cascade operates in vivo. Cell membranes are often rather impermeable to cAMP, but a lipid soluble analogue, *dibutyryl-cyclic AMP*, is often more effective.

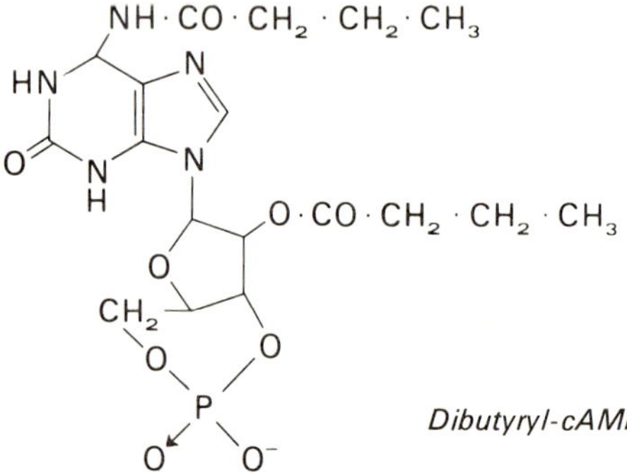

$$NH \cdot CO \cdot CH_2 \cdot CH_2 \cdot CH_3$$

Dibutyryl-cAMP

Although many hormone effects can be reproduced on isolated tissues in vitro, it has characteristically been difficult to obtain hormone effects on broken cell preparations. This has led to the hypothesis that all hormones are extracellular effectors, whose actions within cells are mediated by 'second messengers'. This view is already known to be untrue as stated. Oestrogens, when stimulating the synthesis of specific proteins in sensitive tissues, bind to receptors at the cell surface, but are then themselves carried, as shown by isotopic labelling experiments, by 'cytoplasmic receptor proteins' to the nucleus, where they probably exchange to a 'nuclear receptor protein' before de-repression takes place. Although the binding at the cell surface is significant, the hormone on its way from membrane to nucleus is clearly not a *second* messenger. It is by no means certain that peptide hormones cannot be transferred to intracellular locations in a similar way.

A list of hormone actions known to be mediated by cyclic AMP is given in Table 18.3. It may be pointed out that in some instances one effect of a hormone is cAMP dependent, while another is not (cf. ACTH, p. 380).

Table 18.3

A.	*Hormones acting through adenyl cyclase*
Adrenaline	Phosphorylase (liver; heart)
Noradrenaline	Lipolysis (adipose tissue)
Glucagon	$\begin{cases}\text{Phosphorylase (liver)}\\\text{Insulin release (islets of Langerhans)}\end{cases}$
TSH	Thyroid (hormone production)
ACTH	Steroidogenesis (adrenal cortex)
LH (ICSH)	Steroidogenesis (ovary, testis)
Vasopressin	Water reabsorption (kidney)
Parathormone	Ca^{2+} reabsorption (kidney; ?bone)

B. *Hormones not acting through adenyl cyclase*
Glucocorticoids
Aldosterone
Oestrogens
Androgens
Progesterone
Thyroid hormones
*Insulin
Growth hormone
Prolactin

* Insulin decreases the cAMP level of adipose tissue, but this is not its main function.

Feedback Controls on Hormone Production

In Chapter 16 it was suggested that the concentration of any chemical taking part in a biological system is controlled by the ratio of the production rate, K_1, to the rate constant for destruction, K_2. While this is also true for hormones, K_1 often varies in response to physiological stimuli or to feedback mechanisms, as is natural, since hormones are helping to maintain the constancy of the *milieu intérieur* (homoeostasis). Thus changes in the destruction constant K_2 are often compensated for by a corresponding change in K_1. Nevertheless the plasma level of the hormone is still to some extent controlled by the ratio K_1/K_2. A single example must suffice.

Either hypo- or hyper-thyroidism may lead to secondary adrenal insufficiency, because of the effects on the level of 11-β-hydroxysteroid dehydrogenase which inactivates the main glucocorticoid, cortisol (p. 396). Hypothyroidism leads to diminished 11-β-dehydrogenase activity in liver. The reduced rate of removal (K_2) of cortisol leads to an elevated plasma cortisol level. ACTH secretion is suppressed, and the adrenals become 'refractory'. In hyperthyroidism, on the other hand, there is increased 11-β-dehydrogenase activity. The consequent decrease in plasma cortisol level leads to continuous oversecretion of ACTH and overstimulation of the adrenals, i.e., increase in K_1. The resulting 'thyroid crisis' can be abolished by corticosteroid injections.

Where feedback controls exist, the production of the hormone is not directly controlled by its own concentration. Instead, either the endocrine gland is controlled by the major metabolite affected by the hormone, or the hormone controls production of a *trophic hormone* that regulates the endocrine gland itself. Examples of both these situations are shown in Fig. 18.3.

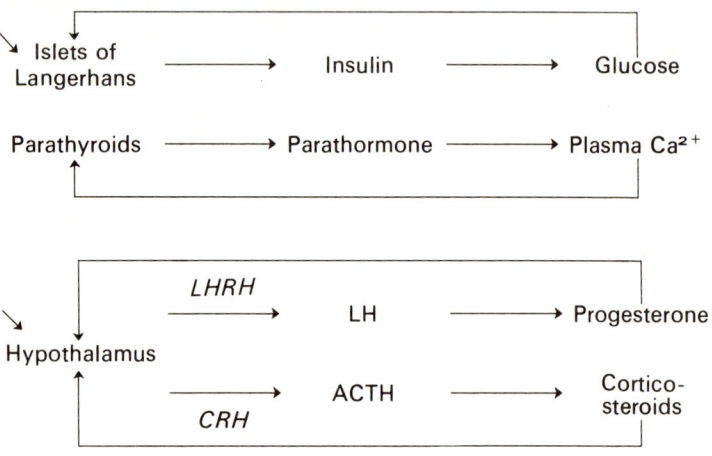

FIG. 18.3. Examples of two-stage feedback mechanisms in the control of hormone secretion. The arrows at the left indicate other controlling influences.

Some hormones do not have feedback control mechanisms of the type shown in Fig. 18.3. Aldosterone secretion is controlled through the effective plasma volume, which is only indirectly related to the metabolite (Na^+) regulated by the hormone. Prolactin secretion is normally inhibited, and the inhibition is released by direct stimuli, such as suckling. Growth hormone disappears from plasma for hours at a time, and its reappearance cannot be uniquely correlated with any known experimental stimuli.

The Pituitary Hormones and Hypothalamic Control Substances

It has been realized for many years that the secretion of anterior pituitary hormones is strongly affected by signals received by the central nervous system, e.g., psychic stress, day length, suckling. Conversely it is known that if the pituitary is removed from its normal site and re-implanted elsewhere in the body, nervous control over its function is lost.

Neurosecretory cells in the hypothalamus synthesize peptide releasing or inhibiting factors in response to signals from other parts of the brain, or as a result of feedback from circulating hormones produced by target organs (Fig. 18.3). These peptides are released from axons terminating in the median eminence, into the hypophysial portal blood system, whence they affect specific secretory cell types in the pituitary itself. Owing to the very considerable difficulties of assaying extracts of the median eminence, the picture is not yet completely clear, but several peptides have been isolated and sequenced. The present situation is outlined below.

Thyrotrophin-releasing hormone. This is a tripeptide of the structure PGA-Pro-GlyNH$_2$ (glycinamide). The hormone has been synthesized and shown to be as effective as the natural one in increasing the circulating TSH level. It is already in wide clinical use for testing thyroid function.

Luteotrophic hormone-releasing hormone. This also releases FSH. It is effective in children, so that the normal absence of gonadotrophin secretion until puberty is due to inhibition of LH-RH secretion by the hypothalamus, and not to functional incompetence in the pituitary. It has the structure PGA-His-Trp-Ser-Tyr-Gly-Leu-Arg-Pro-GlyNH$_2$.

Corticotrophin-releasing hormone. There is strong evidence for the existence of CRH, but it has still not been positively identified.

Growth hormone-releasing hormone. A peptide which stimulates the release of GH from the pituitary in vitro has been identified and indeed sequenced, but it has no effect at all on the plasma growth hormone level. As a GH-*inhibiting* hormone has also been postulated, there is at present complete confusion about the control of GH secretion by the pituitary.

Prolactin-inhibiting hormone. The experimental evidence that prolactin secretion from the pituitary is normally suppressed by a hypothalamic hormone is very clear-cut, but the hormone itself has not yet been identified.

Melanocyte-stimulating hormone-releasing hormone (pars intermedia). This is of no interest in man, but its structure is Pro-Leu-GlyNH$_2$.

Anterior Pituitary Hormones

Growth hormone is a protein. That from cattle pituitaries has a molecular weight of 45 000, but human and monkey growth hormones are single peptide chains, with a molecular weight of 21 500 daltons. The latter are the only growth hormones which are effective in man as well as in animals. Human growth hormone has about 30 amino acids in common with human prolactin, and considerable prolactin activity, but prolactin has very little growth-promoting activity. A protein called *placental lactogen* is secreted by the placenta; it is both growth-promoting and prolactin-like (see p. 383).

Growth hormone injections cause a gain in weight accompanied by a retention of nitrogen, and a drop in plasma amino acid level. This points to an increased rate of tissue protein synthesis. In immature animals, skeletal growth also occurs, so that as a result of prolonged injections, treated animals may be twice as long and more than twice as heavy as normal adult animals. In species in which maturation is accompanied by closure of the epiphyses of the long bones, growth hormone will not cause skeletal growth in the adult. The growth does not exactly mimic normal growth, for example, the brain does not respond at all, and there is no effect on sexual maturation. There is good evidence that growth hormone will not produce nitrogen retention in animals unless insulin is also present.

Growth hormone also markedly affects carbohydrate and fat metabolism. The hormone is almost undetectable in the plasma of fed adults, but the level rises during fasting and drops sharply on the ingestion of glucose. In vivo, the rate of secretion of NEFA (Chapter 10) by adipose tissue is very much increased after a delay of 2–6 hours, and probably as a result of this, the RQ of the whole animal decreases, and ketosis and fatty liver often occur after injections of the hormone. Investigations on adipose tissue in vitro lead to the conclusion that the hormone causes an RNA-mediated increase in lipase activity. The oxidation of carbohydrate is depressed, but the output of glucose by the liver may be increased so that a temporary (or in some species permanent) diabetes may be produced. In muscle, growth hormone stimulates the accumulation and incorporation into protein of amino acids.

In man some types of pituitary dysfunction result in *dwarfism* of a type not improved by thyroxine treatment. Basophil tumours of the pituitary on the other hand cause *gigantism* in the young, and *acromegaly* (a growth of the bones of the face and extremities) in the adult.

Adrenocorticotrophin (ACTH) is extracted as a protein from the pituitary, but the active hormone consists of a peptide containing 39 amino acids (see, however, p. 385). All the species differences in primary structure occur in amino acids 25–39. The hormone is rapidly destroyed in the body, with a half-life of 3 minutes. Slow-acting ACTH preparations contain the peptides in a viscous medium such as gelatin or carboxymethylcellulose.

The biological function of ACTH is complex. It maintains the weight of the adrenal cortex, and stimulates it to produce and secrete adrenal steroid hormones, with the exception of aldosterone (see p. 398). Thus after the removal of the anterior pituitary the adrenals atrophy, but the animal remains in salt balance.

If the hormone is injected over a period of time into a hypophysectomized animal, the atrophied adrenals increase in size, and there is redevelopment of two layers of specialized cells (zona reticularis and zona fasciculata) which characterize the normal adrenal. These effects are relatively slow in onset. They are not mediated by cyclic AMP. Increases in DNA polymerase and thymidylate kinase activities in the adrenal have been observed after ACTH administration. However, ACTH will produce an immediate secretion of steroid hormones even after a single injection into a hypophysectomized animal, or if it is incubated with adrenals in vitro. This is accompanied by a fall in the content of ascorbic acid and cholesterol in the adrenal. There is a rise in the cyclic AMP content of the tissue, and the fast reaction to ACTH can be satisfactorily mimicked by cAMP or analogues. The rate-limiting reactions in adrenal steroid synthesis are hydrolysis of stored cholesterol ester, allowing free cholesterol to be transported to the mitochondria, where it is oxidized to pregnenolone (p. 395). These events are all stimulated by cAMP; it is interesting that stimulation of translation efficiency in the adrenal ribosomes is implied. Fig. 18.4 summarizes the information.

The control of ACTH secretion is complex, as for all the trophic hormones of the pituitary. Secretion is increased by 'stress' (see p. 399), partly through a direct action of adrenaline and histamine on the pituitary, but more importantly because of a nervous reaction, which causes the secretion of an 'ACTH-releasing factor' (CRH) from the hypothalamus. The rate of secretion of CRH, and therefore of ACTH also varies inversely with the concentration of adrenal cortical hormones in the plasma, so that circulating ACTH is virtually absent from the blood of a subject receiving high doses of adrenal steroids (see p. 398).

Thyrotrophin is a glycoprotein with a molecular weight of about 28 000. It has two non-identical sub-units of mol. wt. about 14 000 (see p. 384). Like ACTH, it has an immediate effect and a long-term effect on its target organ, in this case the thyroid. The immediate effect (within 5–15 min) is to increase the rate of secretion of thyroxine, tri-iodothyronine and iodide

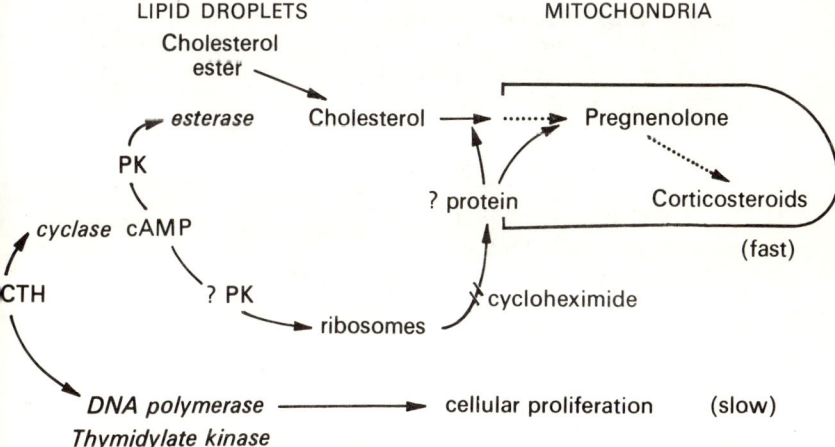

FIG. 18.4. Effects of ACTH on the adrenal gland. PK represents protein kinase.

into the blood. The iodide comes from de-iodination of iodotyrosines. All these effects are secondary to a stimulation by TSH of the hydrolysis of thyroglobulin. TSH also rapidly increases the O_2 consumption, glucose utilization and RNA synthesis in the gland; these effects are presumably secondary to a stimulation of thyroglobulin synthesis and of oxygen-requiring iodinations.

In the long term, TSH maintains the weight of the thyroid and the functional state of the secretory epithelium. The thyroid thus atrophies after hypophysectomy but, as with the adrenals, the gland retains a minimal function and secretion.

Control of thyrotrophin secretion is by a feedback mechanism through a 'thyrotrophin-releasing factor' in the hypothalamus. Unusually, there is an extra feedback of thyroid hormones to the pituitary itself. This is very marked in tadpoles before metamorphosis.

Although human *prolactin*, unlike that of many species, is structurally closely related to human growth hormone, the two are immunologically distinct, and circulating prolactin can be separately detected. Its functions are to promote mammary development and lactation, including the synthesis of milk proteins; to induce luteolysis of the previous crop of corporea lutea during each menstrual cycle; and in early pregnancy, together with LH, to maintain functional corpora lutea. The luteotrophic functions of prolactin in man are, however, small by comparison with many other species. In birds, prolactin stimulates the growth and secretions of the crop gland. Pituitary tumours are often associated with a high plasma prolactin, but galactorrhoea is not an automatic consequence. On the other hand, galactorrhoea is sometimes found with normal prolactin levels. This is not surprising in view of the number of hormones known to be involved in mammary gland development and function (Fig. 18.5).

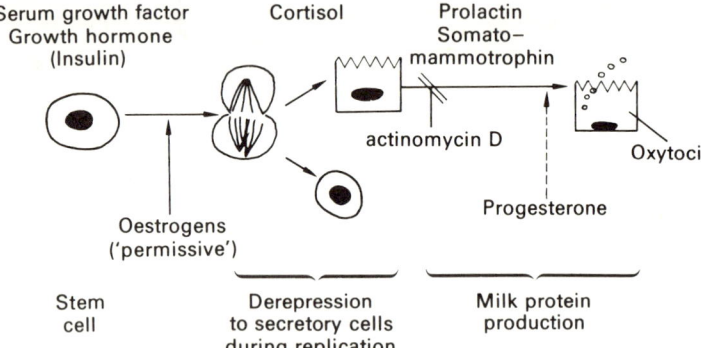

Fig. 18.5. Hormones involved in breast development and milk production.

A polypeptide *somatomammotrophin*, or placental lactogen, whose primary structure much resembles that of growth hormone, is produced by the placenta in late pregnancy.

The *follicle-stimulating hormone* has not yet been completely purified, but it is a glycoprotein with two sub-units. In females it stimulates the growth of ovarian follicles up to ovulation. As with LH, it is secreted continuously during the menstrual cycle; the pre-ovulation 'spike' is much smaller than with LH. Clearly the secretion of FSH and LH is not inhibited by the increasing quantities of oestrogen released into the blood throughout the first half of the cycle.

In the male, FSH by itself has little recognizable effect, but if injected in small quantities with LH into castrated animals it aids in the growth of seminiferous tubules and in spermatogenesis.

The *luteinizing hormone* is a glycoprotein with two sub-units of mol. wt. 13 400 and 14 000 daltons respectively (see also p. 384). A very large 'spike' of this hormone is released just before ovulation, and thereafter it is responsible for the development of the corpus luteum as a temporary endocrine gland. Progesterone is the main hormone secreted by the corpus luteum, and although the details have not been so completely worked out, it is probable that its relationship with LH is very similar to that between ACTH and the adrenal steroids. That is to say, in vitro a very rapid progesterone release can be evoked that may be mimicked by cyclic AMP. Thus the latter probably stimulates an early phase in the conversion of cholesterol to progesterone.

In the male it is the testosterone level in plasma which regulates the release of LH-RH from the hypothalamus. The release of the hormone (often called interstitial cell-stimulating hormone, ICSH) is more or less continuous; there are very rarely any cyclic changes in the circulating level. ICSH stimulates the development of the interstitial (Leydig) cells that secrete testosterone and other androgens, and these in turn are necessary for complete spermatogenesis.

The placenta produces considerable quantities of a glycoprotein *chorionic gonadotrophin* of mol. wt. about 46 000 (see also p. 384), especially during early pregnancy. This is secreted into the mother's blood; some escapes into the urine and forms the basis of some pregnancy diagnosis tests. Human chorionic gonadotrophin has only the properties of LH; its biological role is probably to help make the nutrition and development of the fetus independent of stimuli affecting the mother. In the second half of pregnancy gestation can go to term after complete removal of the pituitary.

Posterior Pituitary

This part of the gland contains a protein to which are linked two peptides, vasopressin and oxytocin, usually in a ratio of about 8:1. These peptides, whose structure has been elucidated (see Fig. 18.6) are synthesized only in

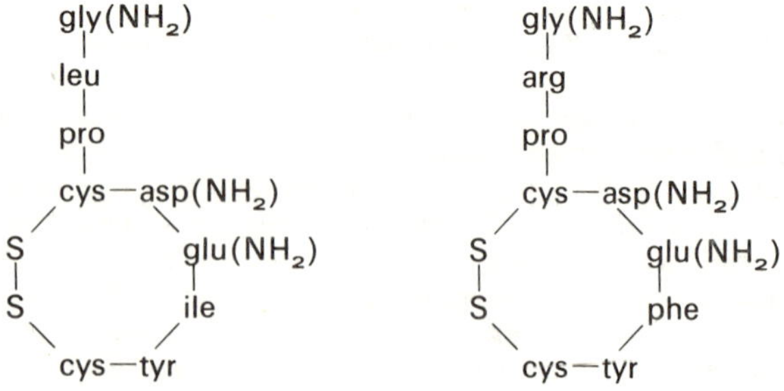

gly(NH₂)
leu
pro
cys—asp(NH₂)
S   glu(NH₂)
S   ile
cys—tyr

Oxytocin

gly(NH₂)
arg
pro
cys—asp(NH₂)
S   glu(NH₂)
S   phe
cys—tyr

Vasopressin

Fig. 18.6. Constitution of the posterior pituitary hormones. The glycine residues are at the *C*-terminal end. The free carboxyl groups of the glutamyl, aspartyl and glycyl residues are amidated.

the hypothalamus and travel down nerve fibres into the posterior pituitary, where they are stored.

Vasopressin is identical with the *antidiuretic hormone* (ADH). It acts directly on the kidney to increase the reabsorption of water in the distal tubule, and is secreted when the osmotic pressure of plasma rises above its normal value. It may act, in the renal tubular cells, through the mediation of cyclic AMP. It is doubtful whether, at physiological concentrations, it increases salt excretion, but it does to some extent act on the kidney in antagonism to the adrenal steroids. Vasopressin has also a vasoconstrictive action on peripheral blood vessels, thus raising blood pressure, but this is thought to be pharmacological rather than physiological.

Oxytocin causes smooth muscle to contract. The uterus immediately before parturition is most sensitive to this effect, but suitable doses will cause other smooth muscle to contract. Oxytocin causes also a rather slowly appearing diuresis.

A *melanophore-expanding* hormone (intermedin) which is active in amphibia can be extracted also from the intermediate lobe of mammalian pituitaries. It is not known to have any function in mammals.

Redundant Information in Polypeptide Hormones

1. Thyroid-stimulating hormone, luteinizing hormone and human chorionic gonadotrophin are all glycoproteins with two sub-units. The α-chain of TSH, the CI chain of LH and A chain of HCG are absolutely identical to one another in molecular weight and amino acid sequence.

2. The placenta secretes a *placental lactogen* (somatomammotrophin). It has 191 amino acids, precisely as many as human growth hormone, and 80 per cent of the residues are homologous with the latter. Human

prolactin is slightly larger than growth hormone and 30 per cent of its amino acid residues are homologous. Placental lactogen has strong growth-promoting *and* prolactin activities, while the biological effects of GH and prolactin overlap considerably.

3. The two posterior pituitary hormones differ only in two of their 9 amino acids (p. 384). The MSH-releasing hormone is identical with the non-cyclic 'tail' of oxytocin, while the peptide calcitonin (p. 388) contains a nine-membered disulphide ring.

4. The α-melanocyte stimulating hormone is a peptide of 13 amino acids, whose sequence is identical with that of the 13 amino acids at the N-terminal end of ACTH. Only the first 20 of the 39 amino acids of ACTH are necessary for full biological activity. Of the 84 amino acids of the para-thyroid hormone, only 29, starting from the N-terminal end, are necessary for full activity.

The remarkable similarities in primary structure within these groups of polypeptide homones, perhaps arising from gene duplication, have interesting implications for the evolutionary development of hormone-sensitive sites and tissues in vertebrates. On the other hand, the presence of long sequences and indeed whole sub-units with no apparent specific function suggests that these may be protective, and may be stripped off before the hormones act.

Neither insulin nor glucagon have any similarity to each other or to other polypeptide hormones.

Thyroid

More than one hormone is produced by the thyroid; L-thyroxine and L-tri-iodothyronine are both found in plasma, although there is much more of the former. Tri-iodothyronine appears to be about 5 times as active as thyroxine probably because it is not so strongly bound to protein in plasma, but in fact the two hormones are equally active. Only L-thyroxine has hormonal activity; the D-isomer lowers blood cholesterol without affecting myxoedema or cardiac activity. Almost all the circulating thyroxine is loosely bound to a specific α-globulin, and is precipitated with it on the addition of deproteinizing agents. The iodine so removed from plasma is known as the *protein bound iodine* (PBI) (as distinct from in-organic I). It amounts to 4–8 μg/100 ml in the normal individual. Both hormones take several days to exert their full effect; little change is seen for 48 hours after the first injection.

$$HO-\underset{I}{\overset{I}{\bigcirc}}-O-\underset{I}{\overset{I}{\bigcirc}}-CH_2 \cdot \underset{NH_2}{\overset{}{CH}} \cdot COOH$$

Thyroxine

The structural requirements for thyroxine-like activity are not very stringent. With reference to the substituted diphenyl ether shown below,

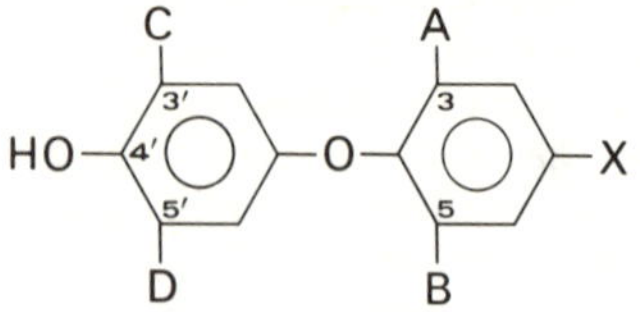

both rings must retain their aromatic character, but otherwise only the —OH group at position 4' is almost essential for activity. X can be alanine (as it is in thyroxine), pyruvic, propionic, acetic or formic acid. The two substituents at A and B need not be iodine; bromine or methyl groups are also effective. In the 'outer' ring, highest activity is obtained if position 5' is free (i.e. D = H, as in tri-iodothyronine). The substituent C also need not be iodine, but is most effective when it is about the same size as an I atom. The compound 3,5-diiodo-3'-isopropyl-L-thyronine is one of the most potent substances known.

Since biologically active thyroxine analogues need not contain any iodine at all, mechanisms of action which depend upon the release of I, I$^-$, or I$^+$ atoms in tissues sensitive to thyroxine are clearly not tenable. Nevertheless, there is no evidence that in the body, thyroid function is carried on by other than iodine-substituted thyronines, synthesized in the thyroid gland.

The chief function of the thyroid is to concentrate iodide from the plasma and to oxidize it to iodine. Enlargement of the thyroid therefore takes place when the diet is deficient in iodide and vice versa. Tyrosine residues in the protein thyroglobulin are iodinated by I_2 and are then condensed to form thyroxine. The thyroxine residues stay in peptide linkage until the thyroglobulin is hydrolysed; normally stimulation of thyroglobulin uptake by thyroid cells and attack on it by a protease is one of the functions of TSH. Iodine itself inhibits the proteolysis of thyroglobulin, and is used, in large doses, as a transitory anti-thyroid agent.

Thyroxine formation is inhibited at several stages by goitrogens, natural ones in *Brassicae* such as allylthiourea, or synthetic ones such as methylthiouracil. Such drugs are used in the treatment of hyperthyroidism; in large doses they cause enlargement of the thyroid, because thyrotrophin secretion is stimulated.

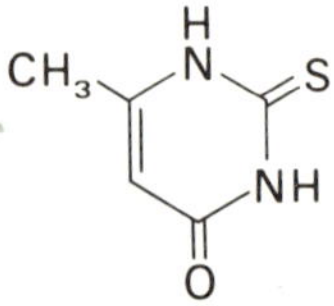

Methylthiouracil

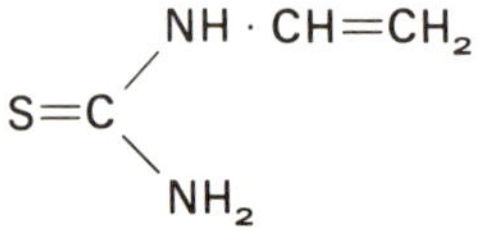

Allylthiourea

Underactivity of the thyroid results in *cretinism* in the young, which is characterized by lowered metabolic rate, dwarfism, and mental deficiency. If it is not treated in infancy it is irreversible. If deficiency develops in adult life the disease is called *myxoedema*, characterized by a puffy dry skin, hypothermia, lowered basal metabolic rate, and lowered intelligence. A raised blood cholesterol is also usual. Overactivity of the thyroid leads to thyrotoxicosis or *Graves's disease*. It is usual to find severe loss in weight, over-excitability and restlessness, and protrusion of the eyeballs (exophthalmos). A raised basal metabolic rate and increased nitrogen excretion are very characteristic. The animal is much more than usually sensitive to catecholamines, and there is usually a high concentration of NEFA. Thyroid hormones particularly affect the heart; there is an increase in its size, and a tachycardia which may be due to a direct effect on the conductive tissue.

The calorigenic effect of thyroxine remains something of a mystery. It does not appear to be identical with the non-shivering calorigenesis caused by catecholamines, although the two phenomena are certainly related. It can be shown that utilization of energy for muscular contraction is less efficient in hypothyroidism, but it has been extremely difficult to establish any effects, e.g., on the uncoupling of oxidative phosphorylation in mitochondria, which cannot be dismissed as artefacts. The concentration of many enzymes connected with oxidation is low in hypothyroidism, particularly cytochrome *c*, but it has not been possible to identify a rate-limiting enzyme whose concentration is increased in hyperthyroidism.

There is always a lag period before any thyroid hormone or analogue becomes biologically effective. Research, particularly into amphibian metamorphosis, suggests that this is due to increased RNA synthesis, preceding an increased rate of protein synthesis. It seems likely that in higher animals the major effect is on the 'export' of ribosomes from the nucleus and their binding to endoplasmic reticulum and to mRNA, i.e. on translation rather than on transcription. In amphibian larvae, thyroid hormones stimulate cell division and differentiation in the limb buds, but it is not clear whether there is any comparable effect in higher animals, for example on the development of the central nervous system.

In thyrotoxicosis endogenous protein breakdown outweighs protein synthesis, but this is a metabolic control which superimposes itself on an enzyme pattern, as in the wasting type of diabetes.

Parathyroids

These four glands secrete a polypeptide hormone which controls Ca metabolism. The concentration of Ca in the plasma is normally maintained very closely at about 10 mg/100 ml; half of the Ca is ionized—this is the part which is essential for the proper contractility of muscle—and the rest is bound to plasma protein. The rate of secretion of the hormone varies inversely with the plasma Ca concentration.

Experiments with injected parathyroid hormone have suggested that the effect is primarily to increase resorption of Ca from bone, and secondarily to diminish phosphate reabsorption by the kidney tubules, so that plasma phosphate falls and urine phosphate rises. The changes in phosphate concentration follow rapidly after injection of the hormone, however, while the rise in plasma Ca only begins after an hour or two. There does not seem to be any direct action of parathormone on Ca excretion by the kidney; the urine level reflects the plasma concentration.

One of the earliest changes seen after administration of parathormone is a rise in blood and urine citrate, which precedes the rise in plasma Ca level by several hours. It may be that liberation of Ca from bone begins by stimulating citrate production (from carbohydrate) by osteoclasts. This citrate could replace phosphate in the bone mineral, and so produce in course of time a soluble Ca citrate complex, which could find its way into the blood stream. The phosphate released from the bone is promptly excreted. It is certainly true that the parathyroid hormone stimulates osteoclast activity.

In overactivity of the parathyroid the plasma Ca may rise to as much as 20 mg/100 ml, with calcification of soft tissue and demineralization of bone. If the condition persists for a long time, excessive remineralization of the skeleton may occur, forming so-called 'marble bone'. Underactivity of the glands may reduce the plasma Ca to about 4 mg/100 ml, and tetany— fatal if uncontrolled—will develop.

The hormone has a mol. wt. of about 9 000. It consists of a single peptide chain of 84 amino acids, of which the 20 at the carboxyl end are essential for biological activity.

Calcitonin

This is sometimes called thyrocalcitonin, but in various species it may also be extracted from parathyroid or thymus. It is a peptide of 32 amino acids, which at the N-terminal end has a nine-membered disulphide ring similar in shape to that of the posterior pituitary hormones (p. 384), although apart from the two Cys residues the amino acids in the ring are different.

Calcitonin inhibits bone resorption, especially the mineral solubilization phase, and hence reduces the plasma Ca^{2+} level. Secretion from the thyroid is stimulated by hypercalcaemia and by gastrointestinal hormones (cf. insulin). It is not clear at present whether calcitonin forms part of the normal physiological mechanisms for regulation of Ca^{2+} metabolism, but it is in clinical use for treating some forms of hypercalcaemia.

Pancreas

The islets of Langerhans in the pancreas are known to contain two peptides with effects on carbohydrate metabolism, *insulin* from the β-cells, and *glucagon* from the α-cells.

Insulin

This hormone is a small protein, mol. wt. about 5 750. There are two peptide chains, held together by two disulphide bonds; the A-chain contains 21 amino acids, the B-chain 30. Neither chain by itself has any biological activity.

It is difficult to demonstrate experimentally a direct effect of insulin upon metabolism in liver; nevertheless injection of insulin into diabetics ameliorates several metabolic imbalances peculiar to that organ. Thus it very rapidly suppresses glucose output, both from glycogen breakdown and from gluconeogenesis. This also happens in the normal subject, and is partly responsible for the severe hypoglycaemia of large doses of insulin. Another effect is a fall in urea output, which seems not to be entirely due to the decrease in plasma amino acid levels. There is a very rapid suppression of ketosis, but this may be secondary to the suppression of FFA release from adipose tissue (see below). A slow effect of insulin in liver is the stimulation of triglyceride synthesis. In diabetes fat synthesis is possible from fructose but not from glucose, demonstrating that in the absence of insulin glucokinase synthesis is repressed. There appears, however, to be a further effect of insulin on fat synthesis, which cannot be identified with any stage before the release of long-chain saturated fatty acids from the fatty acid synthase complex. It may lie in stimulation of triglyceride synthesis, or in the desaturation of stearate or palmitate.

The major effect of insulin is on glucose utilization in peripheral tissues. Insulin increases the rate at which glucose—and other, non-metabolized sugars—are taken into muscle and other tissues. (Liver cells are freely permeable to glucose, while insulin does not affect the transfer of sugars into brain.) The fall in blood sugar which results is extremely rapid and can be fatal if allowed to proceed too far. The hormone appears to increase the activity of 'permeases' at the cell membrane. As a result, both the relative amount of carbohydrate oxidized and the amount stored as glycogen are increased. Insulin directly increases the fraction of glycogen synthetase in muscle which is in the G-6-P-independent (I) form (see Chapter 9); the mechanism is as yet unknown. The fraction of pyruvate dehydrogenase in the active form (p. 365) is increased both in muscle and in adipose cells. In the latter tissue and in the liver, the rate, but not the fraction, of glucose oxidized by the hexose shunt is increased. This makes more $NADPH_2$ available. None of these effects can be shown to depend on cAMP. The hormone also increases the rate of transfer of amino acids into muscle and also their incorporation into protein. The relation between growth hormone and insulin in growth has already been mentioned (p. 380). In vitro, at all events, stimulation of protein synthesis by both these hormones is not blocked by actinomycin D, and they presumably increase in different ways the translational efficiency of the ribosomes.

As already mentioned, the rapid suppression of ketosis by insulin may be more related to the reduction in NEFA output by adipose tissue than a primary effect on the liver. Insulin markedly stimulates the metabolism of fat cells. It increases the glucose taken up by the cell, and increases the rate of conversion of acetyl residues to fatty acids. In addition it makes more glycerol phosphate available for triglyceride synthesis (p. 193) in virtue of stimulating the catabolism of glucose. This is important because adipose tissue contains practically no glycerol kinase, and complete hydrolysis of a triglyceride in the fat cells is therefore automatically followed by loss of the glycerol to the blood stream. It is possible that the anti-lipolytic effect of insulin is partly due to the increased availability of phosphoglycerol, which swings the equilibrium below to the left:

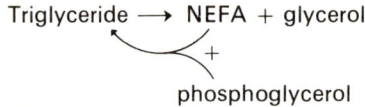

However, the effect of insulin on plasma NEFA levels is very dramatic, and it has been postulated that it directly inhibits the hormone-sensitive lipase. It has not been possible to provide evidence for this, but it is found that insulin will reduce the cyclic AMP concentration in adipose tissue if this has previously been raised by other hormones, e.g., adrenalin. There are certain difficulties in accepting this as the sole mode of action of insulin in suppressing NEFA output by fat cells, but at present no other hypothesis is available.

The islets are stimulated to secrete insulin by a rise in the blood sugar level, but orally ingested glucose is more effective than the same amount given intravenously. The gastrointestinal hormones pancreozymin, secretin and gastrin have been shown to produce a very rapid stimulation of insulin secretion. Several amino acids given orally, among them arginine, leucine, phenylalanine and methionine, stimulate insulin release. Since glucagon release is inhibited by hyperglycaemia, it seems unlikely that it is important in the control of insulin secretion.

Pro-insulin. As shown in Chapter 16 (p. 348) the insulin molecule is synthesized as a single peptide, from which a central nonapeptide is later excised. On stimulation of the islets, true insulin stored in granules is first released, but pro-insulin, which reacts with antibodies to insulin, but less strongly than the latter, makes up from 50–80 per cent of the total immunologically active material in plasma. It has 10 per cent or less of the biological activity of insulin in most tests, but may be more effective than this in suppressing hepatic glucose output. It is not thought to be converted to insulin outside the pancreas.

Oral hypoglycaemic drugs such as 'tolbutamide' are mostly substituted ureas; their formula bears no relation to that of insulin itself. They have very little action on the blood sugar of the normal subject, but can cause a relatively long-lasting fall towards normal levels in the mild diabetic, i.e. one who is still secreting a minimal amount of insulin. The way in which they act seems to be in stimulating the failing islets to produce somewhat more insulin.

Glucagon

This is a peptide, 29 amino acid residues long, mol. wt. about 3 500, formed in the α-cells of the islets of Langerhans. It greatly stimulates glycogenolysis in liver, by activating adenyl cyclase and so triggering the cascade effect described in Chapter 17. It also appears to stimulate gluconeogenesis, and hence urea output, in liver. It also activates the adenyl cyclase of adipose tissue and hence is lipolytic, but it does not affect muscle phosphorylase or muscle glycogen levels.

Glucagon is detectable in plasma, and its secretion is inhibited by hyperglycaemia, so that it must be regarded as a hormone. It is most remarkable, however, that no endocrine disease can be attributed to over- or underproduction of glucagon. Even in a patient suffering from an α-islet cell tumour, it was impossible to detect any symptoms that could with certainty be attributed to hyperglucagonaemia.

An immunologically cross-reacting peptide can be extracted from the gut, but is not thought to have any importance in the control of metabolism.

Regulation of the Blood Sugar Level

It has been explained earlier (Chapter 9) that constancy of blood glucose concentration is essential for adequate body function. Although insulin is continuously secreted, at a rate which rises and falls with changes in blood glucose level, the rate of secretion is very low during fasting. The function of insulin is therefore essentially to promote storage and utilization of food carbohydrate. Other hormones act to maintain the blood glucose level when carbohydrate is not available from food. The only source of *blood* glucose other than the digestive tract is the liver. Glucose can be produced in liver either through the breakdown of glycogen, or through gluconeogenesis. In fasting, liver glycogen stores slowly disappear, becoming exhausted 15–24 hours after the last meal. Glycogenolysis may be under direct hormonal control, but in liver there is always some phosphorylase *a* present, and it is quite likely that the gradual disappearance of liver glycogen is due more to decreasing glycogen synthetase activity than to activation of phosphorylase. Adrenaline is not required for a normal response to fasting, and if glucagon secretion plays any role it may be more important in stimulating gluconeogenesis.

Growth hormone decreases the rate at which carbohydrate is oxidized in peripheral tissues; possibly the actual inhibition of glucose uptake is caused by the raised plasma NEFA levels. The growth hormone effects

Fig. 18.7. Biosynthesis of steroid hormones. In this outline diagram, only one route to each hormone of the several that are usually possible has been shown, and many intermediate compounds have been omitted.

The importance of pregnenolone and progesterone as precursors stands out clearly. Notice, too, the persistence of the 3β-configuration of the androgen precursors on the right-hand side of the diagram.

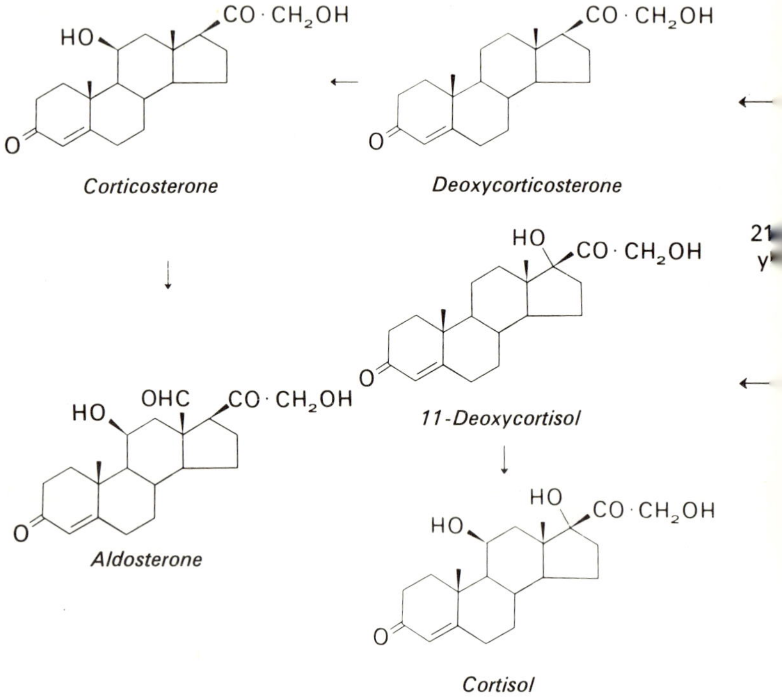

Corticosterone

Deoxycorticosterone

11-Deoxycortisol

Aldosterone

Cortisol

Reactions occurring only in adrenal tissue

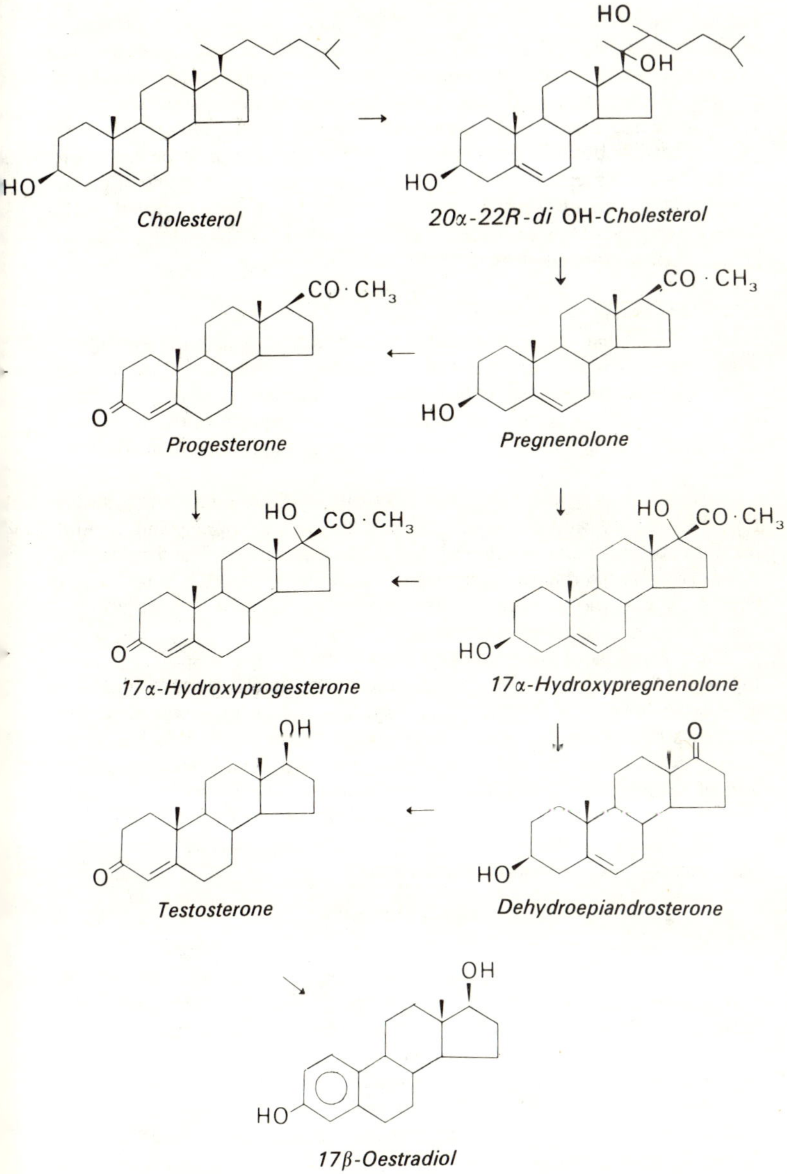

Cholesterol → *20α-22R-di OH-Cholesterol*

Progesterone ← *Pregnenolone*

17α-Hydroxyprogesterone ← *17α-Hydroxypregnenolone*

Testosterone ← *Dehydroepiandrosterone*

17β-Oestradiol

Reactions occurring in gonads and (in upper half of diagram) adrenals require the presence of adrenal cortical hormones, whose secretion is stimulated by ACTH. The steroid hormones also act by increasing the rate of gluconeogenesis from amino acids, and by mobilizing surplus body proteins to the amino acid pool. The glucocorticoids and growth hormone are slow-acting and do not come into full effect until fasting has been prolonged for about 24 hours. Although it is possible in some species to induce hyperglycaemia by massive injections of growth hormone or glucocorticoids, the insulin sensitivity of the hypophysectomized animal or the fasting hypoglycaemia of Addison's disease are more characteristic of the effects of these hormones on carbohydrate metabolism.

Diabetes

It must be stressed that clinical diabetes is an extremely complex syndrome, whose causes are not always well understood. Experimental diabetes is often an oversimplification of the disease; for example, injection of pituitary extracts may cause permanent diabetes in test animals but diabetes has developed in patients who had been suffering from pituitary deficiency for years.

However, two types of diabetes may often be distinguished: a severe type responding well to insulin, appearing in the young; and a mild, insulin-resistant diabetes often appearing in middle age. The former type is almost certainly due to deficiency of insulin secretion; the latter *may* be caused by alterations in metabolism, possibly caused by oversecretion of other hormones.

The symptoms of diabetes are a raised blood sugar level even in fasting, and lowered glucose tolerance, glucosuria accompanied by polyuria and thirst, lowered RQ due to increased fat catabolism, increased gluconeogenesis from protein, and loss of weight. Ketonaemia and ketonuria appear more slowly; they may cause disturbance of electrolyte metabolism and, if severe, reduce cerebral oxygen uptake, causing unconsciousness (diabetic coma). Arteriosclerosis may appear after many years and cause blindness from retinitis or cataract. This may happen even if the patient has been continuously on insulin treatment.

Steroid Hormones

Synthesis

All the steroid hormones are synthesized from cholesterol, which may be manufactured in the gland or taken from the plasma. The first stage, common to all steroid hormones, is double hydroxylation of the side-chain followed by scission to form *pregnenolone.*

The subsequent pathways are complex, because the mixed function oxidases (see Chapter 12) which introduce further —OH groups stereospecifically at different points in the molecule are not absolutely specific for substrates, so that all the possible intermediates may be present, and in addition cholesterol-3-sulphate and other steroids esterified with sulphate at the 3-position are substrates. There are at least 10 hydroxylating systems in the adrenal gland, so the large number of steroids that have been identified in adrenal extracts is not surprising. Many of these enzymes may also be present in the gonads and placenta, but the 21- and 11-β-hydroxylases are restricted to the adrenals, since cortical steroids (p. 48) are not formed elsewhere.

Fig. 18.7 shows some of the more probable pathways to adrenal steroids and sex hormones. It is agreed that oestrogens are formed from androgens and that the essential series of reactions is the hydroxylation of the methyl group on C–10, which is then oxidized to the carboxylic acid and finally lost as CO_2; ring A is then 'aromatized'. However, there are several possible precursor androgens, and the relative importance of the various routes has not yet been worked out. Interestingly, although oestrogen synthesis goes via androgens, the latter are not normally secreted by ovaries or placenta, but the testis normally secretes oestrogens. The adrenal cortex of both sexes secretes some androgens (not testosterone), but these are largely degradation products of corticosteroids.

Inactivation of Steroid Hormones

Over 100 steroid derivatives have been isolated from urine, so what is said here can only be a very rough guide, particularly as steric configuration (Chapter 3) is very important in determining the origin of an excretion product, and is not treated here in any depth. The liver is an important site of steroid catabolism, both in conjugating hydroxyl groups and in oxidation and reduction, so that liver malfunction may lead to apparent adrenal or genital syndromes (see also p. 377).

1. The hydroxylation of oestradiol 17-β at the 16-position to form oestriol may perhaps be regarded as an inactivation mechanism in man, since the latter is a very much weaker oestrogen than the former.

2. All the steroid hormones except the oestrogens have an oxo group at the 3-position, and this is very often reduced to the corresponding —OH group, thus pregnanediol from progesterone, androstanediol from testosterone, and a variety of cortical steroid derivatives. It is characteristic that these inactivation products have the 3α configuration, while precursors of the hormones, such as pregnenolone and dehydroepiandrosterone, have the 3β configuration of the cholesterol from which they are formed (Fig. 18.7). Differentiation of the so-called α-oxosteroids and β-oxosteroids either as a group or individually is therefore of great help in differential diagnosis, particularly of adrenal function. The adrenal corticoids have a Δ4, 5-double bond as well as the oxo group, and this is usually reduced as well, thus tetrahydrocortisol and tetrahydrocortisone are major excretion products of the cortical steroid hormones. Aldosterone, however, is not usually reduced in this way.

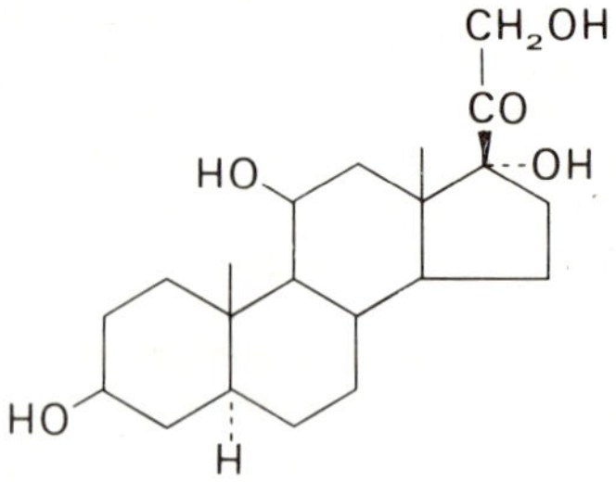

Tetrahydrocortisol

3. Oxidation of the $-OH$ bond at C–11 to $=O$, as in the transformation from cortisol to cortisone, causes loss of biological activity. However, this inactivation is reversible (see below).

4. The characteristic $-CO-CH_2OH$ side-chain at C–17 of the adrenal steroids is very easily oxidized away, both chemically and biologically, particularly if there is also a 17-OH group present, as in cortisol. This is the reason why there is a consistent output of 17-oxosteroids, some of which are androgenic, both in men and women. Those cortical steroid derivatives in urine which have not lost their side-chain may be oxidized in the laboratory after hydrolysis of their conjugating groups; they are known as the '17-oxogenic' steroids (p. 49).

5. In addition to all this, a major method of inactivation is conjugation, which chiefly takes place in liver, although for 3-sulphate esters see p. 393. Apart from sulphate the major conjugating molecule is glucuronide (p. 279); 3-, or 17-, or for aldosterone 18-, glucuronide conjugates may be formed. The solubilized steroids may be partly excreted in bile and re-absorbed from the gut.

Adrenal Cortex

More than 30 steroids in the cortex of the adrenal glands have an effect either on salt metabolism or on carbohydrate/protein metabolism. Only a few of these are normally secreted into the adrenal vein; in man, three compounds, cortisol, corticosterone and aldosterone, make up about 85 per cent of the total secretion (see Table 18.2). As indicated in Fig. 18.7, the usual order in synthesis after 17-hydroxylation is 21-hydroxylation (i.e. in the side-chain) and finally 11β-hydroxylation. Besides this, both androgenic and oestrogenic steroids are secreted but are not necessary for normal sexual function. Most of the adrenal steroid hormones have a hydroxyl group at C–11. Cortisone, which has a carbonyl group, is hormonally inactive, its effectiveness in corticosteroid replacement therapy depending on its reduction to cortisol in the liver. The same is true of the synthetic steroid *prednisone*.

Deoxycorticosterone (DOC), which has no radical at C–11, is quite an active mineralocorticoid, although not nearly so active as aldosterone, which has an —OH group there. The distinction between *glucocorticoids* which are supposed to affect only intermediary metabolism and tissue permeability, and *mineralocorticoids*, affecting only salt and water metabolism, is rather an artificial one. Even aldosterone has some effect on carbohydrate metabolism, and the glucocorticoids affect mineral metabolism quite strongly. Only DOC and some of the synthetic steroids (see Fig. 18.8) can be placed absolutely in one category or the other.

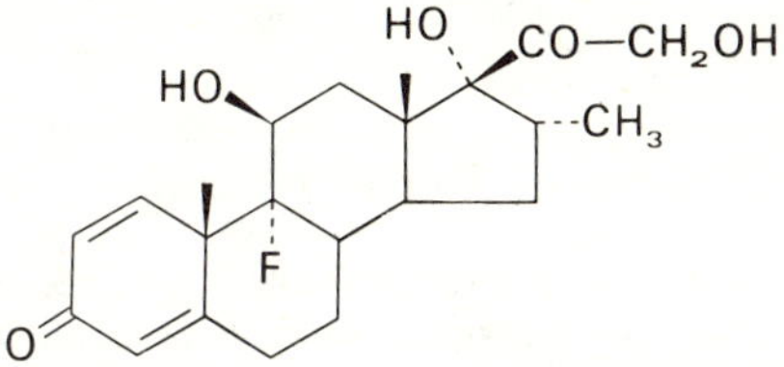

FIG. 18.8. Dexamethasone. Betamethasone is a stereoisomer of this drug. Both have very powerful glucorticoid and anti-inflammatory activity, but no mineralocorticoid activity at all.

Adrenalectomy or hyposecretion of the adrenals (*Addison's Disease*) give rise to severe disturbances of salt metabolism and/or carbohydrate metabolism which may be fatal if untreated. There is excessive excretion of sodium ions and retention of potassium leading to hyperexcitability of muscle and cardiac failure. The loss of sodium is accompanied by a loss of extracellular water. Nevertheless, the kidney is unable to excrete a dilute urine, and a water load will lead to water retention and 'water intoxication'. At the same time gluconeogenesis is inhibited, so that the subject is liable to suffer from hypoglycaemic crises if he goes without food for a few hours. Maintenance of liver glycogen during fasting has in fact been used as a test for activity of adrenal steroid hormones. Appetite is diminished, and with complete adrenalectomy, growth stops. In *primary* adrenal insufficiency, the gland is unresponsive to ACTH. In *secondary* insufficiency the gland is responsive (although it may be atrophic), and the deficiency is essentially one of ACTH secretion.

Oversecretion of adrenal steroid hormones leads to increased excretion of nitrogen (which is not very marked in man), diminution of growth, and inhibition of protein synthesis as in a lowered rate of formation of scar tissue. There is maintenance of liver glycogen even in the fasting state, and diminution of carbohydrate oxidation. Retention of sodium and a loss of potassium also occur. This potassium loss may be serious as it leads to muscular weakness and cardiac dysfuction. Although the adrenal steroid hormones are diuretic, oversecretion may cause sufficient sodium retention for oedema to occur. This may be prevented by a low-salt diet.

The hypothalamic/hypophysial production of ACTH is strongly inhibited by 11–OH steroids, whether they are natural, like cortisol, or synthetic, like dexamethasone (Fig. 18.8).

DOC, on the other hand, has no effect on ACTH production, although it is secreted in traces by the adrenal. The drug *metyrapone* inhibits 11β-

Metyrapone

hydroxylation, and may be used to test the ACTH-secreting capacity of the pituitary, if the responsiveness of the adrenal cortex has first been established with injected ACTH. In the normal person, metyrapone inhibits the 11β-hydroxylase, which first leads to raised plasma DOC and lowered plasma cortisol. This latter fall stimulates ACTH secretion, which brings the cortisol level back to normal, with gross overproduction of DOC. Failure to obtain such an overproduction is evidence of secondary adrenal insufficiency.

Cortisol production at greater than normal rates may be due to neoplasms, stress, or adrenal hyperplasia. In *Cushing's syndrome*, which may be defined as continuous overproduction of cortisol, there is a partial or complete failure of the feedback mechanism by which ACTH secretion is regulated. There are symptoms of muscular weakness, sodium retention, and oedema, together with a characteristic distribution of sub-cutaneous fat and cutaneous striae. Definition in terms of cortisol overproduction distinguishes Cushing's syndrome from the adrenogenital syndrome, in which there has been shown to be a partial deficiency of the 21-hydroxylase. As a result 17–OH progesterone accumulates and is converted to testosterone. At the same time the reduced rate of cortisol production stimulates ACTH secretion and leads to adrenal hyperplasia.

Aldosterone production and secretion is not controlled by ACTH, except in so far as the latter is responsible for maintenance of adrenal size, but by a change in some function of blood volume. As explained on p. 316, the amount of Na^+ in the extracellular fluid space (including the blood) is so great compared to the other cations that a change in total Na^+ means a change in extracellular fluid volume and vice versa. Most investigators think that the secretion of *renin* in response to a reduction in renal blood flow is the primary event in the control of Na^+ balance through aldosterone secretion. Renin leads to the formation of the peptides angiotensin and angiotensin II (p. 65). The latter powerfully stimulates aldosterone secretion from the zona glomerulosa. ACTH and K^+ may potentiate the effect of the peptide.

The stress reaction. Overdosage of steroid hormones, i.e. injection of large doses into the *normal* animal, is followed by the effects described in previous paragraphs and also by a characteristic lymphocytopenia and involution of the lymphatic tissue, including the thymus. Similar effects on lymphocytes, protein metabolism, and salt retention (with potassium excretion) may be produced by subjecting untreated animals to a variety of stresses such as cold, bacterial infections, burns, surgical procedures, or severe muscular exertion. It has been strongly advocated that such stresses cause hypersecretion of adrenal steroid hormones, and that the stress responses are adjustments of the organism induced by these hormones to fight the stress. It is true that the anterior pituitary increases its secretion of ACTH in response to nervous stimulation (both directly and by way of adrenaline), and also that the adrenalectomized animal does not show changes in metabolism in response to stress and succumbs more easily. Nevertheless it can be shown that the adrenalectomized animal given a *constant* dose of steroids will show a normal response to stress. Also the stress response has been demonstrated in patients immediately *after* bilateral adrenalectomy. From this and other similar evidence it seems preferable to say that adrenal steroid hormones have a *permissive* effect on metabolic response to stress, rather than directly causing it. The position is similar with regard to the effect of adrenocortical secretions on growth.

Permeability. Many adrenal steroids, particularly cortisone and cortisol, and synthetic steroids with a 'glucocorticoid-like' action, increase the permeability of connective tissue and of synovial membranes. They may physically reduce the space-filling arrangement of the mucopolysaccharide between the collagen fibrils (p. 27). They also stabilize tissue cells damaged by the inflammatory reaction, and in the long term they stimulate the synthesis of a protease that destroys the collagen itself. This can have a deleterious effect on bone. This effect of cortical steroids may be linked with their stimulation of secretion in general (see, for example, Fig. 18.5). Stimulation of gastric secretion not infrequently induces peptic ulcers during steroid therapy. The effect may be produced by local application of the hormones, and it is frequently used for the relief of oedema and skin reactions to injury (e.g., sunburn) or bacterial infections of connective tissue, or of bursitis and arthritis. Pain is relieved, possibly because oedematous tissue no longer presses on pain receptors. This use of adrenal steroid hormones is very valuable, but the underlying cause of injury is not removed and the spread of bacterial infection may actually be increased by the rise in permeability.

Adrenal Medulla

The chromaffin tissue of the medulla of the adrenal gland contains two hormones:

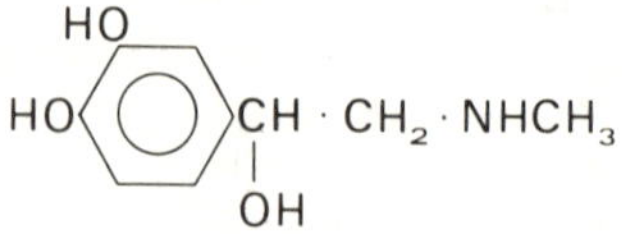

HO—\
HO⟨◯⟩CH · CH$_2$ · NHCH$_3$\
    |\
    OH

Adrenaline

and

HO—\
HO⟨◯⟩CH · CH$_2$ · NH$_2$\
    |\
    OH

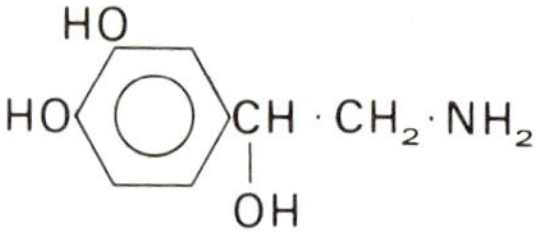

Noradrenaline

In man, 80 per cent of the hormone in the medulla is adrenaline, but noradrenaline is the chief component of the plasma catecholamines (catechol is 1,2-dihydroxybenzene). Noradrenaline is a chemical transmitter at sympathetic ganglia, and at the junctions of nerves and smooth muscle. It therefore has a powerful effect on the peripheral vascular system, while the effects of adrenaline on blood pressure are limited to heart rate and cardiac output.

Both adrenaline and noradrenaline raise the plasma NEFA levels markedly by activating the fat cell lipase, but only adrenaline causes a rise in blood sugar, by causing the conversion of liver phosphorylase *b* to the *a* form. Adrenaline also causes the conversion of muscle phosphorylase *b* to *a*, which produces a rise in blood lactate. The effects both on lipase and phosphorylase are mediated by stimulation of adenyl cyclase, with a subsequent rise in the intracellular concentration of cyclic AMP. There must, however, be differences in the sensitivity of cell receptors towards the two hormones, since biochemical as well as physiological responses to them clearly differ. It is possible that those physiological responses of the catecholamines that are associated with the β-receptors (vasodilation, speeding up and strengthening of the heart's beat) are mediated through cAMP, but that these are much faster in appearing than the rise in blood sugar and the breakdown of glycogen.

The medulla of the adrenal gland is not essential to life, and apart from a certain placidity demedullated animals behave entirely normally. Their blood still contains of course some noradrenaline coming from the preganglionic adrenergic neurones.

An important pathway for the inactivation of 'catecholamines' is methylation at the $-OH$ on C–3 of the benzene ring, followed by removal of the amino group by amine oxidase.

Prostaglandins

These complex carboxylic acids (formula on p. 42) are found in highest concentration in seminal fluid, but they are also widely distributed in tissues such as kidney, pancreas, lung and brain, as well as in blood. There is evidence that some prostaglandins are formed in adipose tissue when it is exposed to lipolytic hormones.

The prostaglandins have many physiological effects, a large percentage of which involve smooth muscle, e.g., lowering the blood pressure, increasing intestinal motility and contractions of the uterus. It is not certain whether any of these are true hormonal control mechanisms. Biochemically, the most important effect of prostaglandins is that they inhibit the release of non-esterified fatty acids by adipose tissue; they are as effective as insulin in this respect. At the same time they cause a rise in the concentration of cyclic AMP in this and other tissues, which is difficult to reconcile with the 'second messenger' theory of hormone action.

Many of the physiological and biochemical effects of prostaglandins can be explained in terms of an antagonism to adrenaline or noradrenaline, but this is not true for all effects.

Sex Hormones

Testosterone (formula p. 46)

This is the main secretion of the testis. The plasma concentration in men is about 15 times higher than in women. It is absolutely necessary for maintenance of the sperm ducts and accessory sex glands, which involute on castration. It is also responsible for the development of the secondary sex characteristics, such as hair pattern, voice and behaviour (sexual drive). The latter two, and some others, are not lost by castration of the adult.

If present very early in fetal life, testosterone can reverse genetic sex by inducing development of the Wolffian ducts and suppressing the Mullerian ducts. The sexual organs so induced may later become functional.

Testosterone and other androgens stimulate the growth of muscle—these are the 'anabolic steroids' familiar in athletics and horse-racing. The increase in net protein synthesis is blocked by actinomycin D, but some protein synthesis seems to precede the synthesis of RNA. The mechanisms described in Chapters 15 and 16 make this intelligible.

Oestradiol and oestrone. Oestradiol is the most important oestrogen; only the 17β-stereoisomer (formula on p. 47) is biologically active. Oestrogens are synthesized in the ovary, but also in testis, adrenal and placenta. The urinary output of oestrogens in men, and in women at the onset of menstruation, is not very different. The human placenta becomes a major source of oestrogens in pregnancy, but in the third trimester much of this is oestriol, which originates in the fetus. The latter synthesizes dehydroepiandrosterone (Fig. 18.7), and 16–OH-dehydroepiandrosterone which is transferred to the placenta. This organ converts it to oestriol, the excretion of which in the last trimester is used as an index of fetal health, especially in high-risk pregnancies.

Oestrogens cause rapid growth to mature size and function of the ovaries, the uterus (both endometrium and myometrium), the vagina and the mammary glands, when injected into immature or castrated animals. Like testosterone, they can reverse genetic sex in the early embryo. They are also responsible for development of the secondary sexual characteristics and behaviour pattern, and promote closure of the epiphysial discs in the long bones.

Progesterone (p. 47) is synthesized in the corpora lutea of the ovary, but during pregnancy the placenta becomes the major source of the hormone in humans. It has little effect on tissues that have not been primed with oestrogens, but in its normal function it thickens the uterine endometrium, and is responsible for the 'decidual reaction', by which the fertilized ovum is buried in the endometrial tissue. It inhibits myometrial contractions, which would otherwise terminate pregnancy. It causes development of the secretory alveoli of breast tissue, but inhibits milk secretion, which does not start until pregnancy is over.

It is progesterone, rather than oestradiol, which inhibits the secretion of the hypothalamic gonadotrophin-releasing hormone (cf. oral contraceptives, below).

Almost all the functions of the sex hormones covered in this brief review are concerned with cell division and differentiation and the de-repression of specific proteins. We know too little of the mechanisms involved, in eukaryote cells, to be able to explain the actions of the hormones in biochemical terms. However, one may observe that the structures of progesterone and to a less extent testosterone, are very similar to those of the adrenal steroid hormones, which have much less effect on genetic material, and it is therefore possible to visualize a rather specific set of receptors in the cells of the various target organs which transfer the bound hormone to the nuclei that can respond to it. Such a mechanism has already been established in outline for oestrogens, as described on p. 376. This transport into the nucleus is only a first stage in the complete answer to the problem of sex hormone action. For example, hepatocytes are able to take up and extensively metabolize almost all steroid hormones without their own nuclei being affected.

Oral Contraceptives

Ovulation is completely inhibited during pregnancy, and this is due to the much increased secretion of progesterone, partly from the corpora lutea and partly from the placenta in late pregnancy, in amounts up to 250 mg per day. It is not practicable to give the large doses of progesterone which would be necessary, and androgens, which are also effective, have undesirable side-effects. It is also important that the agents should be active when taken by mouth, which the natural oestrogens, in particular, are not. Several synthetic steroids, such as norethynodrel and chlormadinone acetate, have been shown to inhibit conception without having considerable oestrogenic or androgenic activity. It is not at all clear how these compounds act; they usually suppress the pre-ovulatory spike of LH and FSH secretion, but with chlormadinone, at all events, ovulation takes place in over half the cycles. Possibly there are also changes in the tissues of the reproductive tract that are unfavourable to fertilization or nidation. Most of these compounds have progestational activity and cause proliferation of the endometrium. They are usually combined with a small quantity of an oestrogen, so that when the inhibitor (which is taken by mouth) is withdrawn, a normal menstruation follows. It is not, however, necessary that this should be so, and some ovulation inhibitors are known which are not progestagens.

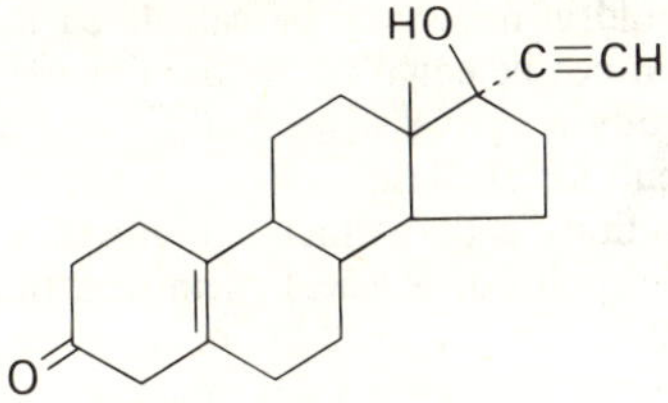

Norethynodrel

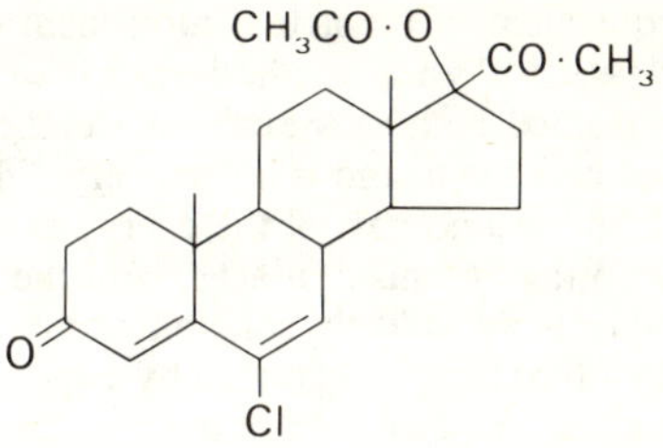

Chlormadinone

Although oral contraceptives have been shown in clinical trials to be reliable, and a normal pregnancy rate has followed when they have been discontinued, several long-term questions remain to be answered, for example, whether their use gives rise to ovarian tumours or to mutations in the ovum itself, and the extent of the definite correlation between these drugs and thrombosis.

19 Energy Requirements and Food Intake

The energy which the human body expends may be divided, from the point of view of an outside observer, into two parts. The first relates to the energy required to move the body or any object in its surroundings. The energy equivalent of the actual movement could, if necessary, be calculated by classical physical methods, but it would not be equal to the total energy expended, because the efficiency of the body as a physical mechanism would not be known, and is indeed very difficult to calculate.

The second part of the energy expenditure is that which is required to keep the organism intact. From the viewpoint of physical chemistry, the body is an *unlikely* conglomeration of essentially unstable compounds, dissolved in or surrounded by a very precise but unusual salt solution, and maintained, as a rule, above the temperature of its surroundings. A good deal of chemical energy must be expended merely to preserve such a combination intact. Some of the energy may be used in physiological ways —in the osmotic work of the kidneys, in the pumping of the blood to transport oxygen and remove wastes, in the contractions of the pulmonary muscles. But it must be remembered that each cell throughout its life must expend energy to maintain its integrity, whether or not it is taking part in the activity of the whole organism. As a striking example, the energy expenditure of the central nervous system has been estimated to be 10 per cent of that of the whole body at rest, but there is very little variation in this energy utilization between directed thought, unconsciousness, or complete imbecility.

All the energy expended to keep the organism intact appears as heat, except for the not inconsiderable amount which is represented by water vapour in expired air or in insensible perspiration. This energy is expressed in joules (J). Most of the energy used in physical work also appears as heat, because of the low efficiency (about 15 per cent) of the body. It must always be remembered that most of the chemical energy of the food, although it is expressed in joules, has performed a number of vital biological functions in the tissues before it reappears in the environment as heat.

The Measurement of Energy Expenditure

As all the body's energy expenditure appears as, or can be converted into, heat units the simplest way of measuring energy expenditure is by using a calorimeter. Although the calorimeters which have been constructed for this purpose are large and complicated to build and use, they need not be described in detail. They consist in essence of a chamber, well insulated, from which the heat is removed by piped water, whose flow rate and temperature rise are accurately measured. Arrangements must be made for gas exchange, and the water formed by the subject must be collected and weighed. The heat used in warming the excreta is small, but must not be neglected.

Such an apparatus is impossible for routine use, or for measuring energy expenditure in physical work, and recourse must be had to *indirect methods*. The guiding principle here is the chemical theorem that *the energy liberated in a chemical reaction is the same, no matter by which intermediate steps it is carried out*. Thus the energy liberated (as heat) in the reaction

$$C_6H_{12}O_6 + 6O_2 \rightarrow 6CO_2 + 6H_2O$$

is exactly the same whether it is carried out on a fire, in a calorimeter, or in the body by way of pyruvic acid and the tricarboxylic acid cycle, so long as the end-products in each case are CO_2 and H_2O. Thus by measuring the heat given off in a calorimeter on burning a given quantity of fat, carbohydrate, or protein, it is possible to say what must be the energy liberated by a body consuming known quantities of the three foodstuffs. The calorimeter usually used is the so-called bomb calorimeter which can be filled with oxygen under pressure and ignited electrically. A correction has to be made for protein, as one of the biological end-products, urea, is different from that obtained in a calorimeter. The figures obtained are 37·6 kJ/g for fat, 17·2 kJ/g for carbohydrate (mixed), and 18 kJ/g for protein.

This method of assessing energy expenditure is only applicable to long-term studies; it cannot be used for measuring the energy used in specific physical tasks. For this, use is made of the fact that the combustion, for example, of 1 gramme-molecule of $C_6H_{12}O_6$ always produces $180 \times 17·2$ kJ, and it always requires 6 moles of O_2. Thus one can state the energy produced when 1 mole, or, more usefully, 1 litre of oxygen is used to combust carbohydrate (or fat, or protein). This comes to 19·7 kJ/litre for fat, 17 kJ/litre for carbohydrate, and 18·6 kJ/litre for protein. The only problem now is to decide the proportions of the three foodstuffs actually being oxidized in any given period. The amount of protein can be estimated by finding the amount of urea excreted, but since the proportion of protein oxidized is not usually more than 10 per cent, it is often ignored completely in the cruder investigations. For the estimation of the proportions of fat and carbohydrate being combusted, the *respiratory quotient* (strictly the non-protein RQ) is used.

$$RQ = \frac{\text{volume of } CO_2 \text{ produced}}{\text{volume of } O_2 \text{ consumed}}$$

The RQ of $\quad C_6H_{12}O_6 + 6O_2 \rightarrow 6CO_2 + 6H_2O$

is 6/6 or 1·0. The RQ of fat oxidation is 0·71. Tables exist showing the joules liberated per litre of oxygen for RQs between 0·71 (19·2 kJ/litre) and 1·00 (21·1 kJ/litre). Since the difference is small, the RQ, to a first approximation, may be neglected. Any of the recognized apparatus for measuring gas exchange may be used for estimating energy expenditure by the indirect method. There are limitations. The subject should not be changing in weight, or synthesizing fat from carbohydrate (the RQ for this > 1), which means that measurements should be made in the post-absorptive period. The subject should not be in oxygen debt, or developing acidosis or alkalosis.

The Basal Metabolic Rate

The energy expended in maintaining the integrity of the organism (see p. 404) may be tentatively identified with the *basal metabolic rate*. The latter is strictly the heat lost when voluntary activity is at a minimum, and is carefully defined. The subject must be at mental and physical rest, 12 hours after the last meal, and in an equable temperature (about 18°C). Even a small departure from these conditions increases the observed heat output considerably. The energy expenditure for adults, in the conditions laid down, is found to depend on sex and surface area. It is 167 kJ/m²/hour for males and 157 kJ/m²/hour for females, and remarkably constant throughout the normal population. The rate is much higher in infants and children, and begins to decrease again after 40 years of age.

The BMR is not fundamentally concerned with maintaining the body temperature, since it is unchanged in subjects living in climates with temperatures above 37°C. The fetus also has a BMR in utero. The basal heat output has, however, been adapted, under the control of the thyroid gland, to be part of the temperature-regulating mechanism. The clinical importance of the BMR is that it is an indicator of thyroid dysfunction. A BMR of − 20 per cent or more indicates myxoedema, of + 20 per cent or more, thyrotoxicosis. Pathological states affecting the BMR are shown in Table 19.1.

Table 19.1
Pathological Conditions Affecting the BMR

Rise in BMR	Fall in BMR
Fever	Starvation
Diabetes insipidus	Hypothalamic obesity
Cardiorenal disease	Addison's disease
Leukaemia, polycythaemia	Hypothyroidism
Hyperthyroidism	

Total Energy Expenditure

The usual way of assessing total energy expenditure is to calculate the BMR, which is usually about 6.7×10^6 J/day for an adult male, and to add to it figures representing the energy used in the various activities of the day. Ten per cent of the total thus obtained is then added to allow for Specific Dynamic Action (see p. 413). Many textbooks have tables of energy expenditure per hour for various activities; they are not very reliable for various reasons. Extended work includes numerous short rest periods which are not allowed for by the figures, and leisure activities account for a great deal of the total energy expenditure. Finally, the efficiency of food utilization varies enormously between individuals; it is therefore point-less to calculate food requirements from an estimation of the activity of a single person. In heavy labour the energy expenditure varies to some extent with the weight of the subject.

A table of daily energy expenditure, obtained from groups of subjects, and checked against food intake, is given in Table 19.2. The figures do not allow for biological variation and differences in leisure activity.

Table 19.2
Daily Energy Expenditure in Kilojoules

	Males	Females
Infants <1 year	460 per kg	
Children, 4–9 years	6 300–8 400	
Children, 10–15 years	up to 21 000	up to 17 000
Clerical worker or student, 24–25 years	11 700–12 500	10 000
Housewife, 30 years		9 200
Miner at coal face	17 000	
Very arduous labour or athletic training	up to 21 000	
Housewife, 50 years		6 300 +
Sedentary worker, 50 years	10 000	

Stress must be laid on the large energy expenditure at 10–13 years, corresponding with the pubertal growth spurt. This is only partly accounted for by the great physical activity of children of this age. The synthesis of new tissue requires much energy, partly because the compounds to be synthesized have chemical energy locked up in them, partly because, in accordance with physicochemical principles, a good deal of the energy supplied is wasted, and appears as heat. Even in rapid synthesis, such as the production of milk, the elaboration of 1 pint of milk (containing 1 550 kJ) requires at least 3 100 kJ of energy. With slower growth, when the formed tissue has to be warmed, oxygenated, and carried about, it has been estimated that the laying down of 1 kg of tissue in a year requires 340 kJ/day extra energy. This principle applies not only to *growth in childhood* but also to *convalescence*. Surgery or feverish infections are almost always accompanied by severe protein loss (see Chapter 18), and loss of weight. It is possible to calculate the energy content of the lost tissue, but in order that it may be replaced, *at least* twice the amount of joules so calculated must be included in the diet, suitably spread out over the recovery period.

Food Intake

Food as an energy supply. From the point of view of useful energy, i.e. that convertible into ATP or otherwise available for biological reactions, it does not matter very much whether the food intake is carbohydrate, fat, protein, or alcohol. Present information suggests that about 65 per cent of the energy combined in fat and carbohydrate can be trapped as ATP. The conversion factor for protein is probably considerably lower, but the problem with most diets is to raise the protein intake above the minimum requirement, so that its inefficiency as a source of useful energy is unimportant. The human body can convert glucose into any other carbohydrate or lipid required by the cells, so the chief limitations on the proportions of foodstuffs eaten are those outlined below:

1. A minimum of 2 500 kJ/day of carbohydrate ($\equiv 150$ g/day) is required to avoid the ketosis which would otherwise arise from fat catabolism (Chapter 10), and to give a protein-sparing effect (Chapter 11).

2. A diet containing no fat is intolerably bulky; 35 per cent fat is recommended by many authorities.

3. Enough fat must be taken to provide the fat-soluble vitamins (see Chapter 20).

4. The minimum protein intake for body maintenance is 40–60 g/day.

5. The contribution made by alcohol is irregular, but should not be underestimated. A study of miners' diets showed that alcohol provided, on the average, 2 500 kJ or 13 per cent of the total calorie intake on Saturdays, and 25 kJ or 0·1 per cent on Tuesdays.

Table 19.3 shows the calorific value, per pound, of some common foodstuffs.

The high energy value of such fatty foods as pork products, chocolate, and nuts should be remarked.

Table 19.3
Calorific Values, per Pound, of Some Foods

	Kilojoules		Kilojoules
Bread	5 000	Eggs (10–16)	3 200
Cakes	7 500	Milk (per pint)	1 715
Potatoes (uncooked)	1 840	Whisky (per pint)	
Rice (cooked)	540	(24 nips)	5 860
Butter	15 000	Beer (per pint)	1 050
Ham	up to 12 550	Chocolate	12 000
Beef	5 400	Nuts	10 700
Cheese	9 000	Sugar	7 800

Fats

It appears now to be established that man, like many other animals, cannot synthesize *linolenic* acid, but this seems to be the only essential fatty acid. The requirement is estimated to be 10 g per day for an adult, but this is normally exceeded in the diet. Only in very young children, and possibly in sufferers from the various sprue syndromes, may a deficiency state arise.

Protein

As described in Chapter 11, a proportion of the body's protein is always being catabolized and the nitrogen excreted. This nitrogen must be replaced (as amino acids) in the daily diet. Whether or not it is being replaced is estimated by the *nitrogen balance* technique. The total nitrogen in the food over a period is estimated, and compared with the total loss by way of urine and faeces, perspiration, hair, and nail clippings, and menstrual flow. A healthy adult should be in *nitrogen equilibrium*. A strong *negative balance* (i.e. net loss of nitrogen) is to be expected during illness. Children should show a *positive balance* (i.e. nitrogen retention associated with growth).

As protein is the most expensive item of the diet, and that most difficult to produce agriculturally, the determination of the minimum amount of protein necessary to maintain an adult in nitrogen equilibrium becomes important. This is very difficult, but a reliable estimate is 44 g/day. There is less tendency now than formerly to multiply this figure by 1·5, as a 'safety factor', because it is realized that in many less-developed countries well-nourished adults may eat less protein than this. The amount of 44 g/day represents about 11 per cent of the energy intake. The proportion of protein rarely rises to 15 per cent of the diet, but has, in certain populations, been known to make up 45 per cent of the total energy.

Children require more protein for their weight than adults. Recent recommendations are 3·5 g/kg/day up to the age of 1 year, falling to 1·5 g/kg/day during puberty.

To state the amount of protein in the diet means nothing if the type of protein is not specified. This is because the body requires certain amino acids, which it cannot make itself, to be supplied in at least minimal amounts. These are the *essential amino acids* (see Chapter 11). Proteins really deficient in one or more amino acids, such as gelatin or zein, will not enable nitrogen equilibrium to be reached if they are the sole dietary protein, no matter how much of them is included in the food intake. An empirical measure of the efficiency of a protein as a supply of essential amino acids is the *biological value*; this is

$$\frac{\text{g protein retained}}{\text{g protein fed}} \times 100$$

The diet must of course be free from proteins other than that under test. Estimates of biological value range from 97 for egg protein to about 63 for gelatin.

Amino acid analysis has shown that most proteins are seriously deficient in only one essential amino acid; usually the deficiency is in *tryptophan, lysine*, or *methionine*. The biological value of a protein tells us that it is deficient in essential amino acids, but does not tell us which is missing. The Food and Agriculture Organization of UNO has worked out a 'protein score' based on the essential amino acid content of proteins. Although the score was worked out a number of years ago, more recent research has not invalidated it, despite the fact that one point (below) is now a matter of controversy. The score is estimated on the following lines. Investigations by various workers have given a tentative picture of the absolute requirements of essential amino acids per kg body weight, which are shown in Table 19.4.

Table 19.4
Minimal Requirements for Essential Amino Acids

	Infants (mg/kg body wt./day)	Adult men (mg/kg body wt./day)	'Pattern' (Tryptophan = 1)
Isoleucine	90	10·4	3·0
Leucine	—	9·9	3·4
Lysine	90	8·8	3·0
Phenylalanine*	90	4·3	2·0
Methionine	85 ⎱85	1·5 ⎱13·1	1·6 ⎱3·0
Cysteine†	0 ⎰	11·6 ⎰	1·4 ⎰
Threonine	60	6·5	2·0
Tryptophan	30	2·9	1·0
Valine	85	8·8	3·0

* Tyrosine present in the diet.

† Cysteine can be replaced by methionine.

This table shows that tryptophan is the amino acid needed in least amount, and the 'pattern' in the last column of the table is arrived at by expressing all the other amino acid requirements as fractions of that of tryptophan. This is the pattern of the composition of an ideal food protein. Cow's milk protein, of high biological value, contains 1.4 g tryptophan per 100 g protein. We can now say that the ideal protein should contain, per 100 g, 1·4 g tryptophan, 2·8 g threonine, 4·2 g lysine, etc. This step in the argument is important, because the concentration of essential amino acids in the diet is significant, as well as the pattern. A protein containing only 0·42 per cent isoleucine, 0·48 per cent leucine, etc., would not be satisfactory. Food proteins are given a score by dividing the concentration of the amino acid in which they are most deficient by the concentration in the ideal protein (Table 19.5).

Table 19.5
Essential Amino Acid Content of Foods in g/100 g Protein

	Pattern	Egg	Beef	White flour	Beans
Isoleucine	4·2	6·7	5·2	4·1	5·6
Leucine	4·8	8·8	8·7	6·9	8·4
Lysine	4·2	6·2	8·4	1·95	7·15
Phenylalanine	2·8	5·7	4·0	5·0	6·4
Total sulphur-amino acids	4·2	5·3	3·7	3·0	1·95
Threonine	2·8	4·8	4·3	2·7	4·3
Tryptophan	1·4	1·6	1·2	1·1	0·9
Valine	4·2	7·2	5·4	4·1	5·9
Protein score	100	100	$\dfrac{1·2}{1·4} \times 100$ $= 86$	$\dfrac{1·95}{4·2} \times 100$ $= 46$	$\dfrac{1·95}{4·2} \times 100$ $= 46$
Biological value		96	77	53	46

This technique makes it possible to provide a proper complement of essential amino acids by mixing incomplete proteins. In general all animal proteins (except gelatin) are *complete*; cereal proteins are *incomplete*, lacking *lysine*, and leguminous proteins are incomplete, lacking *methionine*. A suitable mixture of cereals and legumes can therefore provide a full complement of essential amino acids. (The terms *first class* and *second class* proteins should be avoided, as they give no indication of the possibility of mixing incomplete proteins in the diet.)

The only real point of controversy about this technique is the minimal requirement for lysine (Table 19.4), particularly for infants. The nutritional experiments on which these tables are based were, for obvious reasons, carried out on animals, chiefly rats. Weanling rats grow very fast, and have a much higher lysine requirement, on a weight basis, than adult rats. There is some doubt whether children, who grow much more slowly, need so much lysine. The point is extremely important, because as Table 19.5 shows, cereal proteins, on which malnourished children depend very heavily, are lysine-deficient. Adding free lysine to the diet is inefficient, as well as expensive, since lysine and reducing sugars form inactive condensation products during cooking. Attempts have even been made to breed cereals with a higher lysine content in their proteins. A reliable estimate of the true need of young children for lysine would be a great blessing.

One other facet of protein nutrition must be discussed. There is *no storage* of protein or of amino acids in the body. Therefore all amino acids required for synthesis of replacement protein must be absorbed at the same time. Most work has been done on the omission of tryptophan. If this amino acid is injected into rats fed on a diet which is otherwise complete, the amount of protein synthesized in the liver depends on the interval separating the injection from the food intake. Even a 4-hour interval reduces the rate of protein synthesis very sharply. It is found that injection simultaneously with food not only prevents the accumulation of incomplete peptide chains, but also in some way activates the ribosomal protein-synthesizing machinery. This latter effect appears to be specific, at least in liver, to tryptophan. Thus, if incomplete proteins of different types form part of the diet, they should be mixed at each meal for maximum effectiveness. Similarly, since protein synthesis requires a good deal of energy, carbohydrate or fat should be taken with protein for maximum replacement of tissue protein. If this is not done, part of the absorbed amino acids will be catabolized to supply energy. Some slimming diets have been devised on this principle; a high-protein, low-fat, low-carbohydrate meal allays hunger but leads to net loss of a little endogenous protein (see also next section).

Protein–Calorie Malnutrition

In the light of the information in this chapter, it is understandable that protein deficiency is particularly widespread among children. Two syndromes are prevalent in the literature, *kwashiorkor* (a West African name) and *marasmus* (a word derived from Greek). These are extreme states, and most cases are not so clear-cut, so that the term *protein–calorie malnutrition* (PCM) is more generally useful.

Typically, kwashiorkor attacks children some 3–6 months after the belated weaning customary in many African societies. Apart from failure to gain weight and to thrive, the abdomen is distended, with an enlarged liver, and there may be oedema in the legs. Yet the child may appear well nourished. Marasmus, on the other hand, is characterized by wasting, so that the flesh appears to shrivel and the skin to hang loose. In both diseases there tends to be an intolerance to the disaccharides sucrose and lactose. These are fermented by bacteria in the gut, so that the stools are frothy, copious and fetid.

Kwashiorkor arises when a child which has been breast-fed for 2–4 years is weaned. The milk contains adequate protein, which may be synthesized at the expense of the mother's own protein; but the subsequent diet, although rich in carbohydrate and fat, does not contain enough protein, even when the protein-sparing effect of carbohydrate (Chapter 11) is taken into account. The enlarged liver is fatty and this is not easy to explain; it may be due to choline or phospholipid deficiency. The oedema is due to insufficient synthesis of plasma proteins. The disaccharide intolerance is caused by a deficiency of sucrase and lactase in the mucosa of the small intestine (Chapter 8), attributable to diminished protein synthesis. It may persist for months after other symptoms have been relieved.

Marasmus arises when a child has neither enough protein nor enough carbohydrates, so that its tissue proteins are slowly used for gluconeogenesis, and there is generalized wasting. It is clear that in any given situation, either protein or energy deficiency may predominate.

Much effort has been spent in looking for simple tests which will diagnose these conditions, or better the pre-disease states. Among others, lowered plasma albumin, increased excretion of essential amino acids, increased excretion of prolyl-hydroxyprolyl-peptides (Chapter 13), excretion of short peptides apparently from incomplete chains on ribosomes (especially in marasmus) and reduced hyperglycaemia after a sucrose load have been proposed

Specific Dynamic Action

When a subject is kept in such conditions that his metabolic rate is perfectly steady (not necessarily in BMR conditions), and then allowed to eat, his metabolic rate, measured either as heat output or oxygen consumption, goes up. The increase is proportional to the calorie value of the food taken in, but is not the same for all foodstuffs. For carbohydrate and fat the increase is about 5 per cent, but for protein almost 30 per cent. This rise in heat output which follows eating is known as the *specific dynamic action* (SDA) of foodstuffs. It is of course waste heat from the point of view of cellular energetics. Partly it is due to the chemical energy lost in hydrolysis of the food, and the physical energy lost in intestinal movements, but a considerable SDA has been shown to be produced by the injection of amino acids directly into the bloodstream. It seems likely that the large SDA of protein is a consequence of the facts that amino acid oxidation is not always so efficient a source of useful cellular energy as carbohydrate or fat oxidation, and that urea synthesis from amino acid nitrogen requires considerable energy expenditure (Chapter 11). This view is strengthened by the fact that in periods of rapid growth when amino acid oxidation is suppressed, the SDA of protein may also disappear.

The SDA of a typical meal of a European diet is about 10 per cent. It is usual in calculating energy requirements to find the total energy output of the various daily activities and to add on 10 per cent of this to allow for the SDA. The grand total then represents the joules which must be supplied in the daily diet.

The necessity for an adequate supply of minerals and vitamins (see Chapter 20) must be remembered in preparing diets. The considerations outlined in this chapter are only the basis of proper nutrition.

20   Accessory Food
         Factors

By accessory food factors is meant, in general, the vitamins and minerals which are present in the diet only in traces, but which are essential for the normal function of the body. However, two of the minerals, calcium and phosphorus, are required in more than trace amounts, while a number of the amino acids and one unsaturated fatty acid must be provided in some quantity since the body cannot synthesize them. They are not usually classified as *accessory* food factors.

Minerals

A requirement for *chromium* has been reported, and in animals *selenium* is both required for satisfactory reproduction, and is toxic in excess. Thus it is impossible at the present time to give a list of inorganic materials which will include all the elements required in traces. The most important inorganic requirements are for *Na, K, Ca, Mg, Fe, Cu, Zn, Mo, Cl, I*, and *P. S* as sulphate is essential for the formation of certain body constituents, but it is always organic sulphur, in the form of the sulphur-containing amino acids, which is oxidized to sulphate in vivo. *Co*, as far as is known at present, is only necessary in the form of vitamin B_{12} (q.v.). The cobalt here is not ionized, and is thus not, strictly speaking, an inorganic nutrient. *F* does not form part of the normal body requirements, but it must be mentioned since it may be added regularly in traces to drinking water as a preventative of dental caries.

Potassium is the main inorganic cation present in intracellular fluid, as distinct from plasma and extracellular water. Its concentration is about 3·2 g (80 mEq) per kg (muscle). It also occurs to the extent of about 4 mEq per litre in plasma. Since it is universally abundant in foodstuffs, whether of plant or animal origin, there is usually no difficulty in supplying the daily requirements. These are in fact not precisely stated in many textbooks on nutrition, but are probably of the order of 4 g per day. Disturbances of potassium metabolism may occur, particularly in derangements of the adrenal cortical gland.

Sodium. The average adult requires about 5 g Na$^+$ per day, which is not provided by foodstuffs unless salt is added in cooking. This requirement may be greatly increased in many circumstances, particularly when a hot climate or prolonged exercise have caused a considerable loss of Na$^+$ in perspiration. Insufficient Na$^+$ intake can give rise to heat exhaustion or heat stroke, characterized by muscular weakness, nausea, and fever. The excretion of sodium is carefully regulated by renal mechanisms under the control chiefly of aldosterone, and may be abnormal in diseases of the adrenal gland.

Chloride is the chief anion present in the body, particularly in the extra-cellular fluid. The amounts in food are not sufficient for the normal daily intake, and it is usually added to the diet together with Na$^+$, as common salt. The requirements are about 7·5 g per day of Cl$^-$.

The requirements for *Ca*, *Mg*, and *Fe* are very different, but the absorption of each from the food is complicated by the fact that these three metals form insoluble hydroxides in alkaline conditions. All three also form salts which are insoluble, or in which the metal is poorly ionized. For these reasons a great deal of Ca, Mg, or Fe which is actually ingested with the food may be excreted in the faeces unless the conditions in the intestine are favourable.

Calcium and Phosphate Metabolism

The Structure of Bone

Bones and teeth are distinguished from all other tissues in the body by their high mineral content. The basic formula of this material is that of *hydroxy-apatite*: Ca$_{10}$(PO$_4$)$_6 \cdot$(OH)$_2$, although bone salt as prepared contains considerable amounts of other ions, chiefly Na$^+$, Mg^{2+}, CO$_3^{2-}$, and organic anions. However, like most inorganic molecules, hydroxy-apatite in the solid state does not exist as single molecules, but as a regular crystal lattice (Fig. 20.1). Ions of similar charge and shape can replace Ca^{2+} and PO$_4^{3-}$ and OH$^-$ in the lattice; thus OH$^-$ can be replaced by F$^-$, PO$_4^{3-}$ by CO$_3^{2-}$, and Ca^{2+} by Mg^{2+}, Sr^{2+}, Ba^{2+}, and Ra^{2+}, among others. For this reason the formula of any particular sample of bone salt cannot be precisely specified. Cations which become part of the lattice structure can remain in bone for extended periods, often with serious consequences (see below). The crystals of hydroxy-apatite are very small —never more than 0·1 mm long—and they are surrounded by a hydration shell of water molecules, which is very extensive. For example, 46 per cent of the body's Na$^+$ is in this shell, but outside the crystal structure proper.

Since the crystals are so small they would never form a rigid structure, able to withstand tension, if they were not held together by the matrix.

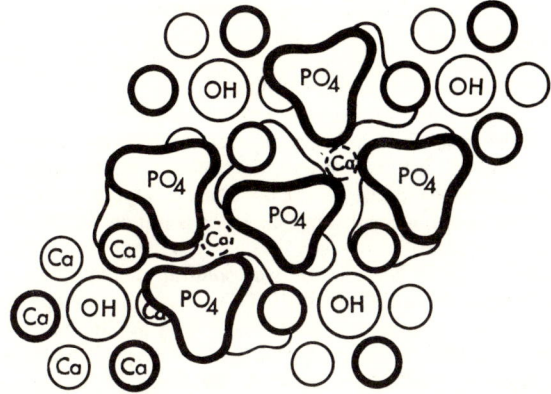

Fig. 20.1. Simplified representation of the hydroxy-apatite lattice. The figure shows the crystallographic 'unit cell', which is repeated in all three dimensions. The ions and interionic distances are roughly to scale, but the individual atoms of the phosphate groups have been replaced by triangular shapes for simplicity. The phosphate groups are in actual fact tetrahedral, not flat.

There are two layers of ions in the unit cell, in the direction at right-angles to the paper. One layer (shown by thick outlines) is at three-quarters of the unit height, the other (shown by thin outlines) is at one-quarter unit height. The calcium ions shown by dashed outlines represent two superimposed atoms, one at zero and the other at half unit height. Two hydroxide ions are superimposed on each other, in the 'tunnel' formed by the Ca^{2+} ions, at one-quarter and three-quarter unit height.

This consists of fibrils of collagen regularly arranged, to which the crystals are attached in a regular fashion. Between the fibrils is the ground substance, a mixture of hyaluronic acid and chondroitin sulphates attached to a carrier protein, see Fig. 2.7, Chapter 2. In the development of bone, the matrix is formed first, in the cartilage, and when bone is resorbed, the matrix disappears together with the bone salt. The proper formation of the matrix, especially of the collagen fibrils, is essential to the formation of tough bone.

In serum the Ca and P concentrations often vary inversely (see p. 433), and it is often said that this is a result of the *solubility product* relationship. This states that for any solution of an ionized salt in contact with the solid (i.e. a saturated solution), the product of the activities of the ions is a constant, equal to the solubility product $K_{s.p.}$. In the simplest case, if the solution is merely saturated with respect to a single uni-univalent salt in it, the concentrations of both ions will be equal, but this need not necessarily be so. In the present case the solubility product would be

$$K_{s.p.} = [Ca^{2+}][PO_4^{3-}]f_{Ca} \cdot f_{PO_4}, \qquad (f_{Ca} \text{ and } f_{PO_4} \text{ are the activity coefficients})$$

The fact that the serum inorganic Ca and P concentrations appear to obey some such relationship has been used as evidence that these ions are in equilibrium with the solid salt, i.e. the bone salt, implying that plasma is a saturated solution of bone salt. This, however, is not so, for the following reasons.

The simple salt which precipitates from solutions containing Ca^{2+} and phosphate at neutral pH is $CaHPO_4,2H_2O$. At the physiological plasma concentrations: 1·3 millimolar (ionized) Ca^{2+}, 1·3 millimolar phosphate, the solubility product for this compound is not reached, i.e. it will not precipitate from normal plasma. However, $CaHPO_4,2H_2O$ is unstable at neutral pH and rapidly transforms into hydroxy-apatite, which is much less soluble. For various reasons this latter compound does not have a true $K_{s.p.}$, but it has been shown by experiment that if bone salt is added to serum, or imitations of serum, the concentrations of dissolved Ca and P will fall, i.e. they will precipitate on the crystals.

We thus have the situation that plasma, and the extracellular fluid, are *supersaturated* with respect to bone salt. Thus variations in the plasma P concentration should have no effect on the Ca level because the product [Ca][P] is already greater than the apparent $K_{s.p.(bone salt)}$.

This approach explains one fact—that calcium phosphate does not precipitate all over the body. It does not normally do so because $K_{s.p.(CaHPO_4,2H_2O)}$ is not exceeded. In hypercalcaemia (plasma Ca > 15 mg per cent) it can and does do so; there is calcification of soft tissues and the formation of renal stones.

It remains to explain why the plasma Ca and P do not precipitate on the already formed bone salt, i.e. why the normal Ca and P concentrations are not much lower than they are. In hypercalcaemia, this precipitation does, in fact, take place, with the formation of 'marble bone'. In the normal person the only explanation can be that the apatite crystals are not in contact with the extracellular fluid, that they are protected from it by a zone (the hydration shell?) in which the activity, but not necessarily the concentration, of Ca is reduced. Of the various compounds which could do this, citrate is the most likely. It has been shown that osteoclasts have a very low concentration of isocitrate dehydrogenase by comparison with other enzymes. The citrate level in bone is higher than that in plasma; in fact bone is the major source of plasma citrate. After parathyroid hormone injection the urinary excretion of citrate rises (see p. 388). In the absence of the parathyroid the plasma Ca level falls to 6–7 mg per 100 ml, which suggests that the protective zone around the apatite crystals has disappeared.

This still leaves the major problem of explaining how the bone salt ever precipitates at all. The answer to this is not known, but the requirement for a matrix, and the regular arrangement of the crystals along the collagen fibrils, both suggest that there are 'nucleation centres' which initiate the crystal growth. These centres do not appear to be the mucopolysaccharides present in large quantities, since they do not bind Ca^{2+} strongly enough, but a specific carbohydrate nucleation centre cannot be ruled out. The initiation of bone salt formation by high local concentrations of phosphate —a theory originally due to Robison—has long since been abandoned in its original form; nevertheless bone growth is associated with high concentrations of plasma alkaline phosphatase. Furthermore, the rare inborn error *hypophosphatasia*, in which there is a partial absence of alkaline phosphatase, is accompanied by the excretion of ethanolamine phosphate (of unknown origin), and is very often associated with defects in bone formation.

In spite of its insolubility, hydroxy-apatite is not metabolically stable. Calcium ions, and presumably phosphate ions also, rapidly exchange with the plasma from the surface of the crystals. The turnover rate is not very large when compared with the total amount of Ca in the skeleton, but it represents complete exchange of the plasma Ca every 2–3 minutes. Even in adult bone there is extensive remodelling of the Haversian canal system by osteocytic activity, which means that apatite crystals may be redissolved, and also that crystals previously adjacent to a canal may be locked away in the interior of the bone structure where the circulation of extracellular fluid is very poor. As a result of this, cations which have replaced Ca ions on the surface of a crystal may remain in the bones more or less permanently. If these ions are long-lived radioactive isotopes (see p. 458), this can be very serious because of their proximity to the bone marrow and the likelihood of blood cell disorders.

The average adult body contains 1·5 kg of calcium, 99 per cent of which is in the skeleton. The concentration in plasma is normally about 10 mg per 100 ml, and is kept very constant by the regulating action of the parathyroid gland. Thus if the absorption of Ca from the intestine is not sufficient to replace that excreted, it may be taken from the bones. This process if continued may lead to serious fragility and weakness of the skeleton, but as a 30 per cent deficiency of bone salt is barely detectable radiologically, Ca imbalance must continue for several years before the effects become serious. Resorption of skeletal Ca is exacerbated by pregnancy and lactation.

The daily excretion of Ca is about 1 g, of which 200 mg is found in the urine. It is not easy to tell how much of the 800 mg found in the faeces has been excreted in the digestive juices, and how much has never been absorbed at all. Absorption of Ca is known to be inefficient if the pH of the duodenum is at all alkaline, e.g., in achlorhydria or emotional disturbances inhibiting gastric secretion; also in the presence of oxalate, citrate, benzoate, or particularly of phytate (a polyphosphoric ester of inositol, occurring in cereal products). The presence in the diet of fat or galactose is said to improve the efficiency of Ca absorption. The most important factor, however, is vitamin D (see p. 430). The importance of phosphate in relation to Ca absorption is less regarded than formerly because, although the efficiency of absorption ought to be proportional to the Ca/P ratio in the diet, it is known that children on low-calcium diets absorb Ca much more efficiently than would be possible if this ratio were a major controlling factor.

Ca requirements are difficult to state exactly as there may be great day-to-day variations in balance. The adult probably requires about 1 g per day. Children require *relatively* more of both because of increasing skeletal size and because the bones are poorly calcified at birth. Pregnant and nursing mothers also require a good deal of extra Ca for the requirements of the fetus and later of the milk. Ca absorption is often poor in the aged, who may require more than the adult in middle life.

There are few good sources of Ca; milk and milk products are by far the most important.

In the same way as calcium, *phosphate* is chiefly required by the body for the formation of bone salt. Unlike calcium, however, it is found in every cell in many different compounds, nucleic acids, coenzymes, phosphate esters of sugars, etc., which are vital for cellular processes. This wide distribution means that it is readily available in the diet, and phosphate deficiency is rarely met with. In agriculture, on the other hand, adequate phosphate in the soil is of prime importance for the growth of plants.

A primary deficiency of *magnesium* probably never occurs because, like K^+, it is so universally distributed in (cellular) foodstuffs. In an average diet 2–400 mg may be present, but as with Ca^{2+}, much of this may be excreted in the faeces unabsorbed. Mg^{2+} deficiency, with symptoms of depression and weakness, with reduced intracellular Mg^{2+} concentrations, may occur as the result of prolonged diarrhoea.

Of the minerals required in trace amounts, *copper* is concerned in some way with the synthesis of haemoglobin, and also occurs in cytochrome oxidase, superoxide dismutase, and several other direct oxygenases. *Zinc* is in the active centre of carbonic anhydrase, and of several peptidases. *Molybdenum* occurs in the prosthetic group of xanthine oxidase. *Chromium* (as Cr^{3+}) is involved in the so-called glucose tolerance factor. *Iodine* is required for the synthesis of *thyroxine* (p. 386). The thyroid gland concentrates iodide, which it oxidizes to iodine. Iodide is supplied in the vegetable part of the diet. Requirements are very low (about 1 mg per day), and are normally easily supplied, *except* in parts of the world (e.g. Derbyshire, districts of Switzerland, the U.S.A., Japan) in which the soil contains little or no iodide. Cretinism was formerly endemic in these areas, but has now been overcome by the addition of iodide to the diet (frequently as iodized table salt).

Fluoride. It was observed that in certain areas of the U.S.A. and Great Britain the teeth of the inhabitants became mottled with an unsightly yellow, but that these teeth were extraordinarily resistant to decay. Mottling was traced to the presence, in the drinking water supply, of an unusually high concentration of fluoride. This ion can substitute for OH^- in the hydroxy-apatite which is the bone salt. It has been shown that there are concentrations of F^- in drinking water (1–2 p.p.m.) at which resistance to decay remains, but mottling does not occur. It has been suggested that suitable amounts of fluoride should be added to drinking water where it is not normally found. A good deal of controversy is at present centred round this proposal.

Iron

The total amount of Fe in the human body is 4–5 g. Of this, 75 per cent is in haemoglobin. Of the other 25 per cent, only a minute proportion is present in such important compounds as the cytochromes; most is in the storage substances *ferritin* and *haemosiderin*, with a very small amount in the plasma iron-transporting protein *transferrin*. Ferritin is soluble, and the Fe is contained inside a hollow shell of the protein, apoferritin. Haemosiderin is an insoluble granular substance of uncertain composition, which is responsible for the histological staining of Fe by the Prussian blue reaction. The important storage organs are the liver, spleen, and red marrow. Liver is quantitatively the most important, although it contains the least haemosiderin.

The life of a red blood corpuscle is approximately 4 months, from which one can calculate that 25 mg Fe per day is needed for new haemoglobin synthesis. The loss of Fe from the body is approximately 1 mg per day in the male (of which only 50 μg (5 per cent) is found in the urine), and on the average 2–3 mg per day in the female of child-bearing age. The absorption of Fe approximately balances the loss; the average normal diet contains 10–15 mg but absorption is poor. The absence of benzidine-reacting substances from normal faeces suggests that dietary haem Fe can be liberated and absorbed if necessary, and there is other evidence for this. So far as non-haem Fe is concerned, experiments have shown that Fe^{2+} is better absorbed than Fe^{3+}. Phosphate and phytate (cf. Ca, p. 420) hinder absorption, while large doses of ascorbate, and possibly also other complexing agents, facilitate it. (This has nothing to do with the vitamin function of ascorbic acid.)

The efficiency of Fe absorption can increase greatly in pregnancy, Fe deficiency, and most kinds of anaemia. Anaemia caused by low dietary iron intake, *in the absence of haemorrhage*, hardly ever occurs. These facts suggest that the intestinal mucosa might be the site of a Fe absorption-regulating mechanism, and this has been elaborated into a mucosal barrier theory. It is true that the absorption of Fe increases according to the body's needs, but the original simple theory is no longer valid. For example, when the Fe content of the diet is raised, the absolute amount of Fe absorbed also rises (even though the proportion absorbed does go down). Further, there is good evidence that apoferritin, which has been proposed as the barrier substance, is rapidly synthesized in response to the presence of non-haem Fe in cells. One suggestion is that a good deal of dietary Fe is actually absorbed and incorporated into ferritin in the intestinal mucosa, but that much of this is later lost (in the absence of abnormally high demand by the body as a whole) by desquamation of the mucosal cells.

There is no relation at all between the rate of Fe absorption and the serum iron level. Normally, 30–40 per cent of the total Fe capacity of the plasma transferrin is occupied. The half-life of plasma Fe is about 90 minutes.

Vitamins

These are chemical compounds of varying complexity which cannot be made by the organism and have to be supplied in small quantities in the diet. Sometimes an organism is able to make the substance in small amounts, not enough for its optimum requirements. Organisms differ in their synthetic abilities, thus what is a vitamin for one species is not for another. In this chapter, only those vitamins at present known to be necessary for human health are dealt with in any detail.

Definition. A definition which will exclude minerals, essential fatty acids, and amino acids is not easy to frame. The following has been suggested. A vitamin must be a substance:

1. Chemically different from the three main nutrients: fats, carbohydrates, proteins.

2. Necessary only in minute quantities in the diet.

3. Its absence must cause a specific deficiency disease.

Many vitamins, particularly those of the B group, are known to be co-enzymes, but it does not appear that all vitamins have a coenzyme function.

Vitamin A

Vitamin A is a highly unsaturated alcohol with the following structure

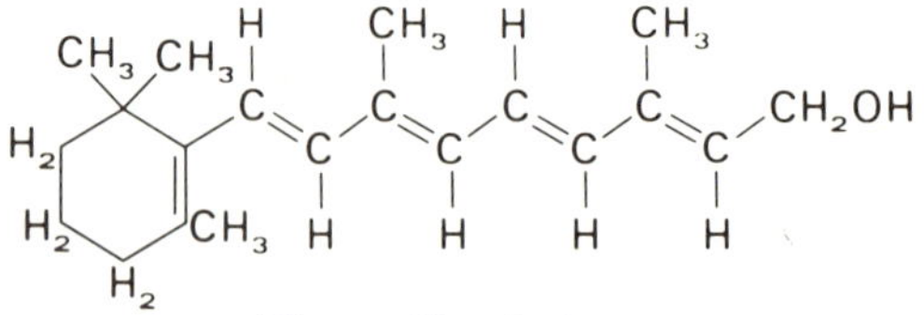

All-trans *Vitamin A*

It is found free or esterified in milk, butter, eggs, and fish liver oils, but equally important sources are certain carotenes found in many plants. The carotenes resemble two molecules of vitamin A joined end to end, and the conversion to vitamin A involves an oxidative fission in the centre of the chain. In order that one of the fission products shall be the vitamin, the structure of the terminal ring must be correct, and although there are a great many carotenes, most of them yellow pigments, only four of them are *provitamins A*. They are found in carrots, tomatoes, and to some extent in grass. Hence the occurrence of provitamins A in milk. Infants and young children can only carry out this transformation to a limited extent; it is almost entirely inhibited in hypothyroidism.

Lack of vitamin A causes cessation of growth and failure of sexual development in rats. In man, however, the deficiency diseases are almost limited to the eyes. Night blindness (see below) is a mild defect, but keratomalacia, giving rise to irreversible corneal opacity, is perhaps the most serious vitamin deficiency disease at the present time, being responsible for thousands of cases of blindness in the tropics. There is a good deal of experimental evidence that vitamin A is necessary to prevent keratinization of mucoid epithelium. Requirement: 5 000–7 000 international units (2 mg) per day.

Vitamin A and vision. The role of vitamin A in visual processes can largely be explained at the molecular level as is shown in the scheme on p.

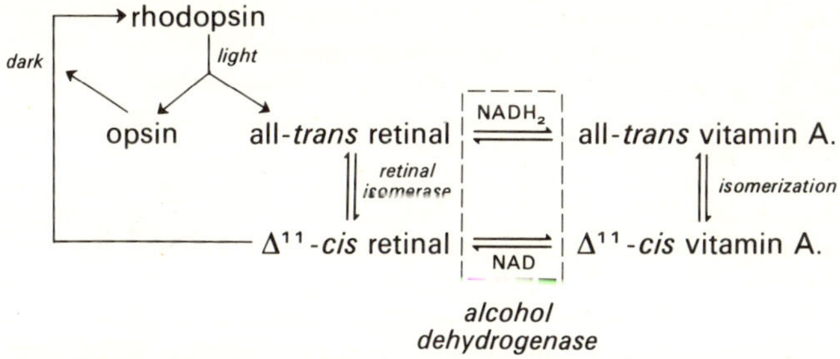

alcohol dehydrogenase

Rhodopsin is a light-sensitive visual pigment consisting of a complex between retinal and a protein called opsin. It occurs in the rods and is important in night vision.

Retinal (retinene) is the aldehyde corresponding to vitamin A. In rhodopsin the double bond between C–11 and C–12 is in the *cis* configuration.

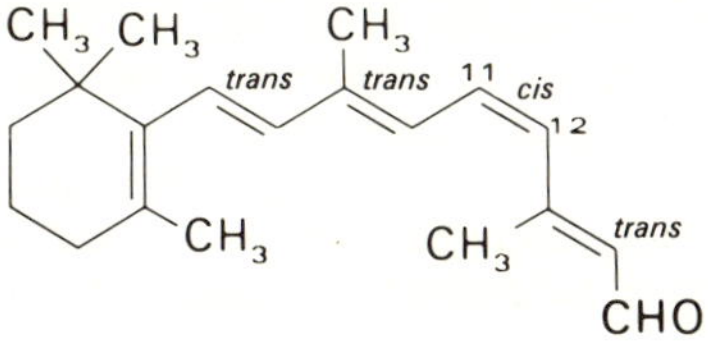

When light falls on rhodopsin the Δ^{11}-*cis* retinal is isomerized to all-*trans* retinal and it becomes detached from the protein moiety, opsin. In the course of these events a nerve impulse is generated by a mechanism which is not yet understood.

The all-*trans* retinal formed can then be isomerized by Δ^{11}-*cis* retinal isomerase. Alternatively, all-*trans* retinal may be reduced by alcohol dehydrogenase and $NADH_2$ to all-*trans* vitamin A which isomerizes to Δ^{11}-*cis* vitamin A. This last compound can be oxidized by alcohol dehydrogenase and NAD to Δ^{11}-*cis* retinal. In the dark Δ^{11}-*cis* retinal and opsin recombine to reform rhodopsin. All-*trans* vitamin A can slowly leak out into the plasma, and so in vitamin A deficiency, the concentrations of retinal and rhodopsin slowly fall, leading to night blindness.

There is evidence that vitamin A has three distinct roles: (1) as the aldehyde retinal in visual processes, (2) as the free alcohol in reproduction, and (3) possibly as the corresponding acid in its other functions. Its metabolism can be summarized as follows:

$$\begin{array}{ccccc}
\text{Vitamin A} & \rightleftharpoons & \text{Vitamin A} & \rightarrow & \text{Vitamin A} & \rightarrow & \text{Metabolites} \\
\text{alcohol} & & \text{aldehyde} & & \text{acid} \\
\updownarrow & & \updownarrow \\
\text{Esters} & & \text{Rhodopsin} \\
\text{(storage)} & & \text{(vision)} & & \text{(? other} \\
& & & & \text{functions)}
\end{array}$$

The Vitamin B Group

This was originally the 'water-soluble vitamin', in distinction to the 'fat-soluble vitamin A', but in fact a large number of different compounds share the functions of the original entity. A number of these compounds are co-enzymes concerned with energy release in cellular oxidations, or with protein synthesis. It is therefore not surprising that B vitamin deficiencies should often manifest themselves as disorders of tissues in which rapid growth normally takes place even in the adult, namely skin and mucous epithelium, digestive glands, and bone marrow. It follows, too, that tissues with high rates of metabolism, including plant seeds, are good sources of these vitamins.

The known coenzyme functions of the B vitamins are shown in Table 20.1.

Thiamine (aneurin, B_1). Found in cereals, in the germ, or in the husk of rice, also in lean meat, liver, kidney, eggs, and seeds. The most serious deficiency disease is *beri-beri*, still endemic in the East. It is characterized by weakness, lassitude, anorexia, and bradycardia, with death usually caused by heart failure. Less serious is generalized *peripheral neuritis*, not

Table 20.1

Name	No.	Enzyme system	See page
Thiamine	B_1	Pyruvate oxidase α-Oxoglutarate oxidase Branched-chain oxo-acid oxidases Transketolase	
Riboflavin	B_2	Flavin mononucleotide (FMN) Flavin adenine dinucleotide (FAD) (Prosthetic groups of many oxidizing enzymes)	
Nicotinamide		Nicotinamide-adenine dinucleotides (NAD and NADP)	
Pyridoxine	B_6	Amino acid transaminases Amino acid decarboxylases	
Pantothenic acid		Coenzyme A	
Folic acid		Transformylation Transhydroxymethylation Transmethylation	
Biotin		Carboxylation of pyruvate acetyl-CoA propionyl-CoA	
Cobalamin	B_{12}	Rearrangement of methylmalonyl-CoA	

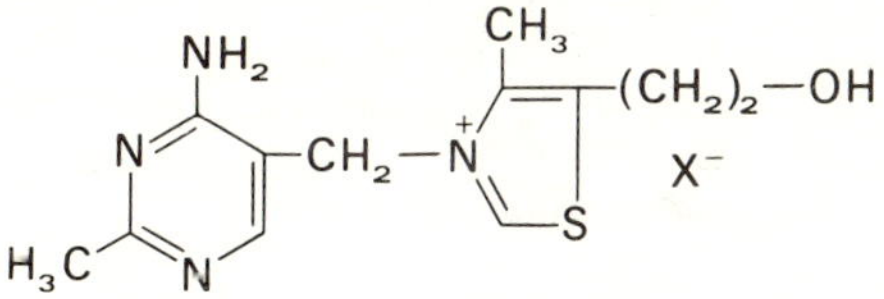

Vitamin B_1 (aneurin, thiamine)

uncommon in chronic alcoholics, because alcoholic drinks are a source of energy, but a poor source of vitamins. Since thiamine is concerned with carbohydrate metabolism, the requirement is related to the carbohydrate intake; it is normally about 2 mg per day. A marked increase in blood pyruvate has been shown to occur in beri-beri, falling to normal on injection of the vitamin. Thiamine pyrophosphate is the prosthetic group of pyruvate decarboxylase, and in its absence pyruvate cannot be oxidized. It is also the prosthetic group of α-oxoglutarate dehydrogenase; reduction in the activity of this enzyme may have the more serious consequences.

Riboflavin (B_2). Found in cereal germ, lean meat, liver, kidney, and in milk. It is unstable to acid and alkali and is also destroyed by light. Although riboflavin deficiency leads to complete cessation of growth in young rats, in humans the known symptoms of deficiency are minor: *cheilitis* is the best known. As riboflavin is a coenzyme or prosthetic group in a large number of systems carrying out cellular oxidations, the effects of deficiency might be expected to be more serious than they are. The reason may possibly be that riboflavin-containing enzymes are present in

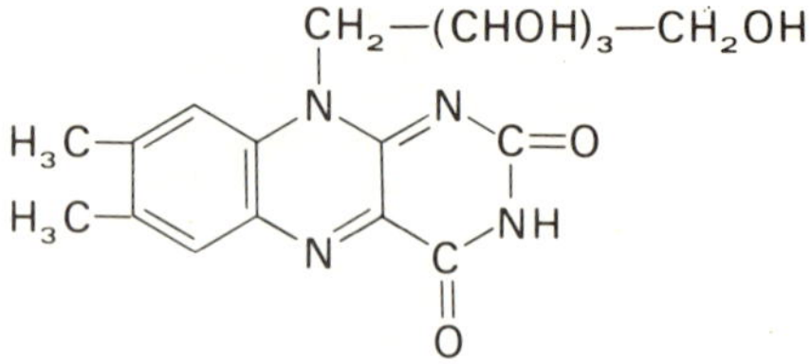

Vitamin B₂ (riboflavin)

considerable excess over the minimum requirements for catalysis so that their 'turnover times' are large, and deficiency must occur for a prolonged period before the cell is denuded of the vitamin. Requirement: about 3 mg per day.

Nicotinic acid (nicotinamide, niacin). This vitamin is required for the synthesis of the oxidative coenzymes NAD and NADP. This is accomplished by the transfer of ribose phosphate (from PRPP, p. 94) to nicotinic acid, and the nucleotide is then amidated. Strictly speaking, therefore, nicotinic acid is the vitamin, but nicotinamide is the form which is obtained from foodstuffs. Hydrolysis of the amide group on the free pyridine derivative is so widespread that the distinction is not important. The most serious deficiency disease is *pellagra* (literally, rough skin), which is

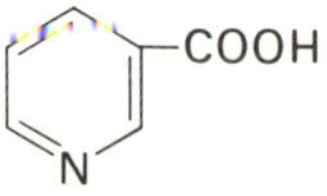

Nicotinic acid

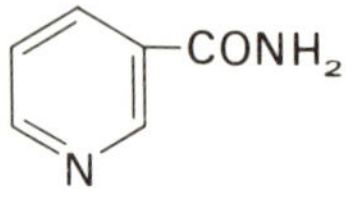

Nicotinamide

characterized by an exfoliative dermatitis particularly on areas of skin exposed to light, anorexia and digestive disturbances, and in its later stages by madness and death. It is a disease particularly of maize-eating areas, partly because most nicotinamide in maize is bound to the seed coat, and therefore unavailable, and partly because a certain amount of nicotinic acid (not enough for all the body's requirements) can be made from tryptophan (p. 241). Maize proteins are particularly deficient in tryptophan, and this cereal is usually eaten in large quantities when it is eaten at all.

Clinical observation and nutritional research have shown that pellagra is not often due to a simple deficiency of nicotinamide; other vitamins, notably riboflavin and pyridoxine, are likely to be deficient in the pellagric diet, and to play some part in the development of the syndrome.

N-methyl nicotinamide is a major urinary metabolite of the vitamin. The parent acid is a very stable compound, resistant both to heat and to acids and alkalis. Requirement: about 20 mg per day.

Pyridoxine (B₆). This has not been shown to cure any specific deficiency disease in man. It is quite probable that the antituberculosis drug iso-nicotinic acid hydrazide (INH) acts by interfering with the function of pyridoxine in the bacillus. There are many enzymes requiring pyridoxine, often as pyridoxal (dehyde) phosphate. These enzymes are concerned with transaminations and oxidative decarboxylations of amino acids. Like riboflavin, pyridoxine is a prosthetic group rather than a coenzyme; it is fairly tightly bound to protein, and this may explain why deficiency rarely occurs.

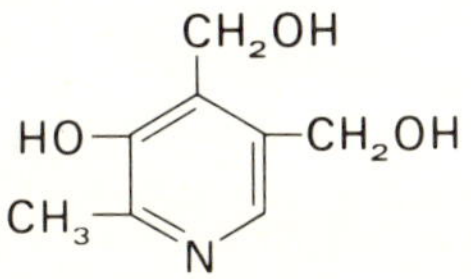

Pyridoxine 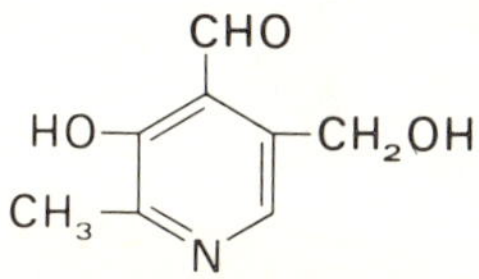   Pyridoxal

Vitamin B₆

Pantothenic acid. This compound again has not been shown to cure any specific deficiency disease in man. It is very widely distributed in foods,

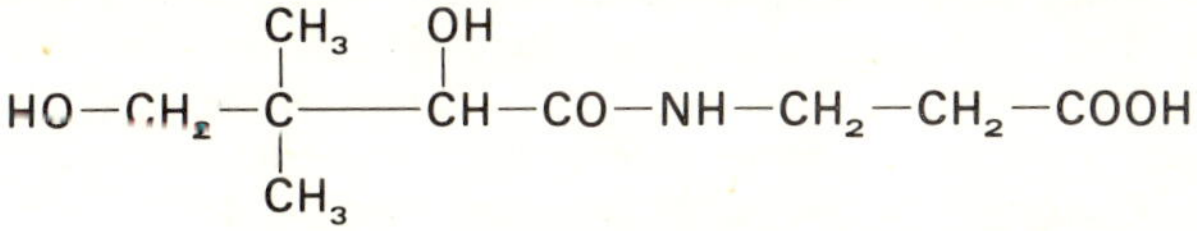

Pantothenic acid

which probably makes a deficiency unlikely to occur. Pantothenic acid is part of coenzyme A, the acyl carrier coenzyme necessary in fat metabolism and in the tricarboxylic acid cycle (see p. 251).

Biotin. Does not cure any specific deficiency disease in man, although a deficiency is sometimes produced by the administration of *avidin*, a biotin-

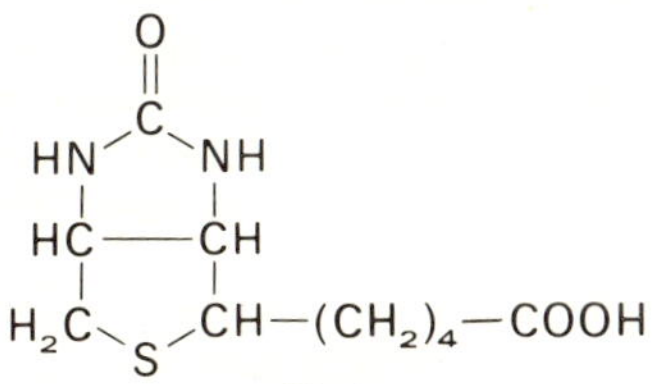

Biotin

binding protein found in raw egg-white (it is destroyed by cooking). Biotin is the prosthetic group of some of the carboxylases. It is attached to the apo-enzymes by an amide link between its carboxyl and the ε-amino group of a lysine residue.

Folic acid. This is found in green leaves and in yeast. Deficiency chiefly shows itself in the development of certain types of anaemia, particularly in unweaned children. A reduced form of the vitamin, tetrahydrofolic acid (formula on p. 146), is a prosthetic group or coenzyme in transfers of one-carbon units, including the formation of thymidylic acid, a constituent of DNA (Chapter 5). It is likely that folic acid anaemia is a deficiency in haemoglobin or erythrocyte synthesis resulting from a lack of RNA. In

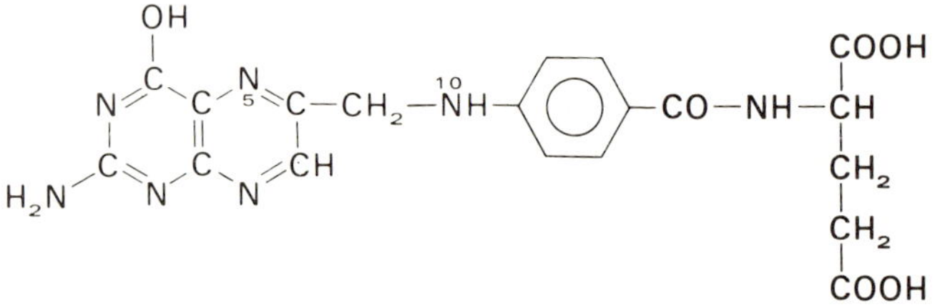

Folic acid (pteroylglutamic acid)

bacteria, folic acid deficiency causes a more widespread cessation of growth. The sulphonamide drugs produce this deficiency by interfering with the incorporation of *p-aminobenzoic acid* into the folic acid molecule. There are several active forms of folic acid: citrovorum factor or folinic acid is N^5-formyl FH_4. Requirement: about 5 mg per day.

Cobalamin (B_{12}). This is not found in plants at all. It can be obtained from liver or lean meat, or from bacteria and moulds. The requirements for the normal adult are very small (about 1 μg per day). Deficiency shows itself in an anaemia, together with degeneration of the spinal cord, which may be fatal in long-standing cases. This is rare; it occurs occasionally after total gastrectomy and in *pernicious anaemia*. In this latter disease the stomach loses, in middle life, the capacity to produce an *intrinsic factor* essential for the absorption of cobalamin (the *extrinsic factor*). The most usual form of vitamin contains one atom of trivalent unionized cobalt, which gives it a deep red colour, and a non-ionized cyanide radical. In the known coenzyme forms of the vitamin, this is replaced by covalently linked organic groups (see p. 150).

The underlying causes of pernicious anaemia are not at present understood, because the only enzyme at present known to require a B_{12} coenzyme in higher animals is methylmalomyl-CoA mutase (p. 204). There is evidence that abnormal branched- and straight chain fatty acids, which could have arisen from methylmalomyl or propionyl residues, accumulate in the neural lipids of B_{12}-deficient animals, but a connection between this and degeneration of the spinal chord has not yet been proved.

Requirements for pernicious anaemics: about 0·01 mg per day.

Choline. This may be regarded as an accessory food factor in some species, because the rate of synthesis *de novo* by methylation of phosphatidyl-ethanolamine is clearly rather slow in many animals. The total daily requirement for man has been estimated to be about 500 mg, but this is normally reached or exceeded in the diet. Some kinds of fatty liver syndromes have been attributed to choline deficiency, e.g. in defects in digestion by pancreatic juice.

Vitamin C (ascorbic acid)

This is one of the water-soluble vitamins. In its chemical properties it is quite strongly acid, owing to ionization of the two hydroxyl groups

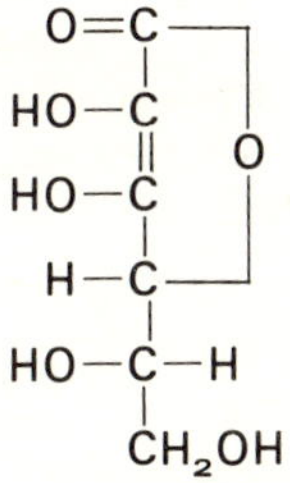

$$
\begin{array}{c}
\text{O=C} \longrightarrow \\
\text{HO--C} \quad | \\
\text{HO--C} \quad \text{O} \\
\text{H--C} \longrightarrow \\
\text{HO--C--H} \\
\text{CH}_2\text{OH}
\end{array}
$$

Vitamin C (L-*ascorbic acid*)

separated by the double bond. It is also oxidized with extraordinary readiness, even by dissolved oxygen at room temperature, particularly if Cu^{2+} ions are present. The first oxidation product is *dehydroascorbic acid*.

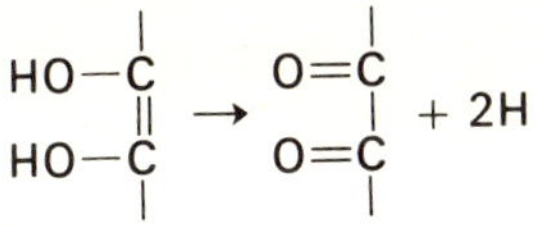

$$
\begin{array}{ccc}
\text{HO--C} & & \text{O=C} \\
\text{HO--C} & \rightarrow & \text{O=C} \quad + 2\text{H}
\end{array}
$$

The disease resulting from vitamin C deficiency is *scurvy*. In its extreme form it is characterized by weakness, loss of teeth and shrinkage of the gums, peripheral haemorrhages, failure of wounds and fractures to heal, and often death. Subclinical scurvy is perhaps not uncommon; symptoms are lassitude, back pains, and hyperkeratosis of, and haemorrhage around, the hair follicles. Almost all the symptoms of ascorbic acid deficiency can be ascribed to lack of collagen, both in scar tissue and in the connective tissue supports of arterioles. Pre-collagen, the precursor of the fibrous form, is not secreted by fibroblasts in the absence of ascorbic acid, and as a result catabolism gradually depletes the connective tissue of its supporting protein. It is not known precisely how ascorbic acid affects pre-collagen formation. It is in some way connected with the hydroxylation of proline residues in procollagen, but is not the oxidizable co-substrate (Chapter 12). It is even possible that it is required for the attachment of carbohydrate chains to hydroxyproline residues, a prerequisite of proper secretion of collagen from fibroblasts. Ascorbic acid is the oxidizable co-substrate for the hydroxylation of dihydroxyphenylethylamine to form noradrenaline, but this cannot be connected with its anti-scorbutic function, nor can its presence in adrenal cortex, whence it disappears on stimulation with ACTH. Ascorbic and dehydroascorbic acids are not hydrogen carriers (e.g., like NAD).

Ascorbic acid is found in citrus fruits, strawberries, blackcurrants, rosehips, potatoes, and green vegetables. The vitamin is readily destroyed by boiling. Daily requirements are 10–30 mg but symptoms may only develop after deprivation for 4–6 weeks because the vitamin is stored in various types of cell. In the U.S.A., daily requirements are put at 100 mg, but this is strongly disputed elsewhere.

Vitamin D

This is a fat-soluble vitamin. Two substances are active when taken by mouth: D_3 (cholecalciferol), a derivative of cholesterol, and D_2 (calciferol), a synthetic derivative of a plant sterol. The vitamins are not strictly sterols since ring B has been opened. The double bonds have the transoid configuration shown. The immediate precursor of D_3 is 7-dehydrocholesterol, which is found in the skin and sebum, and the agent promoting the change is ultra-violet light. If the skin is exposed sufficiently to direct sunlight,

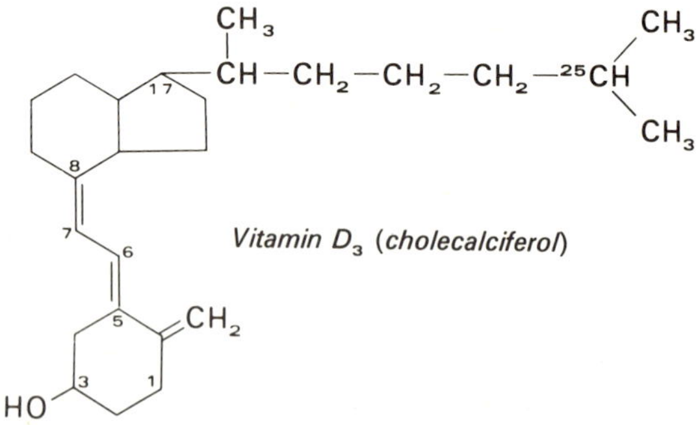

Vitamin D_3 (*cholecalciferol*)

CH_3
CH_3
$\overset{17}{CH}-CH=CH-CH-\overset{25}{CH}$
CH_3
CH_3

Vitamin D$_2$ (calciferol)

therefore, sufficient vitamin will be formed to make its intake in food unnecessary. The biologically active substance in vivo is 1,25-dihydroxycholecalciferol, which is formed by mixed-function hydroxylation of calciferol, as shown in Fig. 20.2.

Absence of vitamin D causes *rickets*, a disease of infants. The bones become soft, the legs unable to support the weight, the flat bones deformed. The rib cage may become so deformed that death from respiratory failure occurs. It is not necessarily a disease of starvation. The skeleton of the newborn child is always deficient in Ca, and any derangement of Ca metabolism quickly makes the bones soft. In adults vitamin D deficiency causes *osteomalacia*, but the calcium reservoir in the adult is so large that the deficiency must be present for several years before softening of the bones is apparent (it is usually only seen associated with pregnancy and lactation).

Vitamin D has three effects: (1) to promote Ca absorption from the gut, (2) to promote mineral deposition in the skeleton, and (3) to inhibit calcium excretion in the urine. Di-OH-cholecalciferol controls function (1) by inducing the synthesis (Chapter 17) of a calcium-binding permease in the intestine. This permease is a protein of mol. wt. 24 000, which binds one Ca^{2+} ion per molecule. It is not yet known how di-OH-cholecalciferol stimulates function (2). A closely related compound *tachysterol* (AT_{10}) can only carry out function (3) and is used for the treatment of tetany.

The best sources are fish liver oils and margarine. The requirement is about 500 iu per day. Because it is stored in the liver a 6 months' requirement can be given at once, but serious overdosage is toxic; it may cause calcification of the renal tubules. Several deaths, mostly of children, from vitamin D overdosage have been reported in the last few years.

Owing to the regulatory action of the parathyroid gland, the serum Ca level in rickets is rarely low, even when the rickets is directly due to Ca deficiency in the diet. Instead, the inorganic phosphate level in serum is usually diminished. This is made clear in Table 20.2, which shows mean values of serum Ca and P in normal subjects, in rickets, tetany, and hyperparathyroidism.

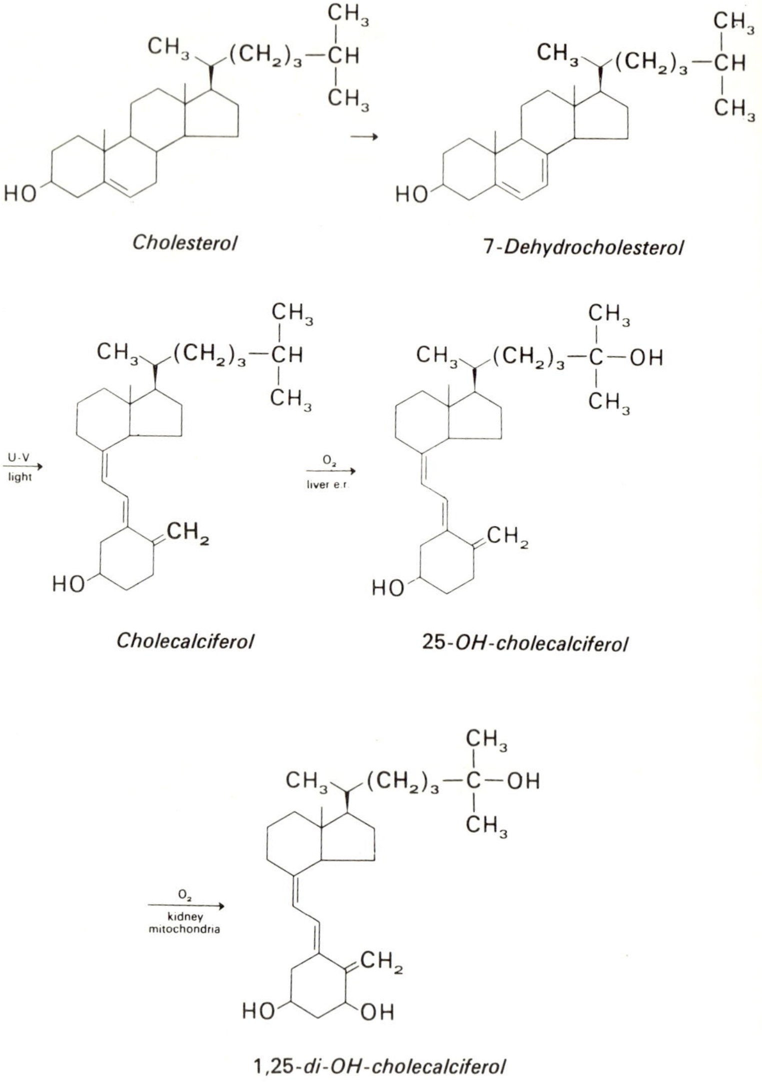

FIG. 20.2. Synthesis of the biologically active form of vitamin D.

Table 20.2
Serum Ca and P Values

	Average values, mg/100 ml	
	Ca	P
Normal*	10·3	4·0
Rickets	9·0	3·0
Tetany	7·9	5·0
Hyperparathyroidism	15·9	2·2

* Adults. Inorganic P in children is normally much higher (about 5·6 mg/100 ml).

Vitamin E (*tocopherol*). This vitamin is not known to be required by man. In animals its absence is said to cause many defects, ranging from

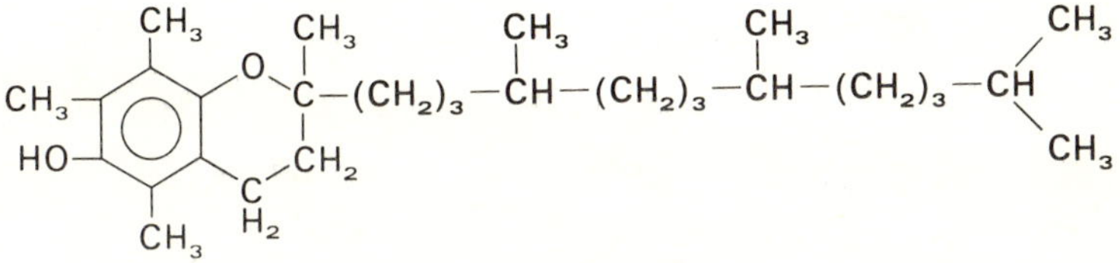

Vitamin E (α-tocopherol)

Several other tocopherols have vitamin E activity

sterility in rats to muscular dystrophy in rabbits. Tocopherol is fat-soluble and easily oxidized and reduced. It is known to protect unsaturated fats, and particularly vitamin A, from oxidation. This it may do by inhibiting the formation of lipid peroxides. It is now widely added to artificially prepared dairy products, e.g., margarine. The human diet contains about 500 mg per day from many plant sources.

In animals, the functions of tocopherol and of selenium, especially in reproduction, are closely associated; a deficiency of one increases the need for the other.

Vitamin K

A fat-soluble vitamin existing in two natural forms. Vitamin K_1 is found in green leafy plants. Vitamin K_2 has a different side-chain (farnesyl instead of phytyl) and is synthesized by bacteria. Various synthetic compounds without either of these side-chains are also potent vitamins. Lack of vitamin K gives rise to haemorrhage, because it is involved, in some way as yet not understood, in the synthesis of prothrombin. The required daily amounts are so small that deficiency rarely arises, except in intestinal ailments or jaundice, which hinder its absorption, or during sulphonamide therapy, since intestinal bacteria may supply the requirement. It is often given as a routine precaution before operations or childbirth, but it is entirely ineffective as an immediate antidote to haemorrhage.

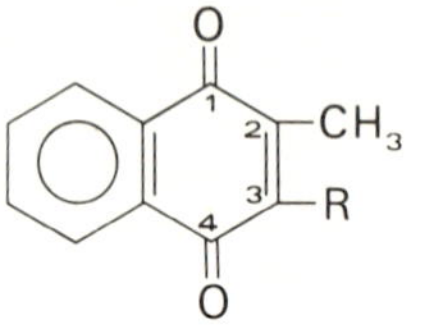

The K vitamins

Vitamin K_1: 2-methyl-3-phytyl-1,4-naphthoquinone, i.e.

R is $-CH_2-CH=\overset{\displaystyle CH_3}{\overset{|}{C}}-(CH_2)_3-\overset{\displaystyle CH_3}{\overset{|}{C}}H-(CH_2)_3-\overset{\displaystyle CH_3}{\overset{|}{C}}H-(CH_2)_3-CH\begin{matrix}/CH\\ \backslash CH_3\end{matrix}$

Vitamin K_2: 2-methyl-3-difarnesyl-1,4-naphthaquinone, i.e.

R is $-CH_2-\left[CH=\overset{\displaystyle CH_3}{\overset{|}{C}}-CH_2-CH_2\right]_5-CH=C\begin{matrix}/CH_3\\ \backslash CH_3\end{matrix}$

Menadione: 2-methyl-1,4-naphthoquinone, i.e.

R is H (menadione is a synthetic analogue)

General Considerations on Vitamin Requirements

It is not easy to give more than rough estimates of vitamin needs for several reasons:

1. The requirements may vary with metabolic rate. This is particularly true of the B vitamins concerned with energy-supplying reactions. These must be supplied in greater amounts as metabolic activity increases.

2. The requirement may depend on the type of foodstuff. This is particularly true of thiamine; the requirement of this vitamin is often expressed as 1 mg per 4 000 kJ carbohydrate oxidized.

3. On subsistence diets, an excess of one vitamin may precipitate a deficiency of another. This is again most true of the B vitamins, for obvious reasons, and it is considered unwise to supplement a diet with, say, nicotinamide or thiamine alone.

4. The rate of destruction of vitamins increases in fevers, so that the vitamin intake is often put up in convalescence. There is little evidence, however, that vitamins *prevent*, or increase resistance to, infections.

5. The requirement depends on the previous dietary history. Not all vitamins are stored to any extent; the most important stores are those of vitamins A, B_{12}, C, and D.

6. An important source of supply of many vitamins is not the food as such, but bacteria living in the alimentary canal. It is difficult to make a quantitative estimate of this supply, but treatment of intestinal infections with insoluble sulphonamides, or other anti-bacterial drugs, has often been found to give rise to more or less serious avitaminoses. All the B vitamins, but particularly nicotinamide and B_{12}, and also vitamin K, are known to be absorbed from autolysing bacteria. Vitamins A, C, and D, however, are not synthesized by intestinal micro-organisms.

In certain circumstances bacteria in the intestine may themselves cause avitaminoses either by competing for the vitamins in food, or by causing conditions unfavourable to absorption (e.g. enteritis.).

Observations on Tissues

It is often desirable to study the metabolic behaviour of an individual organ, or of individual cells, rather than of the body as a whole. This is not difficult for the microbiologist whose organisms are all unicellular; they can be rapidly grown from a small sample of cells in a suitable culture medium, and 'harvested', usually by centrifugation, after a suitable time interval. The cells can then be washed and resuspended in a new medium or analysed as the experiment demands. The metazoan biochemist can readily obtain only erythrocytes or leucocytes (which are rather fragile) or spermatozoa as single cells, and with somewhat more difficulty, ascites tumour cells. Tissue culture does not answer the problem. It is not easy to grow mammalian cells in culture on a large scale, and moreover it is often found that after many generations cells from specialized tissues tend to revert to a more primitive type (de-differentiate). Another approach is to separate the cells in an organized tissue from one another, e.g., by using a proteolytic enzyme to break down the connective tissue. This has been done particularly with adipose tissue, brain and heart, but it is open to question whether the enzymes also attack the plasma membrane of the cells themselves.

For this reason the biochemist is often forced to study either intact organs or preparations from these organs which contain a fairly large proportion of intact cells. He can also take the other approach and deliberately break the tissue up and study the behaviour of a *broken cell preparation*. This will usually contain all the enzymic activities of the intact cell (including some which may have been repressed), but these will no longer be organized. It is also no longer possible to study the ways in which metabolites are transported into, and out of, cells.

Perfusion

The simplest way to study intact cells organized in tissues is to perfuse the whole organ either in situ or in vitro, either with blood, or with a buffered isotonic saline solution. Usually the solution is re-circulated, with some arrangement to re-oxygenate the fluid and to remove CO_2. The whole preparation is kept at 37° by means of a water jacket. The solubility of O_2 is only just enough for the demands of active tissues such as liver to be supplied in this way, and washed erythrocytes, or dissolved haemoglobin, are often used. Many organs have been studied in this way. Brain has to be perfused in situ, and there is no convenient way of perfusing skeletal muscle in vitro, but heart can be perfused quite easily.

In such experiments the tissue itself can be analysed at the end of the perfusion, and changes in particular constituents estimated with reference to the composition of the organ before perfusion. It is common to freeze samples rapidly by compressing them between Wollenberger tongs cooled in liquid N_2, or by dropping small samples in pentane cooled with the same substance. Alternatively, if the perfusion volume is known, the saline can be analysed to determine the rate of uptake of particular metabolites by the tissue, or the rate at which products are secreted from the cells.

Tissue Slices

Much work has been done with this technique in the past, although it is not now so widely used. Oxygen can only enter the cells by diffusion from the medium. The maximum thickness of tissue which can be adequately oxygenated in this way can be calculated; values range from 0·3 mm for liver to 1 mm or more for less active tissues. Slices of this thickness can be cut from soft tissues with a razor blade, either free-hand or in a machine. Liver, kidney, and brain have been studied in this way, and it is also convenient for many plant tissues (e.g., roots, tubers). Many of the cells will of course be damaged, but usually somewhere between 50 and 70 per cent remain intact. It is not possible to slice skeletal muscle without damaging too many cells, but the diaphragm muscle of young rats is less than 1 mm thick. It is often cut away from its attachment to the rib-cage and immersed without further treatment. Adipose tissue is often used, particularly that teased out from the fat pads surrounding the epididymis.

The tissue slices must be suspended in isotonic buffered saline in an airtight vessel (to prevent evaporation). Excess O_2 is provided by having a large air-space above the liquid which may be filled with pure oxygen (or O_2/CO_2). The vessel is usually incubated at 37° with shaking to promote diffusion. Again, changes in cellular composition or changes in the concentration of substituents in the medium can be studied.

Both with this technique and the preceding one the cells are obviously 'dying' from the moment they leave the donor organism, although it may be very difficult to define this term precisely. In some instances it may be possible to use some specialized function of the cell as an index of viability: for example, the regular beating of a perfused heart, or bile secretion by liver.

Homogenates

The logical end to the degradative approach is to break the cell up completely in order to obtain a homogeneous soup which can be adequately oxygenated and whose suspension medium can be varied at will. This is a large gain in exchange for the partial loss of intracellular organization. Nowadays it is usual to homogenize with a Potter–Elvehjem homogenizer, which consists of a glass tube of precise bore into which fits a pestle made of glass or of polytetrafluoroethylene (Teflon). The clearance between pestle and tube is of the order of 0·1–0·15 mm (Fig. 21.1). The pestle is motor-driven at about 2 000 rev/min. Liver may be homogenized by hand. Alternatively, a glass vessel fitted with rotating knives (a blender) is used (Fig. 21.2). In both techniques the tissue is roughly chopped, and suspended in a relatively large volume of medium of suitable composition (see below).

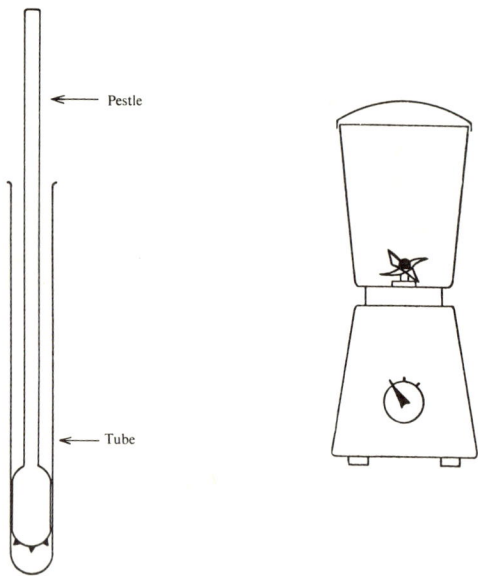

Pestle

Tube

FIG. 21.1. Potter–Elvehjem homogenizer.   FIG. 21.2. Blender.

With both instruments local heating occurs and cooling must be good to prevent degradation of the tissues.

The still-organized cellular structures are best preserved if isotonic or even hypteronic sucrose, or other non-permeating hydrophilic compound, is used instead of isotonic KCl. They can be separated by differential centrifugation, since in this range of dimensions the size of the particles, i.e. their resistance to a gravitational force pulling them through the supporting liquid, matters quite as much as their density. The various fractions so obtained can be identified partly by their biochemical composition and to a limited extent by histological staining, but the intracellular structures are in general too small to be studied by the light microscope. They can, however, be observed by an electron microscope.

Electron Microscopy

Even with theoretically perfect optics, no object which is much shorter than half the wavelength of the light which is used to illuminate it can be resolved. The wavelengths of visible light lie within the range of $0.1–1\ \mu m$. Nuclei and mitochondria can be seen in a light microscope, but the smallest cell particles which can be obtained from a homogenate have a diameter of only 15 nm; even the use of ultra-violet light will not help with these.

Electrons behave in many ways as if they were waves of electromagnetic radiation; their wavelength is, however, much shorter than that of visible light (0.1 nm or less, depending on their speed). Thus a homogeneous beam of electrons (i.e. emitted from a cathode and travelling to the anode under the influence of a precisely regulated electric field) will be absorbed by, or diffracted by, the components of a thin slice of material in exactly the same way as the electromagnetic waves of visible light. For the condensing and focusing lenses it is necessary to use adjustable electromagnets because the path of an electron will 'bend' in an electric field. The whole field of travel—including the specimen—must be in a very high vacuum, since electrons are readily stopped by colliding with gas molecules. The electrons may be visualized by a fluorescent screen or recorded by a photographic plate.

The specimen is embedded in plastic and extremely thin sections are cut. Biological material may be 'stained', by allowing it to react with electron-dense (i.e. highly absorbing) materials such as osmic acid. Negative staining—putting electron-dense phosphotungstic acid in the surrounding medium, which will not penetrate the protein structure—has enabled better definition to be obtained. Distortion caused by heating as the electrons strike the specimen is minimized by using very thin sections ($50\ \mu m$ thick) and by cooling the stage in liquid nitrogen. Three-dimensional structures (e.g., bacteria) are sometimes 'shadowed' by a thin film of metal.

The use of this technique, together with careful fractionation by the biochemist, has enabled a picture of the ultrastructure of the cell to be built up. Fig. 21.3 is an idealized version of this, which must not be literally applied to any cell type in particular.

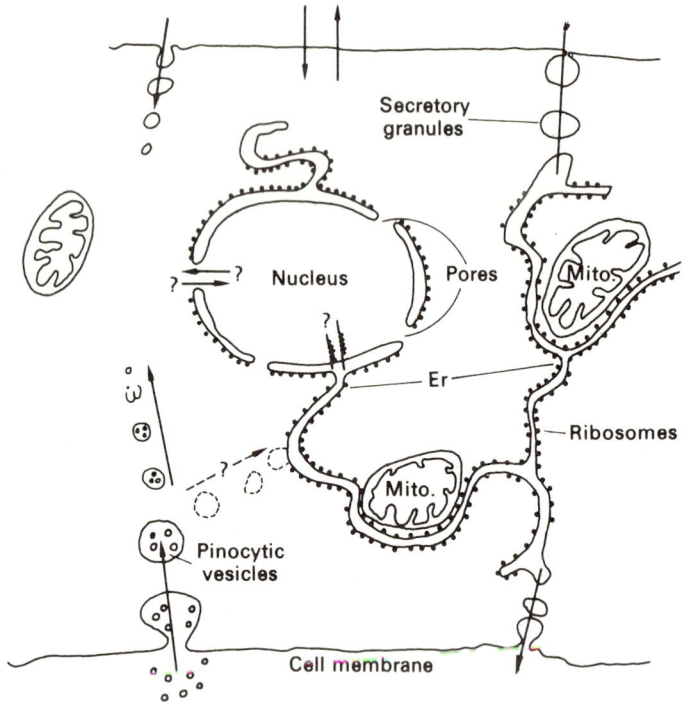

Fig. 21.3. Generalized diagram of the submicroscopic structure of a cell.

Differential Centrifugation

Cell fragments begin to lose their biochemical capacities from the moment of homogenization, so high-speed centrifuges are always refrigerated to keep this rate of loss as small as possible. The centrifugal force exerted on a particle in the solution is expressed in multiples of the force exerted by gravity; it is proportional to the radius of the centrifuge head and to the square of the (angular) velocity. A head about 10 cm in diameter rotating at about 40 000 rev/min. will produce slightly more than 100 000 g. Fields of 200 000 g are now being used routinely. At speeds in excess of 20 000 rev/min it is essential to have the head running in a vacuum to obviate the heat produced by air friction. It is usual to have the tube containing the homogenate held at an angle to the axis of rotation to keep the path of the particles through the solution as short as possible (see Fig. 21.4).

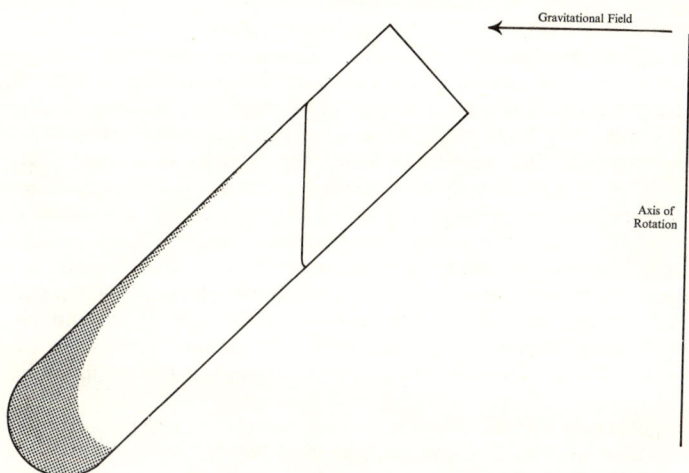

Gravitational Field

Axis of Rotation

Fig. 21.4. Diagram of the tube in an angle centrifuge. The heavier material accumulates against the outside wall of the tube and 'slides' down to the bottom along the wall.

The homogenate, which must be considerably diluted, is centrifuged at a low speed to remove nuclei, etc. The supernatant liquid is then carefully poured off and spun at a higher speed to remove the next fraction. The supernatant is centrifuged at a yet higher speed, and so on. The removal of particles of a particular size depends on the time for which the centrifugal field is applied, as well as on the field itself. Table 21.1 gives an approximate guide.

Table 21.1
Separation of Cellular Structures by Differential Centrifugation

Field and time	Structure
700 g: 10 min.	Nuclei, cell membranes, erythrocytes, etc.
5 000 g: 10 min.	Mitochondria, lysosomes
57 000 g: 60 min.	Microsomes, lysosomes
150 000 g: 30 min.	Ribosomes
Unsedimented	Cell sap and some ribosomes

These centrifugal fields refer to homogenates in 0·25 M sucrose.

A single centrifugation as described produces fractions which are mixtures, although each precipitate contains a predominant structural component. In order to achieve relative purity the precipitates have to be re-suspended and re-centrifuged several times.

Because of convective disturbances, fixed angle rotors are unsuitable for the separation of particles of similar sedimentation behaviour. Swinging-bucket rotors fitted with pivoted buckets, which are vertical when the rotor is stationary and which swing out to a horizontal position when the rotor is spinning, have certain advantages, particularly in gradient-density centrifugation. The sample containing the particles to be separated is layered on to a gradient of concentration of sucrose or salt, established in the centrifuge tube. With proper choice of conditions the components of the mixture will move through the gradient as discrete bands or zones. Large bowl-shaped rotors of high capacity have been developed for large-scale gradient-density centrifugation. These *zonal rotors* permit radial sedimentation without wall effects. They can be adapted to continuous flow. Amongst the applications of zonal rotors is the purification of viruses, which can be used to prepare vaccines of high potency and with minimal side effects.

Application of these techniques has shown that enzymes catalysing particular metabolic pathways are localized in certain sub-cellular structures which can be identified by electron microscopy.

The structure of *mitochondria* cannot be seen with the light microscope. Mitochondria from some tissues do not have the internal membranes or *cristae* shown in Fig. 21.5, but this is characteristic of those from liver and heart. Methods have been devised for disrupting mitochondria (e.g., by ultrasonics or detergents). Fragments of membranes or cristae are obtained which are capable of carrying out all the reactions of the electron transport chain and some phosphorylation (Chapter 12); it is clear that these enzymes are arranged in an orderly way in the mitochondrial walls. Some tissues, e.g., smooth muscle, contain very few mitochondria; they are particularly frequent in heart, pancreas, and liver. A single liver cell may contain 1 000 of them.

The *lysosomes* are probably rather smaller than mitochondria and appear to contain mainly hydrolytic enzymes—cathepsins, phosphatases, etc. These enzymes seem to remain inactive unless the lysosomes are ruptured.

The *microsomes* are, typically, hollow vesicles and on closer examination are found to be fragments of endoplasmic reticulum (see Fig. 21.3), the broken ends of which have joined up. In some tissues, notably pancreas and liver, many ribosomes are attached to the endoplasmic reticulum, and the microsomal vesicles, therefore, have ribosomal activity (e.g., protein synthesis). Certain specialized enzymes (e.g., steroid hydroxylations and drug catabolism) are also associated with microsomes.

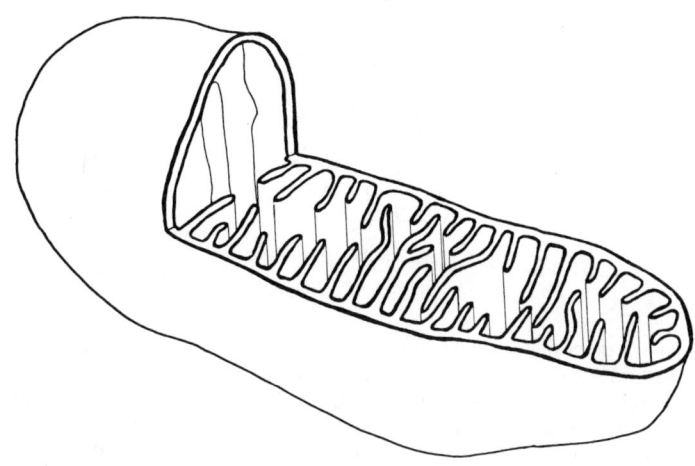

Fig. 21.5. Schematic representation of a mitochondrion. The organelle has been cut open to show the internal structure which consists of a number of internal membranes or cristae.

The *ribosomes* can be prepared by very careful disruption of the cell. They are not frequent in cells from tissues which do not synthesize much protein (e.g., muscle).

The *cell sap* contains all the enzymes of carbohydrate metabolism, other dehydrogenases and the fatty acid synthesizing system and most of the NAD and NADP of the cell. It may also contain very small particles which are not usually centrifuged down, but large single molecules (e.g., glycogen, multi-enzyme complexes and nucleic acids) can also be spun down at these very high speeds, and there is no clear dividing line between organized structures and macromolecules (for *ultracentrifugation*, see separation methods).

Measurements of Gas Exchange

In work with tissue slices and homogenates or sub-cellular particles it is often necessary to measure the oxygen uptake. The classical way of doing this is by measuring the change in volume of the gas-space above the incubation fluid as oxygen disappears from it. The *Warburg manometer* is the best-known of such devices; it actually measures the change in pressure at constant volume (and constant temperature). A special flask is usually used (Fig. 21.6) with a side-arm from which reagents can be added at defined times, and a centre well which is filled with strong alkali to absorb CO_2. The flask is attached by means of a gas-tight joint to a

FIG. 21.6. Warburg flask.

capillary manometer, which contains a device for returning the liquid in it to a standard volume mark. Every time this is done the height of the liquid in the open limb of the manometer, which is subject to atmospheric pressure, is read. The flask, but not the manometer, is immersed in a water bath of carefully controlled temperature and shaken to improve the diffusion of O_2 into the incubation medium. A control flask to allow for variations in bath temperature and atmospheric pressure is necessary. This apparatus, although reliable, is now used only for specialized applications, e.g., the study of photosynthesis.

The *oxygen electrode* consists of a platinum cathode in a salt solution which is connected via a salt bridge to a saturated calomel (or other) electrode. A potential of about 0·6 v is applied across the two electrodes, the platinum being negative. This results in the electro-reduction of molecular oxygen according to the equation:

$$2H_2O + O_2 + 4\varepsilon \longrightarrow 4OH^-$$

The hydroxide ions can then move to the anode and current flows. This electrode consumes the dissolved oxygen and the current flow in the cell is proportional to the concentration of the dissolved molecular oxygen. Thus a microammeter (preferably a recording one) in the circuit will show a deflection proportional to the oxygen tension at the electrode. Such an oxygen electrode is separated from the solution whose oxygen tension is to be measured by a membrane (of polythene or mylar) which is permeable to oxygen and other gases, but not to water or crystalloids and this protects it against most substances that might poison it. If the solution is vigorously stirred oxygen diffuses through the membrane to the electrode and the current flow is proportional to the P_{O_2} in the test solution. The system is very sensitive and well suited to the use of small amounts of sub-cellular fractions. It may also be used in the gas phase and for the continuous monitoring of P_{O_2} in the blood during extended surgical operations.

The *carbon dioxide electrode* is essentially a pH meter (Chapter 1), the glass and reference electrodes of which are in a thin layer of dilute bicarbonate (0·005–0·01 M) in sodium chloride (about 0·1 M). This solution will be in equilibrium with CO_2 and its pH will depend on the $[HCO_3{}^-]/qP_{CO_2}$ ratio (see Chapter 14). The bicarbonate solution is separated from the sample whose P_{CO_2} is to be measured by a thin membrane (of rubber or Teflon) which is permeable to CO_2 and other gases but not to water or crystalloids. The pH of the bicarbonate solution in immediate contact with the electrodes is then dependent on the P_{CO_2} in the test solution. Unlike the oxygen electrode, the CO_2 electrode does not consume the gas being measured. The electrode works quite well but its response is not immediate. It can also be used in the gas phase and to monitor P_{CO_2} in blood during operations.

Analytical Techniques

Spectrophotometry

A great deal of analytical work is done in biochemistry without ever isolating individual compounds, by taking advantage of their light absorption at particular wavelengths. Sometimes this is in the visible region; often in the ultra-violet (short wavelength) region. The specific extinction tends to increase as the wavelength decreases, and thus very small amounts of colourless ultra-violet absorbing substances can be measured with accuracy.

Practical spectrophotometry depends on the fact that most organic substances in dilute solution obey Beer's law,

$$\log_{10} \frac{I_0}{I} = k_\lambda c l$$

where I_0 is the intensity of the incident light, and I that of the light leaving the sample cell. I_0 is measured by putting a sample cell containing water in the light path. The term $\log I_0/I$ is called the *optical density* or *absorbance*. In the equation, l is the length of the path of the light through the solution, and c is the concentration of the absorbing solute. k_λ is a constant; if l is set at 1 cm and c at 1 molar k_λ is equal to the molar extinction coefficient ε. The subscript λ indicates that the extinction coefficient varies with wavelength, i.e. there are absorption maxima. Ideally the absorption should be measured with light of a single wavelength, or at least with a very narrow bandwidth. This demands a very good prism or diffraction grating. For work in the ultra-violet region (below 325 nm) the optics, including the sample cell, must be made of fused quartz, since glass absorbs ultra-violet light.

In the far infra-red (1–2 μm) the absorption of light depends on particular chemical groupings, and spectra in this region are used for determining the chemical constitution of unknown compounds.

Table 21.2
Absorption Maxima of some Compounds of Biological Interest

	Absorption max. nm
Haemoglobin	560 and others
Oxyhaemoglobin	577 and others
Methaemoglobin (in acid)	630 and others
Cytochrome c Fe^{2+}	{ 550 and others
Cytochrome c Fe^{3+}	{ 565 and others
Riboflavin, FMN, FAD	450, 375, 260
NADH$_2$, NADPH$_2$	{ 340, 260
NAD, NADP	{ 260
Ubiquinone	{ 275
Ubiquinone reduced	{ 290
Tyrosine: also in peptide linkage	275
Tryptophan: also in peptide linkage	280
Adenine derivatives	260
(Other purines and pyrimidines very similar)	

A list of some absorption maxima for well-known biochemical compounds is shown in Table 21.2. The appearance of a band at 340 nm on reduction of NAD or NADP is very frequently used for estimating the activity of NAD-linked dehydrogenases or for determining their substrates (see p. 140). It can also be used for estimating the activity of any reaction which can be coupled to a dehydrogenase. Thus in estimating the activity of hexokinase, which catalyses the reaction

$$\text{Glucose} + \text{ATP} \xrightarrow{\text{hexokinase}} \text{G-6-P} + \text{ADP}$$

if glucose-6-P dehydrogenase and NADP are added to the reaction mixture the following reaction takes place:

$$\text{G-6-P} + \text{NADP} \rightarrow \text{6-P-Gluconate} + \text{NADPH}_2$$

and the primary reaction can be followed by measuring the increase in optical density at 340 nm.

Changes in the ratios of compounds listed in Table 21.2 have been measured in intact cells or in mitochondria, using dilute suspensions or very thin slices. Unpredictable light-scattering by colloids and sub-cellular structures is compensated for by using a *double beam spectrophotometer*. This uses two beams of light, one at a wavelength not absorbed by the compound of interest; the diminution in intensity of this beam as it passes through a sample is a measure of light scattering. The other beam is of a wavelength which is absorbed.

Fluorimetry can also be used for analysis. Many substances on irradiation with light of one wavelength emit light of a precise longer wavelength. Since the fluorescent light is emitted in all directions the photocell is placed at right-angles to the path of the incident beam, and it then does not respond to the latter. A refinement of this instrument is the *spectrofluorimeter*; by using one prism to produce light of a given wavelength and another to allow light only of a known wavelength to fall on the photocell, almost any fluorescent compound (which includes almost all aromatic compounds of biological interest) can be determined with precision in the presence of any of the others. The method is remarkably sensitive, but like all fluorimetric techniques it suffers from the quenching phenomenon. This means that many compounds will either absorb the incident or emitted light, or will prevent the sensitive molecule from emitting its characteristic radiation (see scintillation counting).

The usefulness of these optical techniques is commonly extended to measurements of substances which do not absorb light of a convenient wavelength, by carrying out a preliminary chemical reaction, under controlled conditions, which produces a coloured (or fluorescing) substance. This technique is called *colorimetry*.

Flame photometry depends on measuring the radiation *emitted* by a metallic element when a salt of it is vaporized in a flame. (The yellow of a sodium flame will be very familiar.) This method is limited to ions such as Na, K or Li, which emit strongly at the temperature of a gas flame. A more widely applicable technique which also depends on emission spectra is *atomic absorption spectrophotometry*. In this, the radiation coming from a specially constructed bulb with a cathode of the element in question is passed through a vapour containing that element. The amount of radiation absorbed is proportional to the concentration of the species in the vapour. This technique is very sensitive, especially if the sample is vaporized by a 'graphite furnace', and not by a gas flame. It can be used for most metals, in amounts down to 10^{-12} g.

Separation Methods

Many of the techniques which are to be described are used for analysis rather than for preparation on a large scale, but the distinction is an artificial one since the methods may in principle be scaled up or down.

Ultracentrifugation

This is used for proteins and nucleic acids and is a variant of the differential centrifugation already described. Even quite small proteins will sediment in an intense gravitational field; the rate of sedimentation is proportional not only to the molecular weight but also to the shape (e.g., globular, rod-like) and it is expressed as the sedimentation coefficient. This is defined as

$$S = \frac{\text{rate of movement}}{\text{centrifugal force}}$$

and has the dimensions of time. The unit used is the *Svedberg* (S), which is 1×10^{-13} sec. Thus a particle of size 70S (p. 338) has a sedimentation coefficient of 70×10^{-13} sec. The movement of the proteins can be detected by the change in refractive index as the concentration changes at any particular point; a beam of light is directed through the cell as it is being centrifuged and the refracted beam is caused, by an ingenious optical device, to make a record of the refractive index gradients on a photographic plate (schlieren system). The method can be used for testing the homogeneity of protein preparations, in which case several runs in different conditions of pH, ionic strength, etc. are necessary, or, in conjunction with a second method (diffusion, osmosis, etc.), for determining the molecular weight. In *equilibrium* centrifugation the solution is spun for a long time at a low speed, so that centrifugal force is just balanced by diffusion.

Density gradient centrifugation is widely used as a separation technique. In one technique a density gradient of sucrose solution is prepared in a centrifuge tube, and the sample is layered on top of it. On spinning in a horizontal rotor particles of different sizes move at different velocities down the tube. The run may be stopped at any time. In the equilibrium method a uniform solution of 6 M CsCl is used, and the substances to be separated are dissolved in it. After long centrifugation at high speed, the CsCl sediments slightly, producing a density gradient in situ, and the macromolecules come to equilibrium in bands which reflect their sizes. This method is particularly useful for nucleic acids.

Electrophoresis

This technique can be applied to any compound which carries an electric charge, whether positive or negative; it must also be appreciably water-soluble. In practice it is limited to the study of proteins, peptides, amino acids, and phosphate derivatives. The rate of movement of large molecules depends not only on the charge density but also on the size and shape, including the extent of hydration. With dipolar ions (Chapter 1) there is a pH (the isoionic point) at which the positive and negative charges on the molecule balance and the compound will not move in an electric field. To avoid this, proteins are often subjected to electrophoresis in a barbitone buffer of pH 8·6; for other compounds a suitable pH is found by experiment. The advantages of the sodium dodecylsulphate (SDS) technique for inducing a uniform charge density on the protein have already been mentioned (p. 75).

The electrodes to which the voltage is applied must obviously be as far apart as possible, and there must be some arrangement for introducing the sample in as narrow a section as possible, either in the middle or at one end of the cell. The voltage which can be applied ranges from 200 or 300 to 10 000 volts. It is limited by the efficiency of the cooling system, since the heat associated with passage of a current may produce thermal decomposition of the support (and denaturation of proteins).

Free electrophoresis, i.e. in free solution, has largely given way to zone electrophoresis, because of the limited separation it affords, and the difficulty in holding the experimental conditions steady.

Zone electrophoresis. When solid blocks of porous material are used for electrophoresis adsorption becomes important. There may even be a molecular sieve effect if the pores in the material are too small to let large molecules through. The general effect is to improve the separation, e.g., starch gel will enable 22 or more proteins to be demonstrated in plasma instead of the 5 main groups (see Fig. 14.2) which were previously observed with free or paper electrophoresis.

A long and homogeneous block of a gel, or a cylinder in a glass tube, is cast. The cross-section of the gel must be sufficiently low (<1 cm) to make heating effects unimportant. The gel must contain a suitable electrolyte/buffer in the aqueous component. Slots are then cut in the block into which small amounts of sample can be introduced or a disc of filter paper saturated with sample is placed on top of the column, which is run vertically. The surface is covered to prevent evaporation, and contact with the electrodes is made by buffer-saturated paper. Starch grains, partially hydrolysed starch gel, and polyacrylamide gels have been used. In refinements of the method a pH gradient may be introduced into the buffer in the solid support (isoelectric focusing).

A disadvantage of the technique is the limited number of ways in which the components of the sample may be visualized after the run. The gel is distorted by heating or treatment with chemicals so the method is most suitable for proteins, which may be located by dye-binding, or by scanning with ultra-violet light. If gel blocks are used enzymes can be identified after the run by damping the strip or block with a solution containing substrate, a suitable buffer, and some reagent which will give a colour with the product of the reaction.

In *immuno-electrophoresis* a slot is cut in the gel along its length, and after the electrophoretic run a solution of a suitable antibody is poured into it. The antibody and the antigen(s) (component(s) of the sample) diffuse in all directions from their respective positions, and where they interact (see Chapter 14) a white precipitate—a line or arc—is formed in the gel. By this means the presence or absence of particular proteins can be detected.

Paper electrophoresis is very convenient for charged substances like amino acids, because the paper may be dried and treated with reagents like ninhydrin after the run. If the paper is clamped between metal plates, or suspended in a non-aqueous solvent, to dissipate heat, potentials of up to 10 000 v may be applied, and a run may be completed in 20–100 min. Paper electrophoresis may be combined with paper chromatography in a second dimension, to improve separation.

Chromatography

This may be defined as the separation of substances according to their partition coefficients between two immiscible phases. Chromatography may be divided into three main types—adsorption of the substances (to a solid), ion-exchange (using a charged solid), and partition (which may be liquid/liquid or liquid/gas). The chromatogram is 'developed' by causing a liquid or gas containing the substances to be separated to flow past or through the second, supporting phase. The separated substances may be collected as they leave the chromatogram or identified by the time taken for them to leave, or they may be left on the chromatogram and identified by their relative positions.

Paper chromatography. Although this has largely been superseded by thin-layer chromatography (below), it is convenient to explain the general principles by reference to it. It is essentially a form of partition chromatography between two liquid phases, although adsorption to the paper plays some part. The stationary phase is water (or a buffered solution) held between the cellulose fibres of the paper. The mobile phase is usually a mixture of organic solvents, perhaps containing some water. The solution to be analysed is applied to one end of a paper strip as a small spot which is allowed to dry. The paper is then 'equilibrated' by being hung in an airtight tank containing the vapour of the stationary phase, and the end nearest the spot is then immersed in a trough of the mobile phase from which the strip hangs down (see Fig. 21.7). Alternatively, the strip is stood upright in a dish of the mobile phase and the liquid ascends by capillary attraction.

After a suitable time interval the paper is taken out, the distance to which the solvent has run is marked, and the paper is allowed to dry. The separated compounds can then be visualized by any appropriate means. The time of running can be varied according to convenience, but it is found that if the temperature is kept constant the distance run by any particular compound with any given solvent will be a constant fraction of the distance run by the solvent front. This ratio is called the R_F, where

$$R_F = \frac{\text{distance run by compound}}{\text{distance run by solvent}}$$

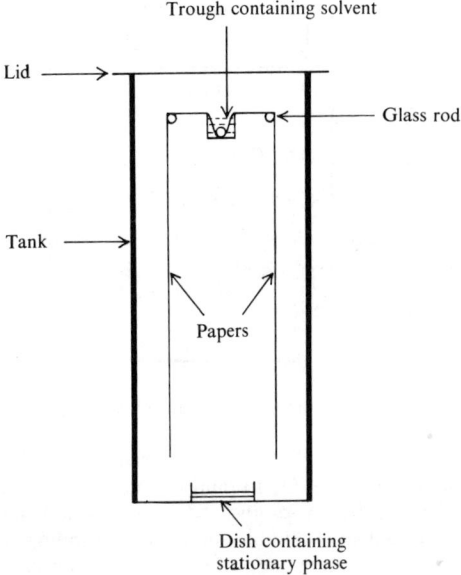

Trough containing solvent

Lid

Glass rod

Tank

Papers

Dish containing
stationary phase

FIG. 21.7. Apparatus for paper chromatography.

Paper chromatography is an extremely versatile analytical method, although proteins cannot be satisfactorily separated by it. Only small quantities of material can be analysed, and it is necessary to ensure that the paper is not overloaded with respect to compounds which are not very soluble in the mobile phase, otherwise streaking will occur. High concentrations of salts, etc. (e.g., from urine) must also be removed. The lower limit of material which can be used is set by the sensitivity of the method of detection.

Even better separations can be made by means of *two-way* paper chromatography (see Fig. 21.8). The paper must be dried before being run in the second dimension.

Thin-layer chromatography. This belongs more to adsorption than to partition chromatography, although the solid support retains some water. Because the support is dry, its preparation does not need to be varied for it to be used with organic solvents to separate hydrophobic substances. This versatility, and the greater sharpness of separation, makes it the method of choice for most applications. A thin (less than 1 mm) layer of inert material (kieselguhr, silica gel) is applied to a flat glass plate which can be as small as a microscope slide. It is very important that

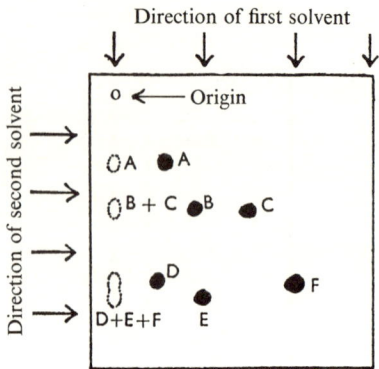

FIG. 21.8. Two-dimensional paper chromatogram.

the layer be of a constant thickness. The material is applied as a thin paste, usually containing some $CaSO_4$ to bind it, and allowed to dry. The spot is applied as for a paper strip, and the plate is stood upright with its bottom end immersed in a suitable solvent. The running time is usually 1–2 hours, separations are good, and very small amounts of material can be detected.

Column chromatography. In this technique a glass tube, typically 1–5 cm in diameter and 10–30 cm high, is packed with a suitable solid in powder form; the interstices between the granules are filled with liquid. It is important that no air bubbles get into the column. The sample, dissolved in a small volume of liquid, is run into the top of the column, which is then developed by running a suitable solvent slowly through it until the separated substances appear in solution one by one in the column effluent. The practical convenience of a chromatographic column depends on the use of a *fraction collector*, a device which will remove a collecting tube from underneath the column when a certain volume has been collected in it (or a certain time has elapsed), and substitute a fresh tube. By this means columns may be allowed to run for hours or days practically without attention. The contents of each tube must then be assayed individually for the substances whose purification has been desired.

Sometimes the solid may be inert, merely a support for a stationary liquid phase, but more usually the substance to be separated is adsorbed to the solid phase. Alternatively, the solid may be acidic or basic, in which case the substance to be separated displaces a soluble ion of the same sign from electrostatic combination. Sometimes all three processes operate, which makes column chromatography difficult to describe but useful for the separation of substances with similar physical properties, e.g., proteins.

Adsorption chromatography. This is typically used for the separation of substances insoluble in water, e.g., fatty acids, sterols, carotenoids, with a support such as silica gel or activated alumina. The sample is applied to the column in a non-polar solvent (e.g., $CHCl_3$), and is then eluted by using successively higher concentrations of a polar solute (e.g., ethanol). The separate substrates in the sample then partition themselves continuously between the adsorbing solid and the liquid, and those which are least strongly adsorbed reach the bottom of the column first.

Ion-exchange chromatography. This works on a different principle. The solid with which the column is packed is an insoluble macromolecule containing free (reactive) acidic or basic groups. There are two important types, synthetic polystyrene resins and modified forms of cellulose. They are used in the salt forms, e.g., $Resin^+Cl^-$ or $Resin^-Na^+$. An example will make the method clear.

Supposing such a column is used to separate anions. An anion-exchanger (i.e. a basic resin) is used as, let us say, the chloride salt. The ions to be separated must have a greater affinity for the resin than chloride; when the sample is applied to the column the anions will displace some chloride ions and themselves remain bound. There are two ways of getting them off the column again. One is *elution development*, in which the column is washed with a solution containing weakly attracted anions, e.g., chloride. These will gradually, by a mass-action effect, displace the bound anions, and the latter will slowly move down the column in the reverse order to their affinities for the resin. When they emerge they will be contaminated by the eluting anions but will be separated from each other. The separation can be improved by carefully controlling the pH of the developing solvent and by gradually increasing the concentration of eluting anions (*gradient elution*).

The second method is *displacement* elution, in which the bound anions are displaced by using an eluting solution containing an anion with a *stronger* affinity for the resin than the bound anions. These are then rapidly displaced and arrive in the effluent in high concentration, free of displacing anion, but somewhat intermixed with each other.

Using a cation-exchanger (i.e. $Resin^-$) and eluting with suitable cations, the separation of cations may be effected. Amino acids, peptides, and proteins can also be separated on ion exchange resins; the separations are better than might be expected, particularly using cellulose derivatives, because adsorption and molecular-sieve phenomena (see below) also play a part.

Molecular sieving. It is possible to join together dextran chains (see Chapter 2) to make granules having pores of fairly precise dimensions. As the solution to be examined passes down the column small molecules will enter the pores and be retarded while large molecules will pass straight on. Molecules of intermediate sizes will be retarded to some extent, thus the material is defined in terms of a minimum and a maximum molecular weight. The pore sizes can be varied over a wide range, so that the technique can be used, e.g., for separating salts from colloids, or for separating small macromolecules from big ones. The use of other porous materials, e.g., glass beads, enables the technique to be extended to separating viruses.

Affinity chromatography. It is possible, by a suitable chemical reaction, to form a covalent bond between many substances (e.g., NAD, proteins, nucleic acids) and a chromatographic support. If a sample containing a species that has affinity with the bound substance is now passed down the column, the elution of that species will be specifically delayed. For example, NAD-linked dehydrogenases will be delayed by NAD, antibodies by antigens, or DNA by complementary RNA. This technique has enormous power, and is widely used for separating antibodies and genetic DNA.

Gas chromatography. A volatile substance, e.g., ammonia, can be removed from a relatively non-volatile solvent, e.g., water, by blowing air through the solution. A less volatile compound, e.g., methylamine, would also be removed, but more slowly, so that there would be a partial separation of the two compounds. A gas–liquid chromatogram uses a column of liquid so that there is a continuous partition between the liquid and the stream of gas flowing through it, and the substances to be separated emerge in the gas in the order of their volatility. In order to prevent mixing of the liquid it is kept as a thin film on an inert solid support (which can be as simple as brick dust). Because most substances of interest are not very volatile, high-boiling liquids (e.g., silicones or paraffins) are used, and the column is kept at a high and carefully controlled temperature (up to 200–300°C). An inert gas, usually argon, is used, and this reduces thermal decomposition. Its flow rate must be carefully controlled. The sample is volatilized directly into the gas.

The chief difficulty in this technique is in detecting the components of the sample as they appear. Changes in the physical properties of the emerging gas, e.g., its density, thermal conductivity, or ionizability, may be used. The effluent may even be connected to a rapid-scanning mass spectrometer (p. 455). It is not easy to collect the compounds to be measured, so their concentration is estimated from the extent of change in, e.g., density, and they are identified by the time taken for them to appear in the effluent. If they are radioactive their activity may be measured while they are still in the gas phase.

This technique has proved invaluable for analysing fatty acids and other lipids, e.g., steroids. Even amino acids may be separated in this way if their polar properties are reduced by esterification.

Isotopes

There are many experiments which can be performed with isotopic tracers. *Dilution analysis* can be used to find the amount of a particular compound in a biological sample if the compound can be obtained pure but not in quantitative yield. A variant of this is the estimation of the extracellular fluid space by means of the dilution of a known amount of isotopic Na^+ or Cl^-.

Isotopes can be used to find out whether compound B is a product of compound A. For this purpose a pure specimen of A, isotopically labelled, must be prepared by synthesis in the laboratory or biologically. If the precise reaction pathway from A to B is being investigated it may be necessary to find out the distribution of the isotope in B, i.e. in which groups of the molecule it is present. This involves degrading the molecule of B in a controlled way, so that for each fragment obtained it is known from precisely which part of B it came.

Isotopes can also be used to measure the rates of reactions in the steady state, i.e. when there is no apparent change in the concentration of reactants or products. Examples are the rate of conversion of blood glucose to CO_2 and the rate of entry of K^+ into cells.

In almost all cases it is necessary to measure the concentration of the isotope in the compound being studied, i.e. the proportion of isotopic atoms relative to the total number of atoms of that element. If, to all intents and purposes, the isotope is not found naturally, the concentration can be expressed as the *specific activity*—the number of isotopic atoms per μmole (or mg) of the compound under study. If the isotope is one that occurs naturally—the most extreme example would be ^{37}Cl, which forms 24 per cent of all naturally occurring chlorine atoms—the concentration of isotope is expressed in *atoms per cent excess*. This is the number of isotopic atoms per 100 total atoms of that element which is present in excess of the normal number occurring naturally.

Stable isotopes. The important stable isotopes of biological interest are ^{15}N and ^{18}O; ^{13}C and ^2H (deuterium) are still sometimes used. (The superscripts refer to the atomic weights of the isotopes; the natural isotopes of the elements quoted are ^{14}N, ^{16}O, ^{12}C, and ^1H.) Except for deuterium, stable isotopes must be estimated by means of a *mass spectrometer*. The samples must first be converted to a gas (N_2 is used for ^{15}N, CO_2 for ^{13}C and ^{18}O) which is introduced at very low pressure into the apparatus. It is there caused to ionize, and the charged ions are projected towards the cathode by the application of a high and steady voltage. On their way they pass through a magnetic field which deflects them to an extent depending on their speed (which is constant) and mass. Thus, $^{12}CO_2$ will pursue a different path from $^{13}CO_2$, and so on. In order to measure the relative proportions of each isotope the ions are collected before they reach the cathode by a metal plate whose change in potential is measured.

This machine is less convenient to use than one for measuring radioactive isotopes, and it is not usually possible to use starting material which is very highly enriched. Thus experiments with stable isotopes have been much less frequent than those with radioactive ones.

Organic substances will become ionized in the conditions prevailing inside a mass spectrometer, and will also partly decompose in a characteristic way. The machine is therefore widely used for detecting and identifying low concentrations of organic compounds (cf. p. 454).

Neutron activation. Most radioactive isotopes are made by bombarding a suitable target with neutrons in an atomic pile. It is sometimes possible to bombard a biological sample directly, thus converting some element into a radioactive isotope which can be detected in very low concentration. The method is usually limited to metallic elements, of medium atomic weight.

Radioactive isotopes. A few definitions are helpful. Radioactive emanations occur when unstable atoms break down, and can consist either of α-particles (helium nuclei), β-particles (electrons or positrons), or γ-rays (electromagnetic radiation very similar to X-rays). Quite often α- and γ-rays, or β- and γ-rays occur together. The number of radioactive atoms in any sample necessarily decreases with time; each isotope has a characteristic instability, which is expressed in the *half-life* $(t_{\frac{1}{2}})$, the time taken for the number of radioactive atoms of a particular isotope, in a given sample, to decrease by one-half. This can vary from a fraction of a second to several thousand years, but only in rare cases can isotopes with half-lives of less than an hour be used experimentally. The half-life is independent of the amount of material present. In two half-lives the number of radioactive atoms will fall to one quarter of the original, and so on.

The *amount* of radioactivity present in any sample would be most logically defined by the number of atoms of each unstable isotope present. In fact, it is defined slightly differently, as the number of disintegrations occurring in unit time. One unit is defined as 3.7×10^{10} disintegrations per second; there are therefore more radioactive atoms in material containing 1 curie of a long-lived isotope than of a short-lived one. In biological experiments a milli- or even micro-curie (μCi) is a more convenient unit; 1 μCi gives approximately 2.2×10^6 disintegrations per minute, which is far more than can be conveniently measured in a small sample. Because of the continual decay in radioactivity, the number of curies in any particular sample is continually decreasing, and in every experiment the specific activity of the products (d.p.m./μmole or mg) has to be corrected to what it would have been at some arbitrary time (the beginning of the experiment).

The radioactive rays from each isotope have a characteristic energy (or range of energies) which depends partly on the mass of the particles and partly on their speeds. A knowledge of the energy values is very useful, both from the point of view of measurement and of biological damage. The range of radioactive rays depends on their energies, but is very different for each type. Heavy α-particles collide with many atoms and are quickly stopped, whereas γ-rays go 'straight through' most atoms and have a much longer range. β-particles are intermediate. In general, radioactive rays of all types are stopped the more easily the higher the atomic weight of the material they are traversing, hence the virtue of shielding with lead. The amount of energy for each ray is measured in electron-volts; the values vary from several million electron-volts (MeV) to about 0·02 MeV (the energy of X-rays can be measured in the same units). The range of a 1 MeV β-particle (electron) in air is about 3 metres.

The total energy, in the form of γ- or X-rays, emitted by any source is measured in *roentgens* (r), which defines their ability to produce ions in gases. This unit is not strictly applicable to α- or β-rays. Time does not enter into the definition of a roentgen, thus 1 r can be produced by a weak source in a year, or by a powerful source in a second.

Biological effects. A given amount of emitted radiation, whether expressed in roentgens or otherwise, will not necessarily be absorbed by any given specimen of matter; if the specimen were in the form of a thin sheet, for example, most of the radiation might go straight through. The unit of radiation actually absorbed is the *rad*, defined in terms of energy absorbed: 1 rad = 100 ergs/g. Note that again no time is specified.

The biological effectiveness of different types of radiation actually absorbed has also to be taken into account, and this depends partly on the density of the ionization caused by the rays. For this reason α-rays, with a short path along which are many ionizations, cause much more damage per rad absorbed than β- or γ-rays. The biological effectiveness or radiation is measured in *rems*, a unit which must vary according to the tissue being studied. Very roughly, 1 rem = 1 rad for β-, γ-, or X-rays, and 1 rem = 0·05 rad for α-rays. As an example of the use of this unit, the maximum occupational exposure to radiation should not exceed 5 rems per year on the average.

It follows from what has been said about the ranges of the various rays that α- and β-emitters are more dangerous inside the body, since all their radiations will be absorbed. Outside the body, their rays will hardly penetrate the skin, and in relatively low dosage will not harm this latter tissue. γ-Emitters, on the other hand, may be relatively harmless inside the body, since most of their radiation may travel through the tissues to the outside air, whereas outside the body their long range makes them dangerous.

All radiation is subject to the *inverse-square law*, which means that the amount of radiation falling on any area is reduced to one-fourth when the distance between it and the source is doubled. Quite apart from the range in air, therefore, the further away from a source of radioactivity the safer, but for high-activity sources, particularly γ-emitters, thicknesses of concrete or lead bricks will be necessary for shielding.

Table 21.3 sets out some properties of important isotopes of biological interest.

Table 21.3

Isotope	Half-life	Emissions, with energy in MeV*	Maximum permissible† body burden in μCi
³H (tritium)	12·26 yr	β(0·006)	1 000
¹⁴C	5 730 yr	β(0·05)	300
²⁴Na	15 hr	β(0·54), γ(1·4, 2·7)	7
³²P	14·2 days	β(0·7)	6
³⁵S	87 days	β(0·05)	100
³⁸Cl	37·3 min	β(1·39), γ(1·6, 2·1)	—
⁴⁰K	1·3 × 10⁹ yr	β(1·3), γ(1·46)	Naturally occurring
⁴²K	12·5 hr	β(1·4), γ(1.5)	10
⁴⁵Ca	153 days	β(0·09)	30
⁵⁹Fe	45 days	β(0·12), γ(1·1, 1·3)	20
⁶⁰Co	5·3 yr	β(0·1), γ(1·17, 1·33)	10
⁹⁰Sr	28 yr	β(1·0)	2
¹³¹I	8·05 days	β(0·2), γ(0·36, 0·7)	0·7
¹⁹⁸Au	2·7 days	β(0·34), γ(0·4)	20
²²⁶Ra	1 622 yr	α(4·8), γ(0·18)	0·1

* The figures give the mean energies of β-particles; the maximum energies are in most cases roughly three times as large.

† The maximum permissible body burden is less when the isotope is concentrated in a particular organ.

Detection and estimation of radioactivity. With the exception of Ra, which is not used as a tracer, the isotopes shown in Table 21.3 do not emit α-particles. The problem is therefore the detection of β- and γ-rays. There are two basic methods: the ionization of gases and phosphorescence (scintillation). Which method is best for a particular isotope depends largely on the penetrating power of its emissions; this depends to some extent on their energy.

An *ionization chamber* consists of a pair of charged plates with an air (or gas) space between them. A *Geiger–Müller tube* is essentially an ionization chamber with a built-in amplifying device. The cathode is a metal cylinder, while the anode is a thin wire running down its axis (Fig. 21.9). The tube is filled with an easily ionized gas, such as argon, at moderately low pressure, together with a trace of an organic molecule as a 'quencher'. The central wire is kept at a high positive potential (300–1 000 v). When a ray enters the tube and ionizes an atom of the gas the electron is accelerated rapidly towards the wire because of the intense field around it. It thus ionizes other atoms of the gas, and the secondary electrons so produced ionize others ('avalanche effect'). The end-result is a substantial 'pulse' (potential drop at the anode) for each entering ray

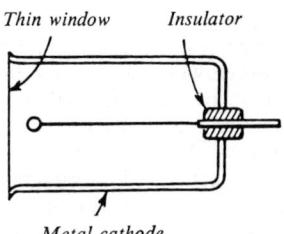

Thin window *Insulator*

Metal cathode

Fig. 21.9. Geiger–Müller tube.

which causes an ionization. The pulse is followed by a 'dead time' (0·5 millisec or less) during which any ray entering the tube will not be detected. The function of the quencher is to prevent the discharge from becoming continuous. Each tube has a certain voltage range over which the count rate is independent of voltage—the 'plateau'. Beyond this, the tube discharges continuously, and is rapidly ruined.

Geiger–Müller counters work best at relatively low count rates ($<10\,000$ per min), since then the dead time is only a small fraction of the total counting time (but see background, below). The chief technical difficulty lies in finding a 'window' which will permit the passage of weak β-rays. For strong β-rays, thin aluminium sheet is satisfactory. A thin mica sheet will permit the passage of some β-rays from a ^{14}C sample, but there is no window that will allow the counting of ^{3}H (see Table 21.3). A type of tube called a 'gas-flow counter' in which argon is continuously passed through an open-ended tube may be used for counting tritium or ^{14}C. Geiger–Müller tubes are very inefficient for counting γ-rays, although a tube with a cylindrical Pb cathode can be used.

Geiger–Müller tubes can be made to dip into a liquid sample, but most samples are solid and are placed on a disc (planchette) at a standard distance from the window. Because of the inverse-square law the 'geometry' (i.e. positioning of the sample) is very important. A problem called *self-absorption* arises with solid samples. β-particles have only a short range in solids, and it frequently happens that rays arising from an isotope at the bottom of the sample are absorbed by the sample itself before they can reach the window of the Geiger–Müller tube. This can be corrected for, but in principle solid samples are best counted as 'infinitely thin' (no self-absorption) or 'infinitely thick' (so that only the top layer of the sample is, in effect, counted; in this case the counts observed are proportional to the *concentration* of the isotope).

Scintillation Counting

Some materials respond to radioactive rays (and suitable light photons) by the excitation of an orbital electron, rather than its complete loss.

When the excited electron returns to a lower energy state a photon is emitted. This phenomenon is known as phosphorescence. Zinc sulphide was used by the early workers with radioactivity; the present-day *phosphors* are either NaI crystals (activated with thallium) or organic materials, of which the type compound is anthracene.

Scintillation counting would be impracticable without a device to detect the tiny flashes of light. A *photomultiplier* (Fig. 21.10) converts the photons into pulses of electrical energy and amplifies them sufficiently to be counted. Each row of the vanes ('dynodes') shown in the diagram is kept at a successively higher voltage. The photon emitted by the phosphor falls on a light-sensitive metal which emits an electron (as in a photoelectric cell). This is accelerated towards the first row of dynodes. When it impinges, three or four secondary electrons are emitted, and these are accelerated towards the next row of dynodes, and so on. An 11-stage photomultiplier will amplify the original impulse by a factor of 10^{11}. The instrument is, of course, sensitive to any visible light, so that it, together with the scintillator and sample, must be kept in the dark. It is a feature of scintillation counting that the intensity of the electrical pulse is proportional to the energy of the original radioactive emission, which makes it possible to identify individual isotopes or to count one in a mixture.

The simplest way of detecting γ-rays is to place the source in contact with a light-shielded NaI crystal which is placed on a photomultiplier. Weak β-rays would be absorbed by the container of such a crystal, and for them *liquid scintillation counting* is used. The phosphor and the sample are both dissolved in a non-aqueous solvent; the emission of photons takes place inside the transparent sample container, which is placed in close contact with the photomultiplier. This is the technique of choice for counting ^{3}H and ^{14}C. Strong β-emitters, such as ^{32}P, are just as easily counted with a Geiger–Müller tube or *Cerenkov* counting.

There are certain difficulties with liquid scintillation counting. Very many chemicals quench the phosphorescence; that is to say, they accept

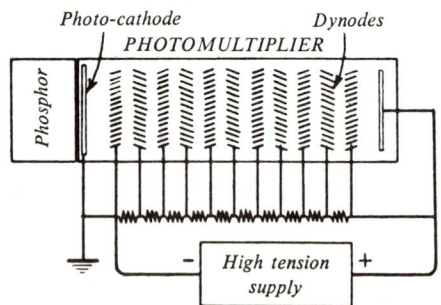

FIG. 21.10. Photomultiplier.

the energy from the excited electron, so abolishing the emission of a photon. Water is a powerful quenching agent, thus a non-aqueous solvent (usually toluene) has to be used. Although up to 10 per cent of water can be tolerated in the final solution, it is sometimes very difficult to dissolve polar labelled compounds (e.g., amino acids) in it. Furthermore, many coloured compounds (particularly yellow ones) which may be present in the sample reduce the efficiency by absorbing photons before they can reach the photomultiplier. It is therefore necessary to pay strict attention to counting efficiency, using internal standards (i.e. isotopes of known activity).

The measurement of counting rate. All the devices which have been described require either a *rate-meter* (to measure the current produced by the ionization) or a *scaler* (to count the individual pulses coming from a Geiger–Müller tube or photomultiplier). Modern machines use a series of *decade* scalers which will record up to 1 000 000 impulses (counts). It is necessary to record the counting time accurately, and this is done by a variant of the same technique (using, for example, the a.c. power as a source of the impulses). The *observed count rate* must be corrected for background count, then for efficiency, dead time, etc., before being used, with a suitable decay correction, for calculating specific activities.

Background. All detectors of radioactivity will show a low rate of impulse production in the absence of a radioactive sample. This is largely due to cosmic rays, partly also to radioactive isotopes in the air and in structural materials, and, in the case of scintillators, to random thermal phosphorescence. The sum total of these is known as the *background rate*, and must be subtracted from the observed rate in the presence of the sample to give the true count rate. It can be shown that the true rate will not be estimated with any accuracy if it is not at least as large as the background, and it is therefore advantageous to keep the latter as low as possible. (For example, γ-ray counting of natural ^{40}K in the body has been used as a method of measuring the proportion of muscle to fat; small variations in a rate of 500 c.p.m. become very important.) The sample and detector are usually shielded from cosmic rays by lead (a 'lead castle'). Other ways of reducing the background are by coincidence counting (which ensures that only events occurring *in the sample* are counted), and for liquid scintillators, cooling the liquid to 0°C or below.

Therapeutic Applications

Radioactive emissions, particularly of γ-rays, are frequently used deliberately to kill unwanted cells. Examples are the radioactive sterilization of surgical accessories, partial destruction of overactive thyroids by ^{131}I (which is concentrated by the thyroid itself), and the irradiation of tumours by ^{60}Co (from outside the body) and by ^{198}Au, ^{226}Ra, or other isotopes implanted in tissues.

Index